HUMAN POLYMORPHIC GENES

HUMAN POLYMORPHIC GENES

WORLD DISTRIBUTION

Arun K. Roychoudhury and Masatoshi Nei

Center for Demographic and Population Genetics
The University of Texas
Health Science Center at Houston

New York · Oxford

OXFORD UNIVERSITY PRESS

1988

Oxford University Press

Oxford   New York   Toronto
Delhi   Bombay   Calcutta   Madras   Karachi
Petaling Jaya   Singapore   Hong Kong   Tokyo
Nairobi   Dar es Salaam   Cape Town
Melbourne   Auckland

and associated companies in
Berlin   Ibadan

Library of Congress Cataloging-in-Publication Data
Roychoudhury, Arun K.
Human polymorphic genes.
Bibliography: p.
1. Human population genetics—Tables. 2. Gene
frequency—tables. 3. Genetic polymorphisms—tables.
I. Nei, Masatoshi. II. Title. [DNLM: 1. Gene Frequency
—tables. 2. Polymorphism (Genetics)—tables.
QH 455 R888h]
QH431.R69   1988   573.215   87-24868
ISBN 0-19-505123-8

2 4 6 8 10 9 7 5 3

Printed in the United States of America
on acid-free paper

To Jack Schull

Preface

Human populations are polymorphic for a large number of genetic loci. These polymorphic loci are useful for constructing linkage maps, genetic counseling, paternity testing, and studying the evolutionary relationships of human populations. For this reason, a large number of human geneticists have studied the gene frequencies at polymorphic loci. Particularly in the last ten years, the number of polymorphic loci (or polymorphic DNA regions) discovered has increased dramatically, and an enormous amount of gene frequency data has accumulated. Unfortunately, these data are published in many different journals and are not easily accessible to research workers. We have therefore decided to compile the gene frequency data for all polymorphic loci so far studied. This book is the product of our efforts in this direction during the last few years.

The purpose of this book is to present gene frequency data for as many loci as possible for various human populations. The total number of loci or genetic systems included is 362, and their names are given in Chapter 2. We have considered a total of about 180 populations from around the world. These populations are chosen either because they represent major ethnic groups or because they are interesting from an anthropological point of view. At many loci, however, gene frequency data are available only for a limited number of populations.

To make this book useful for medical and forensic purposes as well as for evolutionary studies, we have included a chapter on the utility of gene frequency data. We have presented various mathematical formulas that are useful for constructing linkage maps, genetic counseling, and paternity testing. The heterozygosity value (H) and the probability of exclusion of paternity (PE) are presented for each locus in each population whenever meaningful values are obtainable. Information on gene symbols, approximate chromosomal locations, enzyme commission (E.C.) numbers, and polymorphism/monomorphism is also given for each locus.

At the present time, many investigators are studying DNA polymorphisms with the aim of constructing a comprehensive human gene map by using restriction enzymes. These polymorphisms are important for both medical and evolutionary studies. Unfortunately, most of the gene or haplotype frequency data currently available are based on small sample sizes, and their reliability is not very high. Furthermore, the populations and the DNA polymorphisms studied are not always well characterized. Different investigators often use different DNA probes and different restriction enzymes even if they study the same DNA region, so that the data obtained by different investigators are not always comparable. We have therefore included only those data that refer to well-defined DNA polymorphisms and are based on a relatively large sample size. In contrast to the polymorphisms of nuclear DNA, those of mitochondrial DNA have been studied more systematically, and new insight has been obtained into the evolution of human populations. However, all mitochondrial genes are inherited together as a single entity, so that the genetic relationships of populations derived from mitochondrial DNA data are subject to large sampling errors. Furthermore, polymorphism data for mitochondrial DNA are so extensive that it is difficult to tabulate them in a limited space. We have therefore decided not to include them in this book.

We are aware that there are two excellent books on the gene frequency distributions among human populations. One is Mourant et al.'s (1976) The Distribution of the Human Blood Groups and Other Polymorphisms and the other is Tills et al.'s (1983a) The Distribution of the Human Blood Group and Other Polymorphisms, Supplement 1 (both from Oxford University Press). The scope of these books is to list all gene frequency data so far published. Thus, for example, Mourant et al.'s book includes more than 200 different sets of ABO blood group gene frequencies from France alone. They also present phenotype frequencies in addition to gene frequencies. For these reasons, their books are voluminous. However, the number of loci covered in these two books is only 83, which is less than one-fourth of our collection. The above two books include data published up to 1975, but ours includes data up to February 1987. Unlike the above books, ours includes only one set of gene frequency data for each locus and for each population except under special circumstances. Phenotype frequencies are not included because they can be predicted from gene frequencies in most populations. The above two books are useful for studying the microdifferentiation of populations, whereas ours will be useful for studying the genetic differentiation of different ethnic groups. Our book will also be useful for medical and forensic purposes.

As mentioned above, we have included only one set of gene frequency data for each locus for each population. For recently discovered loci, this procedure has created no problems, because there are not many data sets. In the case of well-studied loci such as the ABO blood group system, however, it was not always easy to choose one data set. We used the following criteria for this purpose: (1) large sample size, (2) recent publication, (3) data from experienced laboratories, (4) representative gene frequencies for a population, and (5) data available for many different loci from the same population. However, these criteria served only as guidelines, and the actual choice was often arbitrary. We apologize to many research workers for not including their excellent data; our intention was not to present all published data.

We are indebted to many investigators in producing this book. A large number of authors sent us their reprints and preprints, which were extremely helpful in our compilation of gene frequency data. Since so many authors are involved, we cannot acknowledge our indebtedness to them individually. The books of Mourant et al. (1976) and Tills et al. (1983a), Race and Sanger's (1975) Blood Groups in Man (Blackwell, Oxford), and Steinberg and Cook's (1981) The Distribution of the Human Immunoglobulin Allotypes (Oxford University Press) were very useful in finding early references about blood group and other polymorphic loci. Equally useful were review articles or monographs in which a large number of data sets were compiled. A few examples of such works are Matsunaga's (1962) paper on cerumen, Post's (1971) paper on colorblindness, Baur et al.'s (1984) paper on HLA, Flatz's (1987) paper on lactose tolerance, and Berg's (1974) paper on Scandinavian populations.

Our special thanks go to Ranajit Chakraborty, Aravinda Chakravarti, Peter Smouse, and Clay Stephens who kindly read and commented on the first draft of Chapter 1. Terry Bertin, Stephen Daiger, Robert Ferrell, David Hewett-Emmett, and Clay Stephens were also kind enough to read parts or all of the data tables. We are indebted to Dr. Victor McKusick who not only provided us with the latest information of the human gene map but also kindly checked the list of genetic loci presented in Chapter 2. John Sourdis and Robert Schwartz helped us in computerizing data tables and finding an ingenious way to check references by computer. John Sourdis and Gregory Livshits also helped us in incorporating numerous corrections in the final form of data tables. We also would like to express our gratitude to Jack Schull for his support and encouragement of this project. As a token of our appreciation, we would like to dedicate this book to him on the occasion of his 65th birthday.

The project of compilation of gene frequency data was partially supported by research grants from the United States National Institutes of Health and National Science Foundation. A major part of the compilation was conducted while one of us (A. K. R.) was on leave from the Bose Institute, Calcutta, India.

We hope to continue our work of compiling gene frequency data and would appreciate receiving any reprints concerning genetic polymorphisms in man. We would also appreciate it if any errors contained in this book are brought to our attention. Although we tried to present all numerical values as accurately as possible, some errors may still have crept in.

Contents

PART I

General Guidance

1   Utility of Gene Frequency Data

This book is intended for use by research workers in human genetics, physical anthropology, genetic counseling, and paternity testing. There are four different ways of using the gene frequency data presented.

EVOLUTIONARY AND POPULATION GENETICS STUDIES

Gene frequency data are very useful for studying the genetic relationships and evolution of human populations. In the early days of genetic studies, the genetic relationships among different populations were studied by examining the geographical distribution of gene frequencies at a few loci such as the *ABO* and *Rh* blood group systems. It was later realized that comparison of gene frequencies for one or two loci is not reliable since each locus has a different distribution. Only when a large number of loci are examined do the genetic relationships become clear (Cavalli-Sforza and Edwards 1964). This is partly because the interpopulational genetic variation is very small compared with the intrapopulational variation at the gene level (Nei and Roychoudhury 1972, 1974, 1982; Lewontin 1972). However, if a large number of loci are examined, even small differences can be detected with sufficient accuracy.

The genetic difference between a pair of populations is usually measured by a quantity called the genetic distance, which is a function of gene frequencies. There are several different measures of genetic distance (Jacquard 1975; Nei 1987, Chapter 9). Once genetic distances are estimated for a group of populations, their genetic relationships can be studied by using dendrograms, principal component analysis, etc. (Cavalli-Sforza and Bodmer 1971; Sneath and Sokal 1973; Nei 1987, Chapter 11).

In recent years, intensive studies have been made on human DNA polymorphism. DNA polymorphisms are usually studied by using restriction enzymes. If several different restriction enzymes are used, a restriction-site map can be constructed for each haplotype (DNA sequence). Comparisons of restriction-site maps provide estimates of nucleotide differences between and within populations (see Nei 1987, Chapters 5 and 10). It is therefore possible to study the genetic relationship of populations in terms of the number of nucleotide differences per site. Haplotype frequency data can also be analyzed by using genetic distance measures (Wainscoat et al. 1986). However, DNA polymorphism data obtained from a single DNA segment of the genome (or mitochondrial DNA) are subject to large stochastic errors, so that it is necessary to be cautious about the conclusions obtained from this type of study.

One of the purposes of population genetics is to study the mechanism of maintenance of genetic variability. Noting that the frequency of the thalassemia gene is high in the Mediterranean area, where malaria is endemic, Haldane (1949) suggested that this deleterious gene might be maintained by heterozygote advantage. This suggestion led Allison (1955) to the discovery that the high frequency of the sickle-cell anemia gene in Central Africa is the result of heterozygote advantage, which is caused by higher resistance of sickle-cell heterozygotes to malaria. Similar studies on the association between allele frequencies and environmental factors have been made for a number of polymorphic loci (e.g., Flatz 1987; Flint et al. 1986; Goedde 1986). It seems that this type of association study is an effective way to make an inference about selective mechanisms, although the inference has to be proved later by more direct means. To conduct this type of study, data on the distribution of allele frequencies among various populations are needed.

Gene frequency data are also useful for tracing past gene migration. For example, the gene pool of the American Negroid population is known to contain Caucasian genes because of past interracial mating. The percentage of Caucasian genes has been estimated to be about 20% from gene frequency data for blood group loci (Reed 1969). Gene frequency data also suggest that gene migration has occurred for a long period of time between southern Europe and northern Africa.

To conduct a quantitative study of the evolutionary change of genes or populations, we need statistical methods. There are many such methods for studying various evolutionary and population genetics problems. The reader who is interested in such methods should consult Cavalli-Sforza and Bodmer (1971) and Nei (1987).

LINKAGE STUDIES

In recent years, great effort has been made to construct gene maps for human chromosomes (see Human Gene Mapping 8, 1985). Once gene maps are constructed, they can be used to predict genetic diseases through linked marker genes (see next section). A gene map, however, cannot be constructed without polymorphic loci. In general, highly polymorphic loci are more useful than less polymorphic loci for constructing gene maps. This is because the linkage between two loci can be detected only when there are double heterozygotes. Therefore, the utility of a locus for constructing gene maps can be measured by the frequency of heterozygotes or *heterozygosity* of the locus.

In a random mating population, the heterozygosity of a locus is defined by

$$H = 1 - \sum_{i=1}^{k} x_i^2, \qquad (1)$$

where x_i is the frequency of the ith allele in the population and k is the number of alleles. For example, if there are three alleles, A_1, A_2, and A_3, at a locus and their frequencies are 0.7, 0.2, and 0.1, respectively, H becomes $1 - (0.7^2 + 0.2^2 + 0.1^2) = 0.46$. When the sample size is small, Eq. (1) may give an underestimate of population heterozygosity. In this case, the fol-

lowing formula gives an unbiased estimate:

$$H = 2n/(2n - 1)[1 - \Sigma \hat{x}_i^2], \qquad (2)$$

where $\hat{x}_i$ is the frequency of the ith allele in the sample and n is the number of individuals sampled. In practice, however, Eqs. (1) and (2) are very close to each other unless n is very small.

We computed the heterozygosity values for all populations and for all loci except for those in which the mode of inheritance is not firmly established. They are presented in the tables of gene frequencies. In this computation, we used Eq. (1) because the sample size is usually very large. The heterozygosity varies from locus to locus and from population to population. It is therefore advisable to use highly heterozygous loci and populations for the construction of gene maps.

Chakraborty et al. (1979) proposed a somewhat different criterion by which the usefulness of a polymorphic locus for linkage studies can be judged. This criterion is the probability that a pair of individuals mating at random is capable of producing at least two different genotypes. This probability also varies from locus to locus and from population to population.

GENETIC COUNSELING

The idea of using linked marker genes for genetic counseling is not new. Hoogvliet (1942) used colorblindness for detecting the female carrier of the hemophilia gene. Edwards (1956) and Renwick (1969) suggested the use of linked marker genes for intrauterine diagnosis of genetic diseases when the disease gene is not expressed in the uterus. Similarly, marker genes are useful for predicting late-onset genetic diseases before the diseases are manifested. To evaluate the utility of marker loci for genetic counseling, two different criteria are required. One is the accuracy of prediction or the probability with which an individual with a marker allele contracts the disease in question. This probability depends on the recombination value between the marker and disease genes

4

and on how the linkage phase of the parent is determined. The other is the proportion of individuals or families (informative families) in which a particular marker locus can be used for genetic counseling. This proportion depends on the frequencies and dominance relationships of marker alleles. Obviously, a marker locus must be polymorphic; otherwise it has no utility for genetic counseling.

The linkage phase of a counselee (parent) for X-linked or autosomal dominant diseases can be inferred from information on the phenotypes of his or her relatives. Particularly useful for this purpose is the information from grandparents and offspring. Once the linkage phase of a counselee is determined, the accuracy of predicting a genetic disease depends solely on the recombination value. However, computation of the proportion of informative families is somewhat complicated. Since the mathematical theory of genetic counseling with linked marker genes is rarely discussed in textbooks of human genetics, we present a detailed explanation of the theory, following Nei (1977, 1979) and Chakravarti and Nei (1982).

X-Linked Recessive Diseases

Use of Information from Grandparents

Let us first consider how marker genes are used for predicting genetic diseases. Let D and d be the normal and disease alleles, respectively, at an X-linked locus, and M_1 and M_2 be the alleles at the marker locus. We denote by r the recombination value between the two loci. We consider the case of codominant alleles at the M locus, so that the three possible genotypes in females are identifiable. Since female patients for X-linked recessive genes are very rare, we ignore them. If a mother is heterozygous for d, one-half of her male children are expected to contract the disease. Let the genotype of the mother be DM_1/dM_2 (Fig. I). Then, the proportion $1 - r$ of her sons with allele M_2 is expected to develop the disease. Therefore, if r is small, the probability of predicting the disease before it is expressed is very high. Clearly, marker

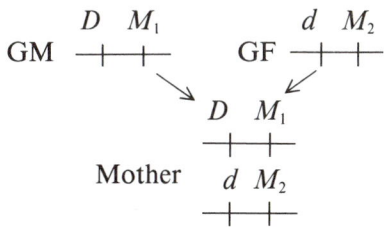

FIG. I. Two gametes producing a double-heterozygous mother for X-linked genes. GM, grandmother; GF, grandfather.

genes are useful for increasing the predictability of genetic diseases. Note that marker genes are also useful for predicting the carrier status of a female offspring. In the above example, if the DM_1/dM_2 mother marries a husband with genotype DM_1, the proportion $1 - r$ of her daughters with genotype M_1M_2 will carry the d gene.

In the above discussion we assumed that the mother is a double heterozygote. Actually, this is essential; if she is not a double heterozygote, the marker locus has no utility for increasing the predictability of genetic diseases. There are two types of double heterozygotes, i.e., DM_1/dM_2 and DM_2/dM_1. Both of these heterozygotes are informative for genetic counseling. However, to make a prediction we must know whether allele d is linked with M_1 or M_2. Knowledge of double heterozygosity alone is not sufficient. In general, we cannot determine the linkage phase of a mother by examining her genotype.

In the case of X-linked genes the linkage phase of a mother can be determined by examining the genotype of the grandfather (see Fig. I). If his genotype is dM_2, then the mother's genotype must be DM_1/dM_2, neglecting new mutation. On the other hand, if his genotype is dM_1, then the mother's genotype is DM_2/dM_1. With certain genetic diseases, the heterozygous condition of the D locus is detectable by biochemical techniques. In this case the linkage phase of the mother can be determined even if the grandfather does not have the disease gene. If the grandfather is normal, the d gene carried by the mother must have come from the grandmother, but the mother's linkage phase can be determined by examining the grandfather. Namely, if his genotype is DM_1, then the mother's genotype must be DM_1/dM_2. On the other hand, if his genotype is DM_2, then the mother's genotype must be DM_2/dM_1.

The question we now face is: How often can we tell the linkage phase of a mother with gene d if she visits a genetic counselor? The answer to this question depends on whether there is dominance at the marker locus or not. We first consider the case of no dominance and then the case of dominance.

Codominant Markers. Let x_1 and x_2 be the frequencies of alleles M_1 and M_2 at the M locus, respectively. The probability that a mother heterozygous for the d gene is also heterozygous at the M locus, given that the grandfather has allele d, is considered. For simplicity, it is assumed that the gene frequencies at the D and M loci have reached the equilibrium value and there is linkage equilibrium between the two loci. [This assumption does not always hold; see Asmussen and Clegg (1982) and Chakravarti (1983) for the effect of linkage disequilibrium.] The grandfather may have either allele M_1 or M_2 at the M locus. The probability that he has M_1 is x_1, and the probability that he has allele M_2 is x_2. The probability that the mother is DM_1/dM_2, given that she is heterozygous for the d gene and the grandfather is dM_2, is x_1. Therefore, the probability that the affected grandfather has allele M_2 at the M locus and the mother is DM_1/dM_2 is x_1x_2. Similarly, the probability that the affected grandfather has allele M_1 at the M locus and the mother is DM_2/dM_1 is x_1x_2. Thus, the total probability that a mother is informative or the proportion of informative mothers is

$$G = 2x_1x_2. \tag{3}$$

We call this the utility index of a polymorphic locus for genetic counseling.

In the above computation, we assumed that there are only two alleles at the M locus, but the above result can easily be extended to the case of multiple alleles. That is, if there are k alleles and the frequency of the ith allele (M_i) is x_i, then the utility index is

$$G = \sum_{i=1}^{k} x_i(1 - x_i) \tag{4}$$
$$= 1 - \Sigma x_i^2.$$

This is identical with Eq. (1). The same formula can be obtained when the grandfather is normal (Nei 1977, 1979). The maximum value of G for a given value of k is obtained when all allele frequencies are equal, i.e., $x_i = 1/k$. In this case, G is $1 - 1/k$ and becomes 0.5, 0.67, 0.75, and 0.80 for $k = 2$, 3, 4, and 5, respectively.

Dominant Markers. If there is dominance at the marker locus, we need information on the genotypes of both grandfather and grandmother to determine the linkage phase of the mother. Figure II shows the pedigrees in which the linkage phase of the mother can be determined. In this figure M and m denote the dominant and recessive alleles at the M locus, and the only thing known about the mother before examination of the M locus is that she is a carrier because her father (grandfather) had the disease. However, the double heterozygosity and linkage phase of the mother can be deter-

$Dm/Dm \times dM$ DM/DM, DM/Dm, $DM/dM \times dm$
$Dm/dm \downarrow$ DM/dm, Dm/dM

Mother dM/Dm Mother dm/DM

(a)   (b)

FIG. II. Pedigrees in which the linkage phase of X-linked genes can be determined. d is an X-linked recessive allele, whereas M (dominant) and m (recessive) are marker alleles.

mined simultaneously if the phenotypes of the grandfather and the grandmother for the M locus are known. In the example given in Fig. IIa, the mother's phenotype for the M locus is M, so that d and M must have come from the grandfather and be present on the same chromosome, whereas D and m must have come from the grandmother. Note that chromosome Dm may come from either the Dm/Dm or Dm/dm grandmother. However, there is no need to distinguish between the two cases, since we are interested in the probability that the mother is a double heterozygote, given that she is heterozygous for the d gene. In the case of Fig. IIa, the genotype of the grandfather for the M locus is M, whereas that of the grandmother is mm. Thus, the probability of observing this combination of grandfather and grandmother is $(1 - x)x^2$, where x is the frequency of allele m. Note that when the grandfather has allele M at the M locus, the grandmother's genotype for this locus must be mm. Otherwise, we cannot determine the linkage phase of the mother. Therefore, the expected frequency of informative mothers is also given by $(1 - x)x^2$.

When the grandfather has allele m, the grandmother's genotype at the M locus can be either MM or Mm (Fig. IIb), but only the gamete with allele M produces a heterozygous mother at the M locus. Therefore, the expected frequency of informative mothers is given by $(1 - x)x$. If we consider both M and m grandfathers, the proportion of informative mothers or utility index is given by

$$G = x(1 - x^2). \qquad (5)$$

This formula holds also for the case in which the grandfather is normal.

Use of Information from Offspring

In the above formulation, we assumed that the phenotypes of grandparents for the marker loci are known. In practice, this assumption does not necessarily hold. When information on the phenotypes of grandparents is not available, one can use information from other relatives to infer the linkage phase of the counselee. The most informative relatives for this purpose are offspring. Indeed, information from offspring is often more useful than that from grandparents (Chakravarti and Nei 1982). When we use information from offspring, however, linkage phase cannot be determined with certainty; it must be estimated.

The principle of using offspring information for predicting the linkage phase is as follows. Suppose that the genotype of a counselee (mother) is DdM_1M_2, where M_1 and M_2 are codominant markers, and that she has two affected sons (d) with genotype M_2 at the marker locus. If the linkage phase of the counselee is coupling (DM_1/dM_2), the probability of having two sons with genotype dM_2 is $P(A|C) = (1 - r)^2/4$. On the other hand, if the linkage phase is repulsion, the probability is $P(A|R) = r^2/4$. It is assumed that the disease and marker loci are in linkage equilibrium, so that the coupling and repulsion phases are equally frequent in the population. Then, the Bayesian posterior probability that the linkage phase is coupling is

$$\begin{aligned} P(C|A) &= \frac{P(A|C)}{P(A|C) + P(A|R)} \\ &= \frac{(1 - r)^2}{(1 - r)^2 + r^2}. \end{aligned} \qquad (6)$$

Similarly, the probability of repulsion of $P(R|A) = r^2/[(1 - r)^2 + r^2]$. Therefore, if $r = 0.01$, $P(C|A) = 0.9999$ and $P(R|A) = 0.0001$. Thus, we can predict the linkage phase with a high probability.

In practice, a counselee may have both affected and normal sons as well as recombinant and nonrecombinant types. Let n_1, n_2, n_3, and n_4 be the numbers of sons with genotypes DM_1, DM_2, dM_1, and dM_2, respectively. If the linkage phase of the mother is coupling, the probability of having n sons with this set of genotypes is

$$P(A|C) = r^{n_2 + n_3}(1 - r)^{n_1 + n_4}/2^n, \qquad (7a)$$

where $n = n_1 + n_2 + n_3 + n_4$. On the other hand, if the linkage phase is repulsion, the probability is

$$P(A|R) = r^{n_1 + n_4}(1 - r)^{n_2 + n_3}/2^n. \qquad (7b)$$

Therefore, the posterior probability that the mother is in coupling phase is

$$P(C|A) = \frac{r^{n_2+n_3}(1-r)^{n_1+n_4}}{r^{n_2+n_3}(1-r)^{n_1+n_4} + r^{n_1+n_4}(1-r)^{n_2+n_3}}$$

$$= \frac{1}{1+\rho^\alpha},$$

(8a)

where $\rho = r/(1-r)$ and $\alpha = n_1 + n_4 - (n_2 + n_3)$. Furthermore, we have

$$P(R|A) = 1 - P(C|A) = 1/(1 + \rho^{-\alpha}). \quad (8b)$$

Formula (8a) indicates that $P(C|A)$ depends on ρ and α. When ρ is small, even $\alpha = 1$ gives a high value of $P(C|A)$. When $\alpha = 0$, $P(C|A) = P(R|A) = \frac{1}{2}$, and information from children is of no use. However, if r is small, this event is expected to occur very rarely, as will be discussed later. The value of $P(C|A)$ for various values of r is given in Table I. It is clear that when r is small and α is large the linkage phase is determined with a high probability.

Prediction of Genetic Diseases. Suppose that a counselee wants to know the probability that her next son will develop the genetic disease in question. Consider the case in which her son's genotype at the marker locus is M_1. In this case, if the mother is in the coupling phase (DM_1/dM_2), the probability that her son will develop the genetic disease is r, whereas if the mother is in the repulsion phase, the probability is $1 - r$, as mentioned earlier. From the information on her previous sons, the probabilities of coupling [$P(C|A)$] and repulsion

[$P(R|A)$] can be computed by Eqs. (8a) and (8b), respectively. Therefore, the Bayesian probability that he will develop the genetic disease is

$$R(M_1) = rP(C|A) + (1 - r)R(C|A). \quad (9a)$$

On the other hand, the risk for a son with genotype M_2 to develop the disease is

$$R(M_2) = (1 - r)P(C|A) + rP(R|A)$$
$$= 1 - R(M_1).$$

(9b)

The values of $R(M_1)$ for various values of r and α are given in Table I. $R(M_1)$ rapidly decreases with increasing α and approaches the value for the case of unambiguous determination of linkage phase when $\alpha \geq 4$ and $r \leq 0.05$. $1 - R(M_1)$ may be called the accuracy of prediction of genetic disease. This accuracy cannot be made close to 1 unless r is close to 0.

So far we have confined ourselves to male offspring. However, Eqs. (9a) and (9b) give the probability of a female offspring being a disease-gene carrier as well, if information on the father's genotype is given. Namely, $R(M_1)$ and $R(M_2)$ give the probabilities of carrier status when the maternally inherited gene is M_1 and M_2, respectively.

Proportion of Informative Families. As mentioned earlier, linked marker genes are useful only when the counselee is a double heterozygote and the linkage phase is known. Consider a counselee who is known to be a disease-gene carrier from her previous sons. She will be a double heterozygote if the marker locus is

TABLE I. Posterior Probabilities of Coupling Phase [$P(C/A)$] and the Risks for a Male Offspring with Marker Gene M_1 to have X-linked Disease Gene d [$R(M_1)$][a]

α	$r = 0.05$		$r = 0.005$		$r = 0.001$	
	$P(C/A)$	$R(M_1)$	$P(C/A)$	$R(M_1)$	$P(C/A)$	$R(M_1)$
0	0.5000	0.5000	0.5000	0.5000	0.5000	0.5000
1	0.9500	0.0950	0.9950	0.0099	0.9990	0.0020
2	0.9972	0.0525	1.0000	0.0050	1.0000	0.0010
3	0.9998	0.0501	1.0000	0.0050	1.0000	0.0010
≥ 4	1.0000	0.0500	1.0000	0.0050	1.0000	0.0010

[a] $\alpha = n_1 + n_4 - (n_2 + n_3)$, where n_1, n_2, n_3, and n_4 are the numbers of DM_1, DM_2, dM_1, and dM_2 males, respectively. r is the recombination value.

8

heterozygous. If the frequency of allele M_1 in the population is x_1, the probability that an individual is heterozygous is $2x_1(1 - x_1)$. If there are multiple alleles at the marker locus, this probability will of course be $1 - \Sigma\, x_i^2$, where x_i is the frequency of the ith allele.

We have seen that the linkage phase of the counselee can be determined whenever $\alpha \neq 0$. If $\alpha = 0$, however, linked marker genes are of no use. Let us now consider the probability of $\alpha = 0$ for the case in which the counselee has n sons. First assume that the counselee is in coupling phase (DM_1/dM_2). In this case, the recombinant genotypes DM_2 and dM_1 appear with probability r among her sons, whereas the nonrecombinants DM_1 and dM_2 appear with probability $1 - r$. Therefore, the probability of $\alpha = 0$ is $\{n!/[(n/2)!]^2\}r^{n/2}(1 - r)^{n/2}$. When the mother is in the repulsion phase, the genotypes DM_2 and dM_1 appear with probability $1 - r$, whereas DM_1 and dM_2 appear with probability r. Therefore, the probability of $\alpha = 0$ is the same as that for the case of coupling. Thus, for computing the proportion of families with $\alpha = 0$, we do not have to consider the linkage phase of the counselee.

The probability of $\alpha = 0$ becomes $2r(1 - r)$ when $n = 2$. This value is very small when r is small. In the case of $n = 4$, the probability is even smaller, i.e., $6r^2(1 - r)^2$. Thus, as long as r is small, say, smaller than 0.01, the probability of $\alpha = 0$ is very small, and it can be assumed that the linkage phase of the counselee can almost always be determined from information on the phenotypes of children. The proportion of informative families is then given by

$$G = 1 - \Sigma\, x_i^2, \qquad (10)$$

which is the same as that for the case of determination of linkage phase through grandpar-

ents [Eq. (4)]. It is clear from Eq. (10) that a locus with high heterozygosity is more useful for genetic counseling than a locus with low heterozygosity.

In the above formulation, we assumed that marker alleles are codominant. Computation of G for dominant markers is a little more complicated, but a relatively simple formula has been obtained by Chakravarti and Nei (1982). The reader who is interested in this problem may refer to their paper.

Autosomal Dominant Diseases

Use of Information from Grandparents

Codominant Markers. In the case of autosomal dominant diseases, information on the genotypes of both parents as well as on those of grandparents is required for linked marker genes to be useful. This can be seen from example (a) in Fig. III, in which D denotes a dominant disease gene. In this example, the genotype of the marker locus is M_1M_2 for the mother and M_1M_1 for the father. Therefore, if r is the recombination value between the D and M loci, the offspring with genotype M_1M_1 is expected to have the dominant disease gene D with probability $1 - r$.

The frequency of dominant disease genes is generally so low that homozygotes for D are virtually nonexistent. Therefore, we assume that all affected individuals are heterozygous. From Fig. III, it is clear that both affected and normal grandparents can be either homozygous or heterozygous for the marker locus. If we consider the case in which the affected grandparent is homozygous for allele M_1, the probability that the affected parent's genotype

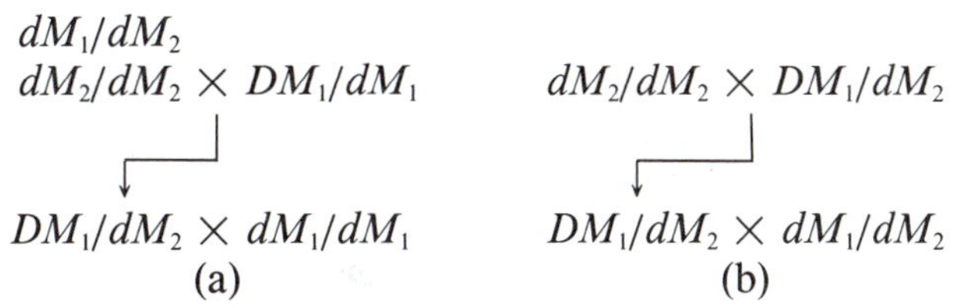

$$dM_1/dM_2$$
$$dM_2/dM_2 \times DM_1/dM_1 \qquad\qquad dM_2/dM_2 \times DM_1/dM_2$$
$$\downarrow\qquad\qquad\qquad\qquad\qquad\downarrow$$
$$DM_1/dM_2 \times dM_1/dM_1 \qquad\qquad DM_1/dM_2 \times dM_1/dM_2$$
$$\text{(a)}\qquad\qquad\qquad\qquad\qquad\text{(b)}$$

FIG. III. Examples of informative families for genetic counseling.

9

(DM_1/dM_2) can be determined is $x_1^2 x_2$. Similarly, when the affected grandparent is M_2M_2, the probability that the affected parent is informative is $x_1 x_2^2$. On the other hand, if the affected grandparent is M_1M_2, the probability is $x_1 x_2(x_1^2 + x_2^2)$. Therefore, the total frequency of informative parents or utility index is

$$G = 2x_1 x_2(1 - x_1 x_2). \qquad (11)$$

In the presence of multiple alleles at the marker locus, the frequency of informative parents may be computed in the following way. We first note that the affected parent must be heterozygous for the marker locus and that one of the genes comes from the grandfather and the other from the grandmother. Obviously, the chance that the affected parent is heterozygous for the ith and jth alleles (M_iM_j) is $2x_i x_j$. However, when both grandfather and grandmother are heterozygous for these alleles, the linkage phase of the affected parent cannot be determined. Therefore, the probability of this event must be subtracted from $2x_i x_j$. The chance that both grandfather and grandmother are heterozygotes is $4x_i^2 x_j^2$, but an affected parent from such a mating becomes heterozygous only with probability ½. Therefore, the probability that an affected parent with M_iM_j is informative is $2x_i x_j - 2x_i^2 x_j^2$. Thus, the total frequency of informative parents or utility index is

$$G = 2 \sum_{i<j} (x_i x_j - x_i^2 x_j^2) \qquad (12)$$
$$= 1 - \Sigma x_i^2 - (\Sigma x_i^2)^2 + \Sigma x_i^4$$

(Nei 1977, 1979). The above equation is identical with Botstein et al.'s (1981) polymorphic information content. Although these authors proposed this quantity as a general measure of utility of a polymorphic locus for genetic counseling, it is valid only for the present case.

The maximum value of G is obtained when all allele frequencies are equal, i.e., $x_i = 1/k$. In this case, G is 0.375, 0.593, 0.703, and 0.768 for $k = 2, 3, 4,$ and 5, respectively.

Dominant Markers with Two Alleles. Dominance reduces the frequency of informative parents considerably, since it makes the distinction of genotypes difficult. Furthermore, if an affected parent mates with a dominant homozygote for a marker locus, marker genes are totally uninformative. In the case in which DM/dm or Dm/dM mates with dM/dm, the family is only partially informative. Namely, in the mating $Dm/dM \times dM/dm$ an offspring with phenotype m (genotype mm) is expected to carry disease gene D with probability $1 - r$, whereas an offspring with phenotype M is expected to carry D with probability $(1 + r)/3$. Thus, marker genes are not very useful if an offspring shows a dominant phenotype. Moreover, the distinction between MM and Mm for a spouse is not possible. Therefore, unless the frequency of m is very high, such "partially informative families" are not really informative. For this reason, we shall not consider these partially informative families in the following.

In the case of two alleles, there are only two types of informative families. They are shown in Fig. IV. It is easy to show that the probability

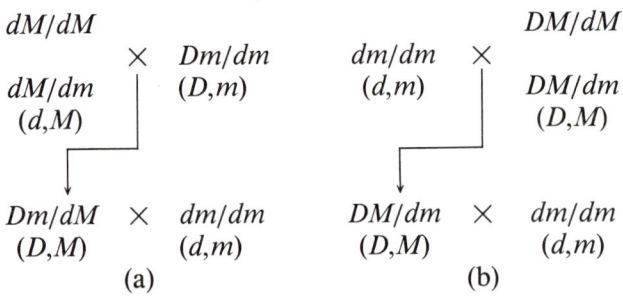

FIG. IV. Pedigrees in which autosomal dominant markers may be used for genetic counseling. Symbols in parentheses stand for the phenotype of each individual.

of obtaining these families or the frequency of informative families on the condition that the affected parent is heterozygous for the D gene and the genotype of his or her spouse is dd is

$$G = 2(1 - x)x^4, \qquad (13)$$

where x is the gene frequency of m.

In the presence of dominance, it is difficult to derive a general formula for the case of multiple alleles, since the dominance relationship among alleles may vary from locus to locus. Nei (1979) has studied the proportion of informative families for the case of the *ABO* blood group locus.

Use of Information from Offspring

We assume that the frequency of disease gene D is so low that the mutant homozygote does not occur, and all affected individuals have genotype Dd, where d is the normal allele. The spouse of a counselee is assumed to be normal with genotype dd. As before, r is the recombination value between the two loci, and we assume that there is linkage equilibrium between the two loci.

Let us assume that the marker locus has two codominant alleles M_1 and M_2. Then, a counselee has the genotype of either DM_1/dM_2 or DM_2/dM_1. The spouse of the counselee can have any of the three genotypes dM_1/dM_1, dM_1/dM_2, and dM_2/dM_2. For codominant markers, all children will give some information for determining linkage phase probabilistically except those from the mating $DdM_1M_2 \times ddM_1M_2$, where only the M_1M_1 and M_2M_2 offspring are useful. This can be seen from Table II, where the expected frequencies of different genotypes in the offspring from the mating $DM_1/dM_2 \times dM_1/dM_2$ are given. It is clear that

if the genotype of an offspring is M_1M_2, it gives no information about the disease gene, since the probability of having gene D is equal to that of having gene d. Note that genotype M_1M_2 appears with probability ½ in the offspring of $DdM_1M_2 \times dM_1/dM_2$.

Consider the mating $DdM_1M_2 \times ddM_1M_1$, and let n_1, n_2, n_3, and n_4 be the numbers of children with genotypes DdM_1M_1, DdM_1M_2, ddM_1M_1, and ddM_1M_2, respectively. Estimation of the probability of linkage phase and the risk of having genetic disease is then exactly the same as that for codominant markers linked with X-linked diseases, and Eqs. (8) and (9) and Table I directly apply. Namely, the probabilities that the offspring with marker genotypes M_1M_1 and M_1M_2 develop the genetic disease, $R(M_1M_1)$ and $R(M_1M_2)$, are given by $R(M_2)$ and $R(M_1)$ in Eqs. (9a) and (9b), respectively. Essentially the same computation can be made by using Eqs. (9a) and (9b) for the mating $DdM_1M_2 \times ddM_2M_2$. The same comment applies to the family $DdM_1M_2 \times ddM_1M_2$, but only the M_1M_1 and M_2M_2 offspring are useful.

Let us now consider the proportion of informative families. We denote by x_1 and x_2 the frequencies of the marker alleles M_1 and M_2, respectively. A counselee who is a carrier of the disease gene is informative if he or she has marker genotype M_1M_2 with a known linkage phase and $\alpha \neq 0$ among his or her offspring. As in the case of sex-linked diseases, the probability of $\alpha = 0$ is very low if r is small, so that the possibility of $\alpha = 0$ can be neglected in practice. As mentioned earlier, all offspring of DdM_1M_2 are informative except when the counselee mates with ddM_1M_2 and the genotypes of the offspring are all M_1M_2 at the M locus. When there are n children born to a counselee, the probability that all of them have genotype M_1M_2 is $(½)^n$. Therefore, the proportion of informative families for n children is

$$G = 2x_1x_2[1 - (½)^{n-1}x_1x_2] \qquad (14)$$

TABLE II. Relative Frequencies of Different Genotypes from the Mating $DM_1/dM_2 \times dM_1/dM_2$

Gamete		dM_1 1/2	dM_2 1/2
DM_1	$(1 - r)/2$	DM_1/dM_1	DM_1/dM_2
DM_2	$r/2$	DM_2/dM_1	DM_2/dM_2
dM_1	$r/2$	dM_1/dM_1	dM_1/dM_2
dM_2	$(1 - r)/2$	dM_2/dM_1	dM_2/dM_2

for $n \geq 1$. Note that if $n = 1$, $G = 2x_1x_2(1 - x_1x_2)$, which is equal to the utility index for the case of determination of linkage phase through grandparents [Eq. (11)]. When $n \geq 2$, G in Eq. (14) is higher than the latter value.

11

When there are multiple alleles at the marker locus, the proportion of informative families is

$$G = 2 \sum_{i<j} x_i x_j [1 - (\tfrac{1}{2})^{n-1} x_i x_j], \qquad (15)$$

where x_i is the frequency of the ith allele.

Autosomal Recessive Diseases

Heterozygotes for autosomal recessive genes are usually identified through their affected offspring. With the advent of biochemical techniques, they may also be identified by carrier detection tests. In the case of autosomal recessive diseases, information from grandparents is useless unless their genotypes for the disease locus are identified by a biochemical test. Therefore, genetic counseling is usually conducted by using information from children. The counselees for this case are usually a couple, both of whom are heterozygous for the disease gene d and its normal allele D. For linked marker genes to be useful in this case, at least one member of the couple must be heterozygous for the marker locus.

Mathematical formulation of the risk estimation and the proportion of informative families for autosomal recessive diseases is somewhat complicated. The reader who is interested in this problem should consult Chakravarti and Nei (1982). Here, we present a formula for computing the proportion of informative families for the case of codominant alleles. It is given by

$$G = 1 - \sum_i x_i^4 - \sum_{i \neq j} x_i^2 x_j^2$$
$$= 1 - \left(\sum_{i=1}^{k} x_i^2 \right)^2. \qquad (16)$$

The maximum value of G for a given value of k is attained when all allele frequencies are equal ($x_i = 1/k$). In this case, G is $1 - (1/k)^2$ and becomes 0.75, 0.89, 0.94, and 0.96 for $k = 2, 3, 4,$ and 5, respectively. Thus, the utility of marker genes for detecting autosomal recessive diseases is very high.

Computation of Utility Indices and Other Remarks

We have seen that the proportion of informative families or the utility index for genetic counseling varies with the mode of inheritance of the disease and marker genes. This makes it difficult to present a general utility index for each marker locus. We have therefore decided not to present utility indices in the data tables in Part II. However, if the mode of inheritance of the disease locus is known, the utility index of a particular marker locus can easily be computed by using one of the formulas presented in this chapter.

It should also be noted that the utility index of codominant markers for an X-linked locus is equal to heterozygosity given by Eq. (1) and that this value is given for all loci. Therefore this heterozygosity value can be used for the utility index for X-linked loci.

Another equation that can be used for computing the proportion of informative families is Eq. (17), which will be presented in the next section. This equation is for computing the probability of exclusion of paternity (PE) for the case of two codominant alleles, but it happens to be half the value given by Eq. (11). The PE value obtained by Eq. (17) is given for all cases of two codominant alleles in the data tables in Part II. Therefore, the proportion of informative families for this case can be obtained by doubling the PE value.

In this book, we were concerned only with the case of a single marker gene. When there are several marker genes linked with a disease locus, the accuracy of prediction of a disease may be increased by using all of them. This problem has been discussed by Chakravarti and Buetow (1985).

PATERNITY TESTING

One of the important areas in which polymorphic loci are effectively used is paternity testing. Since males of certain genotypes cannot be the father of a child born to a particular mother, they can be excluded from paternity when they are accused. Recently, Jeffreys et al.

(1985) initiated the use of highly polymorphic minisatellite DNA for this purpose and suggested that this portion of DNA can be used for identification of paternity rather than exclusion (see also Baird et al. 1986). This is a breakthrough in paternity testing, but its general utility should be examined more carefully because this portion of DNA is often subject to unequal crossing over and the molecular technique used is too sophisticated to be used in general paternity testing laboratories (Jeffreys et al. 1986). Here, we consider the conventional methods of exclusion of paternity and present mathematical formulas for computing the probability of paternity exclusion for several different cases.

Exclusion Probability for Individual Loci

Codominant Genes. Let us consider a codominant locus with alleles A_1 and A_2, whose frequencies are x_1 and x_2, respectively. When the genotype of a mother is A_1A_1, the genotype of her child can be either A_1A_1 or A_1A_2, depending on the genotype of her husband. If the child's genotype is A_1A_1, a male with genotype A_2A_2 cannot be his or her father. Similarly, if the child's genotype is A_1A_2, a male with genotype A_1A_1 can be excluded from parentage. Table III shows all possible combinations of mother and child and the genotype of males who can be excluded from paternity. It also includes the frequency of the genotype of each mother, the probability of each mother having a child with a given genotype, and the frequency of the genotype of the male excluded

from parentage under the assumption of random mating. From this table, one can compute the probability of exclusion of paternity. For example, when the genotypes of a mother and her child are both A_1A_1, a male with genotype A_2A_2 cannot be the father. Males with genotype A_2A_2 occur with a probability of x_2^2 in a randomly mating population. Therefore, the probability of exclusion of paternity for this combination of mother and child (A_1A_1–A_1A_1) is x_2^2. The probabilities of exclusion of paternity for the other mother–child combinations can be obtained in the same way and are presented in column (3) in Table III.

The average probability of paternity exclusion for a locus can be obtained by multiplying the exclusion probability for a given mother–child combination by the probability of occurrence of the mother-child combination in a randomly mating population and taking a sum of the products of the two probabilities over all mother–child combinations. For example, the probability of occurrence of the mother–child combination A_1A_1–A_1A_1 is $x_1^2 \times x_1 = x_1^3$ from Table III, whereas the exclusion probability for this mother–child combination is x_2^2. Therefore, the product of the two probabilities is $x_1^3 x_2^2$. Similarly, the products of the two probabilities can be computed for all other mother–child combinations. Thus, the average exclusion probability for this locus becomes

$$PE = x_1^3 x_2^2 + x_1^4 x_2 + x_1^2 x_2^3 + x_1^3 x_2^2$$
$$+ x_1 x_2^4 + x_1^2 x_2^3 \qquad (17)$$
$$= x_1 x_2 (1 - x_1 x_2)$$

TABLE III. Mother–Child Combination, Putative Fathers Excluded, and Probabilities of Paternity Exclusion

Mother		Child		Father excluded		
Genotype	Frequency (1)	Genotype	Probability (2)	Genotype	Frequency (3)	(1) × (2) × (3)
A_1A_1	x_1^2	A_1A_1	x_1	A_2A_2	x_2^2	$x_1^3 x_2^2$
		A_1A_2	x_2	A_1A_1	x_1^2	$x_1^4 x_2$
A_1A_2	$2x_1x_2$	A_1A_1	$x_1/2$	A_2A_2	x_2^2	$x_1^2 x_2^3$
		A_1A_2	$1/2$			
		A_2A_2	$x_2/2$	A_1A_1	x_1^2	$x_1^3 x_2^2$
A_2A_2	x_2^2	A_1A_2	x_1	A_2A_2	x_2^2	$x_1 x_2^4$
		A_2A_2	x_2	A_1A_1	x_1^2	$x_1^2 x_2^3$

13

(Wiener et al. 1930). This probability can be used as a measure of the utility of a locus for paternity testing.

When multiple codominant alleles exist at a locus, computation of the average probability of paternity exclusion is more complicated (Selvin 1980; Chakravarti and Li 1983; Garber and Morris 1983). Chakravarti and Li (1983), however, produced a relatively simple formula for the case of k alleles. It is given by

$$PE = 1 - 2a_2 + a_3 + 3(a_2a_3 - a_5) - 2(a_2^2 - a_4), \quad (18)$$

where $a_n = \sum_{i=1}^{k} x_i^n$. In the case of three alleles, this reduces to

$$PE = \frac{H}{2}\left(1 - \frac{H}{2}\right) + x_1x_2x_3\left(5 - \frac{3}{2}H\right), \quad (19)$$

where $H = 1 - \sum_{i=1}^{k} x_i^2$ (see also Wiener 1962).

Dominant Genes. Computation of the average probability of paternity exclusion for the case of two alleles with dominance is quite simple. Let M and m be the dominant and recessive alleles at a locus and denote by x_1 and x_2 the frequencies of M and m, respectively. In this case, paternity exclusion is possible only when the genotypes of the mother, child, and putative father are mm, Mm, and mm. If any of the trio has a different genotype, the locus is useless for paternity exclusion. Therefore, the probability of paternity exclusion for the locus is x_2^2 (probability of mm woman) $\times x_1$ (probability of such a woman giving birth to an Mm child) $\times x_2^2$ (probability that the putative father is mm). That is,

$$PE = x_1x_2^4 \quad (20)$$

(Wiener et al. 1930).

For any given gene frequency, Eq. (20) is smaller than Eq. (17). Therefore, codominant loci are more informative than dominant loci for paternity testing.

In the presence of multiple alleles with dominance, computation of the average exclusion probability is quite complicated. In the cases of the ABO and A_1A_2BO system, however, general formulas are worked out. They are as follows.

For the ABO (and also *Duffy*) system,

$$PE = p(1 - p)^4 + q(1 - q)^4 + 2pqr^2 + pq(p + q)r^2, \quad (21)$$

where p, q, and r are the frequencies of alleles A, B, and O (or fy^a, fy^b, and fy in the case of the *Duffy* system), respectively (Wiener et al. 1930). Equation (21) also applies to enzyme or protein loci where two codominant alleles and one null allele exist [e.g., acid phosphatase-A (salivary) locus]. When $r = 0$, Eq. (21) becomes $PE = p(1 - p)^4 + q(1 - q)^4 = pq(p^3 + q^3)$. This is different from Eq. (17), although there are only two codominant alleles. This difference has occurred because different definitions of paternal exclusion are used for Eqs. (17) and (21). For example, in the presence of allele fy at the *Duffy* locus male phenotype fy^a (genotype fy^afy^a or fy^afy) cannot be excluded from paternity for the mother–child phenotype combination fy^b–fy^b because phenotype fy^b may have genotype fy^bfy^b or fy^bfy. However, if we know that there is no fy allele in a population, phenotype fy^a (genotype fy^afy^a) can be excluded. This indicates that in the absence of fy Eq. (17) rather than (21) should be used (see Table 147.1).

For the A_1A_2BO system,

$$\begin{aligned} PE = {} & p_1(1 - p_1)^4 + p_2(1 - p_1 - p_2)^4 \\ & + q(1 - q)^4 + 2(p_1 + p_2)qr^2 \\ & + p_1q(p_1 + q)(p_2 + r)^2 \\ & + p_2q(p_2 + q)r^2 + p_1p_2q(1 - p_1 - p_2)^2 \\ & + 2p_1p_2q(p_2 + q + 2r - p_2q), \end{aligned}$$
$$(22)$$

where p_1, p_2, q, and r are the frequencies of A_1, A_2, B, and O, respectively (Li and Chakravarti 1983).

HLA Loci. *HLA* loci are very useful for paternity testing, because there are many polymorphic alleles. The alleles at these loci are codominant with each other, but some populations contain alleles that are not detectable by available antisera. These alleles may be pooled and treated as a single null or blank allele. This blank allele is recessive to the other alleles. Therefore, the mode of inheritance is similar to that of the *ABO* blood group system. A general formula for the probability of exclusion for *HLA* loci has been worked out by Chakravarti and Li (1983).

Consider one of the *HLA* loci (A, B, C, DR, DQ), and let A_1, A_2, A_3, $\ldots$, and A_k be codominant alleles and A_b be the blank allele. We denote the frequencies of A_1, A_2, $\ldots$, A_k, and A_b by x_1, x_2, $\ldots$, x_k, and x_b. The average probability of paternity exclusion is then given by

$$
\begin{aligned}
PE = &\sum_i x_i(1 - x_i)^4 + 2x_b^2 \sum_{i<j} x_i x_j \\
&+ \sum_{i<j} x_i x_j (x_i + x_j)(1 - x_i - x_j)^2 \\
&+ \sum_{i<j} x_i x_j (x_i + x_b)[(1 - x_b - x_i)^2 \quad (23) \\
&- a_2 - x_i^2] + \sum_i x_i(x_i^2 + 3x_i x_b \\
&+ x_b^2)[(1 - r - x_i)^2 - a_2 + x_i^2],
\end{aligned}
$$

where $a_2 = \sum_i x_i^2$. It should be noted, however,

that when there is no blank allele, Eq. (18) should be used.

Linked Marker Genes. When two loci are closely linked, alleles at the two loci may not be randomly combined. This nonrandom association of alleles complicates the computation of exclusion probability. General treatments of this problem are given by Chakraborty and Hedrick (1983), Chakravarti and Li (1983), and Smouse and Adams (1983). When the recombination value is negligibly small, however, it is possible to have a relatively simple formula.

In the following, we present the formulas only for the cases of the *MNSs* and *Rh* blood group systems under the assumption of no recombination.

For the *MNSs* system,

$$
\begin{aligned}
PE &= P_{MN} + P_{Ss} - P_{MN,Ss} \\
&= mn(1 - mn) + pq(1 - pq) \\
&\quad - [1 - (½)H]\,(a^2d + ad^2 + b^2c \quad (24) \\
&\quad + bc^2) + abcd,
\end{aligned}
$$

where m, n, p, and q are the frequencies of alleles M, N, S, and s, and a, b, c, and d are the chromosome frequencies of MS, NS, Ms, and Ns, respectively (Wiener 1952, Li and Chakravarti 1983).

For the *Rh* system,

$$
\begin{aligned}
PE = P_D + P_C + P_E &- P_{CD} \\
&- P_{DE} - P_{CE} + P_{CDE}, \quad (25)
\end{aligned}
$$

where $P_D = Dd^4$, $P_C = Cc(1 - Cc)$, $P_E = Ee(1 - Ee)$, and

$$
\begin{aligned}
P_{CD} &= (d_1^2 + d_2^2 + d_1 d_2)(d_1^2 D_2 + d_2^2 D_1), \\
P_{DE} &= (d_3^2 + d_4^2 + d_3 d_4)(d_3^2 D_4 + d_4^2 D_3), \\
P_{CE} &= [1 - (H/2)](E_2^2 e_1 + E_2 e_1^2 + E_1^2 e_2 \\
&\quad + E_1 e_2^2) - E_1 E_2 e_1 e_2, \\
P_{CDE} &= (v^2 + w^2 + z^2 + vw + vz + wz) \\
&\quad \times (tz^2 + xw^2 + yv^2) - vw^2xz
\end{aligned}
$$

(Rust 1972; Li and Chakravarti 1983). Here, the mathematical notation represent the following gamete frequencies and their functions.

		c					C		
		E	e				E	e	
D		t	u	D_1			x	y	D_2
d		v	w	d_1			0	z	d_2
		E_1	e_1				E_2	e_2	

$$
\begin{aligned}
t + x &= D_3 & u + y &= D_4 \\
v &= d_3 & w + z &= d_4 \\
c &= t + u + v + w \\
e &= u + w + y + z \\
C &= 1 - c \\
D &= D_1 + D_2 = D_3 + D_4 = t + u + x + y \\
d &= d_1 + d_2 = d_3 + d_4 = v + w + z \\
E &= 1 - e
\end{aligned}
$$

15

The frequency of gamete CdE is assumed to be 0 since the gamete does not exist in most populations (see Table 159).

Exclusion Probability When Many Independent Loci Are Used

Average Exclusion Probabilty. As is clear from the data tables in Part II, the average exclusion probability for a single locus is rather low except for some HLA loci. Therefore, to exclude an accused nonfather with a high probability, we must use many genetic loci. The average exclusion probability when r independent loci are used is given by

$$P = 1 - (1 - PE_1) \times (1 - PE_2) \cdots (1 - PE_r), \quad (26)$$

where PE_i is the PE value for the ith locus.

In the United States and many European countries, it is possible to make P as high as 0.99 by using currently available polymorphic loci. Furthermore, many new DNA polymorphisms are being discovered at the present time. Therefore, in the near future it will be possible to exclude virtually all accused nonfathers by using genetic tests. The only problem with this approach is that some genetic tests, particularly those for DNA polymorphisms, are costly and thus cannot be used routinely. It is therefore important to find polymorphic loci which have a high PE value and of which the genotypes can be assayed inexpensively. This problem has recently been discussed by Smouse and Chakraborty (1986) with respect to DNA polymorphisms.

Exclusion Probability for a Particular Mother–Child Combination. Earlier it was shown that the probability of exclusion of an accused nonfather depends on the genotypes of the mother and child under consideration. Thus, in the case of two codominant alleles (A_1, A_2) at a locus (Table III), the probability of exclusion (q) is x_2^2 for the mother–child combination A_1A_1–A_1A_1 and x_1^2 for the mother–child combination A_1A_1–A_1A_2. Therefore, if x_1 is

smaller than x_2, the probability is higher for the former mother–child combination than for the latter. For example, if $x_1^2 = 0.1$ and $x_2 = 0.9$, the former probability is 81 ($= 0.81/0.01$) times greater than the latter. This indicates that the probability of exclusion that should be considered in courts of law be computed for each mother–child combination. When r independent loci are used, this probability (Q) is given by

$$Q = 1 - (1 - q_1) \times (1 - q_2) \cdots (1 - q_r), \quad (27)$$

where q_i is the q value for the ith locus.

Computation of Exclusion Probability

Using the mathematical formulas presented above, we computed the average exclusion probability for most of the polymorphic loci included in this book. The results obtained are presented in the column of PE in the data tables. The PE value is not presented when the mode of inheritance is not clearly established. The immunoglobulin loci are also excluded from the computation of PE, because there is no general mathematical formula available for these loci. It should be noted, however, that the immunoglobulin loci are very useful for paternity testing. Similarly, no PE values are computed for DNA polymorphisms, although they are also very useful. The reason for this is that the current gene frequency data for these polymorphisms are based on small sample sizes and thus their reliability is low.

As is clear from the data tables, the PE value varies with population as well as with genetic locus. Therefore, to achieve an efficient exclusion of paternity for an accused nonfather, we must use a different set of genetic loci for different populations. For example, the ABO locus is quite useful for European populations but of little use in many American Indian populations. We hope that the PE values in the data tables are useful for finding informative loci for paternity testing in various populations.

REFERENCES

Allison, A. C. (1955) Aspects of polymorphism in man. Cold Spring Harbor Symp. Quant. Biol. 20:239–255.

Asmussen, M. A. and Clegg, M. T. (1982) Use of restriction fragment length polymorphisms for genetic counseling: population genetic considerations. Amer. J. Hum. Genet. 34:369–380.

Baird, M., Balazs, I., Giusti, A., Miyazaki, G. L., Nicholas, L., et al. (1986) Allele frequency distribution of two highly polymorphic DNA sequences in three ethnic groups and its application to the determination of paternity. Amer. J. Hum. Genet. 39:489–501.

Botstein, D., White, R. L., Skolnick, M., and Davis, R. W. (1981) Construction of a genetic linkage map in man using restriction fragment length polymorphisms. Amer. J. Hum. Genet. 32:314–331.

Cavalli-Sforza, L. L. and Bodmer, W. F. (1971) The Genetics of Human Populations. Freeman, San Francisco.

Cavalli-Sforza, L. L. and Edwards, A. W. F. (1964) Analysis of human evolution. In: Genetics Today (Proc. 11th Int. Cong. Genet.), Pergamon Press, Oxford, pp. 923–933.

Chakraborty, R., Fuerst, P. A., and Ferrell, R. E. (1979) Potential information in family studies of linkage. In: Genetic Analysis of Common Diseases: Applications to Predictive Factors in Coronary Heart Disease (Sing, C. F. and Skolnick, M., eds.), Alan R. Liss, New York, pp. 297–303.

Chakraborty, R. and Hedrick, P. W. (1983) Paternity exclusion and the paternity index for two linked loci. Hum. Hered. 33:13–23.

Chakravarti, A. (1983) Utility and efficiency of linked marker genes for genetic counseling. III. Proportion of informative families under linkage disequilibrium. Amer. J. Hum. Genet. 35:592–610.

Chakravarti, A. and Buetow, K. H. (1985) A strategy for using multiple linked markers for genetic counseling. Amer. J. Hum. Genet. 37:984–997.

Chakravarti, A. and Li, C. C. (1983) The effect of linkage on paternity calculations. In: Inclusion Probabilities in Parentage Testing (Walker, R. H., ed.), Amer. Assoc. Blood Banks, Arlington, Virginia, pp. 411–422.

Chakravarti, A. and Nei, M. (1982) Utility and efficiency of linked marker genes for genetic counseling. II. Identification of linkage phase by offspring phenotypes. Amer. J. Hum. Genet. 34:531–551.

Edwards, J. H. (1956) Antenatal detection of hereditary disorders. Lancet 1:i.6922, 579.

Flatz, G. (1987) Genetics of lactose digestion in humans. Adv. Hum. Genet. 16:1–17.

Flint, J., Hill, A. V. S., Bowden, D. K., Oppenheimer, S. J., Sill, P. R., et al. (1986) High frequencies of α-thalassaemia are the result of natural selection of malaria. Nature 321:744–750.

Garber, R. A. and Morris, J. W. (1983) General equations for the average power of exclusion for genetic systems of n codominant alleles in one-parent and no-parent cases of disputed parentage. In: Inclusion Probabilities in Parentage Testing (Walker, R. H., ed.), Amer. Assoc. Blood Banks, Arlington, Virginia, pp. 277–280.

Goedde, H. W. (1986) Ethnic differences in reactions to drugs and other xenobiotics—outlook of a geneticist. In: Ethnic Differences in Reactions to Drugs and Xenobiotics (Kalow, W., Goedde, H. W., and Agarwal, D. P., eds.), Alan R. Liss, New York, pp. 9–20.

Haldane, J. B. S. (1949) The rate of mutation of human genes. Proc. 8th Int. Cong. Genet., Stockholm, pp. 267–273.

Hoogvliet, B. (1942) Genetische en klinische beschouwing naar aanleiding van bloederziekte en kleurenblindheit in dezelfde familie. Genetica 23:93–220.

Jacquard, A. (1974) The Genetic Structure of Populations, Springer-Verlag, Heidelberg.

Jeffreys, A. J., Wilson, V., and Thein, S. L. (1985) Individual-specific "fingerprints" of human DNA. Nature 316:76–79.

Jeffreys, A. J., Wilson, V., Thein, S. L., Wheatherall, D. J., and Ponder, B. A. J. (1986) DNA "fingerprints" and segregation analysis of multiple markers in human pedigrees. Amer. J. Hum. Genet. 39:11–24.

Lewontin, R. C. (1972) The apportionment of human diversity. Evol. Biol. 6:381–398.

Li, C. C. and Chakravarti, A. (1983) On the exclusion and paternity probabilities. In: Inclusion Probabilities in Parentage Testing (Walker, R. H., ed.), Amer. Assoc. Blood Banks, Arlington, Virginia, pp. 609–622.

Nei, M. (1977) Utility and efficiency of linked marker genes for genetic counseling. Bull. Int. Stat. Inst. 47:698–711.

Nei, M. (1979) Proportion of informative families for genetic counseling with linked marker genes. Jpn. J. Human Genet. 24:131–142.

Nei, M. (1987) Molecular Evolutioniary Genetics, Columbia University Press, New York.

Nei, M. and Roychoudhury, A. K. (1972) Gene differences between Caucasian, Negro, and Japanese populations. Science 177:434–436.

Nei, M. and Roychoudhury, A. K. (1974) Genic variation within and between the three major races of man, Caucasoids, Negroids, and Mongoloids. Amer. J. Hum. Genet. 26:421–443.

Nei, M. and Roychoudhury, A. K. (1982) Genetic relationship and evolution of human races. Evol. Biol. 14:1–59.

Reed, T. E. (1969) Caucasian genes in American negroes. Science 165:762–768.

Renwick, J. H. (1969) Widening the scope of antenatal diagnosis. Lancet 2:i.7616, 386.

Rust, P. F. (1972) On the probability of detecting nonpaternity through the Rh blood-group system. Amer. J. Hum. Genet. 24:54–57.

Selvin, S. (1980) Probability of nonpaternity determined by multiple allele codominant systems. Amer. J. Hum. Genet. 32:276–278.

Smouse, P. E. and Adams, J. (1983) The use of closely linked genetic markers for paternity analysis with special consideration of the HLA complex. In: Inclusion Probabilities in Parentage Testing (Walker, R. H., ed.), Amer. Assoc. Blood Banks, Arlington, Virginia, pp. 397–410.

Smouse, P. E. and Chakraborty, R. (1986) The use of restriction fragment length polymorphisms in paternity analysis. Amer. J. Hum. Genet. 38:918–939.

Sneath, P. H. A . and Sokal, R. R. (1973) Numerical Taxonomy, Freeman, San Francisco.

Wainscoat, J. S., Hill, A. V. S., Boyce, A. L., Flint, J., Hernandez, M., et al. (1986) Evolutionary relationships of human populations from an analysis of nuclear DNA polymorphisms. Nature 319:491–493.

Wiener, A. S. (1952) Heredity of the M-N-S blood types: Theoretico-statistical considerations. Amer. J. Hum. Genet. 4:37–53.

Wiener, A. S. (1962) Chances of proving nonpaternity with a system determined by triple allelic codominant genes. Amer. J. Hum. Genet. 20:279–282.

Wiener, A. S., Lederer, M., and Polayes, S. H. (1930) Studies in isohemagglutination. IV. On the chances of proving nonpaternity with special reference to blood groups. J. Immunol. 19:259–282.

2   Genetic Loci Included and Their Properties

This book includes gene frequency data for 362 genetic loci, including both polymorphic and monomorphic loci. The genetic loci are classified into seven different groups: loci for enzymes (131), proteins (69), blood groups and platelet antigens (113), HLA (histocompatibility) (5), immunoglobulins (3), DNA polymorphisms (33), and miscellaneous (8). In each of these groups, genetic loci are arranged in alphabetical order. In the following, we present the list of all the genetic loci included, their chromosomal locations, McKusick numbers, information on polymorphism/monomorphism, and table numbers in which gene frequency data for each locus are given. For enzyme loci, the enzyme commission (E.C.) numbers are also presented. This list is intended to serve as an index for gene frequency data as well as a source of information on some genetic properties of the loci included. The McKusick number refers to the gene number given in McKusick's book Mendelian Inheritance in Man, 7th Edition (Johns Hopkins University Press, 1986). The reader may refer to his book for further information on biochemical and genetic properties of each gene.

The gene symbols used in this book are in accordance with those in Human Gene Mapping 1985 (Karger, Basel), McKusicks's Mendelian Inheritance in Man, or Race and Sanger's (1975) Blood Groups in Man (Blackwell, Oxford, 1975). However, these books do not cover all the genes or genetic systems discussed in our book. We have therefore used our own gene symbols for some loci. These symbols are marked with an asterisk in the following list.

Some readers may want to know the names of the genetic loci studied in each population covered in this book. This information is given in Appendix A at the end of the book. A list of gene symbols and their corresponding names of genes (loci) is also provided (Appendix B). This list may also be used as a subject index.

The chromosomal location of each genetic locus is given by the standard cytogenetic notation. In this notation, the first one or two digits refer to chromosome number, whereas p and q represent the short and long arms of the chromosome. (The X chromosome is represented by X rather than a number.) Each chromosomal arm is divided into one to four regions, and each region is further divided into several chromosome bands. Regions and bands are numbered consecutively from the centromere outward along each chromosome arm. In designating a particular band, four items are required: (1) the chromosome number, (2) the arm symbol, (3) the region number, and (4) the band number within that region. These items are given in order without spacing or punctuation. For example, 4q28 indicates chromosome 4, long arm, region 2, and band 8. A band may be further subdivided into several subbands and subunits of subbands. Thus, 12q24.32 indicates chromosome 12, long arm, region 2, band 4, subband 3, and subband subunit 2. The notation "ter" indicates the terminal (end) of a chromosome arm, whereas "cen" stands for the centromere. When the assignment of the chromosomal location of a locus is provisional, it is marked with (P). When there are inconsistent reports about the location, it is marked with (I). Information on the chromosomal locations of the genes in the following list is based on V. A. McKusick's newsletter on The Human Gene Map, April 15, 1987. The pictorial representation of each chromosome with regions and bands is given in McKusick's book Mendelian Inheritance in Man.

Information on polymorphism is useful in many different areas of human genetics, as discussed in Chapter 1. Although there are several different definitions of genetic polymorphism in the literature, the following one seems to be most reasonable (see Nei 1987 for details). That is, a locus is called polymorphic if the frequency of the most common allele is less than or equal to 0.99. In this book, we have followed this definition except when sample size is very small. It should be noted that there are many loci that are polymorphic in some populations but not in others. In the following list of genetic loci, a locus is called polymorphic (P) if at least one of the populations studied shows polymorphism. The other loci are called monomorphic (M). Note that the genetic loci included in the following list are *not* a random sample of loci from the genome. It seems to us that more polymorphic loci are represented than monomorphic loci simply because the former are studied more extensively than the latter. It should also be noted that the genetic loci for some rare blood groups are not well established and that they may eventually turn out to be identical with some other blood group loci. The reader may refer to Race and Sanger's (1975) book for this problem.

ENZYMES

Genetic locus	Gene symbol	E.C. number	Chromosome location	McKusick number	Poly- morphism	Table
Acetylcholinesterase	ACHE	3.1.1.7		11850	P	1
Acid phosphatase 1	ACP1	3.1.3.2	2p25	17150	P	2
Acid phosphatase-A, salivary	SACPA			17182	P	3
Acid phosphatase-B, salivary	SACPB			17183	P	4
Aconitase 1, soluble	ACO1	4.2.1.3	9p22-p13	10088	P	5
Aconitase 2, mitochondrial	ACO2	4.2.1.3	22q11.21-q13.31	10085	M	75
Adenine phosphoribosyltransferase	APRT	2.4.2.7	16q24	10260	M	75
Adenosine deaminase	ADA	3.5.4.4	20q13.11	10270	P	6
Adenylate kinase 1, soluble	AK1	2.7.4.3	9q34.1-q34.3	10300	P	7
Adenylate kinase 2, mitochondrial	AK2		1p34	10302	M	75
Adenylate kinase 3, mitochondrial	AK3	2.7.4.10	9p24-p13	10303	M	75
Alanine aminotransferase; glutamic pyruvate transaminase	AAT1, GPT1	2.6.1.2	16	13820	P	42
Alcohol dehydrogenase I, alpha polypeptide	ADH1	1.1.1.1	4q21-q25	10370	M	75
Alcohol dehydrogenase I, beta polypeptide	ADH2	1.1.1.1	4q21-q25	10372	P	8
Alcohol dehydrogenase I, gamma polypeptide	ADH3	1.1.1.1	4q21-q25	10373	P	9
Aldehyde dehydrogenase 1, cytosolic	ALDH1	1.2.1.3	9q(P)	10064	M	75
Aldehyde dehydrogenase 2, mitochondrial	ALDH2	1.2.1.3	12	10065	P	10
Aldehyde dehydrogenase 3	ALDH3	1.2.1.3	17(P)	10066	P	11
Aldehyde dehydrogenase 4	ALDH4	1.2.1.3	9q	10067	M	75
Aldolase A	ALDOA	4.1.2.13	16(I)	10385	M	75
Alkaline phosphatase, placental	ALPP, PLAP	3.1.3.1	2(P)	17180	P	12
Aminolevulinate dehydratase, delta	ALAD	4.2.1.24	9q34	12527	P	13
Amylase, alpha; salivary	AMY1	3.2.1.1	1p21	10470	P	14
Amylase, alpha; pancreatic	AMY2	3.2.1.1	1p21	10465	P	15
Arginase, liver	ARG1	3.5.3.1	6q23(P)	20780	M	75
Argininosuccinate lyase	ASL	4.3.2.1	7p21-q22	20790	M	75
Arylesterase; paraoxonase; esterase A	ESA[a], PON	3.1.1.2	7q22	16882	P	55
Arylsulfatase A	ARSA	3.1.6.1	22q13.31-qter	25010	M	75
Beta-N-acetyl-glucosaminidase A; hexosaminidase-A	HEXA	3.2.1.30	15q22-q25.1	27280	M	75
Beta-N-acetyl-glucosaminidase B; hexosaminidase-B	HEXB	3.2.1.30	5q13	26880	M	75
Carbonic anhydrase I	CA1	4.2.1.1	8q13-q22	11480	P	16
Carbonic anhydrase II	CA2	4.2.1.1	8q13-q22	11481	P	17
Carbonic anhydrase III	CA3	4.2.1.1	8q13-q22	11475	P	18
Catalase	CAT	1.11.1.6	11p13	11550	P	19
Cholinesterase (serum) 1; pseudocholinesterase 1	CHE1	3.1.1.8	3q25.2	17740	P	20
Cholinesterase 2; pseudocholinesterase 2	CHE2	3.1.1.8		17750	P	21
Cytidine deaminase	CDA	3.5.4.5		12392	P	22
D-amino acid oxidase	DAMOX	1.4.3.3		12405	M	75
D-aspartate oxidase	DASOX	1.4.3.1		12445	M	75

21

Genetic locus	Gene symbol	E.C. number	Chromosome location	McKusick number	Poly-morphism	Table
Diaphorase NADH	DIA1	1.6.4.3	22q13.31-qter	25080	P	23
Diaphorase 3 (sperm)	DIA3	1.8.1.4		12588	P	24
Diaphorase 4	DIA4	1.6.4.3	16q12-q22	12586	P	25
2,3 Diphosphoglycerate mutase	DPGM	2.7.5.4		22280	P	26
Enolase 1; phosphopyruvate hydratase	ENO1,PPH	4.2.1.11	1pter-p36.13	17243	M	75
Esterase A1; A$_{1-3}$	ESA1	3.1.1.1		13321	P	27
Esterase B, erythrocyte	ESB	3.1.1.1		13326	M	75
Esterase B3, leukocyte	ESB3	3.1.1.1	16(P)	13329	P	28
Esterase D; S-formylglutathione hydrolase	ESD, FGH	3.1.1.1	13q14.11	13328	P	29
Esterase, salivary	SET1			18091	P	30
Formaldehyde dehydrogenase	FDH	1.2.1.1	4q21-q24	13649	M	75
S-formylglutathione hydrolase; esterase D	FGH, ESD	3.1.2.12	13q14.11	13328	P	31
Fucosidase, alpha-L	FUCA	3.2.1.51	1p34	23000	P	32
Fumarate hydratase, soluble	FH1	4.2.1.2	1q42.1	13685	M	75
Fumarate hydratase, mitochondrial	FH2	4.2.1.2		13686	M	75
Gaba transaminase; Gamma-aminobutyric acid transaminase	GABAT	2.6.1.19		13715	P	34
Galactokinase	GALK	2.7.1.6	17q21-q22	23020	M	75
Galactose-1-phosphate uridyl transferase	GALT	2.7.7.12	9p13	23040	P	33
Galactosidase, alpha	GLA	3.2.1.22	Xq22	30150	M	75
Gamma-aminobutyric acid transaminase; Gaba-transaminase	GABAT	2.6.1.19		23015	P	34
Gamma-glutamyl cyclotransferase	GCTG	2.3.2.4	7pter-p14(P)	13717	M	75
Glucose dehydrogenase	GDH	1.1.1.47	1pter-p36.13	13809	P	35
Glucose-6-phosphate dehydrogenase	G6PD	1.1.1.49	Xq28	30590	P	36
Glucose-6-phosphate dehydrogenase, salivary; hexose-6-phosphate dehydrogenase	G6PDS*, H6PD			13810	P	37
Glucose phosphate isomerase; phosphohexose isomerase; phosphoglucose isomerase	GPI, PHI, PGI	5.3.1.9	19cen-q13.2	17240	P	38
Glucosidase, alpha, acid	GAA	3.2.1.20	17q23	23230	P	39
Glutamate dehydrogenase	GLUD	1.4.1.3	10q23-q24(P)	13813	M	75
Glutamic-oxaloacetic transaminase 1, soluble	GOT1	2.6.1.1	10q25.3	13818	P	40
Glutamic-oxaloacetic transaminase 2, mitochondrial	GOT2	2.6.1.1	16cen-q22	13815	P	41
Glutamic pyruvate transaminase; alanine aminotransferase	GPT1, AAT1	2.6.1.2	16	13820	P	42
Glutathione peroxidase	GPX1	1.11.1.9	3p13-q12	23170	P	43
Glutathione reductase	GSR	1.6.4.2	8p21.1	13830	P	44
Glutathione-S-transferase-1	GST1	2.5.1.18		13835	P	45
Glutathione-S-transferase-2	GST2	2.5.1.18	6p12(P)	13836	P	46
Glutathione-S-transferase-3	GST3	2.5.1.18	11q13-q22	13837	M	75
Glyceraldehyde-3-phosphate	GAPD	1.2.1.12	12p13.31-p13.1	13840	M	75

Genetic locus	Gene symbol	E.C. number	Chromosome location	McKusick number	Poly-morphism	Table
dehydrogenase						
Glycerol-3-phosphate dehydrogenase A	GPD1	1.1.1.8	12(P)	13842	M	75
Glycerol-3-phosphate dehydrogenase B	GPD2	1.1.1.8		13843	M	75
Glycolate oxidase	GOX	1.1.3.1			M	75
Glyoxalase I	GLO1	4.4.1.5	6p21.3-p21.2	13875	P	47
Glyoxalase II; hydroxyacyl glutathione hydrolase	GLO2, HAGH	3.1.2.6	16p13	13876	P	49
Guanine deaminase	GDA	3.5.4.3			M	75
Guanylate kinase 1	GUK1	2.7.4.8	1q32.1-q42	13927	M	75
Hexokinase I	HK1	2.7.1.1	10p11.2	14260	M	75
Hexokinase II	HK2	2.7.1.1		14260	M	75
Hexokinase III	HK3	2.7.1.1		14257	P	48
Hexosaminidase-A; beta-N-acetyl-glucosaminidase A	HEXA	3.2.1.30	15q22-q25.1	27280	M	75
Hexosaminidase-B; beta-N-acetyl-glucosaminidase B	HEXB	3.2.1.30	5q13	26880	M	75
Hydroxyacyl glutathione hydrolase; glyoxalase II	HAGH, GLO2	3.1.2.6	16p13	13876	P	49
Hexose-6-phosphate dehydrogenase; glucose-6-phosphate dehydrogenase, salivary	H6PD, G6PDS*			13810	P	37
Hypoxanthine phosphoribosyl transferase	HPRT	2.4.2.8	Xq26-q27.2	30800	M	75
Indophenol oxidase; superoxide dismutase 1, soluble; tetrazolium oxidase	SOD1	1.15.1.1	21q22.1	14745	P	71
Inosine triphosphatase A	ITPA	3.1.6.3	20p	14752	M	75
Isocitrate dehydrogenase, soluble	IDH1	1.1.1.42	2q33.3	14770	M	50
Isocitrate dehydrogenase, mitochondrial	IDH2	1.1.1.42	15q21-qter	14765	M	75
Lactate dehydrogenase A	LDHA	1.1.1.27	11p15-p14	15000	M	51
Lactate dehydrogenase B	LDHB	1.1.1.27	12p12.2-p12.1	15010	M	51
Leucine aminopeptidase, placental	LAPP			15130	M	75
Malate dehydrogenase, soluble	MDH1	1.1.1.37	2p23	15420	M	52
Malate dehydrogenase, mitochondrial	MDH2	1.1.1.37	7p13-q22	15410	M	75
Malic enzyme 1, cytoplasmic	ME1	1.1.1.40	6q12	15425	M	75
Malic enzyme 2, mitochondrial	ME2	1.1.1.40	6pter-p23(P)	15427	P	53
Mannose phosphate isomerase	MPI	5.3.1.8	15q22-qter	15455	M	75
Nucleoside phosphorylase	NP	3.4.2.1	14q13.1	16405	M	54
Ornithine transcarbamylase	OTC	2.1.3.3	Xp21.1	31125	M	75
Paraoxonase; arylesterase; esterase A	PON, ESA[a]	3.1.1.2	7q22	16882	P	55
Peptidase A	PEPA	3.4.11/13	18q23	16980	P	56
Peptidase B	PEPB	3.4.11/13	12q21	16990	P	57
Peptidase C	PEPC	3.4.11/13	1q42	17000	P	58
Peptidase D	PEPD	3.4.13.9	19p13.1-q13.11	17010	P	59
Peptidase E	PEPE		17q23-qter	17020	M	75

ENZYMES (cont'd)

Genetic locus	Gene symbol	E.C. number	Chromosome location	McKusick number	Poly- morphism	Table
Peroxidase, leukocyte	PXL*			17099	M	75
Peroxidase, salivary	PXS*	1.11.1.7		17099	P	60
Phosphofructokinase, liver type	PFKL	2.7.1.11	21q22.3	17186	M	75
Phosphoglucomutase 1	PGM1	2.7.5.1	1p22.1	17190	P	61
Phosphoglucomutase 1 (subtypes)	PGM1	2.7.5.1	1p22.1	17190	P	61.1
Phosphoglucomutase 2	PGM2	2.7.5.1	4p14-q12	17200	P	62
Phosphoglucomutase 3	PGM3	2.7.5.1	6q12	17210	P	63
Phosphogluconate dehydrogenase	PGD	1.1.1.43	1p36.2-p36.13	17220	P	64
Phosphoglycerate kinase	PGK	2.7.2.3	Xq13	31180	P	65
Phosphoglycerate mutase A; phosphoglyceric acid mutase	PGAMA	2.7.5.3	10q25.3	17225	M	75
Phosphoglycolate phosphatase	PGP	3.1.3.18	16p13.31-p13.12	17228	P	66
Phosphohexose isomerase; phosphoglucose isomerase; glucose phosphate isomerase	PHI, GPI, PGI	5.3.1.9	19cen-q13.2	17240	P	38
Phosphoserine phosphatase	PSP	3.1.3.3	7pter-q22	17248	M	75
Pseudocholinesterase 1; cholinesterase (serum) 1	CHE1	3.1.1.8	3q25.2	17740	P	20
Pseudocholinesterase 2; cholinesterase 2	CHE2	3.1.1.8		17750	P	21
Pyridoxine kinase (activity)	PNK	2.7.1.35		17902	P	67
Pyrophosphatase, inorganic	PP	3.6.1.1	10q11.1-q24	17903	M	75
Pyruvate kinase 3	PKM2, PK3	2.7.1.40	15q22-qter	17905	P	68
Rhodanese	RDS			18037	P	69
S-adenosylhomocysteine hydrolase	SAHH	3.3.1.1	20cen-q13.1	18096	P	70
Sorbitol dehydrogenase	SORD	1.1.1.14	15p12-q21	18250	M	75
Succinate dehydrogenase	SDH	1.3.99.1	1p22.1-qter(P)	18547	M	75
Superoxide dismutase 1, soluble; indophenol oxidase; tetrazolium oxidase	SOD1	1.15.1.1	21q22.1	14745	P	71
Superoxide dismutase-A, salivary	SODA				M	75
Superoxide dismutase-B, salivary	SODB				P	72
Tetrazolium oxidase; superoxide dismutase 1, soluble; indophenol oxidase;	SOD1	1.15.1.1	21q22.1	14745	P	71
Triosephosphate isomerase	TPI	5.3.1.1	12p13	19045	M	73
Uridine monophosphate kinase	UMPK	2.7.4	1p32	19171	P	74

[a] Some authors (e.g. Barrantes et al. 1982) have used gene symbol ESA for the ESA1 locus.

PROTEINS

Genetic locus	Gene symbol	Chromosome location	McKusick number	Poly-morphism	Table
Albumin	ALB	4q11-q13	10360	M	76
Alpha-1-antitrypsin; protease inhibitor	PI	14q32.1	10740	P	77
Alpha-1-antitrypsin; protease inhibitor (subtypes)	PI	14q32.1	10740	P	77.1
Alpha-1-acid glycoprotein (locus 1); orosomucoid 1	ORM1	9q34.3-qter	13860	P	114
Alpha-1-acid glycoprotein (locus 2); orosomucoid 2	ORM2			P	115
Alpha-2-HS-glycoprotein	AHSG	3cen-q13	13868	P	78
Anodal tear protein	ATP*		18689	P	79
Antithrombin III	AT3	1q23.1-q23.9	10730	P	80
Apolipoprotein A-IV	APOA4	11q13	10769	P	81
Apolipoprotein E	APOE	19cen-q13.2	20776	P	82
Beta lipoprotein, Ag system	AG		15200	P	83
Beta lipoprotein, Ld system	LD		15210	P	84
Beta lipoprotein, Lp system	LP		15220	P	85
Beta-2-glycoprotein I, Bg system	BG		13870	P	86
Ceruloplasmin	CP	3q21-q25	11770	P	87
Coagulation factor II; prothrombin	F2	11p11-q12(P)			P
Coagulation factor XIII-A	F13A	6pter-p23	13457	P	89
Coagulation factor XIII-B	F13B	7q31.3-q33	13458	P	90
Complement component C1r, subcomponent	C1R	12(P)	21695	P	91
Complement component 2	C2	6p21.3	21700	P	92
Complement component 3	C3	19p13.3-p13.2	12070	P	93
Complement component 4A	C4A	6p21.3	12081	P	94
Complement component 4B	C4B	6p21.3	12082	P	95
Complement component 5	C5		12090	P	96
Complement component 6	C6		21705	P	97
Complement component 7	C7		21707	P	98
Complement component 8, alpha-gamma polypeptide	C8A	1p22(P)	12095	P	99
Complement component 8, beta polypeptide	C8B	1p22(P)	12096	P	100
Double-band parotid salivary protein	DB	12p13.2	16877	P	101
Factor H	HF, CFH	1q32	13437	P	102
Factor I	FI		21703	P	103
Glycine-rich beta-glycoprotein; properdin factor B (subtypes)	GBG, BF	6p21.3	13847	P	127
Glycine-rich beta-glycoprotein; properdin factor B (subtypes)	GBG, BF	6p21.3	13847	P	127.1
Group specific component; vitamin D binding protein	GC, DBP	4q12	13920	P	104
Group specific component; vitamin D binding protein (subtypes)	GC, DBP	4q12	13920	P	104.1
Haptoglobin, alpha	HPA*	16q22.1	14010	P	105
Haptoglobin, alpha (subtypes)	HPA*	16q22.1	14010	P	105.1
Haptoglobin, beta	HPB*		14020	M	139

PROTEINS (cont'd)

Genetic locus	Gene symbol	Chromosome location	McKusick number	Poly-morphism	Table
Hemoglobin, alpha	HBA	16p13.33-p13.11	14180	M	139
Hemoglobin, beta	HBB	11p15.5	14190	P	106
Hemoglobin, delta	HBD	11p15.5	14200	P	107
Hemoglobin, gamma	HBG	11p15.5	14220, 14225	P	108
Hemopexin	HPX	11(P)	14229	P	109
Lymphocyte cytosol 40k polypeptide	LC40P			P	110
Lymphocyte cytosol 49k polypeptide	LC49P			P	111
Lymphocyte cytosol 64k polypeptide	LC64P, LCP1	13q14.1-13q14.3	15343	P	112
Lymphocyte cytosol 100k polypeptide	LC100P		17488	P	113
Orosomucoid 1; alpha-1-acid glycoprotein (locus 1)	ORM1	9q34.3-qter	13860	P	114
Orosomucoid 2; alpha-1-acid glycoprotein (locus 2)	ORM2			P	115
Parotid acidic protein Pa	PA	12p13.2	16873	P	116
Parotid basic protein Pb	PB	12p13.2	16875	P	117
Parotid isoelectric focusing protein	PIF	12p13.2	16872	P	118
Parotid middle-band protein (fast)	PMF*	12p13.2	16878	P	119
Parotid middle-band protein (slow)	PMS*		16878	P	120
Parotid salivary glycoprotein	G1, PRB3	12p13.2	16884	P	121
Parotid size variant	PS	12p13.2	16881	P	122
Pepsinogen	PGA	11q13	16970	P	123
Plasminogen	PLG	6q25-qter	17335	P	124
Proline-rich salivary protein Pr	PR	12p13.2	16879	P	125
Proline-rich salivary protein Pc	PCS	12p13.2	16871	P	126
Properdin factor B; glycine-rich beta-glycoprotein	BF, GBG	6p21.3	13847	P	127
Properdin factor B; glycine-rich beta-glycoprotein (subtypes)	BF, GBG	6p21.3	13847	P	127.1
Protease inhibitor; alpha-1-antitrypsin	PI	14q32.1	10740	P	77
Protease inhibitor; alpha-1-antitrypsin (subtypes)	PI	14q32.1		P	77.1
Prothrombin; coagulation factor II	F2	11p-q12(P)		P	88
Salivary protein CON1	CON1	12p13.2	16887	P	128
Salivary protein CON2	CON2	12p13.2	16888	P	129
Salivary protein Pe	PE	12p13.2	18097	P	130
Salivary protein Po	PO	12p13.2	18099	P	131
Salivary protein SAL I	SAL1		18093	P	132
Salivary protein SAL II	SAL2		18094	P	133
Thyroxine-binding globulin	TBG	Xq28(P)	31420	P	134
Thyroxine-binding prealbumin	TBPA, TTR, PALB	18q11.2-q12.1	17630	M	139
Transcobalamin II	TC2	22q11.2-qter	27535	P	135
Transferrin	TF	3q21	19000	P	136
Transferrin (subtypes)	TF	3q21	19000	P	136.1
Vitamin B$_{12}$ binding (R) protein	VBRP*		19309	P	137
Vitamin D binding protein; group specific component	GC, DBP	4q12	13920	P	104
Vitamin D binding protein;	GC, DBP	4q12	13920	P	104.1

PROTEINS (cont'd)

Genetic locus	Gene symbol	Chromosome location	McKusick number	Poly-morphism	Table
group specific component (subtypes)					
Xm system	XM	Xq28	31490	P	138

BLOOD GROUPS AND PLATELET ANTIGEN SYSTEMS

Genetic locus/ antigen	Gene symbol	Chromosome location	McKusick number	Poly-morphism	Table
ABH secretion	SE	19p13.1-q13.11	18210	P	140
ABO	ABO	9q34	11030	P	141
Ahonen	AN		11035	M	167
Auberger	AU		11040	P	142
August	AUG*			M	167
Bak	BAK		17348	P	168
Batty	BY		11150	M	167
Becker	BEC*		11150	M	167
Berrens	BE		11150	M	167
Biles	BI			M	167
Bishop	BP			M	167
Box	BX			M	167
Caldwell	CL			M	167
Chido	CH*		11043	M	167
Chr	CHR		11150	M	167
Colton	CO	2p(I)	11045	P	143
Cs	CS			P	144
Diego	DI		11050	P	145
Dombrock	DO		11060	P	146
Duch	DH		11065	M	167
Duffy (Fy^a)	FY	1q12-q21	11070	P	147
Duffy (Fy^a, Fy^b)	FY	1q12-q21	11070	P	147.1
Dupuy	DP			M	167
Duzo	DUZ*		17350	P	168
Elridge	EL			M	167
Envelope	EN		11072	M	167
Er	ER*			P	167
Evans	EV*			M	167
Fr	FR			M	167
Gerbich	GE	2q14-q21(P)	11075	P	148
Gonsowski	GN			M	167
Gonzales	GO			M	167
Good	GD*		11150	M	167
Gregory	GY			M	167
Griffths	GF			M	167
Heibel	HEI*			M	167

Genetic locus/ antigen	Gene symbol	Chromosome location	McKusick number	Poly- morphism	Table
Henshaw	HE			P	149
Hey	HEY			M	167
Hov	HOV			M	167
Hunt	HT			M	167
I	I		11080	M	167
Indian antigen	IN			P	167
Jensen	JEN			M	167
Jn	JN			M	167
Jobbins	JO		11150	M	167
Jr	JR			M	167
Kamhuber	KA			M	167
Karhula antigen	UL		11200	P	150
Kell	K, KEL		11090	P	151
Kidd (Jk^a)	JK	2p(I)	11100	P	152
Kidd (Jk^a, Jk^b)	JK	2p(I)	11100	P	152.1
Knops-Helgeson	KN			M	167
Ko	KO		17350	P	168
Lan	LAN		11160	M	167
Lek	LEK			P	168
Levay	LEV*		11150	M	167
Lewis	LE, LES	19p13.1-q13.11	11110	P	153
Lewis II	LS			M	167
Lutheran	LU	19p13.1-q13.11	11120	P	154
LW	LW	19p13.1-q13.11	11125	M	167
Marriott	ZT			M	167
Martin	MT			M	167
McCoy	MCC*			P	155
Mg	MG			M	167
Milne	MIL			M	167
Miltenberger	MI			M	167
Mit	MIT			M	167
MNS, MN	MNS, MN	4q28-q31	11130	P	156
Moen	MO			M	167
Mur	MUR			M	167
Ne	NE*			P	157
Newfoundland	NFLD		11136	M	167
Nyberg	NY			M	167
Ok	OK			M	167
Ol (Oldeide)	OL			M	167
Orriss	ORR*			M	167
Os	OS			M	167
P	P	6(P)	11140	P	158
Pe	PEA*			M	167
Peters	PT			M	167
PL^{A1}, Zw	PLA*		17353	P	168
PL^{E1}	PLE1*		17354	P	168
PL^{E2}	PLE2*		17354	P	168
Raddon	RDD*			M	167

Genetic locus/ antigen	Gene symbol	Chromosome location	McKusick number	Poly- morphism	Table
Radin	RD	1p36.2-p34	11162	M	167
Reid	RE			M	167
Rhesus	RH	1p36.2-p34	11170	P	159
Ridley	RI			M	167
Rm (Romunde)	RM		11150	M	167
Rodgers	RG		11171	P	160
Rosenlund	RL			M	167
Scianna	SC	1p36.2-p34	11175	M	167
Sd	SD		11173	P	161
Skjelbred	SKJ*			M	167
Stobo	ST*			M	167
Stoltzfus	SF	4q28-q31	11180	P	162
Sutter	JS			P	163
Swann	SW		11150	M	167
Tc	TCA*			M	167
Torkildsen	TO			M	167
Traversu	TR		11150	M	167
Tsunoi	TSU*, TS			M	167
V	V			M	164
Vel	VEL		11160	M	167
Ven	VEN*		11150	M	167
Vw (Verweyst)	VW			M	167
Waldner	WAL*, WD		11201	M	167
Webb	WB		11150	M	167
Wright	WR		11205	M	167
Wulfsberg	WU			M	167
Xg	XG	Xpter-p22.32	31470	P	165
Yt (Cartwright)	YT		11210	P	166
Yuk	YUK			P	168
Zd	ZD			M	167
Zw, Pl^{A1}	PLA*		17353	P	168

HLA (HISTOCOMPATIBILITY) SYSTEMS

Genetic locus	Gene symbol	Chromosome location	McKusick number	Poly- morphism	Table
HLA-A histocompatibility type	HLAA	6p21.3	14280	P	169
HLA-B histocompatibility type	HLAB	6p21.3	14283	P	170
HLA-C histocompatibility type	HLAC	6p21.3	14284	P	171
HLA-DQ histocompatibility type	HLADQ	6p21.3	14688	P	172
HLA-DR histocompatibility type	HLADR	6p21.3	14286	P	173

IMMUNOGLOBULIN SYSTEMS

Genetic locus	Gene symbol	Chromosome location	McKusick number	Poly-morphism	Table
Immunoglobulin A2m system	IGA2	14q32.33	14700	P	174
Immunoglobulin Gm1 system	IGHG1	14q32.33	14710	P	175
Immunoglobulin Km (Inv) system	IGKC, KM	2p12	14720	P	176

DNA POLYMORPHISM

Genetic locus	Gene symbol	Chromosome location	McKusick number	Poly-morphism	Table
Albumin	ALB	4q11-q13	10360	P	177
Alpha-1-antitrypsin; protease inhibitor	PI	14q32.1	10740	P	178
Alpha-fetoprotein	AFP	4q11-q13	10415	P	179
Alpha-globin gene cluster	HBAC, AGC	16p13.33-p13.11	14180	P	180
Amylase, alpha, salivary	AMY1	1p21	10470	P	181
Antithrombin III	AT3	1q23.1-q23.9	10730	P	182
Apolipoprotein A-I	APOA1	11q13	10768	P	183
Apolipoprotein A-II	APOA2	1p21-q23	10767	P	184
Apolipoprotein A-IV	APOA4	11q13	10769	P	185
Apolipoprotein B	APOB	2p24	10773	P	186
Apolipoprotein C-II	APOC2	19cen-q13.2	20775	P	187
Apolipoprotein C-III	APOC3	11q13	10772	P	188
Apolipoprotein AI-CIII-AIV	APOA1-APOC3-APOA4	11q13		P	189
Apolipoprotein AI-CIII-AIV	APOA1-APOC3-APOA4	11q13		P	189.1
Beta-globin gene cluster	HBBC, BGC	11p15.5	14190	P	190
Coagulation factor VIIIC; hemophilia A	F8C, HEMA	Xq28	30670	P	191
Coagulation factor IX; hemophilia B	F9, HEMB	Xq27.1-q27.2	30690	P	192
Collagen I alpha-1 polypeptide	COL1A1	17q21.3-q22.05	12015	P	193
Collagen I alpha-2 polypeptide	COL1A2	7q21.3-q22.1	12016	P	194
Collagen II alpha-2 polypeptide	COL2A2			P	194.1
Growth hormone gene cluster	GHC	17q21-q22	13925	P	195
Hemophilia A; coagulation factor VIII	HEMA, F8C	Xq28	30670	P	191
Hemophilia B; coagulation factor IX	HEMB, F9	Xq27.1-q27.2	30690	P	192
Hypoxanthine phosphoribosyltransferase	HPRT	Xq26-q27.2	30800	P	196
Immunoglobulin, heavy chain constant region of IgG	IGHG	14q32.33	14710	P	197
Immunoglobulin, heavy chain constant region of IgM	IGHM	14q32.33	14702	P	198
Insulin	INS	11p15.5	17673	P	199
Insulin receptor	INSR	19p13.3-p13.2	14767	P	200
Nerve growth factor, beta	NGFB	1p22.1	16203	P	201

DNA POLYMORPHISM (cont'd)

Genetic locus	Gene symbol	Chromosome location	McKusick number	Poly-morphism	Table
Parathyroid hormone	PTH	11p15	16845	P	202
Phenylalanine hydroxylase	PAH	12q24.1	26160	P	203
Phosphoglycerate kinase 1	PGK1	Xq13	31180	P	204
Plasminogen	PLG	6q25-qter	17335	P	205
Proopiomelanocortin	POMC	2p25	17683	P	206
Protease inhibitor; alpha-1 -antitrypsin	PI	14q32.1	10740	P	178
Urokinase	URK*	10q24-qter	19184	P	207
DELETION POLYMORPHISM					
Alpha-globin deletion	HBAD*	16q13.33-p13.11	14180	P	208
Alpha-globin deletion (subtypes of -α)	HBAD*	16q13.33-p13.11	14180	P	208.1
Alpha-globin deletion (subtypes of -$\alpha^{3.7}$)	HBAD*	16q13.33-p13.11	14180	P	208.2
Gamma-globin deletion	HBGD*	11p15	14220	P	209

MISCELLANEOUS

Genetic locus	Gene symbol	Chromosome location	McKusick number	Poly-morphism	Table
Acetylator system; isoniazid inactivation	AC		24340	P	210
Beta-aminoisobutyric acid, urinary excretion	BAIB		21010	P	211
Cerumen (ear wax type)	CER*		11780	P	212
Colorblindness	CB*	Xq28	30380	P	213
Debrisoquine oxidation (S-defect); sparteine oxidation (")	DOX*		23685	P	214
Isoniazid inactivation; acetylator system	AC		24340	P	210
Lactase activity	LAA*		22310	P	215
Mephenytoin metabolism (M-defect)	MEP*			P	216
Phenylthiocarbamide taste	PTC		17120	P	217
Sparteine oxidation (S-defect); debrisoquine oxidation (")	DOX*		23685	P	214

Note: Gene symbols assigned by us are marked with *. Symbols (P) and (I) for chromosomal locations indicate "provisional" and "inconsistent", respectively. P and M in the column of "Polymorphism" refer to the polymorphic and monomorphic loci, respectively.

PART II

Tables of Gene Frequencies

3 Explanation of Data Tables

Gene frequency data are presented for different groups of loci separately. For each group, genetic loci are arranged in alphabetical order, excluding those that are monomorphic. Information on variant alleles at monomorphic loci is presented in a single table for each group of loci.

As mentioned earlier, our primary purpose is to present a single set of gene frequency data for each population or each ethnic group at each locus. We have included as many ethnic groups as possible as long as gene frequency data are available. In some countries such as India and New Guinea, however, there is extensive local differentiation of gene frequencies. In these cases, we have included several different populations. Additional populations are also included when the populations are of anthropological interest, even if they are not the major population in the country. For example, the Faroe Island of the North Atlantic belongs to Denmark, but the people there have been isolated for many centuries. For this reason, this population is included in the book.

The populations are classified according to the geographical distribution under the headings of continent, country, and district. This classification is similar to that of Mourant et al.'s (1975) book The Distribution of Blood Groups and Other Polymorphisms. The spelling of the name of each population is also similar to that of Mourant et al.'s book. However, the name of the population belonging to a country is not always described, particularly when (1) there is a single ethnic group in the country, (2) gene frequency data come from the major ethnic group of the country, or (3) there is no clear information. When an ethnic group occupies a large territory, the name of the location at which gene frequency data were obtained is presented in parentheses as long as such information is available. Some ethnic groups such as the San in Africa have two different names. For these ethnic groups, we have used both names.

We have used the standard notations in the literature to designate different alleles at a locus. In the cases of enzyme and protein loci, however, allele notations are often simplified. Alleles at enzyme and protein loci are usually designated by gene symbols with superscripts a, b, c, ... or 1, 2, 3, ... For example, the three common alleles at the acid phosphatase locus ($ACP1$) are designated by $ACP1^a$, $ACP1^b$, and $ACP1^c$. In the present book, these alleses are represented by superscripts only without gene symbols as long as no confusion is introduced.

Some loci contain several rare alleles, and the rare alleles are not always the same for all populations. In this case, the symbol x is used for the heading of rare allele frequencies, and the name of a rare allele existing in a particular population is designated by a superscript for the frequency of the rare allele. For example, the Gujarati population in India has an allele ADA^4 at the adenosine deaminase locus (ADA) with a frequency of 0.004. The allele and its frequency are then given by 0.004^4. When there is an additional rare allele in a population, the allele and its frequency are given under the heading of y (e.g., Table 29). Occasionally, however, the frequencies of two rare alleles are pooled, and the pooled frequency is given with a superscript of two allelic symbols. For example, the sum of the frequencies of alleles $AK1^3$ and $AK1^5$ among the Athabaskan Indians in Alaska is 0.017, and it is given as $0.017^{3,5}$ (Table 7). When a majority of populations contain the same rare allele, the notation of the rare allele is used as the heading for rare allele frequencies, and an exceptional rare allele is represented by the superscript for its frequency (e.g., Table 2).

It should be noted that we have not tried to list all rare alleles for any population. If they happened to be included in the data we cited, they were recorded.

In some genetic systems such as the aldehyde dehydrogenase 2 locus (*ALDH2*), the genetic nature of polymorphism has not been clarified, and the extent of polymorphism is measured by the percentage of individuals with enzyme or protein deficiency. For these loci, we have presented this percentage instead of gene frequencies. Similarly, the polymorphism of lactose tolerance and intolerance is measured by the percentage of individuals with tolerance, although lactose tolerance is apparently controlled by a dominant allele. The reason for this is that this polymorphism is often studied by crude physiological methods and the gene frequency estimates obtained by these methods are not reliable. Actually, even the estimates of phenotype frequencies are not very reliable at this locus, and two different estimates for a given ethnic group are often quite different (Table 215).

Almost all genetic polymorphisms considered in this book are qualitative, and the phenotypes are usually determined unambiguously. However, there are a few loci at which the distinction among different phenotypes is not always clear-cut. A typical example is the polymorphism of the activity level of serum paraoxinase (*PON*). In Caucasian populations, the activity of the enzyme shows a bimodal distribution among different individuals, and low activity individuals are considered to be recessive homozygotes (*BB*). In Oriental and African populations, however, there is no clear-cut bimodal distribution, and consequently gene frequency data for them are not very reliable. We have excluded this type of data from the present book. In the case of the paraoxinase activity polymorphism, La Du et al. (1986) recently showed that the three genotypes (*AA*, *AB*, and *BB*) at this locus are biochemically distinguishable if several different tests are conducted. Applying these tests, they estimated the gene frequencies for a Sudanese population. This set of gene frequency data is included in this book.

Most gene frequencies in the data tables were cited directly from the original papers. Only when the frequencies were apparently mis-printed or computational errors seemed to exist, did we recompute the values. In some cases, we cited gene frequencies from review papers rather than the original papers. In this case, the review papers are identified by (c) in the column of "Source." For the ABO, MNS, Rh, and Xg blood group loci, we also cited gene frequencies from Mourant et al.'s or Tills et al.'s book whenever there were discrepancies between the values in these books and those in the original papers. We did this because the gene frequencies in the two books were estimated by the maximum likelihood method. Similarly we cited GM system gene frequencies from Steinberg and Cook's book when there were discrepancies between them and the estimates given by the original authors.

For the computation of the probability of paternity exclusion (*PE*), we must know the mode of inheritance. At most enzyme or protein loci, different alleles are codominant, so that the *PE* values have been computed by using Eq. (17), (18), or (19) in Chapter 1. When dominance relationships exist among different alleles at a locus, an appropriate formula for the particular case has been used. The dominance relationships for such loci are given in a footnote for the data table. When haplotypes are identified for DNA polymorphism, they are treated as codominant alleles since the recombination value between haplotypes seems to be very small for all DNA polymorphisms included.

Data on restriction-site polymorphisms (restriction fragment length polymorphisms; RFLP) are rapidly accumulating, but the compilation of the data is not easy. This is because different authors often use different DNA probes and different restriction enzymes. The number of individuals or chromosomes studied for a particular DNA segment also often varies with the restriction enzyme used even in the data set obtained by the same authors (e.g., Table 180). Furthermore, the number of chromosomes studied is often very small. We have therefore decided to include only data that are based on relatively large sample sizes and refer to well-defined DNA regions. We have also decided not to include data for "anonymous DNA." This has eliminated a large proportion of RFLP data. However, most of these data are obtainable from the Gene Mapping Laboratory, Howard Hughes Medical Institute, Yale University.

In the tables of DNA polymorphisms, the presence and absence of a polymorphic restriction site are denoted by $+$ and $-$, respectively. The DNA fragment lengths identified by the restriction enzyme method are presented only when such data are available. These lengths are given in units of kilobases (kb). Information on the restriction enzymes used is also provided whenever it is available. At the present time, different authors use different forms of presentation of DNA polymorphisms and do not necessarily provide all information that is useful for population geneticists. We were therefore forced to use various forms of tables for DNA polymorphisms.

In the following tables some general symbols are used. They are as follows.

n	=	Number of individuals examined
n^*	=	Number of chromosomes examined
H	=	Heterozygosity expected under random mating
PE	=	Paternity exclusion probability
kb	=	Kilobases
x,y	=	Rare alleles
+	=	Presence of a restriction site (DNA polymorphisms)
−	=	Absence of a restriction site (DNA polymorphisms)

4 Enzymes

Table 1. Acetylcholinesterase: ACHE

Place/Population	1	2	n	H	PE	Source
NORTH AMERICA						
CANADA						
Whites	.842	.158	70	26.6	11.5	Coates & Simpson 1972

Table 2. Acid phosphatase: ACP1

Place/Population	a	b	c	r	n	H	PE	Source
EUROPE								
AUSTRIA	.370	.570	.060		410	53.5	24.9	Bhasin & Fuhrmann 1972(c)
BELGIUM (Liège)	.337	.611	.052		1,000	51.0	23.5	Brocteur et al. 1980
BULGARIA (Sofia)	.160	.798	.042		119	33.6	16.4	Ananthakrishnan et al. 1972
CZECHOSLOVAKIA	.322	.615	.063		300	51.4	24.4	Salak & Palousova 1971
DENMARK (Copenhagen)	.372	.551	.077		852	55.2	26.6	Sørensen 1973
FAROE IS. (N. Atlantic)	.361	.594	.045		665	51.5	23.2	Tills et al. 1985
FINLAND	.330	.594	.076		2,897	53.2	25.8	Kataja 1975
FRANCE (Paris)	.321	.639	.040		487	48.7	21.9	Bhasin & Fuhrmann 1972(c)
GERMANY (Hamburg)	.353	.582	.065		7,059	53.2	25.1	Brinkmann et al. 1971
GREECE								
Greeks	.322	.633	.045		611	49.4	22.5	Stamatoyannopoulos et al. 1975
Greeks (Plati)	.291	.684	.025		1,038	44.7	19.5	Tills et al. 1983b
HUNGARY	.298	.643	.059		168	49.4	23.4	Goedde et al. 1986b
ICELAND	.374	.562	.063		1,246	54.0	25.3	Tills et al. 1982b
IRELAND	.338	.616	.045		1,787	50.4	22.8	Tills 1977
ITALY (Bologna)	.297	.647	.056		116	49.0	23.1	Beretta et al. 1977
NETHERLANDS (Leiden)	.365	.588	.047		782	51.9	23.5	Fraser et al. 1974
NORWAY								
Norwegians (Oslo)	.381	.560	.059		1,698	53.8	24.9	Lie & Teisberg 1973
Lapps/Saami (Finnmark)	.469	.523	.008		196	50.6	19.7	"
POLAND (Olkusz)	.223	.704	.073		213	44.9	22.4	Wolanski et al. 1983
PORTUGAL (Terras de Bouro)	.358	.522	.121		116	58.5	30.0	Cruz et al. 1973
SPAIN								
Spanish (Madrid)	.324	.624	.052		190	50.3	23.3	Goedde et al. 1972b
Basques (Bermeo/Pamplona)	.257	.717	.026		288	41.9	18.7	"
SWEDEN								
Swedes	.372	.558	.070		517	54.5	25.9	Broman et al. 1971
Lapps/Saami	.507	.462	.031		210	52.9	22.5	Beckman et al. 1971
SWITZERLAND (Bern)	.345	.607	.049		1,365	51.0	23.3	Pflugshaupt et al. 1978
UNITED KINGDOM								
England	.359	.595	.046		1,010	51.5	23.3	Hopkinson & Harris 1969
Scotland (Southwest)	.347	.601	.052		830	51.6	23.7	Mitchell et al. 1976
Isle of Lewis	.452	.500	.048		94	54.3	24.3	Clegg et al. 1985
Isle of Man	.338	.597	.065		325	52.5	24.9	Mitchell et al. 1982
U.S.S.R.								
EUROPEAN PART								
Russians (Moscow)	.322	.654	.024		1,030	46.8	17.9	Altukhov et al. 1981
Abkhazia (Ochamchir Dist.)	.335	.661	.004		254	45.1	17.8	Ferrell et al. 1985
ASIAN PART								
Eskimos (New Chaplino)	.696	.304			102	42.3	16.7	Sukernik et al. 1981
Chukchi (Chukotka & NE Kamchatka)	.570	.430	.001		1,043	49.0	18.6	"

Continued

Table 2. Acid phosphatase: ACP1 (cont'd)

Place/Population	a	b	c	r	n	H	PE	Source
Nganasan-Avam (Taimir Pen.)	.350	.650			203	45.5	17.6	Sukernik et al. 1978
Nentzi (Between Ob R. and Taj R.)	.190	.794	.016		216	33.3	15.0	Sukernik et al. 1980

ASIA

Place/Population	a	b	c	r	n	H	PE	Source
TURKEY (Asia Minor)	.292	.681	.027		274	45.0	19.8	Hummel et al. 1970
ISRAEL								
Ashkenazi Jews	.250	.721	.029		479	41.7	18.8	Goldschmidt 1967
Yemenite Jews	.171	.752	.076		111	39.9	20.3	Tills et al. 1977b
KUWAIT								
Arabs	.203	.778	.019		155	35.3	15.9	Sawhney et al. 1984
SAUDI ARABIA								
Shiites & Sunnites (al-Hasa)	.323	.667	.010		350	45.1	18.4	Goedde et al. 1979b
Tribes (Western)	.098	.846	.056		143	27.2	13.9	Saha et al. 1980
Arabs (Southern)	.159	.841			261	26.7	11.6	Marengo-Rowe et al. 1974
IRAN								
Turkoman (Gonbad, NE Iran)	.371	.606	.023		155	49.5	20.8	Kirk et al. 1977
AFGHANISTAN								
Pushtu & Dari (Kabul)	.297	.683	.020		281	44.5	19.1	Papiha et al. 1977a
INDIA								
Punjabi (Chandigarh)	.318	.675	.007		140	44.3	17.9	Papiha et al. 1972
Gujarati (W. India)	.302	.691	.007		280	43.1	17.5	Papiha et al. 1981
Hindu & Moslems (Madhya Pradesh)	.319	.678	.003		337	43.9	17.4	Roberts et al. 1974
Bhil tr. (Madhya Pradesh)	.199	.801			143	31.9	13.4	Papiha et al. 1978
Hindu (Andhra Pradesh)	.245	.750	.005		210	37.7	15.7	Roberts et al. 1980
Irula tr. (Nilgiri Hills, S. India)	.035	.965			172	6.8	3.3	Saha et al. 1976
Bengali (E. India)	.270	.726	.004		257	40.0	16.3	Das et al. 1970
SRI LANKA								
Sinhalese (Anuradhapura)	.253	.734	.013		156	39.7	17.0	Roberts et al. 1972b
BANGLADESH								
Moslems (Dacca)	.203	.790	.007		200	33.5	14.4	Papiha et al. 1975
NEPAL								
Nepalese	.176	.819	.005		343	29.8	13.0	Sunderland et al. 1979
Sherpas	.330	.670			141	44.2	17.2	Santachiara et al. 1976
BHUTAN	.171	.829			152	28.4	12.2	Mourant et al. 1969
MALAYSIA								
Malays (Singapore)	.326	.672	.002		259	44.2	17.4	Blake et al. 1973a
INDONESIA								
Batak (Samosir Is.)	.241	.759			187	36.6	14.9	McDermid et al. 1973
Balinese (Bali Is.)	.334	.666			148	44.5	17.3	Breguet et al. 1982b
THAILAND	.290	.710			426	41.2	16.4	Edinger et al. 1975
VIETNAM (N. Vietnam)	.260	.740			65	38.5	15.5	Bhasin & Fuhrmann 1972(c)
PHILIPPINES								
Filipino (Germany)	.111	.865	.024		144	23.9	11.6	Windhof & Walter 1983
Negrito (Angeles, C. Luzon)	.144	.856			128	24.7	10.8	Omoto et al. 1978
CHINA								
Mongolians (N. China)	.227	.773			200	35.1	14.5	Goedde et al. 1984b
Zhuang (S. China)	.284	.716			208	40.7	16.2	"
Hui (Ningxia)	.202	.796	.002		218	32.6	13.8	Xu & Du 1984a
Dong (Guangxi)	.216	.784			201	33.9	14.1	"
Tibetans (Switzerland)	.223	.777			110	34.7	14.3	Jeannet et al. 1972
Chinese (Singapore)	.214	.786			378	33.6	14.0	Blake et al. 1973a
JAPAN								
Japanese (Nara & Mie)	.210	.790			1,740	33.2	13.8	Ishimoto 1975
Ainu (Hokkaido)	.274	.726			245	39.8	15.9	Harvey et al. 1978
KOREA								
Koreans (Seoul)	.156	.844			144	26.3	11.4	Goedde et al. 1986c
Koreans (NE China)	.284	.716			214	40.7	16.2	Goedde et al. 1984b

AFRICA

Place/Population	a	b	c	r	n	H	PE	Source
LIBYA (Tripoli & Benghasi)	.140	.845	.015		168	26.6	12.4	Walter et al. 1975

Continued

Table 2. Acid phosphatase: ACP1 (cont'd)

Place/Population	a	b	c	r	n	H	PE	Source
ALGERIA								
Ideles (Ahagger)	.179	.808		.013	151	31.5	14.1	Lefévre-Witier & Vergnes 1977
EGYPT								
Egyptians (Cairo & Anshas)	.246	.719	.035		283	42.1	19.3	Goedde et al. 1980
Nubians (S. Egypt)	.140	.860			154	24.1	10.6	Bertin et al. 1978
SUDAN								
Sudanese (Khartoum)	.174	.789	.035	.002	230	34.6	16.6	Saha et al. 1978b
Beja-Amarar (NE Sudan)	.021	.979			96	4.1	2.0	el Hassan et al. 1968
CHAD								
Sara Majingay	.246	.739	.002	.012	203	39.3	16.9	Hiernaux 1976
MALI								
Kel Kummer Twaregs	.002	.998			285	.4	.2	Lefèvre-Witier & Vergnes
(Menaka)								1977
SENEGAL								
Bedik (E. Senegal)	.257	.740		.003	746	38.6	15.8	Bouloux et al. 1972
LIBERIA	.180	.820			485	29.5	12.6	Goedde et al. 1985a
IVORY COAST								
Tuka	.115	.859		.026	181	24.8	12.1	Vergnes & Cabannes 1976
NIGERIA								
Yoruba (Western)	.182	.773		.045	66	36.7	17.8	Ojikutu et al. 1977
CAMEROON (Bamenda)	.184	.814	.002		282	30.4	13.0	Goedde et al. 1979a
CENTRAL AFRICAN REPUBLIC								
Babinga Pygmies (Western)	.063	.769		.168	167	37.6	18.9	Santachiara et al. 1977
BURUNDI								
Hutu	.124	.859	.014	.002	446	24.7	11.6	Gall et al. 1982
ETHIOPIA								
Amhara tr.	.068	.932			168	12.7	5.9	Harrison et al. 1969
UGANDA								
Baganda (Kampala)	.135	.863		.003	200	23.7	10.7	Roberts et al. 1977
TANZANIA								
Sandawe (Kondoa-Irang Dt.)	.157	.843			210	26.5	11.5	Godber et al. 1976
Nyaturu (")	.153	.847			209	25.9	11.3	"
Hadza (W. Lake Eyasi)	.127	.873			252	22.2	9.9	Tills et al. 1982a
ANGOLA								
Njinga	.193	.807			109	31.0	13.2	Nurse et al. 1979
SOUTHERN AFRICA								
Basters (Namibia)	.344	.572	.008	.076	119	54.9	27.1	Nurse et al. 1982
San/Bushmen								
Tsumkwe !Kung	.203	.574		.223	271	58.0	31.3	Nurse & Jenkins 1977
Khoi/Hottentot								
Keetmanshoop Nama	.164	.617	.003	.218	150	54.5	29.4	Jenkins 1972
Negroes/Dama (Okombahe)	.190	.799	.006	.005	92	32.5	14.4	Nurse et al. 1976
Bantu (Natal)	.130	.834		.036	304	28.6	14.0	Kirk et al. 1971b
Griqua (Campbell,								
Cape Province)	.260	.580	.005	.155	100	57.2	30.8	Nurse & Jenkins 1975
TRISTAN DA CUNHA								
Islanders	.094	.906			159	17.0	7.8	Jenkins et al. 1985

NORTH AMERICA

Place/Population	a	b	c	r	n	H	PE	Source
ALASKA								
Eskimos (Northern)	.601	.399			99	48.0	18.2	Scott et al. 1966
Eskimos (St. Lawrence Is.)	.694	.306			222	42.5	16.7	Ferrell et al. 1981b
Athabaskan Indians								
(Interior valley)	.674	.326			118	43.9	17.1	Scott et al. 1966
CANADA								
Eskimos (Igloolik)	.453	.537	.009		383	50.6	19.8	McAlpine et al. 1974
Ojibwa Indians								
(Pikangikum)	.335	.665			91	44.6	17.3	Szathmary et al. 1974
Cree Indians (Manitoba)	.307	.690	.003		197	43.0	17.1	Lucciola et al. 1974
Dogrib Indians	.462	.538			158	49.7	18.7	Szathmary et al. 1983
(NW Territories)								
UNITED STATES								
Whites (Seattle)	.394	.547	.060		193	54.2	25.1	Giblett 1969
Blacks (")	.250	.720	.014	.015	429	41.9	18.9	"
Whites (Louisiana)	.366	.589	.045		376	51.7	23.3	Fox et al. 1981
Blacks (")	.232	.753	.005	.009	373	37.9	16.5	"
Mexican-Americans (Texas)	.263	.720	.017		668	41.2	17.8	Hewett-Emmett et al. 1986b

Continued

40

Table 2. Acid phosphatase: ACP1 (cont'd)

Place/Population	a	b	c	r	n	H	PE	Source
MIDDLE AMERICA								
MEXICO								
Indians (San Pablo del								
Monte)	.175	.825			103	28.9	12.4	Crawford et al. 1974
Mestizo (Tlaxcala)	.250	.750			72	37.5	15.2	"
GUATEMALA								
Black Caribs (Livingston)	.175	.783		.042	203	35.5	17.2	Crawford et al. 1981
COSTA RICA								
Guaymi	.062	.813		.125[GUA]	286	32.0	1.3	Barrantes et al. 1982
PANAMA								
Blacks (Costa Arriba)	.162	.827		.011	413	29.0	13.1	Ferrell et al. 1978b
WEST INDIES (Haiti)	.185	.810		.005	308	31.0	13.4	Basu et al. 1976
SOUTH AMERICA								
ARGENTINA								
Mapuche	.189	.811			103	30.7	13.0	Haas et al. 1985
BOLIVIA								
Aymara (Western)	.321	.679			316	43.6	17.0	Ferrell et al. 1978a
BRAZIL								
Baniwa (Jandu Cachoeira)	.076	.924			363	14.7	6.8	Salzano et al. 1986
Cayapo (Para & Mato								
Grosso)	.211	.789			83	33.3	13.9	Salzano et al. 1972b
Pacaàs Novos	.032	.968			221	5.8	2.8	Salzano et al. 1985
Macushi (N. Brazil &								
S. Guyana)	.027	.972	.001		499	5.4	2.7	Neel et al. 1977b
Wapishana (")	.061	.939			615	11.5	5.4	"
Ticuna (C. Amazonas)	.062	.827		.111[TIC]	1,763	30.0	15.3	Neel et al. 1980
CHILE								
Atacameño (Toconao,								
N. Chile)	.346	.654			178	45.3	17.5	Goedde et al. 1985b
COLOMBIA								
Noanama	.225	.775			169	34.9	14.4	Kirk et al. 1974
ECUADOR								
Shuara	.167	.833			90	27.8	12.0	Goedde et al. 1977a
Waorani	.099	.901			187	17.8	8.1	Larrick et al. 1985
FRENCH GUIANA								
Wayampi (Trois-Sauts &								
Camopi)	.139	.861			233	23.9	10.5	Tchen et al. 1978
SURINAM								
Trio (S. Surinam)	.170	.830			520	28.2	12.1	Geerdink et al. 1974b
VENEZUELA								
Makiritare (S. Venezuela	.053	.947			535	10.0	4.8	Weitkamp & Neel 1970
& N. Brazil)								
Yanomama (")	.008	.992			2,313	1.6	.8	Weitkamp & Neel 1972
OCEANIA								
AUSTRALIA								
Aborigines/Malag (Elcho								
Is., N. Territory)	.018	.982			641	3.5	1.7	Kirk et al. 1969
Aborigines (Hooker								
Creek, N. Territory)	.005	.995			285	1.0	.5	Blake 1979
Aborigines, (Amoonguna,								
C. Australia)		1.000			104			Kirk et al. 1971a
Aborigines (Kimberley,								
W. Australia)	.015	.985			171	3.0	1.5	Blake & Spargo 1986
MELANESIA								
W. New Guinea								
Asmat (S. Coastal Plain)	.231	.769			650	35.5	14.6	Gajdusek et al. 1978
Papua New Guinea								
Gainj & Kalam (N.								
Central Highlands)	.270	.730			570	39.4	15.8	Long et al. 1986
Jimi Valley								
(W. Highlands)	.249	.751			346	37.4	15.2	Mourant et al. 1981
Yagaria (E. Highlands)	.184	.816			419	30.0	12.8	Mourant et al. 1982

Continued

Table 2. Acid phosphatase: ACP1 (cont'd)

Place/Population	a	b	c	r	n	H	PE	Source
Fuyuge (C. Highlands)	.151	.849			172	25.6	11.2	Woodfield et al. 1974
Daga (Bay Province)	.687	.313			139	43.0	16.9	Jenkins et al. 1983
Karkar Is.								
Takia	.184	.812	.004		277	30.7	13.3	Booth et al. 1982
Manus Is.	.178	.822			138	29.3	12.5	Malcolm et al. 1972
Buka Is.	.188	.812			80	30.5	12.9	McLoughlin et al. 1982a
Solomon Is.	.222	.778			2,723	34.5	14.3	Lai & Bloom 1982
MICRONESIA								
W. Caroline Is.	.060	.940			382	11.3	5.3	Blake et al. 1973b
E. Caroline Is.	.090	.910			824	16.4	7.5	Yamamoto & Fu 1973
Marshall Is.	.234	.766			372	35.8	14.7	Neel et al. 1976
POLYNESIA								
Samoans (Christchurch, New Zealand)	.214	.786			77	33.6	14.0	Booth et al. 1977

Table 3. Acid phosphatase-A, salivary: SACPA

Place/Population	A	A'	A^0	n	H	PE	Source
ASIA							
INDIA							
Indians (Malaysia)	.533	.012	.456	130	50.8	4.0	Tan & Teng 1979
MALAYSIA							
Malays	.469	.001	.530	355	49.9	3.9	"
CHINA							
Chinese (Malaysia)	.436	.010	.555	155	50.2	5.7	"
NORTH AMERICA							
UNITED STATES							
Caucasians (Hawaii)	.698	.009	.293	213	42.7	1.6	Tan & Ashton 1976a
Japanese (")	.731	.014	.255	72	40.0	1.9	"

Note: Alleles A and A' are considered to be codominant to each other and dominant to allele A^0 (null allele).

Table 4. Acid phosphatase-B, salivary: SACPB

Place/Population	B	B^1	B^0	n	H	PE	Source
ASIA							
INDIA							
Indians (Malaysia)	.752		.248	130	37.3	.3	Tan & Teng 1979
MALAYSIA							
Malays	.925		.075	355	13.9	5.2	"
CHINA							
Chinese (Malaysia)	.797	.016	.187	155	33.0	1.8	"
NORTH AMERICA							
UNITED STATES							
Caucasians (Hawaii)	.773		.227	213	35.1	.2	Tan & Ashton 1976a
Japanese (")	.882		.118	72	28.8	6.7	"

Note: Alleles B and B^1 are considered to be codominant to each other and dominant to allele B^0 (null allele).

Table 5. Aconitase 1, soluble: ACO1

Place/Population	1	2	3	4	6	n	H	PE	Source
EUROPE									
Europeans	.992	.003	.004		.001	359	1.6	.8	Slaughter et al. 1975
GERMANY	.991	.005	.004			400	1.8	.9	Schmitt & Ritter 1974
ASIA									
INDIA									
Indians (Malaysia)	1.000					170			Teng et al. 1978b
MALAYSIA									
Malays	.961	.039				155	7.5	3.6	"
CHINA									
Chinese (Malaysia)	.955	.045				165	8.6	4.1	"
JAPAN									
Japanese (Yamanashi)	.951	.049				152	9.3	4.4	Oya et al. 1985
AFRICA									
NIGERIA	.854	.027		.119		359	25.6	12.4	Slaughter et al. 1975
OCEANIA									
AUSTRALIA									
Aborigines									
(N. Australia)	.960	.040				101	7.7	3.7	Blake 1984
Aborigines									
(C. Australia)	.971	.029				119	5.6	2.7	"
MELANESIA									
Papua New Guinea									
E. Highlands	1.000					70			"
Port Moresby	.997	.003				184	.6	.3	"

Table 6. Adenosine deaminase: ADA

Place/Population	1	2	x	n	H	PE	Source
EUROPE							
AUSTRIA (Vienna)	.932	.068	.001[4]	675	12.7	5.9	Wüst 1971
BELGIUM (Liège)	.948	.052		2,074	9.9	4.7	Brocteur et al. 1980
BULGARIA (Sofia)	.862	.138		138	23.8	10.5	Ananthakrishnan et al. 1972
CZECHOSLOVAKIA (Praha)	.957	.043		360	8.2	3.9	Herzog & Bohatová 1973
DENMARK (Copenhagen)	.944	.056		624	10.6	5.0	Sørensen 1972
FAROE IS. (N. Atlantic)	.932	.068		666	12.7	5.9	Tills et al. 1985
FINLAND							
Finns	.921	.079		801	12.8	6.0	Eriksson et al. 1971b
Lapps/Saami	.863	.137		1,154	23.6	10.4	"
FRANCE (Paris)	.959	.041		710	7.9	3.8	Weissmann et al. 1982(c)
GERMANY (Hamburg)	.945	.055		1,070	10.4	4.9	Brinkmann et al. 1971
GREECE							
Greeks	.915	.085		312	15.6	7.2	Stamatoyannopoulos et al. 1975
Greeks (Plati)	.824	.176		1,038	29.0	12.4	Tills et al. 1983b
HUNGARY	.938	.062		1,234	11.6	5.5	Weissmann et al. 1982(c)
IRELAND	.944	.056		1,698	10.6	5.0	Tills 1977
ITALY (Bologna)	.909	.091		276	16.5	7.6	Beretta et al. 1977
NETHERLANDS (Leiden)	.936	.064		798	12.0	5.6	Fraser et al. 1974
NORWAY							
Norwegians (Oslo)	.950	.050		311	9.5	4.5	Camoens et al. 1972
Lapps/Saami (N. Norway)	.848	.152		300	25.8	11.2	"
POLAND (Olkusz)	.941	.059		213	11.1	5.2	Wolanski et al. 1983

Continued

Table 6. Adenosine deaminase: ADA (cont'd)

Place/Population	1	2	x	n	H	PE	Source
PORTUGAL							
Portuguese	.945	.055		571	10.4	4.9	Weissmann et al. 1982(c)
Portuguese (Terras de Bouro)	.920	.080		119	14.7	6.8	Cruz et al. 1973
SPAIN							
Spanish (Madrid)	.966	.034		160	6.6	3.2	Goedde et al. 1972b
Basques (Bermeo/Pamplona)	.972	.028		282	5.4	2.6	"
SWEDEN							
Swedes	.936	.064		342	12.0	5.6	Ageheim & Bergstrom 1972
Lapps/Saami	.893	.107		187	19.1	8.6	Beckman et al. 1971
SWITZERLAND (Bern)	.940	.060		6,604	11.3	5.3	Pflugshaupt et al. 1978
UNITED KINGDOM							
England	.950	.050		1,353	9.5	4.5	Hopkinson et al. 1969
Isle of Lewis	.968	.032		93	6.2	3.0	Clegg et al. 1985
Isle of Man	.943	.057		192	10.8	5.1	Mitchell et al. 1982
Orkney Is. (Westray)	.982	.018		408	3.5	1.7	Welch et al. 1973

U.S.S.R.

Place/Population	1	2	x	n	H	PE	Source
EUROPEAN PART							
Russians (Moscow)	.929	.071		1,047	13.2	6.2	Altukhov et al. 1981
Abkhazia (Ochamchir Dist.)	.893	.107		253	19.1	8.6	Ferrell et al. 1985
Maris (Mari Republic)	.990	.010		316	2.0	1.0	Erickson et al. 1971b

ASIA

Place/Population	1	2	x	n	H	PE	Source
TURKEY (Asia Minor)	.904	.096		566	17.4	7.9	Altay et al. 1974
ISRAEL							
Ashkenazi Jews	.894	.106		437	19.0	8.6	Szeinberg et al. 1971
Yemenite Jews	.911	.089		84	16.2	7.5	Tills et al. 1977b
KUWAIT							
Arabs	.862	.138		87	23.8	10.5	Al-Nassar et al. 1981
SAUDI ARABIA							
Shiites & Sunnites (al-Hasa)	.930	.068	$.002^4$	359	13.0	6.2	Goedde et al. 1979b
IRAN							
Turkoman (Gonbad, NE Iran)	.894	.106		155	19.0	8.6	Kirk et al. 1977
AFGHANISTAN							
Pushtu & Dari (Kabul)	.866	.134		280	23.2	10.3	Papiha et al. 1977a
INDIA							
Punjabi (Chandigarh)	.889	.111		140	19.7	8.9	Papiha et al. 1972
Gujarati (W. India)	.866	.130	$.004^4$	283	23.3	10.5	Papiha et al. 1981
Hindu & Moslems (Madhya Pradesh)	.866	.134		339	23.2	10.3	Roberts et al. 1974
Bhil tr. (Madhya Pradesh)	.921	.079		145	14.6	6.7	Papiha et al. 1978
Hindu (Andhra Pradesh)	.870	.130		211	22.6	10.0	Roberts et al. 1980
Assamese (E. India)	.920	.080		136	14.7	6.8	Goedde et al. 1972a
SRI LANKA							
Sinhalese (Anuradhapura)	.857	.143		154	24.5	10.8	Roberts et al. 1972b
BANGLADESH							
Moslems (Dacca)	.902	.098		200	17.7	8.1	Papiha et al. 1975
NEPAL							
Nepalese	.883	.117		50	20.7	9.3	van den Branden et al. 1971
Sherpas	.976	.024		143	4.7	2.3	Santachiara et al. 1976
MALAYSIA							
Malays (Kuala Lumpur)	.885	.115		325	20.4	9.1	Welch et al. 1975a
INDONESIA							
Batak (Samosir Is.)	.912	.088		188	16.1	7.5	McDermid et al. 1973
Balinese (Bali Is.)	.882	.118		148	20.8	9.3	Breguet et al. 1982b
THAILAND (Northern)	.918	.082		616	15.1	7.0	Sanpitak et al. 1972
PHILIPPINES							
Filipino (Germany)	.957	.043		92	8.2	3.9	Windhof & Walter 1983
Negrito (Angeles, C. Luzon)	.895	.105		129	18.8	8.5	Omoto et al. 1978
CHINA							
Mongolians (N. China)	.953	.047		191	9.0	4.3	Xu et al. 1986
Zhuang (S. China)	.957	.043		211	8.2	3.9	"
Hui (Ningxia)	.939	.061		213	11.5	5.4	Zhao & Du 1984b
Dong (Guangxia)	.955	.045		200	8.6	4.1	"
Tibetans (Switzerland)	.950	.050		110	9.5	4.5	Jeannet et al. 1972

Continued

Table 6. Adenosine deaminase: ADA (cont'd)

Place/Population	1	2	x	n	H	PE	Source
Chinese (Kuala Lumpur)	.939	.061		279	11.5	5.4	Welch et al. 1975a
TAIWAN							
Toroko-Aborigines	.935	.065		124	12.2	5.7	Chen et al. 1985
JAPAN							
Japanese (Tokyo & Mishima)	.969	.031		931	6.0	2.9	Shinoda 1970
Ainu (Hokkaido)	.972	.028		125	5.4	2.6	Omoto & Harada 1972
KOREA							
Koreans (NE China)	.947	.053		216	10.0	4.8	Xu et al. 1986

AFRICA

Place/Population	1	2	x	n	H	PE	Source
LIBYA							
Jews	.909	.091		204	16.5	7.6	Szeinberg et al. 1971
SENEGAL							
Dakar	1.000			200			Weissmann et al. 1982(c)
NIGERIA							
Yoruba (Western)	1.000			65			Ojikutu et al. 1977
CAMEROON (Bamenda)	1.000			284			Goedde et al. 1979a
CENTRAL AFRICAN REPUBLIC							
Pygmies (Western)	1.000			74			Santachiara et al. 1980
BURUNDI							
Hutu	.994	.003	.002[5]	447	1.2	.5	Gall et al. 1982
ZAIRE							
Bantu	1.000			146			Govaerts et al. 1972
UGANDA							
Baganda (Kampala)	.995	.005		199	1.0	.5	Roberts et al. 1977
TANZANIA							
Sandawe	.998	.002		215	.4	.2	van den Branden et al. 1971
Nyaturu	.998	.002		208	.4	.2	"
ANGOLA							
Njinga	.982	.018[Wk]		109	3.5	1.7	Nurse et al. 1979
SOUTHERN AFRICA							
Basters (Namibia)	.975	.025		120	4.9	2.4	Nurse et al. 1982
San/Bushmen							
Tsumkwe !Kung	.998	.002		278	.4	.2	Jenkins 1972
Khoi/Hottentot							
Keemanshoop Nama	.993	.007		134	1.4	.7	"
Negro/Dama (Okombahe)	.995	.005		92	1.0	.5	Nurse et al. 1976
Bantu	.996	.004		138	.8	.4	Weissmann et al. 1982(c)
Griqua (Campbell, Cape Province)	.980	.020		99	3.9	1.9	Nurse & Jenkins 1975
TRISTAN DA CUNHA	.731	.269		156	39.3	15.8	Jenkins et al. 1985

NORTH AMERICA

Place/Population	1	2	x	n	H	PE	Source
GREENLAND							
Eskimos (Augpilagtok)	.974	.026		152	5.1	2.5	Eriksson et al. 1971b
ALASKA							
Eskimos (Anchorage)	1.000			170			Duncan et al. 1974
Eskimos (St. Lawrence Is.)	1.000			222			Ferrell et al. 1981b
Athabaskan Indians	1.000			56[b]			Duncan et al. 1974
CANADA							
Eskimos (Igloolik)	1.000			362			McAlpine et al. 1974
Ojibwa Indians (Pikangikum)	1.000			91			Szathmary et al. 1974
Cree Indians (Manitoba)	.985	.015		197	3.0	1.5	Lucciola et al. 1974
Dogrib Indians (NW Territories)	.997	.003		158	.6	.3	Szathmary et al. 1983
UNITED STATES							
Whites (Seattle)	.952	.048		168	9.1	4.4	Detter et al. 1970b
Blacks (")	.983	.017		184	3.3	1.6	"
Whites (Louisiana)	.956	.044		376	8.4	4.0	Fox et al. 1981
Blacks (")	.995	.005		373	1.0	.5	"
Mexican Americans (Texas)	.963	.037		669	7.1	3.4	Hewett-Emmett et al. 1986b

MIDDLE AMERICA

Place/Population	1	2	x	n	H	PE	Source
GUATEMALA							
Black Caribs (Livingston)	1.000			205			Crawford et al. 1981
COSTA RICA							
Guaymi	1.000			286			Barrantes et al. 1982

Continued

45

Table 6. Adenosine deaminase: ADA (cont'd)

Place/Population	1	2	x	n	H	PE	Source
PANAMA							
Blacks (Bocas del Toro)	.989	.011		N.A.	2.2	1.1	Ferrell et al. 1978b
SOUTH AMERICA							
BOLIVIA							
Aymara (Western)	1.000			316			Ferrell et al. 1978a
BRAZIL							
Baniwa (Jandu Cachoeira)	1.000			363			Salzano et al. 1986
Cayapo (Pará & Mato Grosso)	1.000			87			Salzano et al. 1972b
Caingang	1.000			134			Salzano et al. 1980
Parakana	1.000			93			Black et al. 1980
Pacaás Novos	1.000			152			Salzano et al. 1985
Macushi (N. Brazil & S. Guyana)	1.000			499			Neel et al. 1977b
Wapishana (")	.998	.002		615	.4	.2	"
Ticuna (C. Amazonas)	1.000			1,763[a]			Neel et al. 1980
CHILE							
Atacameño (Toconao, N. Chile)	1.000			178			Goedde et al. 1985b
FRENCH GUIANA							
Wayampi (Trois-Sauts & Camopi)	1.000			237			Tchen et al. 1978
VENEZUELA							
Makiritare (S. Venezuela &							
N. Brazil)	1.000			146			Weitkamp & Neel 1970
Yanomama (")	1.000			669			Weitkamp & Neel 1972
OCEANIA							
AUSTRALIA							
Aborigines/Malag (Elcho Is.,							
N. Territory)	.976	.024		232	4.7	2.3	Omoto 1972
Aborigines (Hooker Creek,							
N. Territory)	.989	.011		280	2.2	1.1	Blake 1979
Aborigines/Walmadjari							
(Kimberley, W.							
Australia)	1.000			171			Blake & Spargo 1986
Aborigines/Yuendumu							
(C. Australia)	1.000			64			Omoto 1972
MELANESIA							
Papua New Guinea							
Gainj & Kalam (N. Central							
Highlands)	.973	.026	.001[b]	570	5.3	2.6	Long et al. 1986
Jimi Valley (W. Highlands)	.864	.136		385	23.5	10.4	Mourant et al. 1981
Yagaria (E. Highlands)	.830	.170		421	28.2	12.1	Mourant et al. 1982
Fuyuge (C. Highlands)	.824	.176		173	29.0	12.4	Woodfield et al. 1974
Daga (Bay Province)	.745	.255		139	38.0	15.4	Jenkins et al. 1983
Karkar Is.							
Takia	.825	.175		289	28.9	12.4	Booth et al. 1982
Manus Is.	.916	.084		137	15.4	7.1	Malcolm et al. 1972
Buka Is.	.956	.044		80	8.4	4.0	McLoughlin et al. 1982a
MICRONESIA							
W. Caroline Is.	.983	.017		381	3.3	1.6	Blake et al. 1973b
Marshall Is.	.935	.065		372	12.2	5.7	Neel et al. 1976
POLYNESIA							
Samoans (Christchurch,							
New Zealand)	.890	.110		77	19.6	8.8	Booth et al. 1977

Note: 'Wk' indicates weak variant showing the same electrophoretic mobility as the common 1-1 type. [a] Two individuals were of genotype (+/2). [b] One individual was of genotype (+/2).

Table 7. Adenylate kinase 1, soluble: AK1

Place/Population	1	2	x	n	H	PE	Source
EUROPE							
AUSTRIA	.966	.034		187	6.6	3.2	Wing 1974(c)
BELGIUM (Liége)	.962	.038		1,000	7.3	3.5	Brocteur et al. 1980
BULGARIA (Sofia)	.964	.036		138	6.9	3.3	Ananthakrishnan et al. 1972
CZECHOSLOVAKIA (Praha)	.959	.041		330	7.9	3.8	Herzog et al. 1972
DENMARK (Copenhagen)	.970	.030		821	5.8	2.8	Sørensen 1972
FAROE IS. (N. Atlantic)	.974	.026		640	5.1	2.5	Tills et al. 1985
FINLAND	.961	.039		804	7.5	3.6	Eriksson et al. 1971a
FRANCE							
French	.963	.037		723	7.1	3.4	Séger 1971
Basques	.976	.024		63	4.7	2.3	Levine et al. 1974
GERMANY (Hamburg)	.966	.034		2,370	6.6	3.2	Brinkmann et al. 1971
GREECE							
Greeks	.960	.039	.001	616	7.7	3.6	Stamatoyannopoulos et al. 1975
Greeks (Plati)	.935	.065		1,026	12.2	5.7	Tills et al. 1983b
HUNGARY	.962	.038		119			Goedde et al. 1986b
ICELAND	.946	.054		1,338	10.2	4.8	Tills et al. 1982b
IRELAND	.966	.034		1,786	6.6	3.2	Tills 1977
ITALY (Bologna)	.970	.030		279	5.8	2.8	Beretta et al. 1977
NETHERLANDS (Leiden)	.966	.034		801	6.6	3.2	Fraser et al. 1974
NORWAY							
Norwegians	.956	.044		377	8.4	4.0	Berg 1969
Lapps/Saami	.987	.013		273	2.6	1.3	"
POLAND (Olkusz)	.948	.052		213	9.9	4.7	Wolanski et al. 1983
PORTUGAL (Terras de Bouro)	.953	.047		118	9.0	4.3	Cruz et al. 1973
SPAIN							
Spanish (Madrid)	.961	.039		203	7.5	3.6	Goedde et al. 1972b
Basques (Bermeo/Pamplona)	.971	.029		286	5.6	2.7	"
SWEDEN							
Swedes	.959	.041		418	7.9	3.8	Beckman et al. 1974
Lapps/Saami	.998	.002		210	.4	.2	Beckman et al. 1971
SWITZERLAND (Bern)	.957	.043		2,269	8.2	3.9	Pflugshaupt et al. 1978
UNITED KINGDOM							
England	.955	.045		1,887	8.6	4.1	Rapley et al. 1967
Scotland (South-west)	.966	.034		412	6.6	3.2	Mitchell et al. 1976
Isle of Lewis	.989	.011		90	2.2	1.1	Clegg et al. 1985
Isle of Man	.966	.034		326	6.6	3.2	Mitchell et al. 1982
Orkney Is. (Westray)	.970	.030		408	5.8	2.8	Welch et al. 1973
YUGOSLAVIA (Serbia)	.961	.039		283	7.5	3.6	Majkic-Singh et al. 1982
U.S.S.R.							
EUROPEAN PART							
Abkhazia (Ochamchir Dist.)	.961	.039		254	7.5	3.6	Ferrell et al. 1985
ASIAN PART							
Eskimos (New Chaplino)	1.000			102			Sukernik et al. 1981
Chukchi (Chukotka & NE Kamchatka)	.999	.001		1,043	.2	.1	"
Nganasan-Avam (Taimir Pen.)	.976	.024		264	4.7	2.3	Sukernik et al. 1978
Nentzi (Between Ob R. & Taj R.)	.998	.002		582	.4	.2	Sukernik et al. 1980
ASIA							
TURKEY (Asia Minor)	.958	.042		274	8.0	3.9	Hummel et al. 1970
ISRAEL							
Ashkenazi Jews	.945	.055		191	10.4	4.9	Szeinberg & Tomashevsky-Tamir 1971
Yemenite Jews	.947	.053		157	10.0	4.8	Tills et al. 1977b
KUWAIT							
Arabs	.972	.028		159	5.4	2.6	Sawhney et al. 1984
SAUDI ARABIA							
Shiites & Sunnites (al-Hasa)	.965	.035		359	6.8	3.3	Goedde et al. 1979b
Arabs (Southern)	.977	.023		261	4.5	2.2	Marengo-Rowe et al. 1974

Continued

Table 7.   Adenylate kinase 1, soluble: AK1 (cont'd)

Place/Population	1	2	x	n	H	PE	Source
IRAN							
Turkoman (Gonbad, NE Iran)	.952	.048		155	9.1	4.4	Kirk et al. 1977
AFGHANISTAN							
Pushtu & Dari (Kabul)	.920	.080		280	14.7	6.8	Papiha et al. 1977a
INDIA							
Punjabi (Chandigarh)	.900	.100		140	18.0	8.2	Papiha et al. 1972
Gujarati (W. India)	.928	.072		283	13.4	6.2	Papiha et al. 1981
Hindu & Moslems							
(Madhya Pradesh)	.902	.098		339	17.7	8.1	Papiha & Chhaparwal 1973
Bhil tr. (Madhya Pradesh)	.959	.041		145	7.9	3.8	Papiha et al. 1978
Hindu (Andhra Pradesh)	.912	.088		211	16.1	7.4	Roberts et al. 1980
Irula tr. (Nilgiri Hills,							
S. India)	.920	.080		175	14.7	6.8	Saha et al. 1976
Bengali (E. India)	.913	.087		271	15.9	7.3	Das et al. 1970
SRI LANKA							
Sinhalese (Anuradhapura)	.917	.083		156	15.2	7.0	Roberts et al. 1972b
BANGLADESH							
Moslems (Dacca)	.913	.087		200	15.9	7.3	Papiha et al. 1975
NEPAL							
Nepalese	.997	.003		320	.6	.3	Sunderland et al. 1979
Sherpas	1.000			142			Santachiara et al. 1976
BHUTAN	1.000			154			Mourant et al. 1969
MALAYSIA							
Malays (Singapore)	.985	.015		259	3.0	1.5	Blake et al. 1973a
INDONESIA							
Batak (Samosir Is.)	.997	.003		188	.6	.3	McDermid et al. 1973
Balinese (Bali Is.)	1.000			148			Breguet et al. 1982b
THAILAND (Northern)	.995	.005		695	1.0	.5	Sanpitak et al. 1972
VIETNAM	.997	.003		150	.6	.3	Bowman et al. 1971
PHILIPPINES							
Filipino (Germany)	.969	.031		144	6.0	2.9	Windhof & Walter 1983
Negrito (Angeles, C. Luzon)	.930	.070		129	13.0	6.1	Omoto et al. 1978
CHINA							
Mongolians (N. China)	.984	.016		191	3.1	1.5	Xu et al. 1986
Zhuang (S. China)	1.000			211			"
Hui (Ningxia)	.998	.002		218	.4	.2	Zhao & Du 1984a
Dong (Guangxi)	1.000			201			"
Tibetans (Switzerland)	.991	.009		110	1.8	.9	Jeannet et al. 1972
Chinese (Singapore)	1.000			378			Blake et al. 1973a
TAIWAN							
Toroko-Aborigines	1.000						Chen et al. 1985
JAPAN							
Japanese (Nagoya City)	1.000			586			Shinoda & Matsunaga 1970
Ainu (Hokkaido)	1.000			104			Harvey et al. 1978
KOREA							
Koreans (Germany)	.933	.067		75	12.5	5.9	Bajatzadeh et al. 1969b
Koreans (NE China)	1.000			216			Xu et al. 1986

AFRICA

Place/Population	1	2	x	n	H	PE	Source
LIBYA (Tripoli & Benghasi)	.982	.018		169	3.5	1.7	Walter et al. 1975
EGYPT							
Egyptians (Cairo & Anshas)	.976	.024		291	4.7	2.3	Goedde et al. 1980
Nubians (S. Egypt)	.952	.048		155	9.1	4.4	Bertin et al. 1978
SUDAN							
Beja-Amarar (NE Sudan)	.995	.005		98	1.0	.5	el Hassan et al. 1968
CHAD							
Sara Majingay	.996		.004[4]	137	.8	.4	Hiernaux 1976
MALI							
Kel Kummer Twaregs							
(Menaka)	.951	.049		285	9.3	4.4	Lefèvre-Witier & Vergnes 1977
SENEGAL							
Bedik (E. Senegal)	1.000			748			Bouloux et al. 1972
LIBERIA	1.000			485			Goedde et al. 1985a
IVORY COAST							
Tuka	1.000			186			Vergnes & Cabannes 1976
NIGERIA							
Yoruba (Western)	1.000			65			Ojikutu et al. 1977

Continued

48

Table 7. Adenylate kinase 1, soluble: AK1 (cont'd)

Place/Population	1	2	x	n	H	PE	Source
CAMEROON (Bamenda)	1.000			284			Goedde et al. 1979a
CENTRAL AFRICAN REPUBLIC							
Babinga Pygmies	1.000			300			Bhasin & Fuhrmann 1972(c)
Pygmies (Western)	1.000			377			Santachiara et al. 1980
BURUNDI							
Hutu	.994	.006		447	1.2	.6	Gall et al. 1982
ETHIOPIA							
Amhara tr.	.987	.013		303	2.6	1.3	Harrison et al. 1969
UGANDA							
Baganda (Kampala)	.997	.003		196	.6	.3	Roberts et al. 1977
TANZANIA							
Sandawe (Kondoa-Irang Dt.)	.993	.007		203	1.4	.7	Godber et al. 1976
Nyaturu (")	.995	.005		203	1.0	.5	"
Hadza (W. Lake Eyasi)	1.000			253			Tills et al. 1982a
ANGOLA							
Njinga	.995	.005		109	1.0	.5	Nurse et al. 1979
SOUTHERN AFRICA							
Basters (Namibia)	.954	.046		120	8.8	4.2	Nurse et al. 1982
San/Bushmen							
!Kung, Dobe	.976	.024		328	4.7	2.3	Nurse & Jenkins 1977
Khoi/Hottentot							
Keetmanshoop Nama	.962	.038		118	7.3	3.5	Jenkins 1972
Negroes/Dama (Okombahe &							
Sesfontein)	.912	.088		119	16.1	7.4	Nurse et al. 1976
Bantu (Natal)	.990	.010		304	2.0	1.0	Kirk et al. 1971b
Griqua (Campbell, Cape	.983	.017		87	3.3	1.6	Nurse & Jenkins 1975
Province)							
TRISTAN DA CUNHA	1.000			153			Jenkins et al. 1985

NORTH AMERICA

GREENLAND							
Eskimos	.984	.016		153	3.1	1.5	Eriksson et al. 1971a
ALASKA							
Eskimos/Inupiak (Northwestern)	1.000			232			Scott & Wright 1978
Eskimos (St. Lawrence Is.)	.980	.002	$.020^6$	222	3.9	1.9	Ferrell et al. 1981b
Athabaskan Indians	.981	.002	$.017^{3,5}$	212	3.7	1.8	Scott & Wright 1978
CANADA							
Eskimos (Igloolik)	1.000			362			McAlpine et al. 1974
Ojibwa Indians (Pikangikum)	1.000			91			Szathmary et al. 1974
Cree Indians (Manitoba)	.995	.005		199	1.0	.5	Lucciola et al. 1974
Dogrib Indians (NW Territories)	1.000			158			Szathmary et al. 1983
UNITED STATES							
Whites (Seattle)	.974	.026		172	5.1	2.5	Giblett 1969
Blacks (")	.993	.007		223	1.4	.7	"
Whites (Louisiana)	.976	.024		376	4.7	2.3	Fox et al. 1981
Blacks (")	.993	.007		372	1.4	.7	"
Mexican-Americans (Texas)	.975	.025		669	4.9	2.4	Hewett-Emmett et al. 1986b

MIDDLE AMERICA

GUATEMALA							
Black Caribs (Livingston)	.988	.012		203	2.4	1.2	Crawford et al. 1981
COSTA RICA							
Guaymi	1.000			286			Barrantes et al. 1982
PANAMA							
Blacks (Bocas del Toro)	.985	.015		N.A.	3.0	1.5	Ferrell et al. 1978b
WEST INDIES (Haiti)	.991	.009		307	1.8	.9	Basu et al. 1976

SOUTH AMERICA

BOLIVIA							
Aymara (W. Bolivia)	.997	.003		316	.6	.3	Ferrell et al. 1978a
BRAZIL							
Baniwa (Jandu Cachoeira)	1.000			363			Salzano et al. 1986
Cayapo (Pará & Mato Grosso)	1.000			206			Salzano et al. 1972b
Caingang	1.000			214			Salzano et al. 1980
Pacaás Novos	1.000			222			Salzano et al. 1985

Continued

Table 7. Adenylate kinase 1, soluble: AK1 (cont'd)

Place/Population	1	2	x	n	H	PE	Source
Parakana	1.000			93			Black et al. 1980
Macushi (N. Brazil & S. Guyana)	1.000			499			Neel et al. 1977b
Wapishana (")	1.000			615			"
Ticuna (C. Amazonas)	1.000			1,762[a]			Neel et al. 1980
CHILE							
Atacameño (Toconao, N. Chile)	1.000			178			Goedde et al. 1985b
COLOMBIA							
Noanama	1.000			170			Kirk et al. 1974
ECUADOR							
Shuara	1.000			90			Goedde et al. 1977a
Waorani	1.000			187			Larrick et al. 1985
FRENCH GUIANA							
Wayampi (Trois-Sauts & Camopi)	.971	.029		238	5.6	2.7	Tchen et al. 1978
VENEZUELA							
Makiritare (S. Venezuela &							
N. Brazil)	1.000			477			Weitkamp & Neel 1970
Yanomama (")	1.000			1,606			Weitkamp & Neel 1972

OCEANIA

AUSTRALIA							
Aborigines/Malag							
(Elcho Is., N. Territory)	1.000			296			Kirk et al. 1969
Aborigines (Hooker Creek,							
N. Territory)	1.000			285			Blake 1979
Aborigines/Amoonguna							
(C. Australia)	1.000			104			Kirk et al. 1971a
Aborigines/Konejandi							
(Kimberley, W. Australia)	1.000			102			Blake & Spargo 1986
MELANESIA							
W. New Guinea							
Asmat (S. Coastal Plain)	1.000			653			Gajdusek et al. 1978
Papua New Guinea							
Gainj & Kalam (N. Central							
Highlands)	1.000			277			Long et al. 1986
Jimi Valley (W. Highlands)	1.000			385[a]			Mourant et al. 1981
Yagaria (E. Highlands)	1.000			424			Mourant et al. 1982
Fuyuge (C. Highlands)	1.000			173			Woodfield et al. 1974
Daga (Bay Province)	1.000			139			Jenkins et al. 1983
Karkar Is.							
Takia	.998	.002		290	.4	.2	Booth et al. 1982
Manus Is.	1.000			183			Malcolm et al. 1972
Solomon Is.	1.000			364			Blake et al. 1983
MICRONESIA							
W. Caroline Is.	1.000			382			Blake et al. 1973b
E. Caroline Is.	1.000			1,005			Yamamoto & Fu 1973
Marshall Is.	1.000			372			Neel et al. 1976
POLYNESIA							
Samoans (Christchurch,							
New Zealand)	1.000			77			Booth et al. 1977

Note: [a] One individual was of genotype (+/2).

Table 8. Alcohol dehydrogenase I, beta polypeptide: ADH2

Place/Population	1	2	n	H	PE	Source
EUROPE						
GERMANY	.956	.044	46	8.4	4.0	Harada et al. 1978
ASIA						
INDIA						
Indians (Malaysia)	1.000		43			Teng et al. 1979
JAPAN						
Japanese (Chiba)	.317	.683	194	43.3	17.0	Yin et al. 1984
NORTH AMERICA						
UNITED STATES						
Whites	1.000		61			Bosron & Li 1981
Blacks	.840	.160[a]	67			"
Navajo Indians (New Mexico)	1.000		46			Rex et al. 1985

Note: [a] A variant allele designated as $ADH_2^{Indianapolis}$.

Table 9. Alcohol dehydrogenase I, gamma polypeptide: ADH3

Place/Population	1	2	n	H	PE	Source
EUROPE						
GERMANY	.576	.424	46	48.8	18.5	Harada et al. 1978
UNITED KINGDOM						
England	.597	.403	314	48.1	18.3	Smith et al. 1972
ASIA						
INDIA						
Indians (Malaysia)	.641	.359	39	46.0	17.7	Teng et al. 1979
CHINA						
Chinese (Malaysia)	.907	.093	86	16.9	7.7	"
JAPAN						
Japanese	.950	.050	40	9.5	4.5	Harada et al. 1980
NORTH AMERICA						
UNITED STATES						
Whites	.535	.465	61	49.8	18.7	Bosron & Li 1981
Blacks	.850	.150	67	25.5	11.1	"
Navajo Indians (New Mexico)	.490	.510	34	50.0	18.7	Rex et al. 1985
SOUTH AMERICA						
BRAZIL						
Bahia	.861	.139	255	23.9	10.5	Azevedo et al. 1975

Table 10. Aldehyde dehydrogenase 2, mitochondrial: ALDH2

Place/Population	Deficient (%)	n	Source
EUROPE			
GERMANY	0	300	Goedde et al. 1986a
ASIA			
TURKEY (Asia Minor)	0	65	"
ISRAEL	0	77	"
INDIA	0	50	"
INDONESIA	39	30	"
THAILAND (N. Thailand)	8	110	"
VIETNAM	53	138	"
PHILIPPINES			
Filipinos	13	110	"
CHINA			
Mongolians (N. China)	30	198	Goedde et al. 1984a
Han (Beijing)	50	120	"
Zhuang (S. China)	45	106	"
Uygur (Xinjiang)	40	207	Chen & Du 1984
Hui (Ningxia)	36	203	"
Dong (Guangxi)	48	201	"
JAPAN			
Japanese	44	184	Goedde et al. 1986a
Ainu	20	80	"
KOREA			
Koreans (NE China)	25	209	Goedde et al. 1984a
Koreans (Seoul)	21	135	Goedde et al. 1986c
AFRICA			
EGYPT	0	260	Goedde et al. 1986a
SUDAN	0	40	"
LIBERIA	0	184	"
KENYA	0	23	"
NORTH AMERICA			
UNITED STATES			
Sioux Indians (N. Dakota)	5	90	"
Navajo Indians (New Mexico)	2	56	"
MIDDLE AMERICA			
MEXICO			
Mestizo (Mexico City)	4	43	"
SOUTH AMERICA			
CHILE			
Atacameño (Toconao, N. Chile)	43	133	Goedde et al. 1984c
ECUADOR			
Shuara Indians	42	99	Goedde et al. 1986a

Note: The mode of inheritance of ALDH2 deficiency is not well established.

Table 11. Aldehyde dehydrogenase 3: ALDH3

Place/Population	Variants		n	Source
	Genotype	Number		
EUROPE				
N. Europeans	+/V[a]	7	180	Hopkinson et al. 1985

Continued

Table 11. Aldehyde dehydrogenase 3: ALDH3 (cont'd)

Place/Population	Variants		n	Source
	Genotype	Number		
ASIA				
INDIA				
Indians (Malaysia)	+/3	1	33	Hopkinson et al. 1985
CHINA				
Chinese (Malaysia)	+/2	4	71	Teng 1981

Note: [a] V stands for variant allele.

Table 12. Alkaline phosphatase, placental: ALPP, PLAP

Place/Population	s1	f1	i1	Others	n	H	PE	Source
EUROPE								
GERMANY (Mainz)	.654	.247	.097		N.A.	50.2	25.5	Ananthakrishnan et al. 1974(c)
GREECE (N. Greece)	.661	.281	.052	.006	407	48.1	23.1	Kouvatsi & Triantaphyllidis 1985
ICELAND	.685	.283	.031		127	45.0	20.0	Beckman & Johannsson 1967
ITALY (Rome)	.661	.256	.075	.007	273	49.2	24.7	Bottini et al. 1970
SWEDEN (Umeå)	.655	.258	.087		2,244	49.7	24.9	Beckman et al. 1972
UNITED KINGDOM								
England	.637	.270	.085	.008	597	51.4	26.2	Robson & Harris 1967
ASIA								
INDIA								
Maratha (W. India)	.701	.217	.066	.016	208	45.7	23.5	Mukherjee et al. 1978
Hindu (Hyderabad)	.740	.186	.068	.006	175	41.3	21.1	Ram Kumar & Rao 1982
Bengali (E. India)	.701	.201	.081	.017	303	46.1	24.4	Das et al. 1974
MALAYSIA								
Malays (Kuala Lumpur)	.758	.109	.121	.012	211	39.9	21.4	Blake et al. 1969c
THAILAND	.746	.081	.165	.008	188	41.0	21.4	Blake et al. 1968
PHILIPPINES								
Filipino (Hawaii)	.759	.078	.163		83	39.1	20.0	Beckman & Beckman 1969
CHINA								
Chinese (Malaysia)	.748	.030	.210	.012	203	39.5	19.1	Blake et al. 1969c
JAPAN								
Japanese (Osaka)	.724	.038	.236	.003	294	41.9	19.7	Blake et al.1969b
AFRICA								
NIGERIA (Ibadan)	.942	.019	.039		130	11.1	5.6	Beckman et al. 1967
NORTH AMERICA								
CANADA								
Eskimos (Ontario)	.556	.142	.296	.006	81	58.3	31.1	Donald 1976
Cree Indians (")	.801	.145	.050	.004	231	33.5	16.9	"
UNITED STATES								
Whites (Connecticut)	.688	.250	.062		150	46.0	22.3	Bottini et al. 1971
Blacks (")	.886	.064	.050		248	20.8	10.7	"
OCEANIA								
MELANESIA								
Papua New Guinea								
Papuans (Port Moresby)	.880	.050	.068	.002	338	21.8	11.2	Blake et al. 1969a
POLYNESIA								
Maori & Polynesian	.696	.125	.179		84	46.8	24.6	Blake et al. 1971a

Note: N.A. - not available.

Table 13. Aminolevulinate dehydratase, delta: ALAD

Place/Population	1	2	n	H	PE	Source
EUROPE						
GERMANY	.889	.111	144	19.7	8.9	Benkmann et al. 1983
ITALY	.897	.103	762	18.5	8.4	Petrucci et al. 1982
SPAIN (Galicia)	.917	.083	500	15.2	7.0	Caeiro & Rey 1985
ASIA						
THAILAND	.970	.030	117	5.8	2.8	Goedde & Benkmann 1987
JAPAN	.942	.058	121	10.9	5.2	Benkmann et al. 1983
KOREA (Seoul)	.969	.031	146	6.0	2.9	Goedde et al. 1986c
AFRICA						
LIBERIA	1.000		296			Benkmann et al. 1983
SOUTH AMERICA						
CHILE Atacameño (Toconao, N. Chile)	.986	.014	175	2.8	1.4	Goedde et al. 1984c

Table 14. Amylase, alpha; salivary: AMY1

Place/Population	1	2	3	4	Others	n	H	PE	Source
EUROPE									
GERMANY	.909	.070	.021			170	16.8	8.3	Kühnl & Tischberger 1980
NETHERLANDS Dutch & Finns	.907[a]	.067[b]	.027			330	17.2	8.7	Pronk et al. 1982
UNITED KINGDOM England	.891	.069	.038	.003		160	20.0	10.3	de Soyza 1978
ASIA									
INDIA Indians (Malaysia)	.980	.008	.008	.004		127	3.9	2.0	Teng et al. 1978a
MALAYSIA Malays (Kuala Lumpur)	.995	.002	.002			406	1.0	.4	"
CHINA Chinese (Malaysia)	.994		.003	.003		157	1.2	.6	"
JAPAN Japanese	.987		.013				2.6	1.3	Ikemoto 1983(c)
AFRICA									
MALI Bozo (W. Mali)	.875[a]	.015[b]			.011	71	22.2	10.5	Pronk et al. 1982
KENYA	.959[a]	.008[b]			.033	200	7.9	3.9	Pronk et al. 1984
NORTH AMERICA									
UNITED STATES Whites	.996	.0005	.003	.0005		961	.8	.3	Merritt et al. 1973
Blacks	.962		.002	.005	.031	208	7.4	3.7	"

Note: Pronk et al. 1982 have postulated that there are chromosomes with a single locus and those with two closely linked duplicate gene loci and that the chromosomes with duplicate loci can be classified into two types, i. e., haplotypes 1-1 and 1-2. a and b in this table represent the frequencies of haplotypes 1-1 and 1-2, respectively.

Table 15. Amylase, alpha; pancreatic: AMY2

Place/Population	1	2	n	H	PE	Source
EUROPE						
HUNGARY	.983	.017	119	3.3	1.6	Goedde et al. 1986b
ITALY (Naples)	.974	.026	476	5.1	2.5	Carfagna et al. 1976
ASIA						
INDIA						
Indians (Malaysia)	.983	.017	121	3.3	1.6	Zarinah et al. 1984
MALAYSIA						
Malays (Kuala Lumpur)	1.000		168			"
CHINA						
Chinese (Malaysia)	1.000		125			"
KOREA						
Koreans (Seoul)	.997	.003	150	.6	.3	Goedde et al. 1986c
NORTH AMERICA						
UNITED STATES						
Whites	.947	.053	673	10.0	4.8	Merritt et al. 1973
Blacks	.986	.014	383	2.8	1.4	"
SOUTH AMERICA						
CHILE						
Atacameño Indians (Tocanao)		1.000	180			Goedde et al. 1985b

Table 16. Carbonic anhydrase I: CA1

Place/Population	1	x	n	H	PE	Source
EUROPE						
Europeans	1.000		434			Hopkinson et al. 1974a
ITALY (Southern)	1.000		394			Sangiorgi et al. 1982
U.S.S.R.						
EUROPEAN PART						
Abkhazia (Ochamchir Dist.)	1.000		262			Ferrell et al. 1985
ASIA						
IRAN (Northern)	1.000		214			Blake 1978
INDIA						
Indians	1.000		1,794[a]			Ghosh 1977b
NEPAL & TIBET	1.000		113			Santachiara et al. 1976
MALAYSIA						
Malays (Singapore)	.989	.011[3]	274	2.2	1.1	Blake 1978
INDONESIA						
Indonesians	.997	.003	357	.6	.3	Blake 1978(c)
Balinese (Bali Is.)	1.000		148			Breguet et al. 1982b
PHILIPPINES						
Negrito/Mamanwa (NE Mindanao)	.783	.217[Guam]	277	34.0	14.1	Omoto et al. 1981
Non-Negrito (Manobo)	.981	.019[3N]	134	3.7	1.8	"
CHINA						
Chinese (Singapore)	1.000		670			Blake 1978
JAPAN	.999	.0005[HIR1]	3,969	.2	5.0	Ueda et al. 1977
AFRICA						
NIGERIA						
Yoruba (Western)	1.000		64			Ojikutu et al. 1977

Continued

Place/Population	1	x	n	H	PE	Source
CENTRAL AFRICAN REPUBLIC						
Pygmies (Western)	1.000		62			Santachiara et al. 1980
RWANDA						
Hutu	1.000		111			Santachiara 1986
SOUTHERN AFRICA						
Ambo (SW Africa)	1.000		490			Marks et al. 1977
San/Bushmen						
G!aokxate	1.000		33			Nurse & Jenkins 1977
Negro/Denasena	1.000		77			"
Bantu (Natal)	1.000		297			Blake 1978
TRISTAN DA CUNHA	1.000		93			Jenkins et al. 1985

NORTH AMERICA

ALASKA						
Eskimos (St. Lawrence Is.)	1.000		222			Ferrell et al. 1981b
CANADA						
Dogrib Indians (NW Territories)	1.000		158			Szathmary et al. 1983

MIDDLE AMERICA

HONDURAS						
Black Caribs	1.000		58			Tashian & Carter 1976
COSTA RICA						
Guaymi	1.000		286			Barrantes et al. 1982

SOUTH AMERICA

BOLIVIA						
Aymara (Western)	1.000		320			Ferrell et al. 1981a
BRAZIL						
Baniwa (Jandu Cachoeira)	1.000		363			Salzano et al. 1986
Xavante	1.000		265			Tashian & Carter 1976
Caingang	1.000		214			Salzano et al. 1980
Macushi (N. Brazil & S. Guyana)	1.000		498			Neel et al. 1977b
Wapishana (")	1.000		614			"
ECUADOR						
Waorani	1.000		187			Larrick et al. 1985
VENEZUELA						
Makiritare (S. Venezuela &						
N. Brazil)	1.000		390			Neel 1978
Yanomama (")	1.000		374			"

OCEANIA

AUSTRALIA						
Aborigines/Malag (Elcho Is., NT)	1.000		142			Blake 1979
Aborigines (Hooker Creek, NT)	.949	.051[9]	287	9.7	4.6	"
Aborigines/Amoonguna						
(C. Australia)	.981	.019[9]	104	3.7	1.8	"
Aborigines/Walmadjari						
(Kimberley, W. Australia)	.983	.017[9]	171	3.3	1.6	Blake & Spargo 1986
MELANESIA						
New Guinea						
New Guineans	1.000		1,980			Blake 1978
Papua New Guinea						
Gainj & Kalam (N. Central						
Highlands)	1.000		274			Long et al. 1986
Daga (Bay Province)	1.000		139			Jenkins et al. 1983
MICRONESIA						
Caroline Is.	1.000		1,271			Blake 1978
Marshall Is.	1.000		351			Neel et al. 1976
POLYNESIA						
Samoans (New Zealand)	1.000		77			Booth et al. 1977

Note: [a] A few individuals with a variant allele were observed in Parsi and Marathi populations in Bombay. A total of 25 electrophoretic variants and a deficiency variant of CA1 locus have been discovered in various human populations. See Tashian et al. (1980) for details.

Table 17. Carbonic anhydrase II: CA2

Place/Population	1	2	x	n	H	PE	Source
EUROPE							
Europeans	1.000			434			Hopkinson et al. 1974a
ITALY (S. Italy)	1.000			394			Sangiorgi et al. 1982
U.S.S.R.							
EUROPEAN PART							
Abkhazia (Ochamchir Dist.)	1.000			262			Ferrell et al. 1985
ASIA							
IRAN (N. Iran)	.998		.002[3]	214	.4	.2	Blake 1978
INDIA							
Indians	1.000			1,794[a]			Ghosh 1977b
NEPAL & TIBET	1.000			113			Santachiara et al. 1976
MALAYSIA							
Malays (Singapore)	1.000			274			Blake 1978
INDONESIA							
Indonesians	1.000			357			Blake 1978(c)
Balinese (Bali Is.)	1.000			148			Breguet et al. 1982b
CHINA							
Chinese (Malaysia)	1.000			670			Blake 1978
JAPAN	1.000			3,969			Ueda et al. 1977
AFRICA							
NIGERIA							
Yoruba (W. Nigeria)	.891	.109		64	19.4	8.8	Ojikutu et al. 1977
CENTRAL AFRICAN REPUBLIC							
Pygmies (Western)	.910	.090		67	16.4	7.5	Santachiara et al. 1980
RWANDA							
Hutu	.940	.060		109	11.3	5.3	Santachiara 1986
CONGO							
Beti Bantu (Cuvette)	.958	.042		95	8.0	3.9	Destro-Bisol et al. 1986a
ETHIOPIA	1.000			171			Moore et al. 1973
UGANDA	1.000			165			Carter 1972
ZAMBIA	.877	.123		333	21.6	9.6	"
SOUTHERN AFRICA							
Ambo (SW Africa)	.880	.120		490	21.1	9.4	Marks et al. 1977
San/Bushmen							
G!aokxate	1.000			33			Nurse & Jenkins 1977
Negro/Denasena	.942	.058		77	10.9	5.2	"
Bantu (Natal)	.924	.076		297	14.0	6.5	Blake 1978
TRISTAN DA CUNHA	1.000			93			Jenkins et al. 1985
NORTH AMERICA							
ALASKA							
Eskimos (Point Hope)	1.000			181			Moore et al. 1973
Eskimos (St. Lawrence Is.)	1.000			222			Ferrell et al. 1981b
CANADA							
Dogrib Indians (NW Territories)	1.000			158			Szathmary et al. 1983
UNITED STATES							
Whites	1.000			171			Moore et al. 1973
Blacks	.899	.101		222	18.2	8.3	Moore et al. 1971
MIDDLE AMERICA							
HONDURAS							
Black Caribs	1.000			58			Tashian & Carter 1976
COSTA RICA							
Guaymi	1.000			286			Barrantes et al. 1982
SOUTH AMERICA							
BOLIVIA							
Aymara (Western)	1.000			320			Ferrell et al. 1981a

Continued

Table 17. Carbonic anhydrase II: CA2 (cont'd)

Place/Population	1	2	x	n	H	PE	Source
BRAZIL							
Baniwa (Jandu Cachoeira)	.947		.053[Bam]	377	9.5	4.5	Salzano et al. 1986
Caingang	1.000			214			Salzano et al. 1980
Pacaás Novos	1.000			221			Salzano et al. 1985
Macushi (N. Brazil & S. Guyana)	1.000			498			Neel et al. 1977b
Wapishana (")	.998	.002		614	.4	.2	"
ECUADOR							
Waorani	1.000			187			Larrick et al. 1985
VENEZUELA							
Makiritare (S. Venezuela &							
N. Brazil)	1.000			390			Neel 1978
Yanomama (")	1.000			306			"

OCEANIA

	1	2	x	n	H	PE	Source
AUSTRALIA							
Aborigines/Malag (Elcho Is.,							
N. Territory)	1.000			142			Blake 1979
Aborigines (Hooker Creek,							
N. Territory)	.983		.017[4]	287	3.3	1.6	"
Aborigines/Amoonguna							
(C. Australia)	.981		.019[4]	104	3.7	1.8	"
Aborigines/Walmadjari							
(W. Australia)	.927		.073[4]	171	13.5	6.3	Blake & Spargo 1986
MELANESIA							
New Guinea							
New Guineans	1.000			1,980			Blake 1978
Papua New Guinea							
Gainj & Kalam (N. Central							
Highlands)	1.000			277			Long et al. 1986
Daga (Bay Province)	.982		.018[4]	139	3.5	1.7	Jenkins et al. 1983
MICRONESIA							
Caroline Is. (East & West)	1.000			1,271			Blake 1978
Marshall Is.	1.000			191			Tashian & Carter 1976
POLYNESIA							
Samoans (Christchurch,							
New Zealand)	1.000			101			Booth et al. 1977

Note: [a] Four heterozygous individuals with a variant allele 3 were observed in 307 Parsi individuals in
Bombay. A total of seven electrophoretic variants of CA2 locus have been discovered in various
human populations. See Tashian et al. (1980) for details.

Table 18. Carbonic anhydrase III: CA3

Place/Population	31-Val	31-Ile	n	H	PE	Source
NORTH AMERICA						
UNITED STATES						
Whites	.22	.78	9	34.3	14.2	Welty & Hewett-Emmett 1986
Blacks	.80	.20	5	32.0	13.4	Hewett-Emmett et al. 1983

Table 19. Catalase: CAT

Place/Population	N	C	n	H	PE	Source
EUROPE						
SWITZERLAND	.99996	.00004	73,661	.0	.0	Aebi 1966
ASIA						
KUWAIT						
Arabs	1.000		193			Al-Nassar et al. 1981
IRAN						
Mazandaranian &						
Guilanian	.9971	.0029	2,036	.6	.3	Ohkura et al. 1984
NEPAL & TIBET	1.000		23			Santachiara et al. 1976
CHINA						
Chinese (Taiwan)	.9985	.0015	20,439	.3	.1	Takahara et al. 1973
JAPAN						
Japanese (Hokkaido)	1.000		225			"
" (Chiba)	.9984	.0016	11,875	.3	.1	"
" (Okyama)	.9994	.0006	10,336	.1	.0	"
" (Hiroshima)	.9995	.0005	10,679	.1	.0	"
" (Nagasaki)	.9980	.0020	4,238	.4	.2	"
" (Okinawa)	1.000		14,681			"
KOREA						
Koreans (Japan)	.9959	.0041	1,603	.8	.4	"

Note: The gene (C) for acatalasemia, i.e., lack of catalase, is recessive to the normal gene (N). However, heterozygotes for the C allele can be detected by biochemical tests.

Table 20. Cholinesterase (serum) 1; pseudocholinesterase 1: CHE1

Place/Population	u	a	f	n	H	PE	Source
EUROPE							
BULGARIA	.986	.009	.005	108	2.8	1.4	Steegmüller 1975
CZECHOSLOVAKIA	.986	.014		312	2.8	1.4	Altland et al. 1969
FINLAND							
Finns	.984	.016		317	3.1	1.5	Singh et al. 1971
Lapps/Saami	.984	.016		805	3.1	1.5	"
FRANCE							
French (Bareges)	.978	.019	.004[s]	693	4.3	2.2	Vergnes et al. 1980a
Basques	.966	.034		141	6.6	3.2	"
GERMANY	.984	.016		8,314	3.1	1.5	Giblett 1969
GREECE	.992	.008		860	1.6	.8	Fraser et al. 1969b
HUNGARY	.991	.009		276	1.8	.9	Walter et al. 1965
ICELAND	.988		.012	128	2.4	1.2	Neumann & Walter 1968
ITALY	.979	.021		382	4.1	2.0	Whittaker 1968
NETHERLANDS (Leiden)	.978	.022		800	4.3	2.1	Fraser et al. 1974
NORWAY	.985	.015		3,143	3.0	1.5	Boman 1981
PORTUGAL							
Portugese	.983	.017		179	3.3	1.6	Kattamis et al. 1962
Portugese (Terras de Bouro)	.987	.013		77	2.6	1.3	Cruz et al. 1973
SPAIN							
Spanish (Madrid)	.991	.009		159	1.8	.9	Goedde et al. 1973
Basques (Bermeo/Pamplona)	.969	.031		224	6.0	2.9	"
SWEDEN	.995	.005		189	1.0	.5	Boman 1981
UNITED KINGDOM							
England	.981	.019		703	3.7	1.8	Kattamis et al. 1962
YUGOSLAVIA	.988	.012		363	2.4	1.2	Fraser et al. 1969a
U.S.S.R.							
EUROPEAN PART							
Maris (Mari Republic)	1.000			295			Singh et al. 1974c

Continued

Table 20. Cholinesterase (serum) 1; pseudocholinesterase 1: CHE1 (cont'd)

Place/Population	u	a	f	n	H	PE	Source
ASIA							
TURKEY (Asia Minor)	.963	.033	.004	725	7.2	3.5	Sayek et al. 1967
ISRAEL							
Ashkenazi Jews	.983	.017		4,196	3.3	1.6	Szeinberg et al. 1972
Yemenite Jews	.980	.020		459	3.9	1.9	"
Arabs	.991	.009		110	1.8	.9	Szeinberg 1974(c)
IRAN							
Kurds	1.000			184			Lehmann et al. 1973
AFGHANISTAN							
Pushtoon (Kabul)	1.000			210			Rahimi et al. 1977
PAKISTAN	.988	.012		121	2.4	1.2	Neumann & Walter 1968
INDIA							
Punjabi (N. India)	.974	.026		500	5.1	2.5	Singh et al. 1974a
Indians (E. India)	.996	.004		139	.8	.4	Steegmüller 1975
MALAYSIA							
Malays	.985	.015		33	3.0	1.5	Arends et al.1967
THAILAND	1.000			723			Altland et al. 1967
VIETNAM (N. Vietnam)	.993	.007		153	1.4	.7	Herzog et al. 1976
PHILIPPINES							
Filipino	.998	.002		427	.4	.2	Motulsky & Morrow 1968
CHINA							
Chinese (Formosa)	.998	.002		340	.4	.2	"
JAPAN							
Japanese (Tokyo)	.997	.003		371	.6	.3	Altland et al. 1967
Ainu (Hokkaido)	1.000			360			Omoto & Harada 1968
KOREA							
Koreans (Germany)	1.000			115			Bajatzadeh & Walter 1969a
AFRICA							
ALGERIA							
Ideles (Ahaggar)	.998	.002		282	.4	.2	Lefèvre-Witier & Vergnes 1977
CHAD							
Sara Majingay	.997	.003		187	.6	.3	Hiernaux 1976
MALI							
Kel Kummer Twaregs (Menaka)	.994	.006		175	1.2	.6	Lefèvre-Witier & Vergnes 1977
SENEGAL							
Bedik (E. Senegal)	1.000			791			Bouloux et al. 1972
IVORY COAST							
Tuka	1.000			186			Vergnes & Cabannes 1976
NIGERIA	1.000			33			Steegmüller 1975
CAMEROON	1.000			284			Goedde et al. 1979a
CENTRAL AFRICAN REPUBLIC							
Babinga Pygmies	1.000			300			Siniscalco 1967
ETHIOPIA							
Harar	.970	.030		150	5.8	2.8	Agarwal et al. 1973
SOUTHERN AFRICA							
Negroes/Dama (Okombahe)	1.000			92			Nurse et al. 1976
MALAWI	.963		.037	191	7.1	3.4	Whittaker & Lowe 1976
MOZAMBIQUE							
Bantu	.937		.063	326	11.8	5.6	Whittaker & Reys 1975
NORTH AMERICA							
GREENLAND							
Eskimos	1.000			146			Singh et al. 1974c
ALASKA							
Eskimos (Northern)	1.000			122			Gutsche et al. 1967
Athabaskan Indians	1.000			141			"
CANADA							
Eskimos (Igloolik)	1.000			362			McAlpine et al. 1974
UNITED STATES							
Whites (Seattle)	.984	.016		246	3.1	1.5	Motulsky & Morrow 1968
Blacks (")	.995	.005		606	1.0	.5	"
Navajo Indians (Arizona)	1.000			357			Garry 1977

Continued

Table 20. Cholinesterase (serum) 1; pseudocholinesterase 1: CHE1 (cont'd)

Place/Population	u	a	f	n	H	PE	Source
MIDDLE AMERICA							
MEXICO							
Mestizo	1.000			321			Lisker et al. 1967
Nahua	.989	.011		355	2.2	1.1	"
SOUTH AMERICA							
BOLIVIA							
Aymara	1.000			90			Arends et al. 1967
BRAZIL							
Xavante	1.000			285			Tashian et al. 1967
CHILE							
Atacameño (Toconao,							
N. Chile)	.984	.016		162	3.1	1.5	Goedde et al. 1984c
ECUADOR							
Shuara	1.000			90			Goedde et al. 1977a
VENEZUELA							
Makiritare (S. Venezuela							
& N. Brazil)	.969	.031		418	6.0	2.9	Arends et al. 1970
OCEANIA							
AUSTRALIA							
Aborigines	1.000			100			Whittaker 1968

Table 21. Cholinesterase 2; pseudocholinesterase 2: CHE2

Place/Population	+	−	n	H	PE	Source
EUROPE						
BULGARIA	.042	.958	109	8.0	3.5	Steegmüller 1975
CZECHOSLOVAKIA	.056	.944	312	10.6	4.4	Altland et al. 1969
FINLAND						
Finns	.018	.982	317	3.5	1.7	Singh et al. 1971
Lapps/Saami	.054	.946	805	10.2	4.3	"
FRANCE						
French (Bareges)	.037	.963	693	7.1	3.2	Vergnes et al. 1980a
Basques	.007	.993	141	1.4	.7	"
GERMANY	.061	.939	952	11.5	4.7	Altland et al. 1969
GREECE	.081	.919	860	14.9	5.8	Fraser et al. 1969b
ICELAND	.017	.983	1,078	3.3	1.6	Tills et al. 1982b
NETHERLANDS (Leiden)	.038	.962	798	7.3	3.3	Fraser et al. 1974
POLAND	.030	.970	445	5.8	2.7	Steegmüller 1975(c)
PORTUGAL (Terras de Bouro)	.040	.960	76	7.7	3.4	Cruz et al. 1973
SPAIN						
Spanish (Madrid)	.045	.955	159	8.6	3.7	Goedde et al. 1973
Basques	.057	.943	224	10.8	4.5	"
UNITED KINGDOM						
England	.050	.950	1,941	9.5	4.1	Robson & Harris 1966
U.S.S.R.						
EUROPEAN PART						
Mari	.049	.951	295	9.3	4.0	Singh et al. 1974c
ASIA						
ISRAEL						
Iraqi Jews		1.000	64			Robson & Harris 1966
INDIA						
Punjabi (Punjab)	.043	.957	500	7.9	3.6	Singh et al. 1974a

Continued

Table 21.   Cholinesterase 2; pseudocholinesterase 2: CHE2 (cont'd)

Place/Population	+	−	n	H	PE	Source
Hindu (Andhra Pradesh)	.020	.980	1,636	3.9	1.8	Rao et al. 1985
Indians (E. India)	.018	.982	550	3.5	1.7	Steegmüller 1975
Moslems (")	.041	.959	99	7.9	3.5	Mukherjee et al. 1974
BHUTAN	.054	.946	266	10.2	4.3	Mourant et al. 1969
THAILAND	.070	.930	81	13.0	5.2	Simpson 1968
VIETNAM (N. Vietnam)	.026	.974	153	5.1	2.3	Herzog et al. 1976
JAPAN						
Japanese (Hokkaido)	.043	.957	237	8.2	3.6	Omoto & Harada 1968
Ainu (")	.029	.971	195	5.6	2.6	"
KOREA	.011	.989	47	2.2	1.1	Steegmüller 1975

AFRICA

Place/Population	+	−	n	H	PE	Source
SUDAN						
Beja-Amarar (Northern)	.005	.995	100	1.0	.5	el Hassan et al. 1968
MALI						
Kel Kummer Twaregs (Menaka)	.027	.973	280	5.3	2.4	Lefévre-Witier & Vergnes 1977
SENEGAL & IVORY COAST	.011	.989	224	2.2	1.1	Steegmüller 1975
NIGERIA	.015	.985	68	3.0	1.4	"
RHODESIA	.016	.984	1,227	3.1	1.5	Whittaker & Lowe 1976
ANGOLA	.007	.993	303	1.4	.7	"
MOZAMBIQUE	.016	.984	162	3.1	1.5	"
SOUTHERN AFRICA						
San/Bushmen						
!Kung, Dobe	.026	.974	98	5.1	2.3	Jenkins & Nurse 1977
Khoi/Hottentot						
Topnaar	.005	.995	97	1.0	.5	"
Negroes/Dama	.040	.960	115	7.7	3.4	Nurse et al. 1976

NORTH AMERICA

Place/Population	+	−	n	H	PE	Source
GREENLAND						
Eskimos	.056	.944	146	10.6	4.4	Singh et al. 1974c
ALASKA						
Eskimos	.030	.970	2,010	5.8	2.7	Scott 1973
CANADA						
Eskimos (Igloolik)	.068	.932	298	12.7	5.1	McAlpine et al. 1974
Cree Indians	.072	.928	589	13.4	5.3	Simpson 1968
UNITED STATES						
Whites (Seattle)	.025	.975	163	4.9	2.3	Ashton & Simpson 1966
Blacks (")	.026	.974	317	5.1	2.3	"

MIDDLE AMERICA

Place/Population	+	−	n	H	PE	Source
DOMINICAN REPUBLIC						
Carib	.046	.954	99	8.8	3.8	Harvey et al. 1969

SOUTH AMERICA

Place/Population	+	−	n	H	PE	Source
BOLIVIA						
Aymara		1.000	87	11.1	4.6	Vergnes et al. 1976a
BRAZIL						
Xavante		1.000	285			Tashian et al. 1967
ECUADOR						
Shuara		1.000	90			Goedde et al. 1977a
VENEZUELA						
Makiritare	.059	.941	418	11.1	4.6	Arends et al. 1970

OCEANIA

Place/Population	+	−	n	H	PE	Source
AUSTRALIA						
Aborigines		1.000	104			Horsfall et al. 1963
MELANESIA						
Papua New Guinea						
Daga (Bay Province)	.003	.997	164	.6	.3	Jenkins et al. 1983
POLYNESIA						
Easter Is.		1.000	497			Simpson 1968

Note: The allele with + sign is dominant to the allele with the − sign.

Table 22. Cytidine deaminase: CDA

Place/Population	1	2	n	H	PE	Source
NORTH AMERICA						
UNITED STATES						
Whites (Seattle)	.651	.349	189	45.4	17.6	Teng et al. 1975

Table 23. Diaphorase NADH: DIA1

Place/Population	1	2	3	4	x	n	H	PE	Source
EUROPE									
FRANCE									
French (Bareges)	1.000					638			Vergnes et al. 1980a
Basques	.993	.007				144	1.4	.7	"
GERMANY	.991	.003		.002	.004[5]	1,008	1.8	.9	Potrafki et al. 1972
GREECE (Cypriot)	.995	.005				101	1.0	.5	Hopkinson et al. 1970
ITALY (Rome)	.998	.002				321	.4	.2	Lucarelli et al. 1972
NETHERLANDS (Leiden)	1.000					806			Fraser et al. 1974
UNITED KINGDOM									
England	.995	.002	.0003	.002	.0008[5,6]	1,975	1.0	.4	Hopkinson et al. 1970
ASIA									
IRAN									
Turkoman (Gonbad, NE Iran)	1.000					155			Kirk et al. 1977
INDIA									
Indians (Delhi)	.992	.007	.001			490	1.6	.8	Blake et al. 1971b
Hindu (Andhra Pradesh)	1.000					85			Naidu et al. 1985
MALAYSIA									
Malays (Singapore)	1.000					256			Blake et al. 1973a
INDONESIA									
Balinese (Bali Is.)	1.000					148			Breguet et al. 1982b
CHINA									
Chinese (Singapore)	.999		.001			377	.2	.1	Blake et al. 1973a
JAPAN									
Japanese (Aichi)	.998	.002	.0003			5,046	.4	.2	Shimizu et al. 1985
Ainu (Hokkaido)	1.000					125			Omoto & Harada 1972
AFRICA									
MALI									
Kel Kummer Twaregs	1.000					284			Lefèvre-Witier & Vergnes 1977
CENTRAL AFRICAN REPUBLIC									
Pygmies (Western)	1.000					67			Santachiara et al. 1980
Pygmies (")	.980				.020[7]	766	3.9	1.9	Meera Khan 1986
BURUNDI									
Hutu	.985	.015				364	3.0	1.5	Gall et al. 1982
NORTH AMERICA									
CANADA									
Eskimos (Igloolik)	1.000					362			McAlpine et al. 1974
Ojibwa Indians (SE Ontario)	1.000					117			Szathmary et al. 1974
UNITED STATES									
Whites (Seattle)	.994				.006[V]	175	1.2	.6	Detter et al. 1970a
Blacks (")	1.000					125			"

Continued

63

Table 23. Diaphorase NADH: DIA1 (cont'd)

Place/Population	1	2	3	4	x	n	H	PE	Source
SOUTH AMERICA									
BOLIVIA									
Siriono (Eastern)	1.000					109			Vergnes et al. 1976b
EQUADOR									
Waorani	1.000					187			Larrick et al. 1985
OCEANIA									
AUSTRALIA									
Aborigines/Malag (Elcho Is., N. Territory)	1.000					266			Blake 1979
Aborigines (Hooker Creek, N. Territory)	1.000					285			"
Aborigines/Cundeelee (W. Australia)	1.000					116			"
MELANESIA									
Papua New Guinea Papuans	1.000					100			Blake & Kirk 1972
Fuyuge (C. Dist. Highlands)	1.000					173			Woodfield et al. 1974
Solomon Is.	1.000					364			Blake et al. 1983
POLYNESIA									
Samoans (Christchurch, New Zealand)	1.000					101			Booth et al. 1977

Table 24. Diaphorase 3 (sperm): DIA3

Place/Population	1	2	3	n	H	PE	Source
EUROPE							
GERMANY (Hessen)	.755	.223	.021	141	38.0	16.9	Kühnl et al. 1977a
UNITED KINGDOM							
England	.760	.230	.010	346	36.9	15.8	Edwards et al. 1979
ASIA							
JAPAN							
Japanese (Miyagi & Yamagata)	.837	.143	.020	263	27.9	13.1	Sebetan et al. 1982
NORTH AMERICA							
UNITED STATES							
Volunteers (more than 90% Whites)	.712	.288		52	41.8	16.4	Caldwell et al. 1976

Table 25. Diaphorase 4: DIA4

Place/Population	1	2	3	4	n	H	PE	Source
EUROPE								
NETHERLANDS (Leiden)	.982	.012	.001	.004	801	3.6	1.7	Fraser et al. 1974

Table 26. 2,3-Diphosphoglycerate mutase: DPGM

Place/Population	1	2	n	H	PE	Source
ASIA						
NEPAL	1.000		103			Chen et al. 1971
AFRICA						
KENYA	1.000		120			"
CONGO	1.000		83			"
BURUNDI						
Hutu	1.000		400			Gall et al. 1982
NORTH AMERICA						
ALASKA						
Eskimos (Anchorage)	1.000		134			Duncan et al. 1974
Eskimos (St. Lawrence Is.)	.980	.020	222	3.9	1.9	Ferrell et al. 1981b
Athabaskan Indians (Anchorage)	1.000		34			Duncan et al. 1974
CANADA						
Eskimos (Igloolik)	.982	.018	137	3.5	1.7	McAlpine et al. 1974
Eskimos (NW Territories)	.996	.004	257	.8	.4	Chen et al. 1971
Cree Indians	1.000		199			Lucciola et al. 1974
Dogrib Indians (NW Territories)	1.000		158			Szathmary et al. 1983
UNITED STATES						
Whites	.998	.002	213	.4	.2	Chen et al. 1971
Blacks	1.000		192			"
Orientals	1.000		101			"
SOUTH AMERICA						
BOLIVIA						
Aymara (Western)	1.000		320			Ferrell et al. 1981a
OCEANIA						
MELANESIA						
Melanesians (New Guinea)	1.000		98			Chen et al. 1971
MICRONESIA						
Marshall Is.	1.000		367			Neel et al. 1976

Table 27. Esterase A1; A_{1-3}: ESA1

Place/Population	A	D_{MAC1}	n	H	PE	Source
MIDDLE AMERICA						
COSTA RICA						
Guaymi	1.000		286			Barrantes et al. 1982
SOUTH AMERICA						
BRAZIL						
Baniwa	1.000		363			Salzano et al. 1986
Cayapo	1.000		404			Neel et al. 1977b
Caingang	1.000		213			Salzano et al. 1980
Parakana	1.000		37			Black et al. 1980
Pacaás Novos	1.000		222			Salzano et al. 1985
Macushi	.951	.049	498	9.3	4.4	Neel et al. 1977b
Wapishana	.976	.024	614	4.5	2.2	"
VENEZUELA						
Makiritare (S. Venezuela						
& N. Brazil)	1.000		382			"
Yanomama (")	1.000		232			"
OCEANIA						
Papua New Guinea						
Gainj & Kalam	1.000		275			Long et al. 1986

Table 28. Esterase B3, leukocyte: ESB3

Place/Population	1	2	n	H	PE	Source
NORTH AMERICA						
UNITED STATES						
Whites	.966	.034	467	6.6	3.2	Coates & Cortner 1986
Blacks	.982	.018	138	3.5	1.7	"

Table 29. Esterase D; S-formylglutathione hydrolase: ESD, FGH

Place/Population	1	2	x	y	n	H	PE	Source
EUROPE								
BELGIUM (Liége)	.866	.134			500	23.2	10.3	Brocteur et al. 1980
DENMARK (Copenhagen)	.901	.099	.0001Cph		3,116	17.8	8.1	Dissing & Eriksen 1984
FAROE IS. (N. Atlantic)	.816	.184			668	30.0	12.8	Tills et al. 1985
FINLAND	.923	.077			317	14.2	6.6	Beckman & Beckman 1977
FRANCE								
French (Paris)	.889	.111			525	19.7	8.9	Séger 1977
Basques	.911	.089			184	16.2	7.5	Vergnes et al. 1980b
GERMANY								
Germans (Dusseldorf)	.861	.109	.021[5]	.010[11]	312	24.6	12.3	Henke et al. 1986
Germans (Cologne)	.883	.117			1,082	20.7	9.3	Köster et al. 1975
GREECE								
Greeks (N. Greece)	.842	.158			404	26.6	11.5	Kouvatsi & Triantaphyllidis 1984
Greeks (Plati)	.847	.153			579	25.9	11.3	Tills et al. 1983b
HUNGARY	.896	.104			178	18.6	8.5	Goedde et al. 1986b
IRELAND	.877	.123			186	21.6	9.6	Welch & Lee 1974
ITALY (Rome)	.845	.138	.017[5]		231	26.7	12.5	Destro-Bisol et al. 1986b
NETHERLANDS (Leiden)	.858	.142			1,018	24.4	10.7	Ebeli-Struijk et al. 1976
NORWAY								
Norwegians (Oslo)	.887	.113			217	20.0	9.0	Olaisen et al. 1976a
Lapps/Saami (Finnmark)	.872	.128			196	22.3	9.9	"
POLAND (Olkusz)	.918	.082			213	15.1	7.0	Wolanski et al. 1983
PORTUGAL (Porto Dist.)	.826	.174			778	28.7	12.3	Amorim & Siebert 1982
SPAIN (Galicia)	.879	.121	.0053[3]		1,086	21.3	9.5	Carracedo & Concheiro 1983
SWEDEN								
Swedes (Vasterbotten)	.915	.085			200	15.6	7.2	Beckman & Beckman 1977
Lapps/Saami (N. Sweden)	.941	.059			219	11.1	5.2	"
SWITZERLAND (Bern)	.867	.133			744	23.1	10.2	Pflugshaupt et al. 1978
UNITED KINGDOM								
England (Nottingham)	.883	.117			998	20.7	9.3	Cartwright et al. 1977
Isle of Man	.845	.155			320	26.2	11.4	Mitchell et al. 1982
U.S.S.R.								
EUROPEAN PART								
Russians (Moscow)	.917	.083			965	15.2	7.0	Altukhov et al. 1981
Abkhazia (Ochamchir Dist.)	.816	.184			239	30.0	12.8	Ferrell et al. 1985
ASIA								
TURKEY (Asia Minor)								
Jews	.869	.131			138	22.8	10.1	Golan et al. 1977
ISRAEL								
Ashkenazi Jews	.900	.100			235	18.0	8.2	"
Yemenite Jews	.788	.212			156	33.4	13.9	"
IRAQ	.792	.208			320	32.9	13.8	Papiha & Al-Agidi 1976
KUWAIT	.803	.197			160	31.6	13.3	Cartwright et al. 1976
SAUDI ARABIA								
Shiites & Sunnites (al-Hasa)	.805	.195			359	31.4	13.2	Goedde et al. 1979b
IRAN								
Turkoman (Gonbad, NE Iran)	.805	.195			87	31.4	13.2	Akbari et al. 1984

Continued

Table 29. Esterase D; S-formylglutathione hydrolase: ESD, FGH (cont'd)

Place/Population	1	2	x	y	n	H	PE	Source
AFGHANISTAN								
Pushtu & Dari (Kabul)	.897	.103			224	18.5	8.4	Papiha & Nahar 1977
INDIA								
Punjabi (N. India)	.797	.203			303	32.4	13.6	Cartwright et al. 1976
Gujarati (W. India)	.829	.171			283	28.4	12.2	Papiha & Nahar 1977
Bhil tr. (Madhya Pradesh)	.738	.262			143	38.7	15.6	Papiha et al. 1978
Hindu (Andhra Pradesh)	.784	.216			201	33.9	14.1	Roberts et al. 1980
Bengali (E. India)	.710	.290			272	41.2	16.4	Ghosh 1977b
SRI LANKA								
Sinhalese (Anuradhapura)	.763	.237			135	36.2	14.8	Papiha & Nahar 1977
BANGLADESH								
Moslems (Dacca)	.771	.229			166	35.3	14.5	"
NEPAL	.618	.382			364	47.2	18.0	Sunderland et al. 1979
MALAYSIA								
Malays (Singapore)	.644	.356			198	45.9	17.7	Blake 1976
INDONESIA								
Toba Batak (Sumatra)	.707	.293			147	41.4	16.4	"
Balinese (Bali Is.)	.682	.318			148	43.4	17.0	Breguet et al. 1982b
THAILAND	.610	.390			116	47.6	18.1	Goedde & Benkmann 1987
PHILIPPINES								
Filipino (Germany)	.729	.271			144	39.5	15.9	Windhof & Walter 1983
Negrito (Angeles, Luzon)	.779	.120	$.101^{3N}$		129	36.9	19.2	Omoto et al. 1978
CHINA								
Mongolians (N. China)	.699	.296	$.005^6$		201	42.4	17.2	Goedde et al. 1984b
Zhaung (S. China)	.601	.391			208	48.6	18.4	"
Hui (Ningxia)	.732	.268			218	39.2	15.8	Xu & Du 1984c
Dong (Guangxi)	.642	.358			201	46.0	17.7	"
Tibetans	.592	.408			114	48.3	18.3	Papiha & Nahar 1977
Chinese (Singapore)	.565	.435			262	49.2	18.5	Blake 1976
TAIWAN								
Toroko-Aborigines	.683	.317			60	43.3	17.0	Chen et al. 1985
JAPAN								
Japanese (Yamaguchi)	.600	.388	$.011^7$	$.001^Y$	504	48.9	19.7	Yuasa et al. 1985b
Japanese (Tokyo)	.658	.342			1,066	45.0	17.4	Omoto et al. 1975
Ainu (Hokkaido)	.681	.319			94	43.4	17.0	"
KOREA								
Koreans (Seoul)	.651	.349			142	45.4	17.6	Goedde et al. 1986c
Koreans (NE China)	.700	.297	$.002^5$		212	42.2	16.8	Goedde et al. 1984b

AFRICA

Place/Population	1	2	x	y	n	H	PE	Source
EGYPT								
Egyptians (Cairo & Anshas)	.762	.238			250	36.3	14.8	Goedde et al. 1980
Nubians (S. Egypt)	.737	.263			154	38.8	15.6	Bertin et al. 1978
GAMBIA	.916	.084			734	15.4	7.1	Welch 1974
LIBERIA	.947	.053			485	10.0	4.8	Goedde et al. 1985a
NIGERIA								
Yoruba (W. Nigeria)	.937	.063			64	11.8	5.6	Ojikutu et al. 1977
CAMEROON (Bamenda)	.974	.026			283	5.1	2.5	Goedde et al. 1979a
CENTRAL AFRICAN REPUBLIC								
Pygmies (Western)	.843	.157			182	26.5	11.5	Santachiara et al. 1980
BURUNDI								
Hutu	.836	.164			363	27.4	11.8	Gall et al. 1982
UGANDA	.890	.110			209	19.6	8.8	Papiha & Nahar 1977
ANGOLA								
Njinga	.947	.053			95	10.0	4.8	Nurse et al. 1979
SOUTHERN AFRICA								
Namibians	.895	.105			57	18.8	8.5	Palmhert-Keller et al. 1983
San/Bushmen, !Xo	.951	.049			51	9.3	4.4	Nurse & Jenkins 1977
Negroes/Denasena	.980	.020			77	3.9	1.9	"
Bantu (Natal)	.972	.028			180	5.4	2.6	Blake 1976
TRISTAN DA CUNHA	.809	.191			157	30.9	13.1	Jenkins et al. 1985

NORTH AMERICA

Place/Population	1	2	x	y	n	H	PE	Source
ALASKA								
Eskimos-Inupiat (NW Alaska)	.919	.081			223	14.9	6.9	Scott & Wright 1978
Eskimos (St. Lawrence Is.)	.935	.065			222	12.2	5.7	Ferrell et al. 1981b

Continued

67

Table 29. Esterase D; S-formylglutathione hydrolase: ESD, FGH (cont'd)

Place/Population	1	2	x	y	n	H	PE	Source
Athabaskan Indians	.694	.306			175	42.5	16.7	Scott & Wright 1978
CANADA								
Eskimos (Igloolik)	.708	.292			336	41.3	16.4	Cox et al. 1978
Cree Indians (Manitoba)	.668	.332			113	44.4	17.3	Lucciola et al. 1974
Dogrib Indians								
NW Territories)	.826	.174			158	28.7	12.3	Szathmary et al. 1983
UNITED STATES								
Whites (Louisiana)	.906	.094			376	17.0	7.8	Fox et al. 1981
Blacks (")	.907	.093			372	16.9	7.7	"
Mexican-Americans (Texas)	.866	.134			666	23.2	10.3	Hewett-Emmett et al. 1986b

MIDDLE AMERICA

Place/Population	1	2	x	y	n	H	PE	Source
GUATEMALA								
Black Caribs (Livingston)	.912	.088			199	16.1	7.4	Crawford et al. 1981
COSTA RICA								
Guaymi	.918	.082			286	15.1	7.0	Barrantes et al. 1982
PANAMA								
Blacks (Costa Arriba)	.905	.095			423	17.2	7.9	Ferrell et al. 1978b

SOUTH AMERICA

Place/Population	1	2	x	y	n	H	PE	Source
BOLIVIA								
Aymara (W. Bolivia)	.682	.318			316	43.4	17.0	Ferrell et al. 1978a
BRAZIL								
Baniwa (Jandu Cachoeira)	.796	.204			358	32.0	13.4	Salzano et al. 1986
Cayapo (Para & Mato Grosso)	.499	.501			393	50.0	18.7	Neel 1978
Caingang	.682	.318			214	43.4	17.0	Salzano et al. 1980
Parakana	.365	.635			37	46.4	17.8	Black et al. 1980
Pacaás Novos	.783	.217			221	34.3	14.2	Salzano et al. 1985
Macushi (N. Brazil &								
S. Guyana)	.687	.313			498	43.0	16.9	Neel et al. 1977b
Wapishana (")	.798	.202			613	32.2	13.5	"
Ticuna (C. Amazonas)	.662	.338			1,293	44.8	17.4	Mestriner et al. 1980
CHILE								
Atacameño (Toconao)	.616	.384			173	47.3	18.1	Goedde et al. 1985b
ECUADOR								
Shuara	.678	.322			90	43.7	17.1	Goedde et al. 1977a
Waorani	.914	.086			187	15.7	7.2	Larrick et al. 1985
FRENCH GUIANA								
Wayampi (Trois-Sauts &								
Camopi)	.884	.116			238	20.5	9.2	Tchen et al. 1978
VENEZUELA								
Makiritare (S. Venezuela								
& N. Brazil)	.797	.203			390	32.4	13.6	Neel 1978
Yanomama (")	.783	.217			306	34.0	14.1	"

OCEANIA

Place/Population	1	2	x	y	n	H	PE	Source
AUSTRALIA								
Aborigines/Malag (Elcho								
Is., N. Territory)	.961	.039			142	7.5	3.6	Blake 1976
Aborigines (Hooker Creek,								
N. Territory)	.830	.170			286	28.2	12.1	"
Aborigines/Walmadjari								
(Kimberley, W. Australia)	1.000				171			Blake & Spargo 1986
MELANESIA								
W. New Guinea								
Asmat (S. Coastal Plain)	.974	.026			115	5.1	2.5	Gajdusek et al. 1978
Papua New Guinea								
Gainj & Kalam (N.								
Central Highlands)	.934	.066			570	12.3	5.8	Long et al. 1986
Aborigines (E. Highlands)	.949	.051			303	9.7	4.6	Blake 1976
Motu (Central Dist.)	.691	.309			204	42.7	16.8	"
Daga (Bay Province)	.939	.061			139	11.5	5.4	Jenkins et al. 1983
Karkar Is.								
Takia (Gamog & Boroman)	.898	.102			366	18.3	8.3	Blake 1976
Buka Is.	.800	.200			80	32.0	13.4	McLoughlin et al. 1982a

Continued

Table 29. Esterase D; S-formylglutathione hydrolase: ESD, FGH (cont'd)

Place/Population	1	2	x	y	n	H	PE	Source
Solomon Is.	.848	.152			253	25.8	11.2	Blake 1976
MICRONESIA								
W. Caroline Is.	.555	.445			273	49.4	18.6	Blake et al. 1982a
E. Caroline Is.	.571	.429			952	49.0	18.5	Blake et al. 1982a
POLYNESIA								
Samoans (Christchurch,								
New Zealand)	.578	.422			77	48.8	18.4	Booth et al. 1977

Note: Esterase D (ESD) and S-formylglutathione hydrolase (FGH) are identical. See Eiberg & Mohr (1986) and
     Apeshiotis & Bender (1986).

Table 30. Esterase, salivary: SET1

Place/Population	F	S		n	H	PE	Source
ASIA							
INDIA							
Indians (Malaysia)	.449	.551		128	49.5	18.6	Teng et al. 1978a
MALAYSIA							
Malays	.601	.399		282	48.0	18.2	"
CHINA							
Chinese (Malaysia)	.497	.503		159	50.0	18.7	"
JAPAN							
Japanese (Hawaii)	.500	.500		53	50.0	18.8	Tan 1976
NORTH AMERICA							
UNITED STATES							
Whites (Hawaii)	.609	.391		96	47.6	18.1	"

Table 31. S-formylglutathione hydrolase; esterase D: FGH, ESD

Place/Population	1	2		n	H	PE	Source
EUROPE							
FINLAND (Helsinki)	.988	.012		242	2.4	1.2	Uotila 1984
ASIA							
INDIA	.802	.198		73	31.8	13.4	Board & Coggan 1986
CHINA	.708	.292		65	41.3	16.4	"
JAPAN	.507	.493		75	50.0	18.7	"
AFRICA							
SOUTHERN AFRICA							
Bantu	.995	.005		93	1.0	.5	"
OCEANIA							
AUSTRALIA							
Aborigines	.975	.025		101	4.9	2.4	"
MELANESIA							
Papua New Guineans	.951	.049		142	9.3	4.4	"
POLYNESIA							
Samoans (Samoa Is.)	.644	.356		101	45.9	17.7	"

Note: S-formylglutathione hydrolase (FGH) and esterase D (ESD) are identical. See Eiberg & Mohr (1986) and
     Apeshiotis & Bender (1986).

Table 32. Fucosidase, alpha-L: FUCA

Place/Population	1	2	n	H	PE	Source
EUROPE						
FRANCE	.640	.360	350	46.1	17.7	Khoi et al. 1979
GERMANY	.756	.244	535	36.9	15.0	Kühnl & Spielmann 1981
POLAND	.653	.347	271	45.3	17.5	Pryzbylski et al. 1982
UNITED KINGDOM						
England (Oxford)	.678	.322	169	43.7	17.1	Gill & Sutton 1984
NORTH AMERICA						
UNITED STATES						
Jews	.754	.246	126	37.1	15.1	Turner et al. 1975
Non-Jews	.750	.250	68	37.5	15.2	"
Blacks	.926	.074	27	13.7	6.4	"

Table 33. Galactose-1-phosphate uridyl transferase: GALT

Place/Population	1	2	0	x	n	H	PE	Source
EUROPE								
DENMARK (Copenhagen)	.923	.077			2,074	14.2	6.6	Eriksen & Dissing 1980
GERMANY	.994		.006		862	1.2	.6	"
ITALY (Central)	.919	.037	.004	.040	1,386	15.4	4.2	Vaccaro et al. 1984
POLAND (Wroclaw)	.947	.053			57	10.0	4.8	Dobosz 1983
SWITZERLAND (Bern)	.920	.080		.001	1,098	14.7	6.8	Pflugshaupt et al. 1978
UNITED KINGDOM								
England	.998		.002		6,415	.4	.2	Tedesco et al. 1975(c)
ASIA								
CHINA								
Chinese	.975	.010	.015		100	4.9	2.5	Xu & Ng 1983
NORTH AMERICA								
ALASKA								
Eskimos (St. Lawrence Is.)	1.000				222			Ferrell et al. 1981b
CANADA								
Dogrib Indians (NW Territories)	.991			.010[Rae]	158	1.8	5.8	Szathmary et al. 1983
MIDDLE AMERICA								
COSTA RICA								
Guaymi	.894	.106			286	19.0	8.6	Barrantes et al. 1982
SOUTH AMERICA								
BRAZIL								
Baniwa (Jandu Cachoeira)	.99	.007			363	1.4	.8	Salzano et al. 1986
Cayapo (Para & Mato Grosso)	.986	.014			396	2.8	1.4	Neel 1978
Macushi (N. Brazil & S. Guyana)	.907	.093			499	16.9	7.7	Neel et al. 1977b
Wapishana (")	.896	.103		.001[WAP]	615	18.7	8.4	"
Ticuna (C. Amazonas)	.956	.044			1,761	8.4	4.0	Neel et al. 1980

Continued

Table 33. Galactose-1-phosphate uridyl transferase: GALT (cont'd)

Place/Population	1	2	0	x	n	H	PE	Source
VENEZUELA								
Makiritare (S. Venezuela								
& N. Brazil)	.965	.035			388	6.8	3.3	Neel 1978
Yanomama (")	.990	.010			973	2.0	1.0	"
OCEANIA								
MELANESIA								
Papua New Guinea								
Gainj & Kalam (N.	1.000				276			Long et al. 1986
Central Highlands)								
MICRONESIA								
Marshall Is.	.999	.001			365	.2	.1	Neel et al. 1976

Note: $Gt^2 = Gt^{Durate}$; $Gt^0 = Galt^G$.

Table 34. Gamma-aminobutyric acid transaminase; gaba transaminase: GABAT

Place/Population	1	2	3	n	H	PE	Source
EUROPE							
Europeans	.558	.442		60	49.3	18.6	Jeremiah & Povey 1981
ASIA							
INDIA							
Indians (Malaysia)	.568	.396	.037	177	52.1	19.4	Bhattacharyya et al. 1985
MALAYSIA							
Malays (Singapore)	.621	.325	.054	140	50.9	19.0	"
CHINA							
Chinese (Malaysia)	.578	.381	.041	289	52.1	19.4	"
JAPAN (Tokyo)	.547	.453		75	49.6	18.6	Ishii et al. 1984

Table 35. Glucose dehydrogenase: GDH

Place/Population	1	2	3	Others	n	H	PE	Source
EUROPE								
ITALY (Rome)	.663	.212	.125		478	50.0	26.2	Scacchi et al. 1985
UNITED KINGDOM								
England	.723	.194	.083		373	43.3	22.0	King & Cook 1981
ASIA								
INDIA								
Indians (Malaysia)	.479	.404		.117	47	59.4	30.2	Saha et al. 1987
MALAYSIA								
Malays (Singapore)	.433	.447		.120	104	59.8	30.5	"
CHINA								
Chinese (Malaysia)	.421	.516		.063	254	55.3	25.7	"
JAPAN								
Japanese (Kyoto)	.522	.478			322	49.9	18.7	Kera et al. 1983a

Table 36. Glucose-6-phosphate dehydrogenase: G6PD

Place/Population	A+	B+	A-	B-	x	n	H	Source
EUROPE								
BULGARIA		.993		.007		16,681	1.4	Toncheva & Tzoneva 1984
FAROE IS. (N. Atlantic)		1.000				655		Tills et al. 1985
GREECE								
Greeks (Almopia)		.977		.023		131	4.5	Kaplanoglou & Triantaphyll-
								idis 1982
Greeks (Plati)		.979		.021		1,034	4.1	Tills et al. 1983b
ICELAND		1.000				1,060		Tills et al. 1982b
IRELAND		1.000				1,800		Tills 1977
ITALY		.986		.014		215	2.8	Corbo et al. 1981
POLAND		.980		.020		50	3.9	Wolanski et al. 1983
SPAIN (Madrid)		.992		.008		629	1.6	Garcia et al. 1979
UNITED KINGDOM								
England (Holy Is.)		1.000				122		Cartwright 1976
Isle of Lewis		1.000				95		Clegg et al. 1985
Orkney Is. (Westray)		1.000				406		Welch et al. 1973
YUGOSLAVIA		.990		.010		493	2.0	Fraser et al. 1966b
U.S.S.R.								
Abkhazia (Ochamchir Dist.)		1.000				262		Ferrell et al. 1985
ASIA								
TURKEY (Asia Minor)		.983		.017[a]		521	3.3	Say et al. 1965
ISRAEL								
Ashkenazi Jews		.996		.004[a]		819	.8	Szeinberg 1974(c)
Yemenite Jews		.971		.029		70	5.6	Tills et al. 1977b
IRAQ		.876		.124		305	21.7	Hamamy & Saeed 1981
JORDAN								
Jordanians (Amman)	.024	.943	.017	.017		543	11.0	Banerjee et al. 1981
Bedouins	.026	.896		.069[a]	.009[BKRT]	115	19.2	Saha & Banerjee 1986
KUWAIT		.786		.214[a]		461	33.6	Shaker et al. 1966
SAUDI ARABIA								
Arabs	.052	.850	.023	.069	.005[BKRT]	822	27.0	Samuel et al. 1986
Tribes (Western)		.942		.058[a]		291	10.9	Bayoumi et al. 1979
UNITED ARAB EMIRATES								
Abu Dhabians		.900		.100[a]		100	18.0	Kamel et al. 1980
IRAN (Caspian Littoral)		.907		.093		237	16.9	Kirk et al. 1977
AFGHANISTAN								
Pushtu & Dari		.975		.025		280	4.9	Papiha et al. 1977a
PAKISTAN		.977		.023		221	4.5	McCurdy & Mehmood 1970
INDIA								
Punjabi (N.India)		.966		.034		322	6.6	Singh et al. 1974b
Gujarati (W.India)		.975		.025		282	4.9	Papiha et al. 1981
Bhil tr. (Madhya Pradesh)		.933		.067		120	12.5	Papiha et al. 1978
Hindu (Andhra Pradesh)		.943		.057		209	10.8	Roberts et al. 1980
Irula tr. (Nilgiri Hills,								
S. India)		.910		.090		89	16.4	Saha et al. 1976
Bengali (E. India)		.957		.043		490	8.2	Ghosh et al. 1981
SRI LANKA								
Sinhalese (Anuradhapura)		.947		.053		132	10.0	Roberts et al. 1972b
BANGLADESH								
Moslems (Dacca)		.960		.040		150	7.7	Papiha et al. 1975
NEPAL (Nepalganji)		.880		.120		137	21.1	Modiano et al. 1986
BHUTAN		1.000				151[b]		Mourant et al. 1969
MALAYSIA								
Malays		.980		.020		346	3.9	Lie-Injo & Ti 1964
Aborigines		.819		.181		504	29.6	"
INDONESIA								
Balinese (Bali Is.)		.892		.108		74	19.3	Breguet et al. 1982b
THAILAND		.875		.125		811	21.9	Flatz & Tantachamroon 1970
VIETNAM		.970		.030[c]		2,034	5.8	Toncheva 1986
PHILIPPINES								
Filipino		.962		.038		293	7.3	Motulsky et al. 1964
CHINA								

Continued

Table 36. Glucose-6-phosphate dehydrogenase: G6PD (cont'd)

Place/Population	A+	B+	A-	B-	x	n	H	Source
Chinese (Malaya)		.976		.024		207	4.7	Lie-Injo & Ti 1964
JAPAN								
Japanese		1.000				458		Yamamoto et al. 1972
Japanese (Hiroshima & Nagasaki)		.997			.003[d]	6,913	.6	Kageoka et al. 1985
Ainu		1.000				125		Omoto & Harada 1972
KOREA (Seoul)		.986		.014		277	2.8	Kang & Lee 1973

AFRICA

Place/Population	A+	B+	A-	B-	x	n	H	Source
LIBYA (Tripoli & Benghasi)	.042	.958				286	8.0	Kamel et al. 1975
ALGERIA								
Ideles (Ahaggar)	.121	.862	.017			116	24.2	Lefèvre-Witier & Vergnes 1977
EGYPT								
Nubians (Southern)	.068	.932				154	12.7	Bertin et al. 1978
SUDAN								
Sudanese (Khartoum)	.106	.826	.021	.038	.009[Kht]	236	30.5	Saha et al. 1978b
Beja-Amarar (NE Sudan)	.060	.940				100	11.3	el Hassan et al. 1968
CHAD								
Sara Majingay	.129	.614	.178	.069		101	57.0	Hiernaux 1976
MALI								
Kel Kummer Twareg (Menaka)	.086	.914				140	15.7	Lefèvre-Witier & Vergnes 1977
SENEGAL								
Bedik (E. Senegal)	.237	.717	.033	.008	.005[Iba]	396	42.9	Bouloux et al. 1972
LIBERIA	.155	.690	.155			71	47.6	Willcox & Beckman 1981
IVORY COAST								
Tuka	.216	.545	.239			134	59.9	Vergnes & Cabannes 1976
NIGERIA								
Yoruba (Western)	.269	.580	.151			63	56.8	Ojikutu et al. 1977
CAMEROON	.258	.676	.064	.002		561	47.2	Bernstein et al. 1980
CENTRAL AFRICAN REPUBLIC								
Babinga Pygmies	.113	.859	.028			71	24.9	Cavalli-Sforza et al. 1969
BURUNDI								
Hutu	.164	.721	.115			262	44.0	Gall et al. 1982
ETHIOPIA								
Amhara tr.	.032	.968				164	6.2	Harrison et al. 1969
TANZANIA								
Sandawe (Kondoa-Irang Dist.)	.093	.847	.060			206	27.0	Godber et al. 1976
Nyaturu (")	.099	.717	.184			207	44.2	"
ANGOLA								
Njinga	.141	.686	.173			106	48.0	Nurse et al. 1979
MOZAMBIQUE								
Bantu (Southeastern)	.170	.640	.190			735	52.5	Reys et al. 1970
SOUTHERN AFRICA								
Basters (Namibia)	.016	.984				119	3.1	Nurse et al. 1982
San/Bushmen								
!Kung, Dobe	.038	.952	.010			204	9.2	Nurse & Jenkins 1977
G/wi (Botswana)		.979	.021			47	4.1	Jenkins et al. 1975
Khoi/Hottentot								
Keetmanshoop Nama		1.000				59		Jenkins 1972
Negroes/Dama	.126	.863	.011			119	23.9	Nurse et al. 1976
Zulu/Bantu	.115	.870	.015			99	23.0	Nurse et al. 1985
Griqua (Campbell, Cape Province)	.146	.829	.024			100	29.1	Nurse & Jenkins 1975
Negroes	.134	.813	.053			981	31.8	Hitzeroth & Bender 1980
TRISTAN DA CUNHA	.054	.946				157	10.2	Jenkins et al. 1985

NORTH AMERICA

Place/Population	A+	B+	A-	B-	x	n	H	Source
ALASKA								
Eskimos (Anchorage)		1.000				74		Duncan et al. 1974
Eskimos (St. Lawrence Is.)		1.000				222		Ferrell et al. 1981b
Athabaskan Indians (Anchorage)		1.000				37		Duncan et al. 1974(c)
CANADA								

Continued

Table 36. Glucose-6-phosphate dehydrogenase: G6PD(cont'd)

Place/Population	A+	B+	A-	B-	x	n	H	Source
Eskimos (Igloolik)		1.000				206		McAlpine et al. 1974
Ojibwa Indians (Pikangikum)		1.000				93		Szathmary et al. 1974
Cree Indians (Manitoba)		1.000				199		Lucciola et al. 1974
Dogrib Indians								
(NW Territories)		1.000				158		Szathmary et al. 1983
UNITED STATES								
White (Louisiana)		1.000				193		Fox et al 1981
Black (Louisiana)	.211	.652	.137			204		Fox et al. 1981

SOUTH AMERICA

BOLIVIA								
Aymara (Western)	.006	.994				155	1.2	Ferrell et al. 1978a
BRAZIL								
Baniwa		1.000				78		Salzano et al. 1986
Cayapo		1.000				156		Salzano et al. 1972b
Caingang		1.000				212		Salzano et al. 1980
Parakana		1.000				98		Black et al. 1980
Pacaás Novos		1.000				222		Salzano et al. 1985
SURINAM								
Trio/Tiriyo (Southern)		.969		.031		255	6.0	Geerdink et al. 1974b
VENEZUELA								
Makiritare (S. Venezuela &								
N. Brazil)		1.000				70		Weitkamp & Neel 1970
Yanomama (")		1.000				788		Weitkamp & Neel 1972

OCEANIA

AUSTRALIA								
Aborigines (Elcho Is.,								
N. Territory)		1.000				90		Blake 1979
Aborigines (Hooker Creek,								
N. Territory)		1.000				285		"
Aborigines (Beagle Bay,								
W. Australia)		1.000				103		"
MELANESIA								
Papua New Guinea								
Gainj & Kalam (N. Central								
Highlands)		1.000				278		Long et al. 1986
Jimi Valley (W. Highlands)		1.000				293		Mourant et al. 1981
Yagaria (E. Highlands)		1.000				233		Mourant et al. 1982
Fuyuge (C. Dist. Highlands)		.989		.011		133	2.2	Woodfield et al. 1974
Daga (Bay Province)		.955		.045		67	8.6	Jenkins et al. 1983
Karkar Is.								
Takia		.854		.146		139	24.9	Booth et al. 1982
Manus Is.		.949		.051		79	9.7	Malcolm et al. 1972
MICRONESIA								
E. Caroline Is.		1.000				1,005		Yamamoto & Fu 1973
POLYNESIA								
Samoans (Christchurch,								
New Zealand)		.982		.018		77	3.5	Booth et al. 1977

Note: [a] It is not clear that the deficiency is due to A or B or both. [b] One individual with enzyme deficiency
was found. [c] Three variants (Canton, Hanoi, and Vin Fu) were found among 13 G6PD deficient males
studied. [d] The combined frequency of 18 variants.

Table 37. Glucose-6-phosphate dehydrogenase, salivary; hexose-6-phosphate dehydrogenase: G6PDS*, H6PD

Place/Population	1	2	n	H	PE	Source
UNITED STATES						
Caucasians (Hawaii)	.755	.245	190	37.0	15.1	Tan & Ashton 1976
Chinese (")	.706	.294	34	41.5	16.4	"
Japanese (")	.659	.341	104	44.9	17.4	"

Table 38. Glucose phosphate isomerase; phosphoglucose/phosphohexose isomerase: GPI, PGI, PHI

Place/Population	1	2	3	x	n	H	PE	Source
EUROPE								
Europeans	.999	.001			1,930	.2	.1	Ritter & Wendt 1971a
FAROE IS. (N. Atlantic)	1.000				666			Tills et al. 1985
GREECE								
Greeks	1.000				215			Schneider et al. 1975
Greeks (Plati)	.997	.003			1,014	.6	.3	Tills et al. 1983b
ITALY (Bologna)	.996	.004			274	.8	.4	Beretta et al. 1977
NETHERLANDS (Leiden)	.999	.001			806	.2	.1	Fraser et al. 1974
UNITED KINGDOM								
Isle of Lewis	1.000				94			Clegg et al. 1985
Orkeny Is. (Westray)	1.000				408			Welch et al. 1973

U.S.S.R.

Place/Population	1	2	3	x	n	H	PE	Source
EUROPEAN PART								
Abkhazia (Ochamchir Dist.)	.996	.004			262	.8	.4	Ferrell et al. 1985

ASIA

Place/Population	1	2	3	x	n	H	PE	Source
ISRAEL								
Yemenite Jews	1.000				157			Tills et al. 1977b
IRAN								
Turkoman (Gonbad, NE Iran)	.990	.010			155	2.0	1.0	Kirk et al. 1977
AFGHANISTAN								
Pushtu & Dari (Kabul)	.998	.002			280	.4	.2	Papiha et al. 1977a
INDIA								
Punjabi (Chandigarh)	.995	.005			100	1.0	.5	Papiha et al. 1972
Gujarati (W. India)	.993	.005		$.002^9$	282	1.4	.7	Papiha et al. 1981
Hindu & Moslems								
(Madhya Pradesh)	.992	.008			338	1.6	.8	Papiha & Chhaparwal 1973
Bhil tr. (Madhya Pradesh)	.993	.007			143	1.4	.7	Papiha et al. 1978
Hindu (Andhra Pradesh)	.998	.002			211	.4	.2	Roberts et al. 1980
Irula tr. (Nilgiri Hills,								
S. India)	1.000				173			Saha et al. 1976
BANGLADESH								
Moslems (Dacca)	.985	.010		$.005^9$	200	3.0	1.5	Papiha et al. 1975
NEPAL								
Sherpas	.988			$.012^{Sh}$	84	2.4	1.2	Santachiara et al. 1976
MALAYSIA								
Malays (Singapore)	.996	.004			259	.8	.4	Blake et al. 1973a
INDONESIA								
Batak (Samosir Is.)	1.000				188			McDermid et al. 1973
Balinese (Bali Is.)	1.000				148			Breguet et al. 1982b
THAILAND	.996	.001		$.003^{4,8}$	441	.8	.4	Detter et al. 1968
PHILIPPINES								
Filipino/Tagalog (Manila)	.998	.002			253	.4	.2	Omoto et al. 1978
Negrito (Angeles,								
C. Luzon)	1.000				129			"
CHINA								
Tibetans	1.000				25			Santachiara et al. 1976
Chinese (Singapore)	.999	.001			378	.2	.1	Blake et al. 1973a
JAPAN								
Japanese (Mie)	.995	.002		$.003^{2,4-6}$	2,676	1.0	.5	Ishimoto & Kuwata 1974a
Ainu (Hokkaido)	1.000				125			Omoto & Harada 1972

AFRICA

Place/Population	1	2	3	x	n	H	PE	Source
EGYPT								
Nubians (S. Egypt)	1.000				155			Bertin et al. 1978
SUDAN								
Sudanese (Khartoum)	1.000				284			Saha et al. 1978b
NIGERIA	1.000				141			Tills et al. 1979
CENTRAL AFRICAN REPUBLIC								
Pygmies (Western)	1.000				316			Santachiara et al. 1980
BURUNDI								
Hutu	.997	.003			447	.6	.3	Gall et al. 1982

Continued

Table 38. Glucose phosphate isomerase; phosphoglucose/phosphohexose isomerase: GPI, PGI, PHI (cont'd)

Place/Population	1	2	3	x	n	H	PE	Source
TANZANIA								
Sandawe (Kondoa-Irang								
Dist.)	.983			.017[V]	212	3.3	1.6	Godber et al. 1976
Nyaturu (")	.964			.036[V]	179	6.9	3.3	"

NORTH AMERICA

ALASKA								
Eskimos (St. Lawrence Is.)	1.000				222			Ferrell et al. 1981b
CANADA								
Eskimos (Igloolik)	1.000				397			McAlpine et al. 1974
Ojibwa Indians								
(Pikangikum)	1.000				93			Szathmary et al. 1974
Cree Indians (Manitoba)	1.000				200			Lucciola et al. 1974
Dogrib Indians								
(NW Territories)	1.000				158			Szathmary et al. 1983

MIDDLE AMERICA

COSTA RICA								
Guaymi	1.000				286			Barrantes et al. 1982
PANAMA								
Blacks	1.000				N.A.			Ferrell et al. 1978b

SOUTH AMERICA

BOLIVIA								
Aymara (Western)	1.000				316			Ferrell et al. 1978a
BRAZIL/GUYANA								
Baniwa (Jandu Cachoeira)	1.000				363			Salzano et al. 1986
Cayapo (Pará & Mato								
Grosso)	.991	.009[2CAY1]			525	1.8	.9	Neel 1978
Macushi (N. Brazil &								
S. Guyana)	1.000				499			Neel et al. 1977b
Wapishana (")	.999		.001[3WAP1]		615	.2	.1	"
Ticuna (C. Amazonas)	1.000				1,765			Neel et al. 1980
ECUADOR								
Waorani	1.000				187			Larrick et al. 1985
FRENCH GUIANA								
Wayampi (Trois-Sauts &								
Camopi)	1.000				219			Tchen et al. 1978
VENEZUELA								
Makiritare (S. Venezuela								
& N. Brazil)	1.000				409			Neel 1978
Yanomama (")	1.000				2,005			"

OCEANIA

AUSTRALIA								
Aborigines/Malag (Elcho								
Is., N. Territory)	1.000				234[a]			Omoto & Blake 1972
Aborigines (Hooker Creek,								
N. Territory)	1.000				285			Blake 1979
Aborigines (C. Australia)	1.000				98			Omoto & Blake 1972
Aborigines (W. Australia)	1.000				102			Blake & Spargo 1986
MELANESIA								
W. New Guinea								
Asmat (S. Coastal Plain)	1.000				653			Gajdusek et al. 1978
Papua New Guinea								
Gainj & Kalam (N.								
Central Highlands)	.967	.033			272	6.4	3.1	Long et al. 1986
Jimi Valley								
(W. Highlands)	1.000				385			Mourant et al. 1981
Yagaria (E. Highlands)	1.000				424			Mourant et al. 1982
Fuyuge (C. Highlands)	1.000				173			Woodfield et al. 1974
Karkar Is.								
Takia	1.000				139			Booth et al. 1982
Manus Is.	1.000				183			Malcolm et al. 1972

Continued

Table 38. Glucose phosphate isomerase; phosphoglucose/phosphohexose isomerase: GPI, PGI, PHI (cont'd)

Place/Population	1	2	3	x	n	H	PE	Source
Solomon Is.	1.000				364			Blake et al. 1983
MICRONESIA								
W. Caroline Is.	1.000				382			Blake et al. 1973b
Marshall Is.	1.000				373			Neel et al. 1976
POLYNESIA								
Samoans (Christchurch,								
New Zealand)	1.000				77			Booth et al. 1977

Note: [a] One individual was of genotype +/4.

Table 39. Glucosidase, alpha, acid: GAA

Place/Population	1	2	3	4	n	H	PE	Source
EUROPE								
UNITED KINGDOM								
England	.969	.031			1,156	6.0	2.9	Swallow et al. 1975
ASIA								
INDIA								
Indians (United Kingdom)	.984	.016			153	3.1	1.5	"
Indians (Malaysia)	.985	.004	.011		136	3.0	1.5	Teng & Tan 1979
MALAYSIA								
Malays (Kuala Lumpur)	.996	.002	.002		236	.8	.4	"
CHINA								
Chinese (Malaysia)	.998	.002			261	.4	.2	"
AFRICA								
Blacks (United Kingdom)	.988	.012			40	2.4	1.2	Swallow et al. 1975
NORTH AMERICA								
CANADA								
Canadians (Manitoba)	.914	.028		.058	633	16.0	8.1	Nickel & McAlpine 1982
OCEANIA								
AUSTRALIA								
Aborigines (N. Territory)	1.000				101			Blake 1984
Aborigines (C. Australia)	.990	.010			155	2.0	1.0	"
MELANESIA								
Papua New Guinea								
E. Highlands	1.000				70			"
Port Moresby	1.000				184			"

Table 40. Glutamic-oxaloacetic transaminase 1, soluble: GOT1

Place/Population	1	2	3	n	H	PE	Source
EUROPE							
GREECE (Northern)	.998	.002		412	.4	.2	Kouvatsi & Triantaphylli-dis 1984
NETHERLANDS (Leiden)	1.000			806			Fraser et al. 1974
UNITED KINGDOM							
England	.999	.0005	.0005	1,195	.2	5.0	Hackel et al. 1972
ASIA							
IRAN							
Iranians (Tehran)	1.000			134			Farhud et al. 1973
INDIA							
Hindu (Andhra Pradesh)	1.000			85			Naidu et al. 1985
Indians (Malaysia)	1.000			190			Teng et al. 1978a
NEPAL							
Sherpas	1.000			56			Santachiara et al. 1976
MALAYSIA							
Malays (Kuala Lumpur)	1.000			311			Teng et al. 1978a
INDONESIA							
Balinese (Bali Is.)	1.000			148			Breguet et al. 1982b
PHILIPPINES							
Filipino/Tagalog (Manila)	.988	.012		253	2.4	1.2	Omoto et al. 1978
Filipino	.991		.009	165	1.8	.9	Chen & Giblett 1971
Negrito	1.000			129			Omoto et al. 1978
CHINA							
Chinese (C. Taiwan)	1.000			160			Chen et al. 1973
Tibetans	1.000			15			Santachiara et al. 1976
Chinese (Malaysia)	.996	.004		283	.8	.4	Teng et al. 1978a
JAPAN							
Japanese (Mie)	.986	.009	.005	2,047	2.8	1.4	Ishimoto & Kuwata 1974b
Japanese (Hiroshima/ Nagasaki)	.983	.014[HR1]	.003[a]	11,737	3.4	1.4	Satoh et al. 1986
AFRICA							
NIGERIA	1.000			233			Hackel et al. 1972
CENTRAL AFRICAN REPUBLIC							
Pygmies (Western)	1.000			43			Santachiara et al. 1980
BURUNDI							
Hutu	1.000			400			Gall et al. 1982
NORTH AMERICA							
ALASKA							
Eskimos (Anchorage)	.985	.015		131	3.0	1.5	Duncan et al. 1974
Athabaskan Indians	.978	.022		45	4.3	2.1	"
CANADA							
Eskimos (Igloolik)	.983	.017		143	3.3	1.6	McAlpine et al. 1974
Ojibwa Indians (Pikangikum)	.929	.071		91	13.2	6.2	Szathmary et al. 1974
Cree Indians (Manitoba)	.980	.015	.005	199	3.9	1.9	Lucciola et al. 1974
UNITED STATES							
Whites	1.000			97			Chen & Giblett 1971
Blacks	1.000			145			"
Yakima Indians	.970	.012	.018	84	5.9	2.9	"
SOUTH AMERICA							
Peruvian Indians	.943	.032	.025	79	10.9	5.5	"
OCEANIA							
AUSTRALIA							
Aborigines (Hooker Creek, N. Territory)	1.000			285			Blake 1979
Aborigines/Cundeelee (W. Australia)	1.000			132			"

Continued

Table 40. Glutamic-oxaloacetic transaminase 1, soluble: GOT1 (cont'd)

Place/Population	1	2	3	n	H	PE	Source
MELANESIA							
New Guinea							
Asmat (S. Coastal Plain)	1.000			118			Gajdusek et al. 1978
Papua New Guinea							
Gainj & Kalam (N. Central							
Highlands)	1.000			277			Long et al. 1986
Fuyuge (C. Highlands)	1.000			173			Woodfield et al. 1974
Solomon Is.	1.000			364			Blake et al. 1983
POLYNESIA							
Samoans (Christchurch,							
New Zealand)	1.000			77			Booth et al. 1977

Note: [a] Pooled frequency of five kinds of rare variants.

Table 41. Glutamic-oxaloacetic transaminase 2, mitochondrial: GOT2

Place/Population	1	2	3	n	H	PE	Source
EUROPE							
GERMANY (Dusseldorf)	.977	.023		310	4.5	2.2	Driesel et al. 1982a
GREECE							
Greeks (Northern)	.989	.011		412	2.2	1.1	Kouvatsi & Triantaphylli-dis 1984
UNITED KINGDOM							
England	.983	.017		710	3.3	1.6	Hackel et al. 1972
ASIA							
INDIA							
Indians (Malaysia)	1.000			177			Teng et al. 1978a
MALAYSIA							
Malays (Kuala Lumpur)	.995	.005		277	1.0	.5	"
CHINA							
Chinese (Malaysia)	.995	.005		284	1.0	.5	"
JAPAN							
Japanese	.996	.004		1,860	.8	.4	Toyomasu et al. 1984
AFRICA							
NIGERIA							
Nigerians	.921	.010	.069	525	14.7	7.1	Hackel et al. 1972

Table 42. Glutamic-pyruvate transaminase; alanine aminotransferase: GPT1, AAT1

Place/Population	1	2	x	n	H	PE	Source
EUROPE							
BELGIUM (Liège)	.506	.494		757	50.0	18.7	Brocteur et al. 1980
DENMARK	.540	.455	.005[3]	902	50.1	19.3	Dissing 1973

Continued

79

Table 42. Glutamic-pyruvate transaminase; alanine aminotransferase: GPT1, AAT1 (cont'd)

Place/Population	1	2	x	n	H	PE	Source
FAROE IS. (N. Atlantic)	.638	.362		666	46.2	17.8	Tills et al. 1985
FRANCE (Pyrenees)	.545	.455		658	49.6	18.6	Vergnes et al. 1980a
GERMANY (Hamburg)	.531	.468	.001[3]	2,026	49.9	18.8	Brinkmann et al. 1972
GREECE	.566	.434		98	49.2	18.6	Chen et al. 1972b
HUNGARY	.424	.576		152	48.8	18.5	Goedde et al. 1986b
ITALY (Ferrara)	.502	.497	.002[3]	294	50.1	19.0	Scozzari et al. 1975
NETHERLANDS (Leiden)	.533	.467	.001	779	49.8	18.8	Fraser et al. 1974
NORWAY							
Norwegians	.537	.461	.002[3,6,7]	4,148	49.9	18.9	Olaisen 1975
Lapps/Saami (C. Finnmark)	.611	.389		198	47.5	18.1	Olaisen & Teisberg 1972
POLAND	.526	.474		497	49.9	18.7	Dobosz 1983
PORTUGAL (Porto Dist.)	.489	.501	.010[1M,3]	619	51.0	20.0	Amorim & Siebert 1982
SPAIN	.505	.495		184	50.0	18.7	Brinkmann et al. 1973
SWITZERLAND (Bern)	.510	.488	.002[3]	1,894	50.2	19.0	Pflugshaupt et al. 1978
UNITED KINGDOM							
England	.528	.471	.001	1,444	49.9	18.8	Welch et al. 1975b
YUGOSLAVIA (Serbia)	.538	.462		277	49.7	18.7	Kalimanovska et al. 1983

U.S.S.R.

Place/Population	1	2	x	n	H	PE	Source
EUROPEAN PART							
Russians (Moscow)	.572	.428		1,006	48.9	18.5	Altukhov et al. 1981
Abkhazia (Ochamchir Dist.)	.562	.438		252	49.2	18.6	Ferrell et al. 1985

ASIA

Place/Population	1	2	x	n	H	PE	Source
TURKEY (Asia Minor)	.549	.448	.002[3]	213	49.8	18.9	Brinkmann et al. 1973
ISRAEL							
Ashkenazi Jews	.600	.400		196	48.0	18.2	Lahav & Szeinberg 1972
Yemenite Jews	.726	.274		190	39.8	15.9	"
SAUDI ARABIA							
Shiites & Sunnites (al-Hasa)	.618	.382		359	47.2	18.0	Goedde et al. 1979b
IRAN							
Turkoman (Gonbad, NE Iran)	.539	.461		155	49.7	18.7	Kirk et al. 1977
AFGHANISTAN							
Pushtu (Kabul)	.455	.545		210	49.6	18.6	Goedde et al. 1977b
INDIA							
Jat Sikhs (Punjab)	.529	.471		140	49.8	18.7	Chahal & Papiha 1981
Hindu (Andhra Pradesh)	.602	.398		49	47.9	18.2	Roberts et al. 1980
NEPAL							
Sherpas	.706	.294		85	41.5	16.4	Santachiara et al. 1976
MALAYSIA							
Malays (Singapore)	.343	.647	.009[6]	214	46.4	18.7	Blake 1976
INDONESIA							
Toba Batak (Sumatra)	.389	.611		148	47.5	18.1	"
THAILAND	.410	.590		116	48.4	18.3	Goedde & Benkmann 1987
PHILIPPINES							
Filipino/Tagalog (Manila)	.324	.671	.005[6]	253	44.5	17.8	Omoto et al. 1978
Filipino	.295	.693	.011[6]	88	43.3	17.9	Chen et al. 1972b
Negrito (Angeles, C. Luzon)	.143	.857		129	24.5	10.8	Omoto et al. 1978
CHINA							
Mongolians (N. China)	.614	.386		200	47.4	18.1	Goedde et al. 1984b
Zhuang (S. China)	.417	.583		209	48.6	18.4	"
Hui (Ningxia)	.537	.463		218	49.7	18.7	Xu & Du 1984b
Dong (Guangxi)	.408	.592		200	48.3	18.3	"
Tibetans	.552	.448		29	49.5	18.6	Santachiara et al. 1976
Chinese (Singapore)	.500	.490	.010[6]	303	51.0	20.0	Blake 1976
JAPAN							
Japanese (Mie)	.623	.376	.001[6]	1,134	47.0	18.1	Ishimoto & Kuwata 1974c
Ainu (Hokkaido)	.455	.545		101	49.6	18.6	Nishigaki et al. 1980(c)
KOREA							
Koreans (Seoul)	.606	.394		137	47.8	18.2	Goedde et al. 1986c
Koreans (NE China)	.597	.403		213	48.1	18.3	Goedde et al. 1984b

AFRICA

Place/Population	1	2	x	n	H	PE	Source
EGYPT (Cairo & Anshas)	.453	.547		190	49.6	18.6	Goedde et al. 1980
NIGERIA							

Continued

80

Table 42. Glutamic-pyruvate transaminase; alanine aminotransferase: GPT1, AAT1 (cont'd)

Place/Population	1	2	x	n	H	PE	Source
Yoruba (Western)	.863	.137		62	23.6	10.4	Ojikutu et al. 1977
CAMEROON (Bamenda)	.869	.131		279	22.8	10.1	Goedde et al. 1979a
CENTRAL AFRICAN REPUBLIC							
Pygmies (Western)	.843	.157		67	26.5	11.5	Santachiara et al. 1980
Hutu (Burundi)	.859	.141		364	24.2	10.6	Gall et al. 1982
CONGO	.841	.259		85	22.6	13.0	Chen et al. 1972b
KENYA	.857	.143		63	24.5	10.8	"
SOUTHERN AFRICA							
San/Bushmen - G!aokxate	.455	.545		33	49.6	18.6	Nurse & Jenkins 1977
Negro/Denasena	.851	.149		56	25.4	11.1	"
Bantu (Natal)	.873	.127		177	22.2	9.9	Blake 1976
TRISTAN DA CUNHA	.545	.455		145	49.6	18.6	Jenkins et al. 1985

NORTH AMERICA

ALASKA							
Eskimos (Anchorage)	.577	.423		124	48.8	18.5	Duncan et al. 1974
Eskimos (St. Lawrence Is.)	.640	.360		222	46.1	17.7	Ferrell et al. 1981b
Athabaskan Indians (Anchorage)	.561	.439		33	49.3	18.6	Duncan et al. 1974
CANADA							
Eskimos (Igloolik)	.621	.369	.010[5]	145	47.8	19.2	McAlpine et al. 1974
Eskimos	.596	.401	.003[5]	146	48.4	18.6	Chen et al. 1972b
Cree Indians (Manitoba)	.430	.570		194	49.0	18.5	Lucciola et al. 1974
Dogrib Indians (NW Territories)	.342	.658		158	45.0	17.4	Szathmary et al. 1983
UNITED STATES							
Whites	.531	.466	.003[3]	294	50.1	19.1	Welch et al. 1975b
Blacks	.810	.190		258	30.8	13.0	"
Whites (Louisiana)	.528	.468	.004[3]	376	50.2	19.2	Fox et al. 1981
Blacks (")	.755	.245		372	37.0	15.1	"
Mexican-Americans (Texas)	.463	.537		567	49.7	18.7	Hewett-Emmett et al. 1986b

SOUTH AMERICA

CHILE							
Aymara	.412	.588		102	48.4	18.3	van der Does et al. 1972
Atacameño (Toconao, N. Chile)	.364	.625	.011[3]	168	47.7	19.2	Goedde et al. 1985b
COLOMBIA							
Am. Indians	.531	.469		227	49.8	18.7	Lanchbury et al. 1984
ECUADOR							
Shuara	.300	.700		90	42.0	16.6	Goedde et al. 1977a
FRENCH GUIANA							
Wayampi (Trois-Sauts & Camopi)	.457	.543		232	49.6	18.7	Tchen et al. 1978
PERU							
Indians	.568	.432		44	49.1	18.5	Chen et al. 1972b

OCEANIA

AUSTRALIA							
Aborigines/Malag (Elcho Is., NT)	.611	.389		144	47.5	18.1	Blake 1976
" /Waljbiri (Hooker Creek, NT)	.930	.070		286	13.0	6.1	"
" /Ngarinjin (Kimberley, WA)	.809	.191		102	30.9	13.1	Blake & Spargo 1986
MELANESIA							
W. New Guinea							
Asmat (S. Coastal Plain)	.890	.110		114	19.6	8.8	Gajdusek et al. 1978
Papua New Guinea							
Gainj & Kalam (N. Central Highlands)	.686	.308	.005[6]	569	43.5	17.5	Long et al. 1986
Aborigines (E. Highlands)	.817	.183		279	29.9	12.7	Blake 1976
Fuyuge (C. Highlands)	.780	.220		173	34.3	14.2	"
Daga (Bay Province)	.795	.205		139	32.6	13.6	Jenkins et al. 1983
Karkar Is.							
Takia (Gamog & Boromon)	.531	.465	.004[6]	357	50.2	19.2	Blake 1976
Buka Is.	.494	.506		80	50.0	18.7	McLoughlin et al. 1982a
Solomon Is. (Eastern Dist.)	.640	.360		250	46.1	17.7	Blake 1976
MICRONESIA							
W. Caroline Is.	.556	.444		266	49.4	18.6	"
E. Caroline Is.	.706	.292	.001[6]	908	41.6	16.6	"
POLYNESIA							
Samoans (New Zealand)	.547	.447	.007[3]	77	50.1	19.4	Booth et al. 1977

Table 43. Glutathione peroxidase 1: GPX1

Place/Population	1	2	3	n	H	PE	Source
EUROPE							
ITALY	1.000			99			Meera Khan et al. 1984
NETHERLANDS	1.000			398			"
NORWAY	1.000			110			"
ASIA							
INDIA							
Punjabi (N. India)	.987	.013		116	2.6	1.3	"
Marathi (W. India)	1.000			47			"
Telugu (S. India)	1.000			198			"
MALAYSIA							
Malays (Singapore)	1.000			100			Board 1983
AFRICA							
NIGERIA							
Yoruba	1.000			65			Ojikutu et al. 1977
CENTRAL AFRICAN REPUBLIC							
Babinga Pygmies	.996	.004		N.A.	.8	.4	Meera Khan et al. 1984
Bantu	.918	.082		N.A.	15.1	7.0	"
RWANDA							
Twa Pygmies	.979	.021		N.A.	4.1	2.0	"
Hutu	.955	.045		122	8.6	4.1	Meera Khan 1986
ZAIRE							
Mbuti	1.000			N.A.			Meera Khan et al. 1984
CONGO							
Beti Bantu (Cuvette)	.942	.058		94	10.9	5.2	Destro-Bisol et al. 1986a
Babinga pygmies	.989	.011		46	2.2	1.1	"
SOUTHERN AFRICA							
Blacks (Durban)	.942	.058		303	10.9	5.2	Board 1983
TRISTAN DA CUNHA	1.000			98			Jenkins et al. 1985
NORTH AMERICA							
UNITED STATES							
Whites (California)	.996	.004		388	.8	.4	Beutler et al. 1974
Blacks (")	.968	.032		392	6.2	3.0	"
Pima Indians	1.000			100			Board 1983
SOUTH AMERICA							
COLOMBIA							
Indians	1.000			40			"
PERU							
Quechua	1.000			149			Meera Khan et al. 1984
SURINAM							
Bush Negroes/Djuka	.945	.054	.001	715	10.4	4.9	Meera Khan et al. 1986
Trio Indians	1.000			504			Meera Khan et al. 1984
Wajana Indians	1.000			312			"
OCEANIA							
AUSTRALIA							
Aborigines	1.000			150			Board 1983
MELANESIA							
Papua New Guinea							
Papuans	1.000			294			"
MICRONESIA							
Nauru	1.000			97			"
POLYNESIA							
Tokelau Is.	1.000			110			"

Table 44. Glutathione reductase: GSR

Place/Population	1	2	n	H	PE	Source
EUROPE						
GREECE	.988	.012	379	2.4	1.2	Stamatoyannopoulos et al. 19[?]
SPAIN (Madrid)	1.000		2,129[a]			Garcia et al. 1979
AFRICA						
SUDAN						
Sudanese (Khartoum)	.949	.051	414	9.7	4.6	Saha 1981
NORTH AMERICA						
CANADA						
Eskimos (Igloolik)	1.000		191			McAlpine et al. 1974
Ojibwa Indians						
(SE Ontario)	.990	.010^V	105	2.0	1.0	Szathmary et al. 1974
SOUTH AMERICA						
Peruvian Indians	1.000		112			Modiano et al. 1972

Note: [a] Deficiency of the enzyme found in two individuals.

Table 45. Glutathione-S-transferase-1: GST1

Place/Population	1	2	3	0	n	H	PE	Source
EUROPE								
Europeans	.130	.230		.640	49	52.1	27.3	Strange et al. 1984
FRANCE								
French (Paris)	.074	.279		.647	56	49.8	24.4	Laisney et al. 1984
ASIA								
INDIA								
Indians (Malaysia)	.161	.279		.560	43	58.3	31.0	Board 1981
CHINA								
Chinese (Singapore)	.156	.112	.005	.727	221	43.5	23.0	Bhattacharyya & Saha 1984b
Chinese (Malaysia)	.171	.065		.764	90	38.3	19.2	Board 1981
JAPAN								
Japanese	.252	.057		.691	168	45.6	21.9	Harada et al. 1986
OCEANIA								
AUSTRALIA								
Whites (Melbourne)	.106	.078		.815	40	31.8	16.4	Board 1981

Table 46. Glutathione-S-transferase-2: GST2

Place/Population	1	2	0	n	H	PE	Source
ASIA							
INDIA							
Indians (Malaysia)	.779	.221		43	34.4	14.3	Board 1981
MALAYSIA							
Chinese (Singapore)	.767	.215	.018	221	36.5	16.2	Bhattacharyya & Saha 1984b
Chinese (Malaysia)	.812	.188		96	30.5	12.9	Board 1981
OCEANIA							
AUSTRALIA							
Whites (Melbourne)	.838	.162		40	27.2	11.7	"

Table 47. Glyoxalase I: GLO1

Place/Population	1	2	n	H	PE	Source
EUROPE						
AUSTRIA (Vienna)	.434	.566	973	49.1	18.5	Pausch et al. 1985
BELGIUM (Liége)	.432	.568	750	49.1	18.5	Hoste et al. 1984
DENMARK	.431	.569	1,220	49.0	18.5	Eriksen 1979
FAROE IS. (N. Atlantic)	.383	.617	658	47.3	18.0	Tills et al. 1985
FINLAND	.359	.641	906	46.0	17.7	Virtaranta-Knowles & Nevanlinna 1982
FRANCE						
French (Toulouse)	.473	.527	386	49.9	18.7	Vergnes et al. 1980b
Basques	.419	.581	167	48.7	18.4	"
GERMANY (Hessen)	.439	.561	1,150	49.3	18.6	Kühnl et al. 1977b
HUNGARY	.422	.578	128	48.8	18.4	Goedde et al. 1986b
ICELAND	.458	.542	178	49.6	18.7	Karlsson et al. 1980a
ITALY (Rome)	.413	.587	259	48.5	18.4	Ranzani et al. 1979
NETHERLANDS (Leiden)	.454	.546	757	49.6	18.6	Meera Khan & Doppert 1976
NORWAY						
Norwegians	.442	.558	216	49.3	18.6	Olaisen et al. 1976b
Lapps/Saami	.304	.696	184	42.3	16.7	"
POLAND (Wroclaw)	.459	.541	648	49.7	18.7	Dobosz 1983
PORTUGAL (Porto Dist.)	.426	.574	631	48.9	18.5	Amorim & Siebert 1982
SPAIN (Galicia)	.423	.577	319	48.8	18.5	Carracedo & Concheiro 1983
SWITZERLAND (Bern)	.444	.556	619	49.4	18.6	Pflugshaupt et al. 1978
UNITED KINGDOM						
England (London)	.440	.560	296	49.3	18.6	Bagster & Parr 1976
YUGOSLAVIA (Serbia)	.384	.616	258	47.3	18.1	Kalimanovska et al. 1985
U.S.S.R.						
EUROPEAN PART						
Russians (Moscow)	.479	.521	1,038	49.9	18.8	Altukhov et al. 1981
Abkhazia (Ochamchir Dist.)	.369	.631	255	46.6	17.9	Ferrell et al. 1985
ASIA						
TURKEY (Asia Minor)						
Jews	.334	.666	151	44.5	17.3	Golan et al. 1979
ISRAEL						
Ashkenazi Jews	.301	.699	191	42.1	16.6	Golan et al. 1979
Yemenite Jews	.332	.668	134	44.4	17.3	"
IRAQ	.423	.577	472	48.8	18.5	Al-Agidi et al. 1980
IRAN						
Turkoman (Gonbad, NE Iran)	.277	.723	47	40.1	16.0	Akbari et al. 1984

Continued

Table 47. Glyoxalase 1: GLO1 (cont'd)

Place/Population	1	2	n	H	PE	Source
INDIA						
N. Indians (Delhi)	.231	.769	505	35.5	14.6	Ghosh 1977b
Gujarati (W. India)	.280	.720	134	40.3	16.1	"
Hindu (Andhra Pradesh)	.297	.703	1,709	41.8	16.5	Char & Rao 1986
Bengali (E. India)	.213	.787	268	33.5	14.0	Ghosh 1977b
Indians (Malaysia)	.287	.713	183	40.9	16.3	Teng et al. 1978b
MALAYSIA						
Malays (Kuala Lumpur)	.196	.804	294	31.5	13.3	"
INDONESIA						
Toba Batak (Sumatra)	.149	.851	148	25.4	11.1	Ghosh 1977a
Balinese (Bali Is.)	.203	.797	148	32.4	13.6	Breguet et al. 1982b
PHILIPPINES						
Negrito (Angeles, C. Luzon)	.244	.756	129	36.9	15.0	Omoto et al. 1978
CHINA						
Mongolians (Inner Mongolia)	.227	.773	207	35.1	14.5	Goedde et al. 1984b
Zhuang (S. China)	.160	.840	209	26.9	11.6	"
Uygur	.247	.753	219	37.2	15.1	Li et al. 1986
Hui	.162	.838	219	27.2	11.7	"
Dong	.187	.813	201	30.4	12.9	"
Chinese (Malaysia)	.200	.800	115	32.0	13.4	Teng et al. 1978b
TAIWAN						
Toroko-Aborigines	.193	.807	122	31.2	13.1	Chen et al. 1985
JAPAN						
Japanese (Tokyo)	.088	.912	572	16.1	7.4	Harada & Misawa 1976
KOREA						
Koreans (Seoul)	.120	.880	138	21.1	9.4	Goedde et al. 1986c
Koreans (NE China)	.111	.889	212	19.7	8.9	Goedde et al. 1984b

AFRICA

Place/Population	1	2	n	H	PE	Source
ALGERIA						
Kabyle	.459	.541	74	49.7	18.7	Bouali et al. 1981
EGYPT						
Egyptians (Cairo & Anshas)	.379	.621	140	47.1	18.0	Goedde et al. 1980
SUDAN						
Tribes	.368	.642	850	45.2	17.5	Laha et al. 1979
LIBERIA	.324	.676	485	43.8	17.1	Goedde et al. 1985a
GAMBIA						
Jali	.279	.721	506	40.3	16.1	Parr et al. 1977
CAMEROON (Bamenda)	.337	.663	280	44.7	17.4	Goedde et al. 1979a
CENTRAL AFRICAN REPUBLIC						
Pygmies (Western)	.391	.609	78	47.6	18.1	Santachiara et al. 1980
Mbugu	.272	.728	92	39.6	15.9	Spedini et al. 1982
RWANDA						
Hutu	.275	.725	122	39.9	16.0	Meera Khan 1986
SOUTHERN AFRICA						
Bantu	.259	.741	843	38.4	15.5	Bender et al. 1977

NORTH AMERICA

Place/Population	1	2	n	H	PE	Source
ALASKA						
Eskimos-Inupiat (Northwestern)	.290	.710	303	41.2	16.4	Scott & Wright 1978
Eskimos (St. Lawrence Is.)	.360	.640	222	46.1	17.7	Ferrell et al. 1981b
Athabaskan Indians	.233	.767	242	35.7	14.7	Scott & Wright 1978
CANADA						
Dogrib Indians (NW Territories)	.127	.873	158	22.2	9.9	Szathmary et al. 1983
UNITED STATES						
Whites	.420	.580	101	48.7	18.4	Weitkamp 1976
Blacks	.280	.720	108			"
Mexican-Americans (Texas)	.378	.622	669	47.0	18.0	Hewett-Emmett et al. 1986b

MIDDLE AMERICA

Place/Population	1	2	n	H	PE	Source
COSTA RICA						
Guaymi	.671	.329	286	44.2	17.2	Barrantes et al. 1982

SOUTH AMERICA

Continued

Table 47. Glyoxalase 1: GLO1 (cont'd)

Place/Population	1	2	n	H	PE	Source
ARGENTINA						
Mapuche	.376	.624	105	46.9	18.0	Haas et al. 1985
BRAZIL						
Baniwa	.215	.785	151	33.8	14.0	Salzano et al. 1986
Pacaás Novos	.213	.787	221	33.2	13.8	Salzano et al. 1985
Macushi (N. Brazil & S. Guyana)	.209	.791	244	33.1	13.8	Salzano et al. 1984
Ticuna (C. Amazonas)	.168	.832	1,762	28.0	12.0	Neel et al. 1980
CHILE						
Atacameño (Toconao, N. Chile)	.420	.580	175	48.7	18.4	Goedde et al. 1985b
COLOMBIA						
Indians	.297	.703	259	41.8	16.5	Ghosh 1977a
ECUADOR						
Waorani	.008	.992	186	1.6	.8	Larrick et al. 1985

OCEANIA

Place/Population	1	2	n	H	PE	Source
AUSTRALIA						
Aborigines/Malag (Elcho Is., N. Territory)		1.000	151			Ghosh 1977a
Aborigines (Hooker Creek, N. Territory)		1.000	153			"
Aborigines/Walmadjari (Kimberley, W. Australia)	.012	.988	171	2.4	1.2	Blake & Spargo 1986
MELANESIA						
Papua New Guinea						
Gainj & Kalam (N. Central Highlands)		1.000	275			Long et al. 1986
Daga (Bay Province)	.083	.917	139	15.2	7.0	Jenkins et al. 1983
Karkar Is.						
Takia	.003	.997	184	.6	.3	Ghosh 1977a
MICRONESIA						
E. Caroline Is.	.046	.954	748	8.8	4.2	"
POLYNESIA						
Samoans (Christchurch, New Zealand)	.233	.767	101	35.7	14.7	"

Table 48. Hexokinase III: HK3

Place/Population	1	2	n	H	PE	Source
EUROPE						
NETHERLANDS (Leiden)	1.000		806			Fraser et al. 1974
UNITED KINGDOM						
England	.985	.015	330	3.0	1.5	Povey et al. 1975
ASIA						
SAUDI ARABIA						
Arabs (Riyadh, Eastern)	.980	.020	784	3.9	1.9	el Hazmi et al. 1986
Arabs (Khaiber, Western)	.980	.020	457	3.9	1.9	"
JAPAN						
Japanese (Nagoya City)	1.000		250			Shinoda & Matsunaga 1970
NORTH AMERICA						
CANADA						
Eskimos (Igloolik)	1.000		192			McAlpine et al. 1974
SOUTH AMERICA						
BOLIVIA						
Aymara	1.000		320			Ferrell et al. 1981a

Table 49. Hydroxyacyl glutathione hydrolase; glyoxalase II: HAGH, GLO2

Place/Population	1	2	n	H	PE	Source
ASIA						
MALAYSIA						
Malays (Singapore)	1.000		13			Board 1980c
CHINA						
Chinese (Malaysia)	1.000		45			"
JAPAN						
Japanese (Miyazaki)	.998	.002	404	.4	.2	Sugita & Takahama 1983
NORTH AMERICA						
UNITED STATES						
Whites (Chicago)	1.000		606			Charlesworth 1972
Blacks (")	1.000		81			"
OCEANIA						
AUSTRALIA						
Aborigines	1.000		96			Board 1980c
MELANESIA						
Papua New Guinea	1.000		96			"
MICRONESIA						
Naru Is.	.984	.016	526	3.1	1.5	"
Caroline Is.	1.000		252			"
POLYNESIA						
Samoans (Samoa Is.)	1.000		96			"

Table 50. Isocitrate dehydrogenase, soluble: IDH1

Place/Population	Variants		n	Source
	Genotype	Number		
EUROPE				
NETHERLANDS (Leiden)			806	Fraser et al. 1974
U.S.S.R.				
EUROPEAN PART				
Abkhazia (Ochamchir Dist.)			252	Ferrell et al. 1985
ASIA				
IRAN				
Turkoman (Gonbad, NE Iran)			155	Kirk et al. 1977
INDIA				
Hindu (Andhra Pradesh)			85	Naidu et al. 1985
Irula tr. (Nilgiri Hills, S. India)			63	Saha et al. 1976
Bhil tr. (Madhya Pradesh)			145	Papika et al. 1978
NEPAL				
Nepalese			63	Chen et al. 1972a
INDONESIA				
Balinese (Bali Is.)			148	Breguet et al. 1982b
PHILIPPINES				
Filipino			88	Chen et al. 1972a
CHINA				
Chinese (C. Taiwan)			160	Chen et al. 1973
Chinese (Malaysia)			86	Blake 1984
JAPAN				
Japanese (Nagoya City)			250	Shinoda & Matsunaga 1970

Continued

Table 50. Isocitrate dehydrogenase, soluble: IDH1 (cont'd)

Place/Population	Variants Genotype	Variants Number	n	Source
AFRICA				
EGYPT				
Nubians (Southern)			155	Bertin et al. 1978
NIGERIA				
Yoruba (Western)			65	Ojikutu et al. 1977
BURUNDI			477	Gall et al. 1982
Hutu			128	Chen et al. 1972a
CONGO			121	"
KENYA	+/3	2	121	"
MOZAMBIQUE	+/4	1	95	"
SOUTHERN AFRICA				
San/Bushmen				
G! aokxate			59	Nurse & Jenkins 1977
Negro/Denasena	+/4	2	77	"
TRISTAN DA CUNHA			157	Jenkins et al. 1985
NORTH AMERICA				
ALASKA				
Eskimos (St. Lawrence Is.)			222	Ferrell et al. 1981b
CANADA				
Eskimos (Igloolik)			200	McAlpine et al. 1974
Eskimos			285	Chen et al. 1972a
Cree Indians (Manitoba)			200	Lucciola et al. 1974
Ojibwa Indians (Pikangikum)			93	Szathmary et al. 1974
Dogrib Indians (NW Territories)			158	Szathmary et al. 1983
UNITED STATES				
Whites (Seattle)			291	Chen et al. 1972a
Blacks (")			247	"
Orientals (")	+/2	1	184	"
MIDDLE AMERICA				
GUATEMALA				
Black Caribs (Livingston)			205	Crawford et al. 1981
COSTA RICA				
Guaymi			286	Barrantes et al. 1982
PANAMA				
Blacks			781	Ferrell et al. 1978b
SOUTH AMERICA				
BOLIVIA				
Aymara (Western)			316	Ferrell et al. 1978a
BRAZIL				
Baniwa (Jandu Cachoeira)			363	Salzano et al. 1986
Cayapo (Pará & Mato Grosso)			521	Neel 1978
Macushi (N. Brazil & S. Guyana)			498	Neel et al. 1977b
Wapishana (")			615	"
ECUADOR				
Waorani			187	Larrick et al. 1985
FRENCH GUIANA				
Wayampi (Trois-Sauts & Camopi)			219	Tchen et al. 1978
VENEZUELA				
Makiritare (S. Venezuela & N. Brazil)			407	Tanis et al. 1973
Yanomama (")			1,912	"
OCEANIA				
AUSTRALIA				
Aborigines (Hooker Creek, (N. Territory)			285	Blake 1979
Aborigines (Beagle Bay, W. Australia)			28	"
Aborigines (Cundeelee, ")			132	"

Continued

Table 50. Isocitrate dehydrogenase, soluble: IDH1 (cont'd)

Place/Population	Variants		n	Source
	Genotype	Number		
MELANESIA				
W. New Guinea				
Asmat (S. Coastal Plain)			118	Gajdusek et al. 1978
Papua New Guinea				
Gainj & Kalam (N. Central				
Highlands)			277	Long et al. 1986
Fuyuge (C. Highlands)			155	Woodfield et al. 1974
Solomon Is.			364	Blake et al. 1983
MICRONESIA				
Marshall Is.			349	Neel et al. 1976
POLYNESIA				
Samoans (Christchurch, New Zealand)			77	Booth et al. 1977

Table 51. Lactate dehydrogenase A & B: LDHA & LDHB

Place/Population	Variants		Locus	n	Source
	Genotype	Number			
EUROPE					
BULGARIA (Sofia)	+/V[a]	1	B	138	Ananthakrishnan et al. 1972
FAROE IS. (N. Atlantic)				666	Tills et al. 1985
FINLAND					
Finns	+/V[a]	2	A	1,769	Nilsson & Eriksson 1972
Lapps/Saami				1,776	"
GREECE					
Greeks				219	Schneider et al. 1975
Greeks (Plati)				1,035	Tills et al. 1983b
ICELAND	+/V[a]	6[b]		1,442	Tills et al. 1982b
IRELAND	+/V[a]	3[b]		1,800	Tills 1977
NETHERLANDS (Leiden)				806	Fraser et al. 1974
POLAND (Olkusz)				213	Wolanski et al. 1983
SWEDEN				3,171	Nilsson & Eriksson 1972
UNITED KINGDOM					
England	+/Memphis-4	2	A	1,015	Davidson et al. 1965
Isle of Lewis				95	Clegg et al. 1985
Orkney Is. (Westray)				408	Welch et al. 1973
U.S.S.R.					
EUROPEAN PART					
Russians (Moscow)				1,894	Altukhov et al. 1981
Abkhazia (Ochamchir Dist.)	+/V[a]	5	A	262	Ferrell et al. 1985
Maris (Mari Republic)				317	Nielsson and Ericksson 1972
ASIA					
ISRAEL					
Yeminite Jews				157	Tills et al. 1976
JORDAN				543	Banerjee et al. 1981
KUWAIT					
Arabs				150	Sawhney et al. 1984
SAUDI ARABIA					
Tribes (Western)				245	Saha et al. 1980
Arabs (Southern)				261	Marengo-Rowe et al. 1974
IRAN					
Turkoman (Gonbad, NE Iran)				155	Kirk et al. 1977

Continued

Table 51. Lactate dehydrogenase A & B: LDHA & LDHB (cont'd)

Place/Population	Variants			n	Source
	Genotype	Number	Locus		
AFGHANISTAN					
Pushtu & Dari (Kabul)				280	Papiha et al. 1977a
INDIA					
Punjabi (Chandigarh)				140	Papiha et al. 1972
Gujarati (W. India)				281	Papiha et al. 1981
Hindu & Moslems					
(Madhya Pradesh)	+/V[a]	1[b]		337	Papiha & Chhaparwal 1973
Hindu (Andhra Pradesh)				219	Roberts et al. 1980
Irula tr. (Nilgiri Hills, S. India)	+/Calcutta-1	1	A	175	Saha et al. 1976
Bengali (E. India)	+/Calcutta-1	5	A	269	Das et al. 1970
SRI LANKA					
Sinhalese (Anuradhapura)				156	Roberts et al. 1972b
BANGLADESH					
Moslems (Dacca)	+/Calcutta-1	1	A	200	Papiha et al. 1975
NEPAL					
Sherpas				15	Santachiara et al. 1976
BHUTAN				152	Mourant et al. 1969
MALAYSIA					
Malays (Singapore)	+/Malay-1	1	A	259	Blake et al. 1973a
	+/Malay-2	1[b]			
INDONESIA					
Batak (Samosir Is.)				188	McDermid et al. 1973
Balinese (Bali Is.)	+/Calcutta-1	4	A	147	Breguet et al. 1982b
PHILIPPINES					
Filipino/Tagalog (Manila)	+/Calcutta-1	2	A	253	Omoto et al. 1978
Negrito (Angeles, C. Luzon)				129	"
CHINA					
Tibetans				8	Santachiara et al. 1976
Chinese (Singapore)	+/Chinese-1	1	A	378	Blake et al. 1973a
	+/Chinese-2	1[b]			
JAPAN					
Japanese (Hiroshima/Nagasaki)	+/Nagasaki	1	B	4028	Ueda et al. 1977
Ainu (Hidaka)				125	Omoto & Harada 1972

AFRICA

Place/Population	Variants			n	Source
	Genotype	Number	Locus		
ALGERIA					
Ideles (Ahaggar)		1[b]		55	Lefèvre-Witier & Vergues 1977
EGYPT					
Nubians (S. Egypt)				155	Bertin et al. 1978
SUDAN					
Sudanese (Khartoum)				294	Saha et al. 1978b
Beja-Amarar (NE Africa)				100	el Hassan et al. 1968
MALI					
Kel Kummer Twaregs				285	Lefèvre-Witier & Vergnes 1977
NIGERIA					
Yoruba	+/V[a]	1	B	200	Boyer et al. 1963a
CENTRAL AFRICAN REPUBLIC					
Pygmies (Western)				922	Meera Khan (1986)
Bantu				43	"
BURUNDI					
Hutu	+/V[a]	1	B	447	Gall et al. 1982
KENYA			2[b]		474	Herzog et al. 1970
UGANDA					
Baganda (Kampala)				197	Roberts et al. 1977
TANZANIA					
Sandawe (Kondoa-Irang Dist.)	+/V[a]	3[b]		209	Godber et al. 1976
Nyaturu (")				209	"
Hadza (W. Lake Eyasi)				195	Tills et al. 1982a
SOUTHERN AFRICA					
San/Bushmen					
Tsumkwe !Kung				65	Nurse & Jenkins 1977
Bantu (Natal)				304	Kirk et al. 1971b

NORTH AMERICA

Continued

Table 51. Lactate dehydrogenase A & B: LDHA & LDHB (cont'd)

Place/Population	Variants			n	Source
	Genotype	Number	Locus		
GREENLAND					
Eskimos				153	Nilsson & Eriksson 1972
ALASKA					
Eskimos (St. Lawrence Is.)				222	Ferrell et al. 1981b
CANADA					
Eskimos (Igloolik)				362	McAlpine et al. 1974
Ojibwa Indians (Pikangikum)				52	Szathmary et al. 1974
Dogrib Indians (NW Territories)				158	Szathmary et al. 1983
UNITED STATES					
Whites, (Memphis)	+/Mem-4	1	A	330	Kraus & Neely 1964
Blacks, (")	+/Mem-1	3	A	610	"
	+/Mem-2	2	A		
	+/Mem-3	2	B		

MIDDLE AMERICA

Place/Population	Genotype	Number	Locus	n	Source
COSTA RICA					
Guaymi			A	286	Barrantes et al. 1982
PANAMA					
Guaymi	GUA1/GUA1	4	B	484	Neel 1978
	+/GUA1	61	B		
Blacks		3	A	750	Ferrell et al. 1978b
WEST INDIES (Haiti)				307	Basu et al. 1976

SOUTH AMERICA

Place/Population	Genotype	Number	Locus	n	Source
BOLIVIA					
Aymara (W. Bolivia)				429	Ferrell et al. 1978a
BRAZIL					
Baniwa (Jandu Cachoeira)				363	Salzano et al. 1986
Cayapo (Pará & Mato Grosso)				87	Salzano et al. 1972b
Macushi (N. Brazil & S. Guyana)				499	Neel et al. 1977b
Wapishana (")				615	"
Ticuna (C. Amazonas)				1,765	Neel et al. 1980
CHILE					
Aymara	+/M1	4	A	1,379	Ferrell et al. 1980
	+/M2	7	A		
COLOMBIA					
Noanama				170	Kirk et al. 1974
EQUADOR					
Waorani			A	187	Larrick et al. 1985
FRENCH GUIANA					
Wayampi (Trois-Sauts & Camopi)				219	Tchen et al. 1978
VENEZUELA					
Makiritare (S. Venezuela & N. Brazil)				326	Weitkamp & Neel 1970
Yanomama (")				1,019	Weitkamp & Neel 1972

OCEANIA

Place/Population	Genotype	Number	Locus	n	Source
AUSTRALIA					
Aborigines/Malag (Elcho Is., N. Territory)				595	Kirk et al. 1969
Aborigines (Hooker Creek, N. Territory)				286	Blake 1979
Aborigines/Amoonguna (C. Australia)				104	Kirk et al. 1971a
Aborigines/Walmadjari (Kimberley, W. Australia)				171	Blake & Spargo 1986
MELANESIA					
W. New Guinea					
Asmat (S. Coastal Plain)				653	Gajdusek et al. 1978
Papua New Guinea					
Gainj & Kalam (N. Central Highlands)				589	Long et al. 1986
Jimi Valley (W. Highlands)				385	Mourant et al. 1981
Yagaria (E. Highlands)				424	Mourant et al. 1982
Fuyuge (C. Dist. Highlands)				173	Woodfield et al. 1974

Continued

Table 51. Lactate dehydrogenase A & B: LDHA & LDHB (cont'd)

Place/Population	Variants			n	Source
	Genotype	Number	Locus		
Karkar Is.					
Takia		1		290	Booth et al. 1982
Manus Is.				183	Malcolm et al. 1972
Buka Is.				80	McLoughlin et al. 1982a
Solomon Is.				364	Blake et al. 1982
MICRONESIA					
W. Caroline	+/Calcutta-1	1	A	382	Blake et al. 1973b
Marshall Is.		4	A	372	Neel et al. 1976
POLYNESIA					
Samoans (Christchurch, New Zealand)				77	Booth et al. 1977

Note: [a] V stands for a variant allele. [b] It is not specified whether the variant belongs to the locus A or B.

Table 52. Malate dehydrogenase, soluble: MDH1

Place/Population	Variants		n	Source
	Genotype	Number		

EUROPE

FAROE IS. (N. Atlantic)			666	Tills et al. 1985
FINLAND				
Lapps/Saami			148	Leakey et al. 1972
FRANCE				
French (Bareges)			600	Vergnes et al. 1980a
Basques (SW France)			144	"
GREECE				
Greeks			215	Schneider et al. 1975
Greeks (Plati)			1,035	Tills et al. 1983b
ICELAND			1,442	Tills et al. 1982b
IRELAND			1,800	Tills 1977
NETHERLANDS (Leiden)	+/V[a]	1	806	Fraser et al. 1974
POLAND (Olkusz)			213	Wolanski et al. 1983
SWEDEN				
Swedes (Uppsala)			1,004	Beckman & Christodoulou 1974
UNITED KINGDOM				
England (Northeast)	+/3	1	600	Papiha et al. 1977b
Isle of Lewis			94	Clegg et al. 1985
Orkney Is. (Westray)			408	Welch et al. 1973

U.S.S.R.

EUROPEAN PART				
Russians (Moscow)		2	1,965	Altukhov et al. 1981
Abkhazia (Ochamchir Dist.)			262	Ferrell et al. 1985

ASIA

ISRAEL				
Yemenite Jews			204	Leakey et al. 1972
KUWAIT				
Arabs			150	Sawhney et al. 1984
SAUDI ARABIA				
Tribes (Western)			93	Saha et al. 1980
Arabs (Southern)			261	Marengo-Rowe et al. 1974

Continued

Table 52. Malate dehydrogenase, soluble: MDH1 (cont'd)

Place/Population	Variants Genotype	Variants Number	n	Source
IRAN				
Turkoman, NE Iran)			155	Kirk et al. 1977
AFGHANISTAN				
Pushtu & Dari (Kabul)			280	Papiha et al. 1977a
INDIA				
Punjabi (N. India)			182	Mukherjee et al. 1974
Gujarati (W. India)	+/5	1	331	Papiha et al. 1977b
Bhil tr. (Madhya Pradesh)			145	Papiha et al. 1978
Hindu (Andhra Pradesh)			209	Roberts et al. 1980
Irula tr. (S. India)			175	Saha et al. 1976
Bengali (E. India)			108	Das et al. 1970
BANGLADESH				
Muslims (Dacca)	+/5	1	200	Papiha et al. 1977b
NEPAL			179	Leakey et al. 1972
BHUTAN			236	"
MALAYSIA				
Malays (Singapore)			259	Blake et al. 1973a
INDONESIA				
Batak (Samosir Is.)			188	McDermid et al. 1973
Balinese (Bali Is.)			147	Breguet et al. 1982b
PHILIPPINES				
Filipino/Tagalog (Manila)			253	Omoto et al. 1978
Negrito (Angeles, C. Luzon)			129	"
CHINA				
Tibetans			23	Santachiara et al. 1976
Chinese (Singapore)			378	Blake et al. 1973a
JAPAN				
Japanese (Hiroshima & Nagasaki)	+/7 HIR1		4,029	Ueda et al. 1977
Ainu (Hokkaido)			125	Omoto & Harada 1972

AFRICA

Place/Population	Variants Genotype	Variants Number	n	Source
EGYPT				
Nubians (Southern)			155	Bertin et al. 1978
SUDAN				
Sudanese (Khartoum)			152	Saha et al. 1978b
Beja			96	Leakey et al. 1972
NIGERIA			144	"
CENTRAL AFRICAN REPUBLIC				
Pygmies (Western)			32	Santachiara et al. 1980
BURUNDI				
Hutu			447	Gall et al. 1982
ETHIOPIA				
Amhara tr.	+/2	3	298	Leakey et al. 1972
UGANDA				
Baganda (Kampala)			197	Roberts et al. 1977
TANZANIA				
Sandawe (Konda-Irang Dist.)			209	Godber et al. 1976
Nyaturu (")			209	"
Hadza (W. Lake Eyasi)			239	Tills et al. 1982a
SOUTHERN AFRICA				
San/Bushmen -!Kung, Tsumkwe			65	Nurse & Jenkins 1977
Bantu (Natal)			304	Kirk et al. 1971b

NORTH AMERICA

Place/Population	Variants Genotype	Variants Number	n	Source
ALASKA				
Eskimos (St. Lawrence Is.)			222	Ferrell et al. 1981b
CANADA				
Dogrib Indians (NW Territories)			158	Szathmary et al. 1983
UNITED STATES				
Whites			1,440	Davidson & Cortner 1967
Blacks	+/2	1	1,470	"

MIDDLE AMERICA

Table 52. Malate dehydrogenase, soluble: MDH1 (cont'd)

Place/Population	Variants Genotype	Variants Number	n	Source
GUATEMALA				
Black Caribs (Livingston)			205	Crawford et al. 1981
COSTA RICA				
Guaymi			286	Barrantes et al. 1982
PANAMA				
Blacks			781	Ferrell et al. 1978b
WEST INDIES (Haiti)			307	Basu et al. 1976

SOUTH AMERICA

Place/Population	Genotype	Number	n	Source
BOLIVIA				
Aymara (Western)			316	Ferrell et al. 1978a
BRAZIL				
Baniwa (Jandu Cachoeira)			363	Salzano et al. 1986
Cayapo (Pará & Mato Grosso)			87	Salzano et al. 1972b
Macushi (N. Brazil & S. Guyana)	$+/2_{MAC1}$	1	499	Neel et al. 1977b
Wapishana (")			615	"
Ticuna (C. Amazonas)			1,765	Neel et al. 1980
COLOMBIA				
Noanama			170	Kirk et al. 1974
ECUADOR				
Waorani			187	Larrick et al. 1985
FRENCH GUIANA				
Wayampi (Trois-Sauts & Camopi)			219	Tchen et al. 1978
VENEZUELA				
Makiritare (S. Venezuela & N. Brazil)			308	Weitkamp & Neel 1970
Yanomama (")			975	Weitkamp & Neel 1972

OCEANIA

Place/Population	Genotype	Number	n	Source
AUSTRALIA				
Aborigines/Malag(Elcho Is., N. Territory)			35	Kirk et al. 1971a
Aborigines (Hooker Creek, N. Territory)			286	Blake 1979
Aborigines (Amoonguna, C. Australia)			104	Kirk et al. 1971a
Aborigines/Walmadjari (Kimberley, W. Australia)			171	Blake & Spargo 1986
MELANESIA				
W. New Guinea				
Asmat (S. Coastal Plain)	+/3	22	653	Gajdusek et al. 1978
Papua New Guinea				
Gainj & Kalam (NC Highlands)	+/3	4	571	Long et al. 1986
Jimi Valley (W. Highlands)	+/NG1	13	385	Mourant et al. 1981
Yagaria (E. Highlands)	+/3	21	423	Mourant et al. 1982
Fuyuge (C. Highlands)			173	Woodfield et al. 1974
Daga (Bay Province)			139	Jenkins et al. 1983
Karkar Is.	+/3, 3/3	73, 2	1,899	Leakey et al. 1972
Manus Is.			183	Malcolm et al. 1972
Buka Is.			80	McLoughlin et al. 1982a
Solomon Is.			364	Blake et al. 1983
MICRONESIA				
W. Caroline Is.			382	Blake et al. 1973b
Marshall Is.			375	Neel et al. 1976
POLYNESIA				
Samoans (Christchurch, New Zealand)			77	Booth et al. 1977

Note: [a] V stands for a variant allele.

Table 53. Malic enzyme 2, mitochondrial: ME2

Place/Population	1	2	n	H	PE	Source
EUROPE						
GERMANY (SW. Germany)	.668	.332	184	44.4	17.3	Abe & Akiyama 1983(c)
UNITED KINGDOM						
England	.651	.349	409	45.4	17.6	Saha et al. 1978a
ASIA						
INDIA						
Bengali (E. India)	.560	.440	182	49.3	18.6	Ghosh et al. 1980
MALAYSIA						
Malays (Singapore)	.677	.323	113	43.7	17.1	Bhattacharyya & Saha 1984a
CHINA						
Chinese (Malaysia)	.711	.289	161	41.1	16.3	"
JAPAN						
Japanese (Tokyo)	.556	.444	126	49.4	18.6	Abe & Akiyama 1983
NORTH AMERICA						
UNITED STATES						
Whites	.693	.307	132	42.6	16.7	Cohen & Omenn 1972
Blacks	.825	.175	20	28.9	12.4	"

Table 54. Nucleoside phosphorylase: NP

Place/Population	Variants		n	Source
	Genotype	Number		
EUROPE				
GREECE				
Greeks	+/2	1	44	Edwards et al. 1971
UNITED KINGDOM				
England	+/3	1		
	+/4	1	1,542	"
ASIA				
INDIA				
Hindu (Kerala, S. India)			192	Saha et al. 1976
Irula tr. (Nilgiri Hills, S. India)			77	"
Indians (England)	+/2	1	197	Edwards et al. 1971
JAPAN				
Japanese (Nagoya City)		1	586	Shinoda & Matsunaga 1970
AFRICA				
NIGERIA				
Negroes (England)			238	Edwards et al. 1971
NORTH AMERICA				
ALASKA				
Eskimos (St. Lawrence Is.)	+/Eskimo	10	222	Ferrell et al. 1981b
CANADA				
Ojibwa Indians (Pikangikum)			93	Szathmary et al. 1974
Cree Indians (Manitoba)			200	Lucciola et al. 1974
Dogrib Indians (NW Territories)			158	Szathmary et al. 1983
MIDDLE AMERICA				

Continued

Table 54. Nucleoside phosphorylase: NP (cont'd)

Place/Population	Variants		n	Source
	Genotype	Number		
COSTA RICA				
Guaymi			286	Barrantes et al. 1982

SOUTH AMERICA

BRAZIL				
Baniwa (Jandu Cachoeira)			363	Salzano et al. 1986
Macushi (N. Brazil & S. Guyana)			498	Neel et al. 1977b
Wapishana (")			614	"
Ticuna (C. Amazonas)			1,764	Neel et al. 1980
VENEZUELA				
Makiritare (S. Venezuela & N. Brazil)			390	Neel 1978
Yanomama (")			342	"

OCEANIA

MELANESIA				
Papua New Guinea				
Gainj & Kalam (NC Highlands)	+/4	2	270	Long et al. 1986
MICRONESIA				
Marshall Is.			371	Neel et al. 1976

Table 55. Paraoxonase; arylesterase; esterase A: PON, ESA

Place/Population	A	B	n	H	PE	Source
EUROPE						
DENMARK	.726	.274	1,664	39.8	15.9	Eiberg & Mohr 1981
FRANCE	.683	.317	N.A.	43.3	17.0	Vincent-Viry et al. 1986
GERMANY	.760	.240	799	36.5	14.9	Mallinckrodt 1978
UNITED KINGDOM						
England	.703	.297	190	41.8	16.5	Playfer et al. 1976
ASIA						
INDIA						
N. Indians (Malaysia)	.463	.537	70	49.7	18.7	"
AFRICA						
SUDAN	.531	.469	129	49.8	18.7	La Du et al. 1986
NORTH AMERICA						
CANADA						
Whites	.733	.267	82	39.1	15.7	Carro-Ciampi et al. 1981
UNITED STATES						
Whites (Ann Arbor)	.685	.315	215	43.2	16.9	Eckerson et al. 1983
SOUTH AMERICA						
CHILE						
Atacameño (Toconao, N. Chile)	.802	.198	171	31.8	13.4	Goedde et al. 1984c

Note: The three genotypes at this locus are identifiable by using several biological tests (see La Du et al. 1986). Some authors (e.g. Barrantes et al. 1982) have used gene symbol ESA for the ESA1 locus.

Table 56. Peptidase A: PEPA

Place/Population	Genotypes							n	Source
	1/1	1/1w	2/1	2/2	8/1	8/8	0		

EUROPE

BULGARIA (Sofia)	121						16	137	Ananthakrishnan et al. 1972
NETHERLANDS (Leiden)							1[v]	806	Fraser et al. 1974
UNITED KINGDOM									
England	61				41	7		109	Lewis 1973

U.S.S.R.

EUROPEAN PART									
Abkhazia (Ochamchir Dist.)	262							262	Ferrell et al. 1985

ASIA

IRAN									
Turkoman (Gonbad, NE Iran)	151		3	1				155	Kirk et al. 1977
INDIA									
Indians (N. India)	389	97	1				6	493	Blake et al. 1971a
Hindu (Andhra Pr.)	66	12	1[a]				6	85	Naidu et al. 1985
Irula tr. (Nilgiri Hills, S. India)	170							172	Saha et al. 1976
Bengali (E. India)	175						2	176	Das et al. 1970
NEPAL									
Sherpas	80							80	Santachiara et al. 1976
MALAYSIA									
Malays (Singapore)	156	71					32	259	Blake et al. 1973a
INDONESIA									
Batak (Samosir Is.)	164	21					2	187	McDermid et al. 1973
Balinese (Bali Is.)	130	10					8	148	Breguet et al. 1982b
CHINA									
Tibetans	27							27	Santachiara et al. 1976
Chinese (C. Taiwan)	160							160	Chen et al. 1973
Chinese (Singapore)	252	108					18	378	Blake et al. 1973a
JAPAN									
Japanese (Hiroshima & Nagasaki)	4,006						6[b]	4,009	Tanis et al. 1978
Japanese (Mie)	1,069						1[c]	1,070	Ishimoto 1975
Ainu (Hokkaido)	125							125	Omoto & Harada 1972

AFRICA

EGYPT									
Nubians (Southern)	143		9	1				153	Bertin et al. 1978
NIGERIA									
Yoruba (Western)	48		17	1				66	Ojikutu et al. 1977
Nigerians (UK)	27		5	1	6	1[d]		40	Lewis 1973
CENTRAL AFRICAN REPUBLIC									
Pygmies (Western)	347	32	23				12	414	Santachiara et al. 1980
ANGOLA									
Njinga	84		18	2			1[v]	105	Nurse et al. 1979
SOUTHERN AFRICA									
San/Bushmen									
!Kung,Tsumkwe	277		1					278	Nurse & Jenkins 1977
Khoi/Hottentot									
Keetmanshoop Nama	168		7					175	Jenkins 1972
Negroes/Dama (Okombahe)	88		4					92	Nurse et al. 1976
Griqua (Cape Province)	95		4	1				100	Nurse & Jenkins 1975
Bantu (Natal)	249		50	4				303	Kirk et al. 1971b
TRISTAN DA CUNHA	158							158	Jenkins et al. 1985

NORTH AMERICA

Continued

97

Table 56. Peptidase A: PEPA (cont'd)

Place/Population	1/1	1/1w	2/1	2/2	8/1	8/8	0	n	Source
ALASKA									
Eskimos									
(St. Lawrence Is.)	222							222	Ferrell et al. 1981b
CANADA									
Eskimos (Igloolik)	361							361	McAlpine et al. 1974
Cree Indians (Manitoba)	200							200	Lucciola et al. 1974
Ojibwa Indians (N. Ontario)	93							93	Szathmary et al. 1974
Dogrib Indians (NW Territories)	158							158	Szathmary et al. 1983
UNITED STATES									
Whites (Louisiana)	376							376	Fox et al. 1981
Blacks (")	320		39	5			8	373	"

MIDDLE AMERICA

	1/1	1/1w	2/1	2/2	8/1	8/8	0	n	Source
COSTA RICA									
Guaymi	286							286	Barrantes et al. 1982
WEST INDIES									
Negroes	254		30	3	1[e]			294	Lewis & Harris 1967

SOUTH AMERICA

	1/1	1/1w	2/1	2/2	8/1	8/8	0	n	Source
BOLIVIA									
Aymara (W. Bolivia)	316							316	Ferrell et al. 1978a
BRAZIL									
Baniwa (Jandu Cachoeira)	363							363	Salzano et al. 1986
Cayapo (Para & Mato Grosso)	521							521	Neel 1978
Pacaás Novos	221							221	Salzano et al. 1985
Macushi (N. Brazil & S. Guyana)	499							499	Neel et al. 1977b
Wapishana (")	594		20[f]					614	"
Ticuna (C. Amazonas)	1,765							1,765	Neel et al. 1980
COLOMBIA									
Noanama	170							170	Kirk et al. 1974
ECUADOR									
Waorani	146	26	15					187	Larrick et al. 1985
FRENCH GUIANA									
Wayampi	214							214	Tchen et al. 1978
VENEZUELA									
Makiritare (S. Venezuela & N. Brazil)	407							407	Neel 1978
Yanomama (")	1,949							1,949	"

OCEANIA

	1/1	1/1w	2/1	2/2	8/1	8/8	0	n	Source
AUSTRALIA									
Aborigines (Elcho Is., N. Territory)	35	8						43	Kirk et al. 1971a
Aborigines (Hooker Creek, N. Territory)	224	60					2	286	Blake 1979
Aborigines (Amoonguna, C. Australia)	63	26					17	106	Kirk et al. 1971a
Aborigines/Walmadjari (Kimberley, W. Australia)	171							171	Blake & Spargo 1986
MELANESIA									
Papua New Guinea									
Gainj & Kalam (N. Central Highlands)	269						1[g]	270	Long et al. 1986
Fuyuge (C. Dist. Highlands)	173							173	Woodfield et al. 1974
Daga (Bay Province)	127	12						139	Jenkins et al. 1983
Manus Is.	87	39					12	138	Malcolm et al. 1972

Continued

Table 56. Peptidase A: PEPA (cont'd)

Place/Population	Genotypes							n	Source
	1/1	1/1w	2/1	2/2	8/1	8/8	0		
MICRONESIA									
W. Caroline Is.	329	53						382	Blake et al. 1973b
Marshall Is.	373							373	Neel et al. 1976
POLYNESIA									
Samoans (Christchurch,									
New Zealand)	77							77	Booth et al. 1977

Note: [v] Variant individual. [a] Genotype (1/2Brahmin). [b] Three individuals with genotype $(+/4_{HIR1})$ were found in each of Hiroshima and Nagasaki. [c] One individual of genotype (+/4) was found. [d] Genotype (2/8). [e] Genotype (3/1). [f] Individuals of genotype $(1/2_{WAP1})$. [g] Variant is electrophoretically similar to the type-3 variant of Lewis & Harris (1967). [h] Genotype (1/7).

Table 57. Peptidase B: PEPB

Place/Population	1	2	3	x	n	H	PE	Source
EUROPE								
Europeans	.998	.001	.0004	$.0002^4$	2,197	.4	.1	Lewis & Harris 1967
BULGARIA (Sofia)	1.000				105[a]			Ananthakrishnan et al. 1972
NETHERLANDS (Leiden)	.999	.0007			806	.2	.1	Fraser et al. 1974
U.S.S.R.								
EUROPEAN PART								
Abkhazia (Ochamchir Dist.)	1.000				262			Ferrell et al. 1985
ASIA								
IRAN								
Turkoman (Gonbad, NE Iran)	.997	.003			155	.6	.3	Kirk et al. 1977
INDIA								
Indians (N.India)	1.000				493			Blake et al. 1971a
Hindu (Andhra Pradesh)	1.000				85			Naidu et al. 1985
Irula tr. (Nilgiri Hills, S. India)	1.000				171			Saha et al. 1976
Bengali (E. India)	1.000				264			Das et al. 1970
NEPAL								
Sherpas	1.000				84			Santachiara et al. 1976
MALAYSIA								
Malays (Singapore)	1.000				259			Blake et al. 1973a
INDONESIA								
Batak (Samosir Is.)	1.000				187			McDermid et al. 1973

Continued

Table 57. Peptidase B: PEPB (cont'd)

Place/Population	1	2	3	x	n	H	PE	Source
Balinese (Bali Is.)	1.000				148			Breguet et al. 1982b
CHINA								
Tibetans	.982			$.018^{Tibet}$	28	3.5	1.7	Santachiara et al. 1976
Chinese (C. Taiwan)	1.000				160			Chen et al. 1973
Chinese (Singapore)	1.000				378			Blake et al. 1973a
JAPAN								
Japanese (Mie)	.999			.001	1,070	.2	.1	Ishimoto 1975
Ainu (Hokkaido)	1.000				125			Omoto & Harada 1972

AFRICA

Place/Population	1	2	3	x	n	H	PE	Source
EGYPT								
Nubians (Southern)	1.000				155			Bertin et al. 1978
NIGERIA								
Yoruba (Western)	1.000				66			Ojikutu et al. 1977
CENTRAL AFRICAN REPUBLIC								
Pygmies (Western)	1.000				464			Santachiara et al. 1980
RWANDA								
Hutu	1.000				105			Santachiara 1986
ANGOLA								
Njinga	1.000				105			
SOUTHERN AFRICA								
San/Bushmen								
G/wi & G//ana	1.000				129			Jenkins et al. 1975
Negroes/Dama (Okombahe)	1.000				92			Nurse et al. 1976
Griqua (Cape Province)	1.000				100			Nurse & Jenkins 1975
Bantu (Natal)	1.000				304			Kirk et al. 1971b
TRISTAN DA CUNHA	1.000				158			Jenkins et al. 1985

NORTH AMERICA

Place/Population	1	2	3	x	n	H	PE	Source
ALASKA								
Eskimos (St. Lawrence Is.)	.982			$.018^{6ESK}$	222	3.5	1.7	Ferrell et al. 1981b
CANADA								
Eskimos (Igloolik)	1.000				361			McAlpine et al. 1974
Cree Indians (Manitoba)	1.000				200			Lucciola et al. 1974
Ojibwa Indians (N. Ontario)	1.000				93			Szathmary et al. 1974
Dogrib " (NW Territory)	1.000				158			Szathmary et al. 1983

MIDDLE AMERICA

Place/Population	1	2	3	x	n	H	PE	Source
COSTA RICA								
Guaymi	1.000				286			Barrantes et al. 1982
PANAMA								
Blacks	1.000				781			Ferrell et al. 1978b
WEST INDIES	1.000				294			Lewis & Harris 1967

SOUTH AMERICA

Place/Population	1	2	3	x	n	H	PE	Source
BOLIVIA								
Aymara (Western)	1.000				316			Ferrell et al. 1978a
BRAZIL								
Baniwa (Jandu Cachoeira)	.999	$.001^{2Ban1}$			507	.2	.1	Salzano et al. 1986
Cayapo (Para & Mato								
Grosso)	1.000				527			Neel 1978
Pacaás Novos	1.000				221			Salzano et al. 1985
Macushi (N. Brazil &								
S. Guyana)	1.000				499			Neel et al. 1977b
Wapishana (")	1.000				615			"
Ticuna (C. Amazonas)	1.000				1,765			Neel et al. 1980
COLOMBIA								
Noanama	1.000				170			Kirk et al. 1974
ECUADOR								
Waorani	1.000				187			Larrick et al. 1985
FRENCH GUIANA								
Wayampi	1.000				215			Tchen et al. 1978
VENEZUELA								

Continued

Table 57. Peptidase B: PEPB (cont'd)

Place/Population	1	2	3	x	n	H	PE	Source
Makiritare (S. Venezuela & N. Brazil)	1.000				409			Neel 1978
Yanomama (")	1.000				2,005			"

OCEANIA

AUSTRALIA

Aborigines/Malag (Elcho Is., N. Territory)	1.000				35			Kirk et al. 1971a
Aborigines (Hooker Creek, N. Territory)	.993		.007[6]		286	1.4	.7	Blake 1979
Aborigines (Amoonguna, C. Australia)	.995		.005[7]		104	1.0	.5	Kirk et al. 1971a
Aborigines/Walmadjari (Kimberley, W. Australia)	1.000				171			Blake & Spargo 1986
Aborigines/Ngarinjin (Kimberley, W. Australia)	.986		.014[6]		109	2.8	1.4	"

MELANESIA

W. New Guinea Asmat (S. Coastal Plain)	.992	.008			648	1.6	.8	Gajdusek et al. 1978
Papua New Guinea								
Gainj & Kalam (N. Central Highlands)	1.000				275			Long et al. 1986
Fuyuge (C. Highlands)	1.000				173			Woodfield et al. 1974
Daga (Bay Province)	1.000				139			Jenkins et al. 1983
Manus Is.	1.000				183			Malcolm et al. 1972
Solomon Is.	1.000				364			Blake et al. 1983

MICRONESIA

W. Caroline Is.	1.000				382			Blake et al. 1973b
Marshall Is.	1.000				375			Neel et al. 1976

POLYNESIA

Samoans (Christchurch, New Zealand)	1.000				77			Booth et al. 1977

Note: [a] Phenotype PepB 0 is ignored.

Table 58. Peptidase C: PEPC

Place/Population	1	3	x	n	H	PE	Source
EUROPE							
Europeans	.997	.003		4,012	.6	.3	Povey et al. 1972
NETHERLANDS (Leiden)	1.000			806			Fraser et al. 1974
U.S.S.R.							
EUROPEAN PART							
Abkhazia (Ochamchir Dist.)	1.000			262			Ferrell et al. 1985
ASIA							

Continued

Table 58. Peptidase C: PEPC (cont'd)

Place/Population	1	3	x	n	H	PE	Source
IRAN							
Turkoman (Gonbad, NE Iran)	1.000			155			Kirk et al. 1977
INDIA							
Hindu (Andhra Pradesh)	1.000			85			Naidu et al. 1985
Tribes (S. India)	1.000			59			Saha et al. 1976
NEPAL							
Sherpas	.994	.006[5Sh]		83	1.2	.6	Santachiara et al. 1976
INDONESIA							
Balinese (Bali Is.)	1.000			148			Breguet et al. 1982b
CHINA							
Tibetans	1.000			28			Santachiara et al. 1976
JAPAN							
Japanese	1.000			580			Ishimoto 1975

AFRICA

Place/Population	1	3	x	n	H	PE	Source
EGYPT							
Nubians (Southern)	1.000			155			Bertin et al. 1978
NIGERIA							
Yoruba (Western)	1.000			66			Ojikutu et al. 1977
CENTRAL AFRICAN REPUBLIC							
Pygmies (Western)	.759	.011[2]	.230[0]	332	37.1	16.0	Santachiara et al. 1980
ANGOLA							
Njinga	.995	.005[wk]		105	1.0	.5	Nurse et al. 1979
SOUTHERN AFRICA							
San/Bushmen (Botswana)							
G/wi & G//ana	1.000			129			Jenkins et al. 1975
Griqua (Campbell, Cape							
Province)	1.000			97			Nurse & Jenkins 1975
TRISTAN DA CUNHA	1.000			158			Jenkins et al. 1985

NORTH AMERICA

Place/Population	1	3	x	n	H	PE	Source
ALASKA							
Eskimos (St. Lawrence Is.)	1.000			222			Ferrell et al. 1981b
CANADA							
Eskimos (Igloolik)	1.000			360			McAlpine et al. 1974
Cree Indians (Manitoba)	1.000			200			Lucciola et al. 1974
Ojibwa Indians (Pikangikum)	1.000			93			Szathmary et al. 1974
Dogrib Indians							
(NW Territories)	1.000			158			Szathmary et al. 1983

MIDDLE AMERICA

Place/Population	1	3	x	n	H	PE	Source
COSTA RICA							
Guaymi	1.000						Barrantes et al. 1982
PANAMA							
Blacks	1.000			781			Ferrell et al. 1978b

SOUTH AMERICA

Place/Population	1	3	x	n	H	PE	Source
BOLIVIA							
Aymara (Western)	1.000			316			Ferrell et al. 1978a
BRAZIL							
Baniwa	1.000			153			Salzano et al. 1986
Pacaàs Novos	1.000			221			Salzano et al. 1985
Ticuna (C. Amazonas)	1.000			1,763			Neel et al. 1980

OCEANIA

Place/Population	1	3	x	n	H	PE	Source
AUSTRALIA							
Aborigines/Konejandi							
(Kimberley, W. Australia)	1.000			102			Blake & Spargo 1986
MELANESIA							
W. New Guinea							
Asmat (S. Coastal Plain)	1.000			118			Gajdusek et al. 1978
Papua New Guinea							
Gainj & Kalam (N. Central							

Continued

Table 58. Peptidase C: PEPC (cont'd)

Place/Population	1	3	x	n	H	PE	Source
Highlands)	1.000			271			Long et al. 1986
Fuyuge (C. Dist. Highlands)	1.000			173			Woodfield et al. 1974
Solomon Is.	1.000			364			Blake et al. 1983
POLYNESIA							
Samoans (Christchurch, New Zealand)	1.000			77			Booth et al. 1977

Table 59. Peptidase D: PEPD

Place/Population	1	2	3	n	H	PE	Source
EUROPE							
Europeans	.989	.011		728	2.2	1.1	Lewis & Harris 1969
ASIA							
IRAN							
Turkoman (Gonbad, NE Iran)	1.000			155			Kirk et al. 1977
INDIA							
Hindu (Andhra Pradesh)	1.000			85			Naidu et al. 1985
Irula tr. (Nilgiri Hills, S. India)	1.000			78			Saha et al. 1976
INDONESIA							
Balinese (Bali Is.)	1.000			148			Breguet et al. 1982b
JAPAN							
Japanese (Mie)	.997	.001	.002	824	.6	.3	Ishimoto 1975
Ainu (Hokkaido)	1.000			125			Omoto & Harada 1972
AFRICA							
NIGERIA							
Yoruba (Western)	1.000			66			Ojikutu et al. 1977
ANGOLA							
Njinga	.971	.019[1wk]	.010[v]	105	5.7	1.9	Nurse et al. 1979
SOUTHERN AFRICA							
San/Bushmen							
!Kung, Tsumkwe	.997		.003	278	.6	.3	Nurse & Jenkins 1977
Khoi/Hottentot	.985		.015	68	3.0	1.5	"
Negroes/Dama	1.000			92			"
Griqua (Campbell, Cape Province)	.985		.015	98	3.0	1.5	Nurse & Jenkins 1975
TRISTAN DA CUNHA	1.000			158			Jenkins et al. 1985
NORTH AMERICA							
ALASKA							
Eskimos (St. Lawrence Is.)	1.000			222			Ferrell et al. 1981b
CANADA							
Eskimos (Igloolik)	1.000			361			McAlpine et al. 1974
Dogrib Indians (NW Territories)	1.000			158			Szathmary et al. 1983
UNITED STATES							
Whites (Lousiana)	.985	.012	.013	376	2.9	2.4	Fox et al. 1981
Blacks (")	.942	.035	.023	372	11.1	5.6	"
MIDDLE AMERICA							

Continued

Table 59. Peptidase D: PEPD (cont'd)

Place/Population	1	2	3	n	H	PE	Source
COSTA RICA							
Guaymi	1.000			286			Barrantes et al. 1982
WEST INDIES	.956	.024	.020	146	8.5	4.3	Lewis & Harris 1969

SOUTH AMERICA

BRAZIL							
Ticuna (C. Amazonas)	1.000			1,763			Neel et al. 1980

OCEANIA

AUSTRALIA							
Aborigines/Konejandi							
(Kimberley, W. Australia)	1.000			102			Blake & Spargo 1986
MELANESIA							
Papua New Guinea							
Gainj & Kalam (N. Central							
Highlands)	1.000			281			Long et al. 1986
Fuyuge (C. Highlands)	.991	.006	.003	173	1.8	.9	Woodfield et al. 1974
Solomon Is.	.990	.010		364	2.0	1.0	Blake et al. 1983
POLYNESIA							
Samoans (Christchurch, NZ)	1.000			77			Booth et al. 1977

Table 60. Peroxidase, salivary: PXS*

Place/Population	1	2	3	n	H	PE	Source

ASIA

JAPAN							
Japanese	.755	.245			37.0	8.0	Ikemoto 1983(c)

AFRICA

NIGERIA	.860	.140			24.1	7.7	"

NORTH AMERICA

UNITED STATES							
Whites	.787	.208	.005	101	33.7	8.8	Azen 1977

Note: Allele PXS^2 and PXS^3 are codominant to each other and dominant to PXS^1 (Azen 1977). The phenotypes at this locus are highly correlated with those at the PA locus (see Azen 1977; Table 116)

Table 61. Phosphoglucomutase 1: PGM1

Place/Population	1	2	x	y	n	H	PE	Source
EUROPE								
AUSTRIA	.791	.209			220	33.1	13.8	Wing 1974(c)
BELGIUM (Liége)	.768	.232			1,431	35.6	14.6	Brocteur et al. 1980
BULGARIA (Sofia)	.835	.165			127	27.6	11.9	Ananthakrishnan et al. 1972
CZECHOSLOVAKIA	.781	.219			320	34.2	14.2	Herzog & Bohatová 1969
DENMARK (Copenhagen)	.793	.207			309	32.8	13.7	Sørensen 1972
FAROE IS. (N. Atlantic)	.792	.208			667	32.9	13.8	Tills et al. 1985
FINLAND	.782	.218			804	34.1	14.1	Eriksson et al. 1971c
FRANCE								
French (Paris)	.736	.264			457	38.9	15.7	Séger 1971
Basques	.706	.294			63	41.5	16.4	Levine et al. 1974
GERMANY (Hamburg)	.783	.217			4,403	34.0	14.1	Brinkmann et al. 1971
GREECE								
Greeks	.681	.319			631	43.4	17.0	Stamatoyannopoulos et al. 1975
Greeks (Plati)	.685	.315			1,036	43.2	16.9	Tills et al. 1983b
ICELAND	.821	.179			1,338	29.4	12.5	Tills et al. 1982b
IRELAND	.737	.257	.006[6]		1,785	39.1	16.2	Tills 1977
ITALY (Bologna)	.720	.280			279	40.3	16.1	Beretta et al. 1977
NETHERLANDS (Leiden)	.785	.215			801	33.8	14.0	Fraser et al. 1974
NORWAY								
Norwegians	.776	.224	.001[7]		2,674	34.8	14.4	Monn 1969b
Lapps/Saami (Finnmark)	.513	.487			303	50.0	18.7	Monn 1969a
POLAND (Olkusz)	.742	.258			213	38.3	15.5	Wolanski et al. 1983
PORTUGAL								
(Porto Dist.)	.756	.244			544	36.9	15.0	Amorim & Siebert 1982
(Terras de Bouro)	.643	.357			119	45.9	17.7	Cruz et al. 1973
SPAIN								
Spanish (Madrid)	.754	.244	.002[6]		213	37.2	15.3	Goedde et al. 1972b
Basques	.762	.238			277	36.3	14.8	"
SWEDEN								
Swedes	.772	.228			412	35.2	14.5	Beckman et al. 1971
Lapps/Saami	.586	.414			210	48.5	18.4	"
SWITZERLAND (Bern)	.759	.241			2,635	36.6	14.9	Pflugshaupt et al. 1978
UNITED KINGDOM								
England	.765	.235			2,115	36.0	14.7	Hopkinson & Harris 1968
Scotland (Southwest)	.758	.241	.001[7]		828	36.7	15.1	Mitchell et al. 1976
Isle of Lewis	.848	.152			92	25.8	11.2	Clegg et al. 1985
Isle of Man	.741	.252	.007[7]		311	38.7	16.1	Mitchell et al. 1982
Orkney Is. (Westray)	.820	.180			402	29.5	12.6	Welch et al. 1973
U.S.S.R.								
EUROPEAN PART								
Russians (Moscow)	.709	.291			1,559	41.2	16.3	Altukhov et al. 1981
Abkhazia (Ochamchir Dist.)	.718	.282			252	40.5	16.1	Ferrell et al. 1985
ASIAN PART								
Eskimos (New Chaplino)	.941	.059			102	11.1	5.2	Sukernik et al. 1981
Chukchi (Chukotka & NE Kamchatka)	.921	.079			1,043	14.6	6.7	"
Nganasan-Avam (Taimir Penn)	.833	.167			264	27.8	12.0	Sukernik et al. 1978
Nentzi (Between Ob R. & Taj R.)	.655	.345			582	45.2	17.5	Sukernik et al. 1980
ASIA								
TURKEY (Asia Minor)	.677	.323			274	43.7	17.1	Hummel et al. 1970
ISRAEL								
Ashkenazi Jews	.792	.208			185	32.9	13.8	Szeinberg & Tomashevsky-Tamir 1971
Yemenite Jews	.678	.322			157	43.7	17.1	Tills et al. 1977b
KUWAIT								
Arabs	.704	.296			147	41.7	16.5	Sawhney et al. 1984

Continued

105

Table 61. Phosphoglucomutase 1: PGM1 (cont'd)

Place/Population	1	2	x	y	n	H	PE	Source
SAUDI ARABIA								
Shiites/Sunnites (al-Hasa)	.696	.298	.006[3,6]		359	42.7	17.3	Goedde et al. 1979b
Arabs (Southern)	.780	.218	.002[7]		261	34.4	14.4	Marengo-Rowe et al. 1974
IRAN								
Turkoman (Gonbad, NE Iran)	.600	.400			155	48.0	18.2	Kirk et al. 1977
AFGHANISTAN								
Pushtu & Dari (Kabul)	.733	.267			275	39.1	15.7	Papiha et al. 1977a
INDIA								
Punjabi (Chandigarh)	.668	.329	.003[7]		140	44.6	17.6	Papiha et al. 1972
Gujarati (W. India)	.668	.325	.007[7]		282	44.8	18.0	Papiha et al. 1981
Hindu & Moslems								
(Madhya Pradesh	.725	.274	.001[5]		338	39.9	16.1	Roberts et al. 1974
Bhil tr. (Madhya Pradesh)	.704	.296			142	41.7	16.5	Papiha et al. 1978
Hindu (Andhra Pradesh)	.654	.344	.002[5]		211	45.4	17.7	Roberts et al. 1980
Irula tr. (Nilgiri Hills,								
S. India)	.514	.486			175	50.0	18.7	Saha et al. 1976
Bengali (E. India)	.699	.299	.002[6]		269	42.2	16.8	Das et al. 1970
SRI LANKA								
Sinhalese (Anuradhapura)	.739	.261			155	38.6	15.6	Roberts et al. 1972b
BANGLADESH								
Moslems (Dacca)	.722	.272	.006[6,7]		200	40.5	16.7	Papiha et al. 1975
NEPAL								
Nepalese	.802	.196	.002[6]		248	31.8	13.5	Sunderland et al. 1979
Sherpas	.674	.326			138	43.9	17.1	Santachiara et al. 1976
BHUTAN	.770	.224	.006[6]		154	35.7	15.1	Mourant et al. 1969
MALAYSIA								
Malays (Singapore)	.762	.234	.004[3]		259	36.5	15.2	Blake et al. 1973a
INDONESIA								
Batak (Samosir Is.)	.782	.218			188	34.1	14.1	McDermid et al. 1973
Balinese (Bali Is.)	.682	.318			148	43.4	17.0	Breguet et al. 1982b
THAILAND (Northern)	.738	.249	.013[6,7]		587	39.3	16.8	Sanpitak et al. 1972
PHILIPPINES								
Filipino (Germany)	.747	.253			144	37.8	15.3	Windhof & Walter 1983
Negrito (Angeles,								
C. Luzon)	.500	.496	.004[6]		129	50.4	19.3	Omoto et al. 1978
CHINA								
Tibetans	.681	.306	.014[6Ti]		36	44.3	18.1	Santachiara et al. 1976
Tibetans (Switzerland)	.715	.285			107	40.8	16.2	Jeannet et al. 1972
Chinese (Singapore)	.732	.263	.005[8]		378	39.5	16.3	Blake et al. 1973a
TAIWAN								
Toroko-Aborigines	.934	.066			106	12.3	5.8	Chen et al. 1985
JAPAN								
Japanese (Nara & Mie)	.772	.225	.002[7]	.001[4,5,6]	1,740	35.3	14.8	Ishimoto 1970
Japanese (Hiroshima &								
Nagasaki)	.760	.223	.014[7]	.003[a]	11,823	37.2	16.5	Satoh et al. 1984b
Ainu (Hokkaido)	.870	.130			104	22.6	10.0	Harvey et al. 1978
KOREA								
Koreans (Seoul)	.775	.225			142	34.9	14.4	Goedde et al. 1986c

AFRICA

Place/Population	1	2	x	y	n	H	PE	Source
ALGERIA								
Ideles (Ahaggar)	.724	.276			214	40.0	16.0	Lefèvre-Witier & Vergnes 1977
LIBYA (Tripoli & Benghasi)	.793	.207			169	32.8	13.7	Walter et al. 1975
EGYPT								
Egyptians (Cairo & Anshas)	.716	.284			266	40.7	16.2	Goedde et al. 1980
Nubians (S. Egypt)	.771	.229			155	35.3	14.5	Bertin et al. 1978
SUDAN								
Sudanese (Khartoum)	.606	.391	.003[6]		174	48.0	18.5	Saha et al. 1978b
Beja-Amarar (NE Sudan)	.719	.276	.005[7]		98	40.7	16.6	el Hassan et al. 1968
CHAD								
Sara Majingay	.814	.186			175	30.3	12.8	Hiernaux 1976
MALI								
Kel Kummer Twaregs (Menaka)	.455	.545			264	49.6	18.6	Lefèvre-Witier & Vergnes 1977
SENEGAL								
Bedik (E. Senegal)	.819	.181			755	29.6	12.6	Bouloux et al. 1972
IVORY COAST								
Tuka	.829	.171			176	28.4	12.2	Vergnes & Cabannes 1976

Continued

106

Table 61. Phosphoglucomutase 1: PGM1 (cont'd)

Place/Population	1	2	x	y	n	H	PE	Source
NIGERIA								
Yoruba (Western)	.769	.231			65	35.5	14.6	Ojikutu et al. 1977
CAMEROON (Bamenda)	.745	.255			284	38.0	15.4	Goedde et al. 1979a
CENTRAL AFRICAN REPUBLIC								
Pygmies (Western)	.860	.140			732	24.1	10.6	Santachiara et al. 1980
BURUNDI								
Hutu	.814	.186			447	30.3	12.8	Gall et al. 1982
ETHIOPIA								
Amhara tr.	.696	.298	.006[6]		163	42.7	17.3	Harrison et al. 1969
UGANDA								
Baganda (Kampala)	.842	.154	.005[6]		202	26.7	11.9	Roberts et al. 1977
KENYA	.840	.160			309	26.9	11.6	Herzog et al. 1970
TANZANIA								
Sandawe (Kondoa-Irang Dt.)	.804	.191	.005[7]		202	31.7	13.7	Godber et al. 1976
Nyaturu (")	.837	.160	.003[6]		203	27.4	12.0	"
Hadza (W. Lake Eyasi)	.746	.254			215	37.9	15.4	Tills et al. 1982a
ANGOLA								
Njinga	.849	.151			109	25.6	11.2	Nurse et al. 1979
SOUTHERN AFRICA								
Basters (Namibia)	.786	.214			119	33.6	14.0	Nurse et al. 1982
San/Bushmen								
!Kung, Dobe	.980	.020			328	3.9	1.9	Nurse & Jenkins 1977
Khoi/Hottentot								
Keetmanshoop Nama	.803	.163	.034[6]		150	32.7	15.7	Jenkins 1972
Negroes/Dama (Okombahe &								
Sesfontein)	.752	.248			119	37.3	15.2	Nurse & Jenkins 1977
Bantu (Natal)	.855	.145			304	24.8	10.9	Kirk et al. 1971b
Griqua (Campbell, Cape								
Province)	.848	.152			99	25.8	11.2	Nurse & Jenkins 1975
TRISTAN DA CUNHA	.767	.233			159	35.7	14.7	Jenkins et al. 1985

NORTH AMERICA

Place/Population	1	2	x	y	n	H	PE	Source
GREENLAND								
Eskimos	.655	.345			152	45.2	17.5	Eriksson et al. 1971c
ALASKA								
Eskimos (Northern)	.815	.185			108	30.2	12.8	Scott et al. 1966
Eskimos (St. Lawrence Is.)	.878	.122			222	21.4	9.6	Ferrell et al. 1981b
Athabaskan Indians								
(Interior valley)	.894	.106			127	19.0	8.6	Scott et al. 1966
CANADA								
Eskimos (Igloolik)	.758	.242			397	36.7	15.0	McAlpine et al. 1974
Ojibwa Indians								
(Pikangikum)	.835	.165			95	27.6	11.9	Szathmary et al. 1974
Cree Indians (Manitoba)	.927	.073			199	13.5	6.3	Lucciola et al. 1974
Dogrib Indians								
(NW Territories)	.915	.085			158	15.6	7.2	Szathmary et al. 1983
UNITED STATES								
Whites (Seattle)	.752	.248			508	37.3	15.2	Giblett 1969(c)
Blacks (")	.809	.191			654	30.9	13.1	"
Whites (Louisiana)	.783	.217			376	34.0	14.1	Fox et al. 1981
Blacks (")	.787	.213			373	33.5	14.0	"
Mexican-Americans (Texas)	.792	.208			647	32.9	13.8	Hewett-Emmett et al. 1986b

MIDDLE AMERICA

Place/Population	1	2	x	y	n	H	PE	Source
MEXICO								
Indians (San Pablo del								
Monte)	.784	.216			104	33.9	14.1	Crawford et al. 1974
Mestizo (Tlaxcala)	.799	.201			72	32.1	13.5	"
GUATEMALA								
Black Caribs (Livingston)	.879	.120			199	21.3	9.5	Crawford et al. 1981
COSTA RICA								
Guaymi	.937	.063			286	11.8	5.6	Barrantes et al. 1982
PANAMA								
Blacks (Costa Arriba)	.795	.205			422	32.6	13.6	Ferrell et al. 1978b
WEST INDIES (Haiti)	.776	.224			308	34.8	14.4	Basu et al. 1976

Continued

107

Table 61. Phosphoglucomutase 1: PGM1 (cont'd)

Place/Population	1	2	x	y	n	H	PE	Source
SOUTH AMERICA								
ARGENTINA								
Mapuche	.589	.411			101	48.4	18.3	Haas et al. 1985
BOLIVIA								
Aymara (Western)	.832	.168			316	28.0	12.0	Ferrell et al. 1978a
BRAZIL								
Baniwa (Jandu Cachoeira)	.826	.174			363	28.2	12.1	Salzano et al. 1986
Cayapo (Para & Mato								
Grosso)	.765	.235			204	36.0	14.7	Salzano et al. 1972b
Caingang	.931	.069			211	12.8	6.0	Salzano et al. 1980
Parakana	.984	.016			93	3.1	1.5	Black et al. 1980
Pacaás Novos	.669	.331			222	44.2	17.2	Salzano et al. 1985
Macushi (N. Brazil &								
S. Guyana)	.824	.170	.006[10]		499	29.2	12.9	Neel et al. 1977b
Wapishana (")	.772	.228			615	35.2	14.5	"
Ticuna (C. Amazona)	.829	.171			1,765	28.4	12.2	Neel et al. 1980
CHILE								
Atacameño (Toconao, N.								
Chile)	.906	.094			172	17.0	7.8	Goedde et al. 1985b
COLOMBIA								
Noanama	.769	.231			169	35.5	14.6	Kirk et al. 1974
ECUADOR								
Shuara	.789	.211			90	33.3	13.9	Goedde et al. 1977a
Waorani	.679	.321			187	43.6	17.0	Larrick et al. 1985
FRENCH GUIANA								
Wayampi (Trois-Sauts &								
Camopi)	.871	.123	.006[4,10]		237	22.6	10.3	Tchen et al. 1978
SURINAM								
Trio (Southern)	.814	.186			411	30.3	12.8	Geerdink et al. 1974b
VENEZUELA								
Makiritare (S. Venezuela &								
N. Brazil)	.850	.150			535	25.5	11.1	Weitkamp & Neel 1970
Yanomama (")	.957	.043			2,351	8.2	3.9	Weitkamp & Neel 1972
OCEANIA								
AUSTRALIA								
Aborigines/Malag								
(Elcho Is., NT)	.847	.153			255	16.5	7.6	Blake & Omoto 1975
Aborigines (Hooker Creek, NT)	.913	.087			286	15.9	7.3	Blake 1979
" (Amoonguna, CA)	.909	.091			104	16.5	7.6	Kirk et al. 1971a
" /Walmadjari (Kimberley								
W. Australia)	.909	.091			171	16.5	7.6	Blake & Spargo 1986
MELANESIA								
W. New Guinea								
Asmat (S. Coastal Plain)	.942	.026	.032[3]		653	11.1	5.6	Gajdusek et al. 1978
Papua New Guinea								
Gainj & Kalam (N. Central								
Highlands)	.983	.017			571	3.3	1.6	Long et al. 1986
Jimi Valley								
(W. Highlands)	.957	.043			317	8.2	3.9	Mourant et al. 1981
Yagaria (E. Highlands)	.954	.046			419	8.8	4.2	Mourant et al. 1982
Fuyuge (C. Highlands)	.934	.055	.012[3]		173	12.4	6.2	Woodfield et al. 1974
Karkar Is.								
Takia	.863	.130	.008[3]		288	23.8	11.0	Booth et al. 1982
Manus Is.	.917	.083			138	15.2	7.0	Malcolm et al. 1972
Buka Is.	.587	.413			80	48.5	18.4	McLoughlin et al. 1982
Solomon Is.	.736	.214	.018[3]	.032[7]	364	36.4	16.2	Blake et al. 1983
MICRONESIA								
W. Caroline Is.	.707	.166	.068[3]	.059[7]	382	46.4	26.1	Blake et al. 1973b
E. Caroline Is.	.898	.102			1,005	18.3	8.3	Yamamoto & Fu 1973
Marshall Is.	.891	.109			371	19.4	8.8	Neel et al. 1976
POLYNESIA								
Samoans (Christchurch,								
New Zealand)	.669	.331			77	44.3	17.2	Booth et al. 1977

Note: [a] Pooled frequency of 13 different kinds of rare alleles.

108

Table 61.1   Phosphoglucomutase 1 (subtypes): PGM1

Place/Population	1+	1−	2+	2−	Others	n	H	PE	Source

EUROPE

Place/Population	1+	1−	2+	2−	Others	n	H	PE	Source
CZECHOSLOVAKIA (Prague)	.639	.118	.180	.063		495	54.1	31.6	Ranzani et al. 1985
FINLAND	.531	.220	.180	.069		639	63.2	38.2	Lukka et al. 1985
FRANCE									
French (Bareges)	.650	.034	.306	.009		164	48.3	22.1	Vergnes & Sevin 1981
Basques	.609	.109	.241	.041		160	55.7	31.3	"
GERMANY (Hessen)	.630	.132	.184	.053		1,678	54.9	31.9	Kühnl et al. 1978
GREECE	.562	.166	.234	.036		250	60.1	34.5	Kouvatsi & Triantaphyllidis 1984
HUNGARY	.647	.182	.112	.059		179	53.2	30.8	Goedde et al. 1986b
ITALY (Tuscany)	.601	.106	.250	.043		519	56.3	31.6	Bargagna & Abbagnale 1982
POLAND (Wroclaw)	.602	.124	.198	.076		743	57.7	34.3	Dobosz 1983
SPAIN (Galicia)	.621	.114	.211	.054		1,086	55.4	31.9	Carracedo & Concheiro 1982
SWEDEN	.614	.164	.156	.066		2,135	56.7	33.6	Svensson & Wetterling 1982
SWITZERLAND (Bern)	.628	.130	.194	.049		501	54.9	31.7	Scherz et al. 1981a
UNITED KINGDOM									
England	.693	.097	.150	.060		207	48.4	27.8	Papiha et al. 1982a

ASIA

Place/Population	1+	1−	2+	2−	Others	n	H	PE	Source
IRAN									
Zoroastrians	.672	.072	.198	.058		109	50.1	28.0	Papiha et al. 1982b
INDIA									
Jat Sikhs (Punjab)	.601	.125	.219	.055		144	57.2	33.2	Papiha et al. 1982a
Indians (Delhi)	.570	.140	.228	.060	.002	498	60.0	35.5	Kamboh & Kirk 1984
Indians (Madras)	.523	.139	.246	.080	.013	238	64.0	39.5	"
Soliga tr. (S. India)	.467	.066	.320	.147		61	65.4	38.9	"
INDONESIA									
Lesser Sunda Is.	.600	.209	.089	.102		225	57.8	34.3	"
THAILAND									
Thais	.615	.098	.197	.082	.008	61	56.7	33.8	"
Thais	.610	.144	.119	.127		118	57.7	35.1	Goedde & Benkmann 1987
CHINA									
Mongolians (N. China)	.681	.081	.186	.052		202	49.2	27.7	Goedde et al. 1984b
Han	.642	.146	.127	.085		107	54.3	32.4	Chen & Yan 1984
Zhuang (S. China)	.569	.138	.169	.124		202	61.3	37.8	"
Hui (Ningxia)	.512	.171	.178	.129		213	66.0	41.2	Zhao et al. 1984
Dong (Guangxi)	.555	.190	.165	.090		200	62.1	37.9	"
JAPAN									
Japanese (W. Japan)	.651	.120	.164	.052	.014	504	53.2	31.5	Yuasa et al. 1986a
KOREA									
Koreans (Seoul)	.682	.186	.089	.036	.007	140	49.1	27.6	Goedde et al. 1986c
Koreans (NE China)	.690	.076	.189	.045		210	48.0	26.6	Goedde et al. 1984b

AFRICA

Place/Population	1+	1−	2+	2−	Others	n	H	PE	Source
LIBERIA	.787	.039	.147	.027		485	35.7	18.6	Goedde et al. 1985a
SOUTHERN AFRICA									
San/Bushmen									
Ai/ai !Kung	.732	.250	.012		.006	84	40.1	17.5	Tipler et al. 1982
Khoi/Hottentot									
Nama (Rehboth)	.680	.134	.109	.051	.026	78	50.4	30.2	"
Bantu/Zulu	.730	.142	.098	.029		102	43.6	24.1	"

NORTH AMERICA

Place/Population	1+	1−	2+	2−	Others	n	H	PE	Source
ALASKA									
Eskimos	.669	.126	.047	.158		332	50.9	29.2	Dykes et al. 1983a
UNITED STATES									
Whites (Minnesota)	.633	.139	.171	.057		8,662	50.9	30.9	"
Blacks (")	.651	.145	.160	.044		633	52.8	30.3	"
Pima Indians	.485	.342	.020	.153		101	62.4	34.5	"
Apache Indians (San Carlos)	.648	.182	.061	.109		381	53.1	30.8	"
Mexican-Americas (Texas)	.566	.226	.141	.067		647	60.4	35.9	Hewett-Emmett et al. 1986a

Continued

Table 61.1 Phosphoglucomutase 1 (subtypes): PGM1 (cont'd)

Place/Population	1+	1−	2+	2−	Others	n	H	PE	Source
MIDDLE AMERICA									
GUATEMALA									
Black Caribs (Livingston)	.645	.230	.122	.003		176	51.6	27.0	Dykes et al. 1983a
SOUTH AMERICA									
CHILE									
Atacameño (Toconao, N. Chile)	.301	.594	.032	.073		172	55.0	28.9	Goedde et al. 1985b
OCEANIA									
AUSTRALIA									
Aborigines (C. Australia)	.704	.212	.072	.011		313	45.4	23.3	Kamboh & Kirk 1984
MELANESIA									
Papua New Guinea									
E. Highlands	.685	.252	.050	.012		200	46.5	22.8	"
Port Moresby	.771	.157	.065	.007		201	37.7	19.4	"
Daga (Bay Province)	.863	.050	.061	.026		166	24.8	13.3	Jenkins et al. 1983
Fiji	.445	.245	.145	.160		200	69.5	44.3	Kamboh & Kirk 1984
MICRONESIA									
Kiribati	.692	.165	.022	.120		200	47.9	26.4	"
POLYNESIA									
Samoa Is. (Western)	.432	.182	.155	.230		200	70.3	45.4	"
Cook Is.	.645	.175	.080	.100		200	53.7	31.5	"

Table 62. Phosphoglucomutase 2: PGM2

Place/Population	1	2	x	10	n	H	PE	Source
EUROPE								
BELGIUM (Liége)	1.000				1,431[b]			Brocteur et al. 1980
FAROE IS. (N. Atlantic)	1.000				666			Tills et al. 1985
FINLAND								
Finns	.990	.010			406	2.0	1.0	Kirjarinta et al. 1969
Lapps/Saami	1.000				49			"
FRANCE								
French (Bareges)	1.000				593			Vergnes et al. 1980a
Basques	1.000				144			"
GERMANY (Berlin)	1.000				860			Fielder & Pettenkofer 1968
GREECE								
Greeks (Plati)	1.000				1,036			Tills et al. 1983b
ICELAND	1.000				1,338			Tills et al. 1982b
IRELAND	1.000				1,785			Tills 1977
ITALY (S. Italy)	1.000				192			Sangiorgi et al. 1982
NETHERLANDS (Leiden)	1.000				806			Fraser et al. 1974
NORWAY								
Lapps/Saami	.987		.013		303	2.6	1.3	Monn 1969a
POLAND (Olkusz)	1.000				213			Wolanski et al. 1983
UNITED KINGDOM								
England	1.000				2,115			Hopkinson & Harris 1966
Isle of Lewis	1.000				94			Clegg et al. 1985

U.S.S.R.

EUROPEAN PART

Continued

Table 62. Phosphoglucomutase 2: PGM2 (cont'd)

Place/Population	1	2	x	10	n	H	PE	Source
Russians (Moscow)	.999		.001		1,695	.2	.1	Altukhov et al. 1981
Abkhazia (Ochamchir Dist.)	1.000				252			Ferrell et al. 1985

ASIA

TURKEY								
Turks (Cyprus)	1.000				243			Hopkinson & Harris 1966
IRAN								
Turkoman (Gonbad, NE Iran)	.987		.013[9]		155	2.6	1.3	Kirk et al. 1977
AFGHANISTAN								
Pushtu & Dari (Kabul)	1.000				281			Papiha et al. 1977a
INDIA								
Punjabi (Chandigarh)	1.000				140			Papiha et al. 1972
Gujarati (W. India)	1.000				281			Papiha et al. 1981
Hindu & Moslems (Madhya Pradesh)	.999	.001			337	.2	.1	Roberts et al. 1974
Bhil tr. (Madhya Pradesh)	1.000				142			Papiha et al. 1978
Hindu (Andhra Pradesh)	1.000				211			Roberts et al. 1980
Irula tr. (Nilgiri Hills, S. India)	1.000				174			Saha et al. 1976
Bengali (E. India)	.998		.002[6]		269	.4	.2	Das et al. 1970
SRI LANKA								
Sinhalese (Anuradhapura)	1.000				155			Roberts et al. 1972b
BANGLADESH								
Moslems (Dacca)	1.000				200			Papiha et al. 1975
NEPAL								
Sherpas	1.000				138			Santachiara et al. 1976
MALAYSIA								
Malays (Singapore)	1.000				259			Blake et al. 1973a
INDONESIA								
Batak (Samosir Is.)	1.000				188			McDermid et al. 1973
Balinese (Bali Is.)	.993		.007[9]		148	1.4	.7	Breguet et al. 1982b
PHILIPPINES								
Filipino/Tagalog (Manila)	1.000				253			Omoto et al. 1978
Negrito (Angeles, C. Luzon)	1.000				129			"
CHINA								
Tibetans	1.000				36			Santachiara et al. 1976
Chinese (Singapore)	1.000				378			Blake et al. 1973a
JAPAN								
Japanese (Hiroshima & Nagasaki)	1.000				11,833[a]			Satoh et al. 1984a
Ainu (Hokkaido)	1.000				191			Omoto & Harada 1972

AFRICA

EGYPT								
Nubians (S. Egypt)	1.000				155			Bertin et al. 1978
SUDAN								
Sudanese (Khartoum)	1.000				174			Saha et al. 1978b
CHAD								
Sara Majingay	.997	.003			174	.6	.3	Hiernaux 1976
MALI								
Kel Kummer Twaregs (Menaka)	1.000				264			Lefèvre-Witier & Vergnes 1977
SENEGAL								
Bedik (E. Senegal)	.999	.001			755	.2	.1	Bouloux et al. 1972
IVORY COAST								
Tuka	.972	.028			176	5.4	2.6	Vergnes & Cabannes 1976
NIGERIA								
Yoruba (Western)	.992		.008[3]		65	1.6	.8	Ojikutu et al. 1977
CENTRAL AFRICAN REPUBLIC								
Pygmies (Western)	.932	.001	.067[4,6]		732	12.7	6.0	Santachiara et al. 1980
BURUNDI								
Hutu	.992	.001	.007[6]		447	1.6	.8	Gall et al. 1982

Continued

111

Table 62. Phosphoglucomutase 2: PGM2 (cont'd)

Place/Population	1	2	x	10	n	H	PE	Source
UGANDA								
Baganda (Kampala)	.998	.002			202	.4	.2	Roberts et al. 1977
TANZANIA								
Sandawe (Kondoa Irang Dt.)	.993	.007			202	1.4	.7	Godber et al. 1976
Nyaturu (")	1.000				203			"
Hadza (W. Lake Eyasi)	.991	.009			215	1.8	.9	Tills et al. 1982a
ANGOLA								
Njinga	.972	.028			109	5.4	2.6	Nurse et al. 1979
SOUTHERN AFRICA								
Basters (Namibia)	.996	.004			119	.8	.4	Nurse et al. 1982
San/Bushmen								
Dobe	.974	.026			328	5.1	2.5	Nurse & Jenkins 1977
Khoi/Hottentot								
Keetmanshoop Nama	.996	.004			149	.8	.4	Jenkins 1972
Negroes/Dama (Okombahe &								
Sesfontein)	1.000				119			Nurse & Jenkins 1977
Bantu (Natal)	.956	.044			304	8.4	4.0	Kirk et al. 1971b
Griqua (Campbell, Cape								
Province)	.960	.040			99	7.7	3.7	Nurse & Jenkins 1975
TRISTAN DA CUNHA	1.000				159			Jenkins et al. 1985

NORTH AMERICA

Place/Population	1	2	x	10	n	H	PE	Source
GREENLAND								
Eskimos (Western)	1.000				152			Kirjarinta et al. 1969
ALASKA								
Eskimos								
(St. Lawrence Is.)	1.000				222			Ferrell et al. 1981b
CANADA								
Eskimos (Igloolik)	1.000				397			McAlpine et al. 1974
Ojibwa Indians								
(Pikangikum)	1.000				93			Szathmary et al. 1974
Cree Indians (Manitoba)	1.000				199			Lucciola et al. 1974
Dogrib Indians								
(NW Territories)	1.000				158			Szathmary et al. 1983
UNITED STATES								
Whites (Louisiana)	1.000				376			Fox et al. 1981
Blacks (")	.992	.008			373	1.6	.8	"
Blacks (Seattle)	1.000				654			Giblett 1969(c)

MIDDLE AMERICA

Place/Population	1	2	x	10	n	H	PE	Source
MEXICO								
Huastec	1.000				233			Lisker & Giblett 1967
GUATEMALA								
Black Caribs (Livingston)	1.000				200			Crawford et al. 1981
COSTA RICA								
Guaymi	1.000				286			Barrantes et al. 1982
PANAMA								
Blacks	1.000				780			Ferrell et al. 1978b
WEST INDIES (Haiti)	.992		.008[v]		308	1.6	.8	Basu et al. 1976

SOUTH AMERICA

Place/Population	1	2	x	10	n	H	PE	Source
ARGENTINA								
Mapuche	.849		.151[11]		93	25.6	11.2	Haas et al. 1985
BOLIVIA								
Aymara (W. Bolivia)	1.000				316			Ferrell et al. 1978a
BRAZIL								
Baniwa (Jandu Cachoeira)	1.000				363			Salzano et al. 1986
Cayapo (Pará & Mato								
Grosso)	1.000				204			Salzano et al. 1972b
Caingang	1.000				211			Salzano et al. 1980
Parakana	.995	.005			93	1.0	.5	Black et al. 1980
Pacaás Novos	1.000				222			Salzano et al. 1985
Macushi (N. Brazil &								
S. Guyana)	1.000				499			Neel et al. 1977b
Wapishana (")	1.000				615			"

Continued

Table 62. Phosphoglucomutase 2: PGM2 (cont'd)

Place/Population	1	2	x	10	n	H	PE	Source
Ticuna (C. Amazonas)	1.000				1,765			Neel et al. 1980
COLOMBIA								
Noanama	1.000				169			Kirk et al. 1974
ECUADOR								
Waorani	1.000				187			Larrick et al. 1985
FRENCH GUIANA								
Wayampi (Trois-Sauts & Camopi)	.945	.055			238	10.4	4.9	Tchen et al. 1978
SURINAM								
Trio (Southern)	.902		.098[5Trio]		240	17.7	8.1	Geerdink et al. 1974a
VENEZUELA								
Makiritare (S. Venezuela & N. Brazil)	1.000				535			Weitkamp & Neel 1970
Yanomama (")	1.000				2,352			Weitkamp & Neel 1972

OCEANIA

Place/Population	1	2	x	10	n	H	PE	Source
AUSTRALIA								
Aborigines/Malag (Elcho Is., N. Territory)	1.000				255			Blake & Omoto 1975
Aborigines (Hooker Creek, N. Territory)	.947		.053[3]		286	10.0	4.8	Blake 1979
Aborigines (Amoonguna, C. Australia)	.976		.024[3]		104	4.7	2.3	Blake & Omoto 1975
Aborigines/Walmadjari (Kimberley, W. Australia)	.950		.050[3]		171	9.5	4.5	Blake & Spargo 1986
MELANESIA								
W. New Guinea								
Asmat (S. Coastal Plain)	.901		.071[9]	.028	653	18.2	9.1	Gajdusek et al. 1978
Papua New Guinea								
Gainj & Kalam (N. Central Highlands)	.915		.085[9]		571	15.6	7.2	Long et al. 1986
Jimi Valley (W. Highlands)	.912		.088[9,10]		385	16.1	7.4	Mourant et al. 1981
Aborigines (E. Highlands)	.940		.036[9]	.024	449	11.5	5.8	Blake & Omoto 1975
Fuyuge (C. Highlands)	.896		.064[9]	.040	173	19.1	9.7	Woodfield et al. 1974
Daga (Bay Province)	.904		.097[9]		139	17.3	8.0	Jenkins et al. 1983
Karkar Is.								
Takia	.944		.051[9]	.005	403	10.6	5.1	Blake & Omoto 1975
Manus Is.	1.000				183			Malcolm et al. 1972
Buka Is.	.987		.013[9]		80	2.6	1.3	McLoughlin et al. 1982a
Solomon Is.	.999			.001	364	.2	.1	Blake et al. 1983
MICRONESIA								
W. Caroline Is.	1.000				382			Blake et al. 1973b
E. Caroline Is.	1.000				1,005			Yamamoto & Fu 1973
Marshall Is.	1.000				374			Neel et al. 1976
POLYNESIA								
Samoans (Christchurch, New Zealand)	1.000				77			Booth et al. 1977

Note: [a] One individual of genotype (+/2HR1), one individual of (+/5NG2), and six individuals of (+/9NG1) were were observed. [b] A variant similar but not identical to the genotype (+/5) was observed. [v] Five individuals with a variant allele were observed.

Table 63. Phosphoglucomutase 3: PGM3

Place/Population	1	2	3	n	H	PE	Source
EUROPE							
CZECHOSLOVAKIA (Praha)	.764	.236		146	36.1	14.8	Herzog & Drdová 1971
DENMARK	.745	.255		1,031	38.0	15.4	Lamm 1970
GERMANY (Dusseldorf)	.739	.261		462	38.6	15.6	Günther 1982
ITALY	.780	.220		559	34.3	14.2	Corbo et al. 1980
NORWAY	.734	.266		660	39.0	15.7	Monn & Gjønnaess 1971
UNITED KINGDOM							
England	.744	.256		583	38.1	15.4	Hopkinson & Harris 1968
ASIA							
INDIA							
Indians (Malaysia)	.755	.245		198	37.0	15.1	Teng et al. 1978a
MALAYSIA							
Malays (Kuala Lumpur)	.788	.212		343	33.4	13.9	"
CHINA							
Chinese (Malaysia)	.745	.255		296	38.0	15.4	"
JAPAN							
Japanese	.810	.190		370	30.8	13.0	Ishimoto 1969
AFRICA							
NIGERIA							
Nigerians (England)	.340	.660		235	44.9	17.4	Hopkinson & Harris 1968
NORTH AMERICA							
CANADA							
Eskimos (James & Hudson, Ontario)	.884	.116		82	20.5	9.2	Donald 1977
Cree Indians (")	.765	.235		230	36.0	14.7	"
OCEANIA							
AUSTRALIA							
Aborigines (N. Territory)	.589	.411		101	48.4	18.3	Wierst et al. 1973
MELANESIA							
Papua New Guinea							
Papuans (Port Moresby)	.511	.484	.005	191	50.5	18.9	"

Table 64. Phosphogluconate dehydrogenase: PGD

Place/Population	A	C	x	n	H	PE	Source
EUROPE							
BELGIUM (Liège)	.983	.017		500	3.3	1.6	Brocteur et al. 1980
BULGARIA (Sofia)	.989	.011		138	2.2	1.1	Ananthakrishnan et al. 1972
CZECHOSLOVAKIA (Praha)	.988	.012		330	2.4	1.2	Herzog et al. 1972
DENMARK	.980	.020		1,574	3.9	1.9	Berg 1974
FAROE IS. (N. Atlantic)	.986	.014		666	2.8	1.4	Tills et al. 1985
FINLAND							
Lapps/Saami	.968	.032		282	6.2	3.0	Tills et al. 1971(c)
FRANCE							
French (Bareges)	.995	.005		608	1.0	.5	Vergnes et al. 1980a
Basques	.997	.003		144	.6	.3	"
GERMANY (Hamburg)	.980	.020		1,162	3.9	1.9	Brinkmann et al. 1971
GREECE							
Greeks	.958	.039	.003[V]	649	8.1	3.9	Stamatoyannopoulos et al. 1975
Greeks (Plati)	.974	.024	.002[R]	1,035	5.1	2.5	Tills et al. 1983b

Continued

114

Table 64. Phosphogluconate dehydrogenase: PGD (cont'd)

Place/Population	A	C	x	n	H	PE	Source
HUNGARY	.987	.013		116	2.6	1.3	Goedde et al. 1986b
ICELAND	.979	.021		1,069	4.1	2.0	Tills et al. 1982b
IRELAND	.985	.014	.001R	1,797	3.0	1.5	Tills 1977
ITALY (Southern)	.946	.054		193	10.2	4.8	Sangiorgi et al. 1982
NETHERLANDS (Leiden)	.979	.021		801	4.1	2.0	Fraser et al. 1974
POLAND (Olkusz)	.974	.026		213	5.1	2.5	Wolanski et al. 1983
SPAIN							
Spanish (Madrid)	.984	.016		188	3.1	1.5	Goedde et al. 1972b
Basques (Bermeo/Pamplona)	.988	.012		287	2.4	1.2	"
SWEDEN							
Swedes	.981	.019		412	3.7	1.8	Beckman et al. 1971
Lapps/Saami	.869	.131		210	22.8	10.1	"
UNITED KINGDOM							
England	.978	.021		4,557	4.3	2.0	Parr 1966
Isle of Lewis	.989	.011		94	2.2	1.1	Clegg et al. 1985
Isle of Man	.987	.013		189	2.6	1.3	Mitchell et al. 1982
Orkney Is. (Westray)	.985	.015		408	3.0	1.5	Welch et al. 1973

U.S.S.R.

Place/Population	A	C	x	n	H	PE	Source
EUROPEAN PART							
Russians (Moscow)	.962	.038		1,553	7.3	3.5	Altukhov et al 1981
Abkhazia (Ochamchir Dist.)	.976	.024		255	4.7	2.3	Ferrell et al. 1985
ASIAN PART							
Eskimos (New Chaplino)	.946	.054		102	10.2	4.8	Sukernik et al. 1981
Chukchi (Chukotka/N. Kamchatka)	.944	.056		1,042	10.6	5.0	"
Nganasan-Avam (Taimir Penn)	.957	.043		264	8.2	3.9	Sukernik et al. 1978
Nentzi (Between Ob R. &							
Taj R.)	.957	.043		581	8.2	3.9	Sukernik et al. 1980

ASIA

Place/Population	A	C	x	n	H	PE	Source
TURKEY (Asia Minor)	.999	.001		500	.2	.1	Altay et al. 1974
ISRAEL							
Yemenite Jews	.928	.072		111	13.4	6.2	Tills et al. 1977b
JORDAN							
Jordanians	.986	.014		538	2.8	1.4	Banerjee et al. 1981
KUWAIT							
Arabs	.967	.033		150	6.4	3.1	Sawhney et al. 1984
SAUDIA ARABIA							
Shiites & Sunnites (al-Hasa)	.888	.112		354	19.9	9.0	Goedde et al. 1979b
Tribes (Western)	.970	.030		149	5.8	2.8	Saha et al. 1980
Arabs (Southern)	.885	.115		261	20.4	9.1	Marengo-Rowe et al. 1974
IRAN							
Turkoman (Gonbad, NE Iran)	.948	.052		155	9.9	4.7	Kirk et al. 1977
AFGHANISTAN							
Pushtu & Dari (Kabul)	.954	.045	.002H	280	8.8	4.3	Papiha et al. 1977a
INDIA							
Punjabi (Chandigarh)	.979	.021		140	4.1	2.0	Papiha et al. 1972
Gujarati (W. India)	.977	.023		281	4.5	2.2	Papiha et al. 1981
Hindu & Moslems							
(Madhya Pradesh)	.984	.014	.002	339	3.2	1.6	Papiha & Chhaparwal 1973
Bhil tr. (Madhya Pradesh)	.966	.034		145	6.6	3.2	Papiha et al. 1978
Hindu (Andhra Pradesh)	.976	.024		209	4.7	2.3	Roberts et al. 1980
Irula tr. (Nilgiri Hills,							
S. India)	.977	.023		175	4.5	2.2	Saha et al. 1976
Bengali (E. India)	.980	.020		271	3.9	1.9	Das et al. 1970
SRI LANKA							
Sinhalese (Anuradhapura)	.984	.016		156	3.1	1.5	Roberts et al. 1972b
BANGLADESH							
Moslems (Dacca)	.960	.040		200	7.7	3.7	Papiha et al. 1975
NEPAL	.914	.086		208	15.7	7.2	Sunderland et al. 1979
BHUTAN	.770	.230		154	35.4	14.6	Mourant et al. 1969
MALAYSIA							
Malays (Singapore)	.938	.060	.002R	259	11.7	5.5	Blake et al. 1973a
INDONESIA							
Batak (Samosir Is.)	.953	.047		182	9.0	4.3	McDermid et al. 1973
Balinese (Bali Is.)	.983	.017		147	3.3	1.6	Breguet et al. 1982b

Continued

115

Table 64. Phosphogluconate dehydrogenase: PGD (cont'd)

Place/Population	A	C	x	n	H	PE	Source
THAILAND	.956	.044		3,185	8.4	4.0	Tuchinda et al. 1968
VIETNAM	.946	.054		259	10.2	4.8	Bowman et al. 1971
PHILIPPINES							
Filipino (Germany)	.920	.077	.003[R]	144	14.8	6.9	Windhof & Walter 1983
Negrito (Angeles, C. Luzon)	.977	.023		128	4.5	2.2	Omoto et al. 1978
CHINA							
Mongolians (N. China)	.872	.128		199	22.3	9.9	Goedde et al. 1984b
Zhuang (S. China)	.936	.064		180	12.0	5.6	"
Hui (Ningxia)	.939	.061		218	11.5	5.4	Xu & Du 1984d
Dong (Guangxi)	.928	.072		201	13.4	6.2	"
Chinese (Singapore)	.933	.066	.001[S]	378	12.5	5.9	Blake et al. 1973a
JAPAN							
Japanese (Nara & Mie)	.910	.090		1,740	16.4	7.5	Ishimoto 1970
Ainu (Hokkaido)	.928	.072		104	13.4	6.2	Harvey et al. 1978
KOREA							
Koreans (Seoul)	.899	.094	.007[Ko]	148	18.3	8.6	Benkmann et al. 1986
Koreans (NE China)	.886	.114		211	20.2	9.1	Goedde et al. 1984b

AFRICA

Place/Population	A	C	x	n	H	PE	Source
LIBYA (Tripoli)	.952	.048		146	9.1	4.4	Kamel et al. 1975
ALGERIA							
Ideles (Ahaggar)	.987	.013		224	2.6	1.3	Lefèvre-Witier & Vergnes 1977
EGYPT							
Egyptians (Cairo & Anshas)	.962	.038		239	7.3	3.5	Goedde et al. 1980
Nubians (S. Egypt)	.831	.169		148	28.1	12.1	Bertin et al. 1978
SUDAN							
Sudanese (Khartoum)	.976	.024		287	4.7	2.3	Saha et al. 1978b
Beja-Amarar (NE Sudan)	.845	.155		100	26.2	11.4	el Hassan et al. 1968
CHAD							
Sara Majingay	.995	.005		207	1.0	.5	Hiernaux 1976
MALI							
Kel Kummer Twaregs (Menaka)	.836	.164		283	27.4	11.8	Lefèvre-Witier & Vergnes 1977
SENEGAL							
Bedik (E. Senegal)	.952	.048		756	9.1	4.4	Bouloux et al. 1972
LIBERIA	.941	.059		485	11.1	5.2	Goedde et al. 1985a
IVORY COAST							
Tuka	1.000			186			Vergnes & Cabannes 1976
NIGERIA							
Yoruba (Western)	.945	.055		64	10.4	4.9	Ojikutu et al. 1977
CAMEROON (Bamenda)	.954	.042	.004[V]	284	8.8	4.3	Goedde et al. 1979a
BURUNDI							
Hutu	.949	.049	.002[R]	527	9.7	4.7	Gall et al. 1982
ETHIOPIA							
Amhara tr.	.865	.135		100	23.4	10.3	Harrison et al. 1969
UGANDA							
Baganda (Kampala)	.964	.036		195	6.9	3.3	Roberts et al. 1977
TANZANIA							
Sandawe (Kondoa-Irang Dt.)	.941	.054	.005[R]	211	11.2	5.4	Godber et al. 1976
Nyaturu (")	.966	.034		209	6.6	3.2	"
Hadza (W. Lake Eyasi)	.998		.002[R]	203	.4	.2	Tills et al. 1982a
ANGOLA							
Njinga	.917	.083		109	15.2	7.0	Nurse et al. 1979
SOUTHERN AFRICA							
Basters (Namibia)	.966	.034		119	6.6	3.2	Nurse et al. 1982
San/Bushmen							
!Kung, Dobe	1.000			328			Nurse & Jenkins 1977
Khoi/Hottentot							
Keetmanshoop Nama	.996	.004		118	.8	.4	Jenkins 1972
Negroes/Dama (Okombahe &							
Sesfontein)	.970	.017	.013[R]	119	5.9	3.0	Nurse et al. 1976
Bantu (Natal)	.893	.104	.003[R]	304	19.2	8.8	Kirk et al. 1971b
Griqua (Campbell, Cape Province)	.964	.036		96	6.9	3.3	Nurse & Jenkins 1975
TRISTAN DA CUNHA	1.000			159			Jenkins et al. 1985

NORTH AMERICA

ALASKA

Continued

Table 64. Phosphogluconate dehydrogenase: PGD (cont'd)

Place/Population	A	C	x	n	H	PE	Source
Eskimos (Anchorage)	1.000			170			Duncan et al. 1974
Eskimos (St. Lawrence Is.)	.998	.002		222	.4	.2	Ferrell et al. 1981b
Athabaskan Indians (Anchorage)	1.000			37			Duncan et al. 1974(c)
CANADA							
Eskimos (Igloolik)	.990	.010		136	2.0	1.0	McAlpine et al. 1974
Ojibwa Indians (Pikangikum)	.967	.033		91	6.4	3.1	Szathmary et al. 1974
Cree Indians (Manitoba)	.970	.030		199	5.8	2.8	Lucciola et al. 1974
Dogrib Indians (NW Territory)	1.000			158			Szathmary et al. 1983
UNITED STATES							
Whites (Seattle)	.982	.018		647	3.5	1.7	Giblett 1969(c)
Blacks (")	.945	.055		506	10.4	4.9	"
Whites (Louisiana)	.976	.024		376	4.7	2.3	Fox et al. 1981
Blacks (")	.965	.035		373	6.8	3.3	Fox et al. 1981
Mexican-Americans (Texas)	.981	.019		566	3.7	1.8	Hewett-Emmett et al. 1986b

MIDDLE AMERICA

MEXICO							
Indians (Yucatan)	1.000			85			Bowman et al. 1966
GUATEMALA							
Black Caribs (Livingston)	.983	.017		205	3.3	1.6	Crawford et al. 1981
COSTA RICA							
Guaymi	.897	.103		286	18.5	8.4	Barrantes et al. 1982
PANAMA							
Blacks (Costa Arriba)	.946	.054		402	10.2	4.8	Ferrell et al. 1978b
WEST INDIES (Haiti)	.938	.058	.003^V	308	11.7	5.5	Basu et al. 1976

SOUTH AMERICA

BOLIVIA							
Aymara (W. Bolivia)	.986	.011	.003^{Ay}	316	2.8	1.4	Ferrell et al. 1978a
BRAZIL							
Baniwa (Jandu Cachoeira)	1.000			363			Salzano et al. 1986
Cayapo (Pará & Mato Grosso)	1.000			206			Salzano et al. 1972b
Caingang	.990		.010^{Cai}	211	2.0	1.0	Salzano et al. 1980
Parakana	1.000			98			Black et al. 1980
Pacaás Novos	1.000			222			Salzano et al. 1985
Macushi (N. Brazil & S. Guyana)	.998		.002^V	499	.4	.2	Neel et al. 1977b
Wapishana (")	.990		.010^V	615	2.0	1.0	"
Ticuna (C. Amazonas)	1.000			1,764			Neel et al. 1980
CHILE							
Atacameño (Toconao, N. Chile)	.959	.041		176	7.9	3.8	Goedde et al. 1985b
COLOMBIA							
Noanama	.988	.012		170	2.4	1.2	Kirk et al. 1974
ECUADOR							
Shuara	.989	.011		90	2.2	1.1	Goedde et al. 1977a
Waorani	1.000			186			Larrick et al. 1985
FRENCH GUIANA							
Wayampi (Trois-Sauts & Camopi)	1.000			237			Tchen et al. 1978
SURINAM							
Trio (S. Surinam)	1.000			523			Geerdink et al. 1974b
VENEZUELA							
Makiritare (S. Venezuela & N. Brazil)	.990	.005	.005^{Mak}	532	2.0	1.0	Weitkamp & Neel 1970
Yanomama (")	1.000			2,218			Weitkamp & Neel 1972

OCEANIA

AUSTRALIA							
Aborigines/Malag (Elcho Is., N. Territory)	.938	.043	.019^{Elcho}	594	11.8	5.9	Kirk et al. 1969
Aborigines (Hooker Creek,, N. Territory)	.958	.042		286	8.0	3.9	Blake 1979
Aborigines (Amoonguna, C. Australia)	.942	.058		104	10.9	5.2	Kirk et al. 1971a
Aborigines/Walmadjari (Kimberley, W. Australia)	.977	.023		171	4.5	2.2	Blake & Spargo 1986
MELANESIA							

Continued

117

Table 64. Phosphogluconate dehydrogenase: PGD (cont'd)

Place/Population	A	C	x	n	H	PE	Source
W. New Guinea							
Asmat (S. Coastal Plain)	.784	.216		654	33.9	14.1	Gajdusek et al. 1978
Papua New Guinea							
Gainj & Kalam (N. Central							
Highlands)	.941	.057	.003^W	571	11.1	5.4	Long et al. 1986
Jimi Valley (W. Highlands)	.875	.125		385	21.9	9.7	Mourant et al. 1981
Yagaria (E. Highlands)	.889	.111		423	19.7	8.9	Mourant et al. 1982
Fuyuge (C. Highlands)	.858	.142		173	24.4	10.7	Woodfield et al. 1974
Daga (Bay Province)	.745	.255		139	38.0	15.4	Jenkins et al. 1983
Karkar Is.							
Takia	.933	.067		290	12.5	5.9	Booth et al. 1982
Manus Is.	.815	.185		138	30.2	12.8	Malcolm et al. 1972
Buka Is.	.869	.131		80	22.8	10.1	McLoughlin et al. 1982a
Solomon Is.	.799	.201		364	32.1	13.5	Blake et al. 1983
MICRONESIA							
W. Caroline Is.	.938	.062		382	11.6	5.5	Blake et al. 1973b
E. Caroline Is.	.974	.026		1,005	5.1	2.5	Yamamoto & Fu 1973
Marshall Is.	.962	.038		372	7.3	3.5	Neel et al. 1976
POLYNESIA							
Samoans (Christchurch,							
New Zealand)	.877	.123		77	21.6	9.6	Booth et al. 1977

Table 65. Phosphoglycerate kinase: PGK

Place/Population	1	2	x	n	H	Source
EUROPE						
NETHERLANDS (Leiden)	1.000			806		Fraser et al. 1974
ASIA						
IRAN						
Turkoman (Gonbad, NE Iran)	1.000			155		Kirk et al. 1977
INDIA						
Indians (N. India)	1.000			490		Blake et al. 1971a
Irula tr. (Nilgiri Hills,						
S. India)	1.000			173		Saha et al. 1976
NEPAL	1.000			58		Chen & Giblett 1972
MALAYSIA						
Malays (Singapore)	1.000			259		Blake et al. 1973a
INDONESIA						
Batak (Samosir Is.)	1.000			188		McDermid et al. 1973
Balinese (Bali Is.)	1.000			148		Breguet et al. 1982b
PHILIPPINES						
Filipino/Tagalog (Manila)	1.000			253		Omoto et al. 1978
Negrito (Angeles, C. Luzon)	1.000			129		"
CHINA						
Chinese (C. Taiwan)	1.000			160		Chen et al. 1973
Chinese (Singapore)	1.000			378		Blake et al. 1973a
JAPAN						
Japanese (Mie)	1.000			876		Ishimoto 1975
AFRICA						
BURUNDI						
Hutu	.998		.002^6	364	.4	Gall et al. 1982

Continued

Table 65. Phosphoglycerate kinase: PGK (cont'd)

Place/Population	1	2	x	n	H	Source
KENYA	1.000			184		Chen & Giblett 1972
MOZAMBIQUE	1.000			94		"
SOUTHERN AFRICA						
San/Bushmen						
!Kung, Tsumkwe	1.000			65		Nurse & Jenkins 1977

NORTH AMERICA

Place/Population	1	2	x	n	H	Source
ALASKA						
Eskimos (St. Lawrence Is.)	1.000			222		Ferrell et al. 1981b
CANADA						
Eskimos (Igloolik)	1.000			217		McAlpine et al. 1974
Cree Indians (Manitoba)	1.000			200		Lucciola et al. 1974
Ojibwa Indians (N. Ontario)	1.000			93		Szathmary et al. 1974
Dogrib Indians (NW Territories)	1.000			158		Szathmary et al. 1983

SOUTH AMERICA

Place/Population	1	2	x	n	H	Source
BOLIVIA						
Aymara (Western)	1.000			320		Ferrell et al. 1981a
ECUADOR						
Waorani	1.000			187		Larrick et al. 1985
PERU						
Amerindians	1.000			142		Chen & Giblett 1972

OCEANIA

Place/Population	1	2	x	n	H	Source
AUSTRALIA						
Aborigines/Malag (Elcho Is., N. Territory)	.998	.002		234		Omoto & Blake 1972
Aborigines (Hooker Creek, N. Territory)	1.000			285		Blake 1979
Aborigines (Areyonga, C. Australia)	1.000			50		Omoto & Blake 1972
Aborigines/Konejandi (Kimberley, W. Australia)	1.000			102		Blake & Spargo 1986
MELANESIA						
W. New Guinea						
Asmat (S. Coastal Plain)	1.000			408		Gajdusek et al. 1978
Papua New Guinea						
Gainj & Kalam (N. Central Highlands)	.996	.004[4]		551	.8	Long et al. 1986
Western Highlands	.949	.051[4]		214	9.1	Omoto & Blake 1972
Eastern Highlands	.963	.037[4]		81	7.1	"
Fuyuge (C. Highlands)	1.000			173		Woodfield et al. 1974
Karkar Is.	1.000			170		Omoto & Blake 1972
Manus Is.	1.000			183		Malcolm et al. 1972
Buka Is.	1.000			78		McLoughlin et al. 1982a
Solomon Is.	.901	.099		293	17.8	Blake et al. 1983
MICRONESIA						
W. Caroline Is.	.918	.082		380	15.1	Blake et al. 1973b
Ulthi Atoll	.922	.078		385	14.1	Omoto & Blake 1972
POLYNESIA						
Samoans (New Zealand)	.964	.036		77	6.9	Booth et al. 1977

119

Table 66. Phosphoglycolate phosphatase: PGP

Place/Population	1	2	3	n	H	PE	Source
EUROPE							
Europeans	.826	.129	.045	656	29.9	14.9	Barker & Hopkinson 1978
GERMANY							
Germans (Schleswig-Holstein)	.842	.117	.041	886	27.6	13.7	Christiansen & Sachs 1981
HUNGARY	.851	.105	.044	124	26.3	13.2	Goedde et al. 1986b
ITALY							
Italians (Rome)	.923	.056	.021	501	14.4	7.2	Santolamazza et al. 1986
Italians (Sardinia)	.987	.013		303	2.6	1.3	"
SPAIN							
Spanish (Galicia)	.927	.047	.026	500	13.8	7.0	Caeiro & Rey 1985
SWITZERLAND	.889	.089	.022	900	20.1	9.9	Scherz et al. 1981b
U.S.S.R.							
EUROPEAN PART							
Abkhazia (Ochamchir Dist.)	.939	.061		256	11.5	5.4	Ferrell et al. 1985
ASIA							
ISRAEL							
Ashkenazi Jews	.949	.041	.010	255	9.8	4.8	Golan et al. 1981
Yemenite Jews	.972	.019	.009	108	5.5	2.7	"
Arabs	.937	.049	.014	213	11.9	5.9	"
IRAN	.928	.065	.007	200	13.5	6.5	Blake & Hayes 1980
INDIA							
Indians (Bombay)	.954	.046		120	8.8	4.2	"
Indians (Malaysia)	.965	.035		116	6.8	3.3	"
MALAYSIA							
Malays (Singapore)	.995	.005		109	1.0	.5	"
INDONESIA							
Sumatra	1.000			100			"
Bali	1.000			100			"
THAILAND	1.000			110			Goedde & Benkmann 1987
CHINA							
Chinese (Malaysia)	1.000			121			Zarinah et al. 1984
KOREA (Seoul)	1.000			148			Goedde et al. 1986c
AFRICA							
CENTRAL AFRICAN REPUBLIC							
Pygmies (Western)	1.000			34			Meera Khan 1986
SOUTHERN AFRICA							
Bantu	1.000			100			Blake & Hayes 1980
NORTH AMERICA							
ALASKA							
Eskimos (St. Lawrence Is.)	.990	.010		295	2.0	1.0	Ferrell et al. 1981b
CANADA							
Dogrib Indians (NW Territories)	.658	.342		158	45.0	17.4	Szathmary et al. 1983
UNITED STATES							
Pima Indians (Arizona)	.690	.310		205	42.8	16.8	Blake & Hayes 1980
Mexican-Americans (Texas)	.871	.124	.005	667	22.6	10.3	Hewett-Emmett et al. 1986b
SOUTH AMERICA							
BRAZIL							
Amazon	.841	.159		129	26.7	11.6	Blake & Hayes 1980
COLOMBIA							
Noanama	.672	.299	.029	154	45.8	20.2	"
Cofan	.667	.333		66	44.4	17.3	"
OCEANIA							
AUSTRALIA							
Aborigines (C. Australia)	1.000			120			"

Continued

Table 66. Phosphoglycolate phosphatase: PGP (cont'd)

Place/Population	1	2	3	n	H	PE	Source
Aborigines (Kimberley, WA)	1.000			200			Blake & Hayes 1980
MELANESIA							
Papua New Guinea							
Eastern Highlands	1.000			160			"
Port Moresby	1.000			100			"
Karkar Is.	1.000			100			"
MICRONESIA							
W. Caroline Is.	1.000			98			"
POLYNESIA							
Samoa Is. (Western)	.996	.004		130	.8	.4	"
Cook Is.	1.000			100			"

Table 67. Pyridoxine kinase (activity): PNK

Place/Population	H	L	n	H	PE	Source

NORTH AMERICA

UNITED STATES						
Whites	.81	.19	51	30.8	13.0	Chern & Beutler 1976
Blacks	.35	.65	52	45.5	17.6	"

Note: Alleles H and L refer to high and low activity respectively.

Table 68. Pyruvate kinase 3: PKM2, PK3

Place/Population	1	2	x	n	H	PE	Source

EUROPE

GERMANY	.993	.007		214	1.4	.7	Blume et al. 1968
POLAND	.980	.010	.010	78	3.9	2.0	Dobosz 1983
SPAIN							
Spanish (Madrid)	.999	.001		1,636	.2	.1	Garcia et al. 1979

ASIA

SAUDI ARABIA							
Arabs (Riyadh, Eastern)	.940	.060		784	11.3	5.3	el-Hazmi et al. 1986
Arabs (Khaiber, Western)	.934	.066		457	12.3	5.8	"
JAPAN							
Japanese	1.000			230[a]			Nakashima 1974

NORTH AMERICA

CANADA							
Eskimos (Igloolik)	1.000			198			McAlpine et al. 1974

SOUTH AMERICA

BOLIVIA							
Aymara (western)	1.000			320			Ferrell et al. 1981a

Note: [a] One variant allele was observed in a patient with acute hepatitis.

121

Table 69. Rhodanese: RDS

Place/ Population	1	2	n	H	PE	Source
NORTH AMERICA						
ALASKA						
Eskimos-Yupik	1.000		480			Scott & Wright 1980
Athabaskans-Ahtna	.821	.179	N.A.	29.4	12.5	"
Athabaskans-Upper Tanana	.870	.130	N.A.	22.6	10.0	"
Athabaskans-Others	1.000		257			"
Whites	1.000		237			"
Asians	1.000		50			"

Table 70. S-adenosylhomocysteine hydrolase: SAHH

Place/Population	1	2	3	n	H	PE	Source
EUROPE							
GERMANY (SW Germany)	.961	.039		114	7.5	3.6	Bissbort et al. 1983
ITALY (Rome)	.968	.024	.008	248	6.2	3.1	Corbo 1986
ASIA							
JAPAN	.953	.047		214	9.0	4.3	Akiyama et al. 1984

Table 71. Superoxide dismutase 1; indophenol oxidase; tetrazolium oxidase: SOD1

Place/Population	1	2	n	H	PE	Source
EUROPE						
FAROE IS. (Atlantic)	1.000		666			Tills et al. 1985
FINLAND						
Finns	.990	.010	406	2.0	1.0	Eriksson 1974
Lapps/Saami	.998	.002	949	.4	.2	"
FRANCE						
French	1.000		2,012			Vergnes et al. 1980a
Basques	1.000		144			"
GERMANY (Tübingen/Marburg)	.999	.001	4,100	.2	.1	Ritter & Wendt 1971b
GREECE						
Greeks	1.000		229			Schneider et al. 1975
Greeks (Plati)	1.000		1,027			Tills et al. 1983b
ICELAND	1.000		350			Eriksson 1974
IRELAND (Southern)	1.000		260			Welch & Mears 1972
ITALY	.999	.001	738	.2	.1	Crosti et al. 1976
NETHERLANDS (Leiden)	1.000		806			Fraser et al. 1974
PORTUGAL	1.000		572			Weissmann et al. 1982(c)
SWEDEN						
Swedes	.990	.010	2,045	2.0	1.0	Beckman 1973
Lapps/Saami	.998	.002	210	.4	.2	"
UNITED KINGDOM						
England	.9997	.0003	11,237	.2	5.0	Harris et al. 1974
Isle of Lewis	1.000					Clegg et al. 1985
Orkney Is. (Westray)	.985	.015	94	3.0	1.5	Welch et al. 1973

Continued

Table 71. Superoxide dismutase 1; indophenol oxidase; tetrazolium oxidase: SOD1 (cont'd)

Place/Population	1	2	n	H	PE	Source
ASIA						
IRAQ	.997	.003	320	.6	.3	Papiha & Al-Agidi 1976
SAUDI ARABIA						
Tribes (Western)	1.000		149			Saha et al. 1980
IRAN						
Turkoman (Gonbad, NE Iran)	.987	.013	155	2.6	1.3	Kirk et al. 1977
AFGHANISTAN	1.000		280			Papiha et al. 1977a
INDIA						
Indians (Delhi)	1.000		493			Blake et al. 1971a
Indians (W. India)	1.000		648			Blake et al. 1970
Bhil tr. (Madhya Pradesh)	1.000		145			Papiha et al. 1978
Hindu (Andhra Pradesh)	1.000		211			Roberts et al. 1980
Irula tr. (Nilgiri Hills,						
S. India)	1.000		175			Saha et al. 1976
Bengali (E. India)	1.000		269			Das et al. 1970
BANGLADESH						
Moslems (Dacca)	1.000		200			Papiha et al. 1975
NEPAL	1.000		15			Santachiara et al. 1976
MALAYSIA						
Malays (Singapore)	1.000		259			Blake et al. 1973a
Senoi-Aborigines	1.000		123			Tan & Teng 1978
INDONESIA						
Balinese (Bali Is.)	1.000		148			Breguet et al. 1982b
PHILIPPINES						
Filipino	.997	.003[F]	146	.6	.3	Teng & Lie-Injo 1977
CHINA						
Tibetans	1.000		8			Santachiara et al. 1976
Chinese (C. Taiwan)	1.000		160			Chen et al. 1973
Chinese (Singapore)	1.000		378			Blake et al. 1973a
JAPAN						
Japanese (Nagoya City)	1.000		586			Shinoda & Matsunaga 1970
AFRICA						
SUDAN						
Sudanese (Khartoum)	1.000		174			Saha et al. 1978b
CENTRAL AFRICAN REPUBLIC						
Mbugu	.995	.005	92	1.0	.5	Spedini et al. 1982
ETHIOPIA	1.000		160			Welch & Mears 1972
SOUTHERN AFRICA						
Bantu (Natal)	1.000		304			Kirk et al. 1971b
NORTH AMERICA						
GREENLAND						
Eskimos	1.000		152			Weissmann et al. 1982(c)
ALASKA						
Eskimos (St. Lawrence Is.)	1.000		222			Ferrell et al. 1981b
CANADA						
Eskimos (Igloolik)	1.000		359			McAlpine et al. 1974
Dogrib Indians (NW Territories)	1.000		158			Szathmary et al. 1983
MIDDLE AMERICA						
WEST INDIES (Haiti)	1.000		307			Basu et al. 1976
SOUTH AMERICA						
BRAZIL						
Cayapo	1.000		119			Salzano et al. 1972b
COLOMBIA						
Noanama	1.000		170			Kirk et al. 1974
ECUADOR						
Waorani	1.000		187			Larrick et al. 1985
FRENCH GUIANA						
Wayampi (Trois-Sauts & Camopi)	1.000		219			Tchen et al. 1978

Continued

Table 71. Superoxide dismutase 1; indophenol oxidase; tetrazolium oxidase: SOD1 (cont'd)

Place/Population	1	2	n	H	PE	Source
VENEZUELA						
Makiritare (S. Venezuela &						
N. Brazil)	1.000		535			Weitkamp & Neel 1970
Yanomama (")	1.000		2,352			Weitkamp & Neel 1972

OCEANIA

AUSTRALIA						
Aborigines (Mainorou, N.						
Territory)	1.000		51			Kirk et al. 1971a
Aborigines (Hooker Creek, N.						
Territory)	1.000		286			Blake 1979
Aborigines (Walmadjari, Kimber-						
ley, W. Australia)	1.000		171			Blake & Spargo 1986
MELANESIA						
W. New Guinea						
Asmat (S. Coastal Plain)	1.000		653			Gajdusek et al. 1978
Papua New Guinea						
Gainj & Kalam (N. Central						
Highlands)	1.000		588			Long et al. 1986
Fuyuge (C. Highlands)	1.000		173			Woodfield et al. 1974
Manus Is.	1.000		183			Malcolm et al. 1972
Buka Is.	1.000		80			McLoughlin et al. 1982a
Solomon Is.	1.000		77			Blake et al. 1983
MICRONESIA						
W. Caroline Is.	1.000		382			Blake et al. 1973b
POLYNESIA						
Samoans (Christchurch,						
New Zealand)	1.000		77			Booth et al. 1977

Table 72. Superoxide dismutase-B, salivary: SODB

Place/Population	1	2	n	H	PE	Source
ASIA						
INDIA						
Indians (Malaysia)	.992	.008	124	1.6	.8	Tan & Teng 1978
MALAYSIA						
Malays (Kuala Lumpur)	.998	.002	288	.4	.2	"
Senoi (Gombak)	.988	.012	123	2.4	1.2	"
CHINA						
Chinese (Malaysia)	1.000		140			"

Table 73. Triosephosphate isomerase: TPI

| Place/Population | Variants | | n | Source |
	Genotype	Number		
EUROPE				
Europeans	+/2	1	1,705	Peters et al. 1973
	+/3	1		
ASIA				
INDIA			183	"
JAPAN (Hiroshima & Nagasaki)			10,864[a]	Asakawa et al. 1984
AFRICA				
NIGERIA (Ibadan)			417	Peters et al. 1973
MIDDLE AMERICA				
COSTA RICA				
Guaymi			286	Barrantes et al. 1982
SOUTH AMERICA				
BRAZIL/GUYANA				
Baniwa (Jandu Cachoeira)			363	Salzano et al. 1986
Macushi (N. Brazil & S. Guyana)			499	Neel et al. 1977b
Wapishana (")			615	"
Ticuna (C. Amazonas)			1,765	Neel et al. 1980
VENEZUELA				
Makiritare (S. Venezuela & N. Brazil)			390	Neel 1978
Yanomama (")			343	"
OCEANIA				
MELANESIA				
Papua New Guinea				
Gainj & Kalam (N. Central Highlands)			277	Long et al. 1986
MICRONESIA				
Marshall Is.			370	Neel et al. 1976

Note: [a] Four kinds of new variants were observed.

Table 74. Uridine monophosphate kinase: UMPK

Place/Population	1	2	3	n	H	PE	Source
EUROPE							
GERMANY	.949	.051		351	9.7	4.6	Kuhn et al. 1975
ITALY (Rome)	.972	.028		486	5.4	2.6	Ranzani et al. 1977
ASIA							
INDIA							
Indians (Malaysia)	.942	.058		121	10.9	5.2	Zarinah et al. 1984
SRI LANKA							
Sinhalese	1.000			112			Giblett et al. 1975
MALAYSIA							
Malays (Kuala Lumpur)	.851	.149		168	25.4	11.1	Zarinah et al. 1984
PHILIPPINES							
Negrito (Angeles, C. Luzon)	.655	.345		129	45.2	17.5	Omoto et al. 1978

Continued

125

Table 74. Uridine monophosphate kinase: UMPK (cont'd)

Place/Population	1	2	3	n	H	PE	Source
CHINA							
Chinese (Malyasia)	.880	.120		125	21.1	9.4	Zarinah et al. 1984
JAPAN							
Japanese (Tokyo)	.947	.053		635	10.0	4.8	Harada et al. 1975

AFRICA

KENYA & SOUTH AFRICA							
Negroes	1.000			203			Giblett et al. 1975

NORTH AMERICA

ALASKA							
Eskimos-Inupiat (Northwestern)	.778	.022	.200	228	35.4	16.1	Scott & Wright 1978
Athabaskan Indians	.862	.007	.131	213	24.0	10.9	"
CANADA							
Cree Indians	.846	.040	.114	136	27.0	13.4	Lucciola et al. 1974
UNITED STATES							
Whites (Seattle)	.953	.045	.001	390	9.0	4.2	Giblett et al. 1974
Blacks (")	.989	.011		174	2.2	1.1	"

SOUTH AMERICA

VENEZUELA							
Mestizo	.979	.020	.001	442	4.1	2.0	Gallango & Suinaga 1978

Table 75. Enzymes with no variants or rare variants

Enzyme: Symbol	Populations studied (n) & remarks	Source
Acetylesterase: ESA	Europeans (2600): One individual with a variant allele was found.	Tashian 1969
	US Blacks (600): One individual with a variant allele was found.	"
	North American Indians (366)	"
	Micronesians (490)	"
Aconitase 2, mitochondrial: ACO2	Europeans (359)	Slaughter et al. 1975
	Nigerians (545): Three individuals of genotype (+/2) were found.	"
	Germans in Tubingen/Stuttgart (400)	Schmitt & Ritter 1974
	Caucasians (227), Chinese in Malaysia (86),	Blake 1984
	Australian Aborigines (101),	"
	Papua New Guineans in Port Moresby (184)	"
Adenine phosphoribosyl-transferase: APRT	Caucasians (685)	Mowbray et al. 1972
	Nigerians (601): Five indivduals of genotype (+/2) were found.	"
Adenylate kinase 2, mitochondrial: AK2	Europeans (145)	Wilson et al. 1976
Adenylate kinase 3, mitochondrial: AK3	Europeans (50)	"
Alcohol dehydrogenase I, alpha polypeptide: ADH1	English (56)	Smith et al. 1971
	Bahia Indians in Brazil (300)	Azevedo et al. 1975

Continued

Table 75. Enzymes with no variants or rare variants (cont'd)

Enzyme: Symbol	Populations studied (n) & remarks	Source
Aldehyde dehydrogenase 1, cytosolic: ALDH1	Northern Europeans (7,200)	Hopkinson et al. 1985
Aldehyde dehydrogenase 4: ALDH4	Northern Europeans (7,200)	"
Aldolase A: ALDOA	Eskimos in Igloolik, Canada (198)	McAlpine et al. 1974
	US Whites (878): Two individuals with a variant allele were found.	Charlesworth 1972
	US Blacks (875)	"
	Aymara, W. Bolivia (320)	Ferrell et al. 1981a
	Micronesians in Marshall Is. (366)	Neel et al. 1976
Arginase, liver: ARG1	Northern Europeans (237)	Nelson et al. 1977b
Argininosuccinate lyase: ASL	Northern Europeans (100)	"
Arylsulfatase A: ARSA	Israelis (250): Seven individuals with enzyme deficiency were found.	Herz & Bach 1984
	Habbanite Jews (75): One individual with enzyme deficiency was found.	Zlotogora et al. 1980
D-amino acid oxidase: DAMOX	Northern Europeans (163)	Barker & Hopkinson 1977
	Burmese (20)	
D-aspartate oxidase: DASOX	Northern Europeans (1963): One indivdual with a variant allele was found.	"
	Burmese (20)	"
Enolase 1; phosphopyruvate hydratase: ENO1, PPH	Eskimos in Igloolik, Canada (200)	McAlpine et al. 1974
	Cree Indians in Canada (199): Four individuals of genotype (+/2) were found.	Lucciola et al. 1974
Esterase B, erythrocyte: ESB	Gainj & Kalam in Papua New Guinea (273)	Long et al. 1986
Formaldehyde dehydrogenase: FDH	Australian Caucasians (127)	Castle & Board 1982
	Indians (42)	"
	Chinese (74)	"
Fumarate hydratase, soluble: FH1	Europeans (776): One individual of genotype (+/2) was found.	Edwards & Hopkinson 1979
	US Blacks (1,061)	"
	Nigerians (100)	"
	Ethiopians (955)	"
	Japanese (586): One variant found.	Shinoda & Matsunaga 1970
	Australian Aborigines (101)	Blake 1984
	Papua New Guineans in E. Highlands (70), and in Port Moresby (184)	"
Fumarate hydratase, mitochondrial: FH2	Australian Aborigines (101)	Blake 1984
	Papua New Guineans in E. Highlands (70), and in Port Moresby (184)	"
Galactokinase: GALK	Italians (620): Three individuals with enzyme deficiency were found.	Magnani et al. 1982
	US Whites (642): Six individuals with enzyme deficiency were found.	Mayes & Guthrie, 1968
Galactosidase, alpha: GLA	US Whites (197)	Beutler & Kuhl 1972
	US Blacks (14), US Orientals (5)	"
Gamma-glutamyl-cyclo-transferase: GCTG	Caucasians, Australia (200)	Board 1980a
Glutamate dehydrogenase: GLUD	Caucasians (200)	Nelson et al. 1977a
Glutathione-S-transferase-3: GST3	Indians in Malaysia (43)	Board 1981
	Chinese in Malaysia (96)	"

Continued

127

Table 75. Enzymes with no variants or rare variants (cont'd)

Enzyme: Symbol	Populations studied (n) & remarks	Source
Glyceraldehyde-3-phosphate dehydrogenase: GAPD	Dutch in Leiden (806): No variants found. Eskimos in Igloolik, Canada (200) Dogrib Indians in NW Territories, Canada, (158): No variants found. US Whites (1,061): One indivdual with a variant was found. Aymara, W. Bolivia (320)	Fraser et al. 1974 McAlpine et al. 1974 Szathmary et al. 1983 Charlesworth 1972 Ferrell et al. 1981a
Glycerol-3-phosphate dehydrogenase A: GPD1	Northern Europeans (707): One individual of genotype (+/2) was found. Eskimos in St. Lawrence Is., Alaska (222)	Hopkinson et al. 1974b Ferrell et al. 1981b
Glycerol-3-phosphate dehydrogenase B: GPD2	Northern Europeans (707): Two individuals of genotype (+/2) were found.	Hopkinson et al. 1974b
Glycolate oxidase: GOX	Europeans (96)	Harris & Hopkinson 1978(c)
Guanine deaminase: GDA	Europeans & Africans (1,000)	Harris & Hopkinson 1978(c)
Guanylate kinase 1: GUK1	Europeans (1,152) Indians (189) Individuals, USA (385)	Jamil et al. 1975 " Monn & Christiansen 1972
Hexokinase I : HK1	Gainj & Kalam in Papua New Guinea (275)	Long et al. 1986
Hexokinase II: HK2	Gainj & Kalam in Papua New Guinea (273)	"
Hexosaminidase A; beta-N-acetyl-glucosaminidase A: HEXA	Am. Jews by region of origin & heterozygote frequencies. Poland & Russia (24,572) - .033 Middle Europe (2,928) - .048 W. Europe (148) - .020 Mediterranean (108) - .037	Petersen et al. 1983
Hexosaminidase B; beta-N-acetyl-glucosaminidase B: HEXB	See Note.[a]	Cantor & Kaback 1985
Hypoxanthine phosphoribosyl transferase: HPRT	US Whites (325) US Blacks (54)	Kaloustian et al. 1969 "
Inosine triphophatase A: ITPA	Europeans, Africans & Indians studied; no variants detected.	Harris & Hopkinson 1978(c)
Isocitrate dehydrogenase, mitochondrial: IDH2	Caucasians (204), Kuala Lumpur, Chinese (86), Australian Aborigines (101) Papua New Guineans in Port Moresby (184)	Blake 1984 " "
Leucine aminopeptidase, placental: LAPP	Swedes in Uppsala (700)	Beckman et al. 1966
Malate dehydrogenase, mitochondrial: MDH2	Swedes (1,754): One individual of genotype (+/3) was found. Chinese (67) Japanese (348): One individual of genotype (+/4) was found. Filipinos (83) US Whites (523): Five individuals with a variant allele were found. Chinese in Malaysia (86): Two individuals of genotype (+/3) were found. Australian Aborigines (100), Papua New Guineans in E. Highlands (70) and in Port Moresby (191). Polynesians (23)	Beckman & Christodoulou 1974 " " " Davidson & Cortner 1967 Blake 1984 " " Beckman & Christodoulou 1974
Malic enzyme 1, cytoplasmic: ME1	Chinese (161), Malays (113), and Indians (150) living in Singapore: no variant found.	Battacharyya & Saha 1984a

Continued

128

Table 75. Enzymes with no variants or rare variants (cont'd)

Enzyme: Symbol	Populations studied (n) & remarks	Source
	US Whites/Blacks (154): One individual with a variant allele was found.	Cohen & Omenn 1972
Mannose phosphate isomerase: MPI	Germans in Tubingen/Stuttgart (400): Five individuals of genotype (+/2) were found.	Ritter et al. 1974
	Caucasians (227): Five individuals of genotype (+/2) were found.	Blake 1984
	Chinese (86) in Malaysia: One individual of genotype (+/3) and another individual of genotype (+/4) were found.	"
	Australian Aborigines (101)	"
	Papua New Guineans in Port Moresby (184): Six individuals of genotype (+/3) were found.	"
Ornithine transcarbamylase: OTC	Europeans (162)	Harris & Hopkinson 1978(c)
	Ticuna Indians in Brazil (1,760)	Neel et al. 1980
Peptidase E: PEPE	Peruvian Indians (98)	Modiano et al. 1972
Peroxidase, leukocyte: PXL*	US Whites (45)	Azen 1977
Phosphofructokinase, liver: PFKL	Eskimos in Igloolik, Canada (194)	McAlpine et al. 1974
	Ojibwa Indians in Canada (93)	Szathmary et al. 1974
	US Whites (195)	Niessner & Beutler 1974
	Aymara, W. Bolivia (320)	Ferrell et al. 1981a
Phosphoglycerate mutase A phosphoglyceric acid mutase: PGAMA	Greeks (98), Nepalese (84)	Chen et al. 1974
	Filipinos (88), East Africans (380),	"
	Congolese (292), Eskimos (348),	"
	Canadian Indians (165),	"
	Eskimos in Igloolik, Canada (200)	McAlpine et al. 1974
	Cree Indians in Canada (200)	Lucciola et al. 1974
	US Whites (530), US Blacks (273),	Chen et al. 1974
	US Orientals (232), Peruvian Indians (44),	"
	New Guineans (290)	"
Phosphoserine phosphatase: PSP	Europeans (378): Four individuals of genotype (+/2) and one individual of genotype (+/3) were found.	Moro-Furlani et al. 1980
Pyrophosphatase, inorganic: PP	Europeans (970)	Fisher et al. 1974
	Negroes (69)	"
	Asian Indians (104)	"
	Ticuna, C. Amazonas, Brazil (1,760)	Neel et al. 1980
	Caucasians (227), Chinese in Malaysia (86),	Blake 1984
	Australian Aborigines (101), Papua New Guineans in Port Moresby (184).	"
Sorbitol dehydrogenase: SORD	US Whites (586): One individual with a variant allele was found.	Charlesworth 1972
	US Blacks (79)	"
Succinate dehydrogenase: SDH	Japanese in Nagoya city, Japan (200)	Shinoda & Matsunaga 1970
Superoxide dismutase A: SODA (salivary)	Malays in Malaysia (287)	Tan & Teng 1978
	Senoi " (132)	"
	Chinese " (140)	"
	Indians " (122)	"

Note: + stands for the wild-type allele. [a] The deficiency of this enzyme is apparently responsible for Sandhoff disease. The frequency of the deficiency allele has been estimated to be 1/1000 in the Jews and 1/600 in Non-Jews in the United States.

5   Proteins

Table 76.   Albumin: ALB

Place/Population	Variants		n	Source
	Name	No. of heterozygotes		

EUROPE

FINLAND
  Finns ... 2,682 ... Melartin 1967
  Lapps/Saami ... 62 ... Laurell & Niléhn 1966
FRANCE ... B ... 7 ... 10,000 ... Weitkamp 1974(c)
GREECE ... 606 ... Parisi et al. 1980
ITALY ... 5 ... 12,000 ... Bonazzi 1968
NETHERLANDS (Leiden) ... 806 ... Fraser et al. 1974
NORWAY ... 950 ... Efremov & Braend 1964
POLAND ... 500 ... Shrivastava 1969
SWEDEN ... 6 ... 4,750 ... Laurell & Niléhn 1966
UNITED KINGDOM
  England ... 12,000 ... Cooke et al. 1961

Place/Population	Name	No. of heterozygotes	n	Source
EUROPE				
FINLAND				
Finns			2,682	Melartin 1967
Lapps/Saami			62	Laurell & Niléhn 1966
FRANCE	B	7	10,000	Weitkamp 1974(c)
GREECE			606	Parisi et al. 1980
ITALY		5	12,000	Bonazzi 1968
NETHERLANDS (Leiden)			806	Fraser et al. 1974
NORWAY			950	Efremov & Braend 1964
POLAND			500	Shrivastava 1969
SWEDEN		6	4,750	Laurell & Niléhn 1966
UNITED KINGDOM				
England			12,000	Cooke et al. 1961
U.S.S.R.				
EUROPEAN PART				
Russians (Moscow)			1916	Altukhov et al. 1981
Abkhazia (Ochamchir Dist.)			293	Ferrell et al. 1985
ASIA				
JORDAN			537	Banerjee et al. 1981
SAUDI ARABIA				
Tribes (Western)			191	Saha et al. 1980
IRAN				
Turkoman (Gonbad, NE Iran)			155	Kirk et al. 1977
AFGHANISTAN				
Pushtoon (Gawargin)	Afghanistan	10	251	Weitkamp & Buck 1972
	Pushtoon	7		
INDIA				
Hindu (New Delhi)	Gainsville	2	518	McDermid 1971a
Hindu (Andhra Pradesh)			86	Naidu et al. 1985
Irula tr. (Nilgiri Hills,				
S. India)			171	Saha et al. 1976
MALAYSIA				
Malays (Kuala Lumpur)	Gombak	1	410	Welch & Lie-Injo 1972
	Medan	1		
Aborigines	Gombak	2	165	Lie-Injo et al. 1971
INDONESIA				
Batak (Samosir Is.)			168	McDermid et al. 1973
CHINA				
Chinese (Malaysia)	Kuala Lumpur	1	392	Welch & Lie-Injo 1972
JAPAN				
Japanese	Nagasaki-1	10	4,029	Ferrell et al. 1977
Ainu			113	Weitkamp 1974
AFRICA				
EGYPT				
Nubians (S. Egypt)			155	Bertin et al. 1978
SUDAN				
Sudanese (Khartoum)			284	Saha et al. 1978b
SOUTHERN AFRICA				
Bantu (Natal)			304	McDermid & Vos 1971a
NORTH AMERICA				

Continued

Table 76. Albumin: ALB (cont'd)

Place/Population	Variants Name	Variants No. of heterozygotes	n	Source
GREENLAND				
Eskimos			491	Persson et al. 1971
ALASKA				
Eskimos			437	Melartin 1967
Eskimos (St. Lawrence Is.)			222	Ferrell et al. 1981b
Athabaskan Indians	Naskapi	11	185	Melartin 1967
CANADA				
Eskimos (Igloolik)			356	McAlpine et al. 1974
Cree Indians	Naskapi	37	610	Weitkamp 1974
Ojibwa " (Pikangikum)	Naskapi	1	92	Szathmary et al. 1974
Dogrib " (NW Territory)	Naskapi	2	158	Szathmary et al. 1983
UNITED STATES				
Blackfeet Indians (Montana)	Naskapi	6	180	Rokala et al. 1977
Papago " (Arizona)	Mexico	2	179	Melartin 1967
Navajo " (California)	Naskapi	30	468	Johnston et al. 1969a
	Mexico	6		
Papago " (")	"	3	115	"
Zuni " (New Mexico)	"	16	655	Workman et al. 1974

MIDDLE AMERICA

MEXICO				
Indians (San Pablo del Monte)	"	1	258	Crawford et al. 1974
Mestizo (Tlaxcala)	"	4	138	"
GUATEMALA				
Black Caribs (Livingston)	Mexico	2	205	Crawford et al. 1981
Am. Indians	"	4	386	Johnston et al. 1973
COSTA RICA				
Guaymi			286	Barrantes et al. 1982
PANAMA				
Blacks			781	Ferrell et al. 1978b
WEST INDIES (Haiti)	Cayemite	2	308	Basu et al. 1976

SOUTH AMERICA

ARGENTINA				
Mapuche			71	Haas et al. 1985
BOLIVIA				
Aymara (Western)	Aymara	4	305	Ferrell et al. 1978a
BRAZIL				
Baniwa (Jandu Cachoeira)			377	Salzano et al. 1986
Cayapo (Para & Mato Grosso)			708	Salzano et al. 1972b
Caingang			214	Salzano et al. 1980
Parakana			129	Black et al. 1980
Pacaás Novos			222	Salzano et al. 1985
Macushi (N. Brazil & S. Guyana)			509	Neel et al. 1977b
Wapishana (")	Maku	23[a]	623	"
Ticuna (C. Amazonas)			761	Neel et al. 1980
COLOMBIA				
Noanama			155	Kirk et al. 1974
FRENCH GUIANA				
Wayampi (Trois-Sauts & Camopi)			226	Tchen et al. 1978
SURINAM				
Trio (S. Surinam)	Makiritare-1	11	413	Geerdink et al. 1974b
VENEZUELA				
Makiritare (S. Venezuela &				
N. Brazil)	Makiritare	13	534	Arends et al. 1970
Yanomama (")	Maku	4	2,261	Weitkamp et al. 1972
	Yanomama	1		

OCEANIA

AUSTRALIA				
Aborigines (Elcho Is.)			130	McDermid 1971b
MELANESIA				
Papua New Guinea				

Continued

Table 76. Albumin: ALB (cont'd)

Place/Population	Variants		n	Source
	Name	No. of heterozygotes		
New Guineans		1	595	McDermid 1971b
Gainj & Kalam (N. Central Highlands)			594	Long et al. 1986
Fuyuge (C. Highlands)			145	Woodfield et al. 1974
Manus Is.			183	Malcolm et al. 1972
Solomon Is.	New Guinea	5	308	Blake et al. 1983
MICRONESIA				
Marshall Is.		1	372	Neel et al. 1976
POLYNESIA				
Samoans (Christchurch, New Zealand)			77	Booth et al. 1977

Note: [a] One individual was homozygous.

Table 77. Alpha-1-antitrypsin; protease inhibitor: PI

Place/Population	M	S	Z	F	I	V	Others	n	H	PE	Source
EUROPE											
DENMARK	.946	.022	.023	.006			.003	909	10.4	5.4	Thymann 1986
FINLAND											
Finns	.986	.005	.007	.001		.001	.001	1,037	2.8	1.5	Arvilommi 1972
Lapps/Saami	.996	.003	.001					468	.8	.4	Fagerhol et al. 1969
FRANCE	.902	.071	.014	.004	.004		.003	1,653	18.1	9.0	Arnaud et al. 1977
GERMANY	.969	.018	.004	.008				1,441	6.1	3.0	Schmechta & Geserick 1972
GREECE	.960	.029	.003	.006		.002		504	7.8	3.9	Fertakis et al. 1974
HUNGARY	.892	.017	.014	.070	.003			172	19.9	9.9	Kellermann & Walter 1970
ICELAND	.872	.117	.011					94	22.6	10.5	"
IRELAND	.936	.039	.020	.002	.003			1,000	12.2	6.2	Blundell et al. 1975
ITALY (Rome)	.952	.028	.010	.003	.006		.001[w]	513	9.3	4.8	Pascali & de Marcurio 1981
NETHERLANDS	.956	.030	.005	.006	.004			708	8.5	4.4	Klasen et al. 1977
NORWAY											
Norwegians	.946	.023	.016	.013	.001	.0004		2,830	10.4	5.3	Fagerhol 1968
Lapps/Saami	.992		.008					302	1.6	.8	Fagerhol et al. 1969
PORTUGAL											
Portuguese	.865	.115	.018		.002			330	23.8	11.4	Martin et al. 1976
Portugeuse (Terras de Bouro)	.929	.055	.004			.013		119	13.4	6.8	Cruz et al. 1973
SPAIN											
Spanish	.866	.112	.012	.003	.001	.005		378	23.7	11.4	Fagerhol & Tenfjord 1968
Spanish (Madrid)	.903	.097						170	17.5	8.0	Goedde et al. 1973
Basques (Bermeo/ Pamplona)	.877	.116	.007					146	21.7	10.0	"
SWEDEN											
Swedes (Northern)	.981	.010	.008	.0003				1,869	3.7	1.8	Beckman et al. 1980a
Lapps/Saami	.995	.005						217	1.0	.5	"
UNITED KINGDOM											
England	.930	.050	.014	.003	.002	.0006		4,565	13.2	6.6	Cook 1975
ASIA											
SAUDI ARABIA											
Shiites & Sunnites (al-Hasa)	.983	.010	.007					357	3.4	1.7	Goedde et al. 1979b
IRAN											

Continued

Table 77. Alpha-1-antitrypsin; protease inhibitor: PI (cont'd)

Place/Population	M	S	Z	F	I	V	Others	n	H	PE	Source
Turkoman (Gonbad, NE Iran)	1.000							155			Kirk et al. 1977
AFGHANISTAN											
Pushtoon (Kabul)	.964	.014	.019	.002				210	7.0	3.5	Rahimi et al. 1977
PAKISTAN	.953		.009	.038				53	9.0	4.5	Kellermann & Walter 1970
INDIA											
North Indians (Kumaon)	.994	.006						430	1.2	.6	"
Parsi (W. India)	1.000							363			Undevia et al. 1973
Bengali (E. India)	.990	.001	.001	.001		.007		982	2.0	1.0	Walter et al. 1972b
NEPAL	1.000							144			Yuasa et al. 1983b
MALAYSIA											
Malays (Kuala Lumpur)	.979	.015					.007	908	4.1	2.2	Lie-Injo et al. 1978
INDONESIA											
Batak (Samosir Is.)	.973	.027						166	5.3	2.6	McDermid et al. 1973
THAILAND	.963	.018	.011	.006	.001			852	7.2	3.6	Pongpaew & Schelp 1980
PHILIPPINES											
Negrito (Angeles, C. Luzon)	1.000							129			Omoto et al. 1978
CHINA											
Chinese (Kuala Lumpur)	.981	.019						371	3.7	1.8	Lie-Injo et al. 1978
JAPAN											
Japanese	.983	.003		.013		.001		965	3.4	1.7	Harada & Omoto 1969
Ainu	.979	.002		.019				238	4.1	2.0	"
KOREA	.989		.011					90	2.2	1.1	Kellermann & Walter 1970

AFRICA

Place/Population	M	S	Z	F	I	V	Others	n	H	PE	Source
EGYPT											
Egyptians (Cairo & Anshas)	.976	.012	.012					202	4.7	2.4	Goedde et al. 1980
SOMALIA	.967	.014	.007	.004		.001		347	6.5	2.6	Massi & Vecchio 1977
CAMEROON (Bamenda)	.985	.011	.004					275	3.0	1.5	Goedde et al. 1979a
MOZAMBIQUE	.982	.002		.016				274	3.5	1.8	Kellermann & Walter 1970
SOUTH AFRICA											
Bantu (Natal)	.965	.002		.033				226	6.8	3.3	McDermid & Vos 1971b

NORTH AMERICA

Place/Population	M	S	Z	F	I	V	Others	n	H	PE	Source
CANADA											
Eskimos (Igloolik)	.994	.006						170	1.2	.6	Cox et al. 1978
UNITED STATES											
Whites (St. Louis)	.948	.036	.012			.004		1,933	10.0	5.0	Pierce et al. 1975
Blacks	.980	.010	.005			.005		204	3.9	2.0	"

SOUTH AMERICA

Place/Population	M	S	Z	F	I	V	Others	n	H	PE	Source
ECUADOR											
Shuara	.889	.005	.078	.028				90	20.3	10.3	Goedde et al. 1977a
FRENCH GUIANA											
Indian tribes	.939	.039	.004	.011		.006		230	11.7	5.9	Vandeville et al. 1972

OCEANIA

Place/Population	M	S	Z	F	I	V	Others	n	H	PE	Source
MELANESIA											
Manus Is.	.996			.004				138	.8	.4	Malcolm et al. 1972

Table 77.1. Alpha-1-antitrypsin (subtypes); protease inhibitor (subtypes): PI

Place/Population	M_1	M_2	M_3	M_4	S	Z	Others	n	H	PE	Source
EUROPE											
AUSTRIA (Tyrol)	.706	.148	.104		.022	.014	.006	868	46.8	26.7	Böhme et al. 1983
DENMARK	.728	.136	.082		.022	.023	.009	909	44.4	25.5	Thymann 1986
FINLAND											
Finns	.787	.122	.081				.010	136	35.9	18.1	Frants & Eriksson 1978
Finns (Newborns)	.665	.185	.125		.018	.005	.002	200	50.8	28.2	Rantala et al. 1982
Lapps/Saami	.958	.042						48	8.0	3.9	Frants & Eriksson 1976
FRANCE											
French (Rouen)	.734	.156	.110					853	42.5	22.2	Charlionet et al. 1981
French (Toulouse)	.626	.092	.104	.037	.141			163	56.8	35.5	Constans et al. 1980b
GERMANY	.782	.065	.153					497	36.1	18.3	Genz et al. 1977
HUNGARY	.794	.135	.059		.008	.004		119	34.8	18.1	Goedde et al. 1986b
ICELAND	.881	.119						42	21.0	9.4	Frants & Eriksson 1976
ITALY (Northern)	.676	.156	.082	.037	.030	.010	.009	202	50.9	30.5	Klasen et al. 1982
NETHERLANDS	.679	.147	.081	.048	.029	.013	.003	357	50.7	30.8	"
SPAIN (Central)	.665	.170	.068	.015	.082			103	51.7	30.5	"
SWEDEN											
Swedes (Northern)	.839	.142			.011	.009	.001	1,869	27.6	13.2	Beckman et al. 1980a
Lapps/Saami	.919	.076			.055			217	15.0	7.1	"
UNITED KINGDOM											
England (Southern)	.819	.119	.062					792	31.1	15.9	Arnaud et al. 1979
U.S.S.R.											
EUROPEAN PART (Moscow)	.820	.145	.019				.016	171	30.5	15.5	Ikramov et al. 1987
ASIA											
NEPAL	.608	.212	.181					144	55.3	29.7	Yuasa et al. 1983b
INDONESIA											
Balinese (Bali Is.)	.706	.286	.004				.004	131	42.0	17.3	Constans et al. 1986
THAILAND	.602	.301	.097					113	53.8	27.0	Goedde & Benkmann 1987
VIETNAM											
Vietnamese (Australia)	.738	.190	.064				.008	63	41.5	21.1	Clark 1982
CHINA											
Mongolians (Hohhot, Inner Mongolia)	.818	.137	.040				.005	211	31.0	15.5	Xu et al. 1986
Han (Beijing)	.737	.162	.091	.002			.008	274	42.2	22.3	Ying et al. 1985b
Uigur (Xinjiang)	.750	.169	.061	.017			.002	204	40.5	21.2	Ying et al. 1985a
Zhuang (S. China)	.780	.172	.041		.007			207	36.0	17.8	Xu et al. 1986
Hui (Ninxia)	.862	.118	.017				.005	217	24.3	11.8	Niu et al. 1984
Dong (Guangxi)	.878	.098	.002				.022	201	21.9	10.7	"
JAPAN											
Japanese (Osaka)	.719	.233	.042	.002	.004			341	42.7	20.4	Shibata 1983
Japanese (Yamaguchi & Izumo)	.706	.239	.048	.002			.005	1,000	44.2	21.4	Yuasa et al. 1984b
KOREA											
Koreans (Seoul)	.768	.139	.073		.003		.016	151	38.5	20.6	Goedde et al. 1986c
Koreans (NE. China)	.811	.094	.091				.005	214	32.5	17.1	Xu et al. 1986
AFRICA											
MALI											
Bozo (Bani-Niger)	.932	.040	.024				.005	102	12.9	6.7	Frants & Eriksson 1978
LIBERIA	.968	.020	.006		.004		.002	485	6.3	3.2	Goedde et al. 1985a
CONGO & CAMEROON											
Bi-Aka Pygmies	.846	.073	.081					296	27.2	14.1	Constans et al. 1981b
KENYA											
Bantu	.979	.021						48	4.1	2.0	Frants & Eriksson 1976
SOUTHERN AFRICA											
San/Bushmen											
!Kung	.768				.222		.010	97	36.1	15.5	Dunn et al. 1986
Negroes/Dama	.959	.008			.033			60	7.9	3.9	"
NORTH AMERICA											

Continued

Table 77.1. Alpha-1-antitrypsin (subtypes); protease inhibitor (subtypes): PI(cont'd)

Place/Population	M_1	M_2	M_3	M_4	S	Z	Others	n	H	PE	Source
GREENLAND											
Eskimos	.887	.113						31	20.0	9.0	Frants & Eriksson 1976
UNITED STATES											
Whites (Minnesota)	.724	.137	.095		.023	.014	.007	904	44.7	25.5	Dykes et al. 1984
Whites (Philadelphia)	.640	.192	.108		.042	.013	.006	240	54.0	31.7	Kueppers & Christopher-son 1978
Blacks	.903	.028	.054		.005		.010	304	18.1	9.4	"

SOUTH AMERICA

CHILE											
Atacameño (Toconao, N. Chile)	.710	.015	.275					169	42.0	17.9	Goedde et al. 1984c

OCEANIA

AUSTRALIA											
Aborigines (Elcho Is., N. Territory)	.284	.546	.044		.114		.011	133	60.6	35.0	Clark 1982

Table 78. Alpha-2-HS-glycoprotein: AHSG

Place/Population	1	2	3	x	n	H	PE	Source
EUROPE								
GERMANY	.654	.322	.015	.009[4]	166	46.8	20.2	Weidinger et al. 1984c
ASIA								
NEPAL	.757	.243			140	36.8	15.0	Yuasa et al. 1985a
JAPAN (Yamagata)	.735	.264		.0005[5]	2,050	39.0	15.7	Umetsu et al. 1984
AFRICA								
EGYPT	.858	.142			95	24.4	10.7	Abe et al. 1986
NORTH AMERICA								
CANADA								
Whites (Toronto)	.642	.353	.005		215	46.3	18.3	Cox et al. 1986
Blacks (Caribbean)	.690	.261		.049[B]	71	45.3	21.3	"
UNITED STATES								
Mexican-Americans (Texas)	.645	.355			76	45.8	17.7	Hewett-Emmett et al. 1986a

Table 79. Anodal tear protein: ATP*

Place/Population	1	2	3	4	5	n	H	PE	Source
NORTH AMERICA									
UNITED STATES									
Whites	.986		.014			108	2.8	1.4	Azen 1976
Blacks	.972	.028				53	5.4	2.6	"
Chinese	.984			.008	.008	60	3.2	1.6	"

135

Table 80. Antithrombin III: AT3

Place/Population	1	2	3	4	n	H	PE	Source
EUROPE								
GERMANY								
Germans (Southern)	.984	.012	.002	.002	212	3.2	1.6	Weidinger et al. 1983
UNITED KINGDOM								
England	1.000				200[a]			Milner et al. 1985
ASIA								
JAPAN								
Japanese (Kyoto)	1.000				370[b]			Kera et al. 1983b
NORTH AMERICA								
UNITED STATES								
Whites (Pittsburgh)	.878	.103	.068		268	21.4	14.2	Kamboh & Ferrell 1987a
Blacks (")	.916	.068	.016		95	15.6	7.7	"

Note: [a] One individual appeared to have a variant type. [b] Three new variants, each found in heterozygous state.

Table 81. Apolipoprotein A-IV: APOA4

Place/Population	1	2	Others	n	H	PE	Source
EUROPE							
GERMANY							
Germans	.923	.075	.002	1,000	14.2	6.7	Menzel et al. 1982
NORWAY							
Norwegians	.950	.050		124	9.5	4.5	Schamaun et al. 1985
NORTH AMERICA							
UNITED STATES							
Whites (Pittsburgh)	.909	.088	.003	159	16.6	7.7	Kamboh & Ferrell 1987b
Blacks (")	.961	.035	.004	127	7.5	3.7	"

Table 82. Apolipoprotein E: APOE

Place/Population	E_2	E_3	E_4	Others	n	H	PE	Source
EUROPE								
FINLAND	.041	.733	.227		615	41.0	19.3	Ehnholm et al. 1986
FRANCE (Nancy)	.130	.742	.128		223	41.6	21.9	Boerwinkle et al. 1987
GERMANY (Marburg)	.078	.783	.139		1,000	36.2	18.6	Menzel et al. 1983
UNITED KINGDOM								
Scotland (Grampian)	.080	.770	.150		400	37.8	19.4	Cumming & Robertson 1984

Continued

Table 82. Apolipoprotein E: APOE (cont'd)

Place/Population	E_2	E_3	E_4	Others	n	H	PE	Source
ASIA								
INDIA								
Indians (Malaysia)	.070	.875	.055		137	22.6	11.6	Boerwinkle et al. 1986
MALAYSIA								
Malays (Singapore)	.115	.770	.115		117	38.1	19.9	"
CHINA								
Chinese (Malaysia)	.106	.794	.100		175	34.8	18.2	"
JAPAN (Tokyo)	.038	.843	.111	.008[5,6]	197	27.6	14.0	Tsuchiya et al. 1985
NORTH AMERICA								
CANADA (Ottawa)	.078	.770	.152		102	37.8	19.4	Davignon et al. 1984
UNITED STATES								
Whites (Washington, DC)	.095	.756	.149		74	39.7	20.6	Ghiselli et al. 1982
SOUTH AMERICA								
BRAZIL								
Amerindians		.816	.184		107	30.0	12.8	Asakawa et al. 1985
OCEANIA								
NEW ZEALAND								
Blood donors								
(Christchurch)	.120	.720	.160		426	44.2	23.2	Wardell et al. 1982

Table 83. Beta lipoprotein, Ag system: AG

Place/Population	Ag^x	Ag	n	H	PE	Source
EUROPE						
BELGIUM (Liege)	.217	.783	577	34.0	8.2	Brocteur et al. 1980
BULGARIA (Sofia)	.257	.743	107	38.2	7.8	Müller et al. 1974
FINLAND						
Finns	.306	.694	27	42.5	7.1	Berg & Eriksson 1973a
Lapps/Saami	.475	.525	639	49.9	3.6	"
GERMANY (Mainz)	.199	.801	312	31.9	8.2	Müller et al. 1974
GREECE (Athens)	.400	.600	75	48.0	5.2	"
ICELAND	.218	.782	340	34.1	8.2	Berg & Eriksson 1973c
IRELAND	.210	.790	1,317	33.2	8.2	Tills 1977
ITALY	.273	.727	152	39.7	7.6	Hirschfeld et al. 1968
NORWAY						
Norwegians	.202	.798	3,162	32.2	8.2	Solaas 1970
Lapps/Saami (Finnmark)	.417	.583	294	48.6	4.8	Monn et al. 1971
SWEDEN	.213	.787	250	33.5	8.2	Hirschfeld 1968
SWITZERLAND (Bern)	.209	.791	249	33.1	8.2	Morganti et al. 1970
UNITED KINGDOM						
England	.206	.794	956	32.7	8.2	Bradbrook et al. 1971
Isle of Man	.172	.828	111	28.5	8.1	Mitchell et al. 1982

Continued

Table 83. Beta lipoprotein, Ag system: AG (cont'd)

Place/Population	Ag^x	Ag	n	H	PE	Source
ASIA						
ISRAEL						
Yemenite Jews	.113	.887	84	20.0	7.0	Tills et al. 1977b
IRAN						
Kurds	.263	.737	184	38.8	7.8	Lehmann et al. 1973
INDIA (Southern)	.738	.262	42	38.7	.3	Hirschfeld & Okochi 1967
THAILAND	.694	.306	54	42.5	.6	Hirschfeld & Okochi 1967
CHINA						
Tibetans (Switzerland)	.631	.369	110	46.6	1.2	Jeannet et al. 1972
JAPAN						
Japanese	.731	.269	1,205	39.3	.4	Hirschfeld & Okochi 1967
Ainu	.698	.302	96	42.2	.6	Misawa et al. 1971
AFRICA						
NIGERIA						
Yoruba (Southwestern)	.108	.892	426	19.3	6.8	Heiken et al. 1974
SENEGAL						
Negroes	.140	.860	238	24.1	7.7	Müller et al. 1974
ANGOLA						
Negroes	.086	.914	232	15.7	6.0	"
SOUTHERN AFRICA						
Zulu	.072	.928	97	13.4	5.3	Hitzeroth et al. 1980
NORTH AMERICA						
GREENLAND						
Eskimos (Augpilagtok Is.)	.596	.404	153	48.2	1.6	Berg & Eriksson 1971
UNITED STATES						
Whites (Southern)	.580	.420	227	48.7	1.8	Blumberg et al. 1964
Blacks (")	.433	.567	149	49.1	4.5	"
SOUTH AMERICA						
PERU						
Peruvian	.252	.748	121	37.7	7.9	Johnston et al. 1969b
VENEZUELA						
Yanomama	.085	.925	372	15.7	6.0	Weitkamp et al. 1972
OCEANIA						
AUSTRALIA						
Aborigines	.468	.532	166	49.8	3.7	Müller et al. 1974

Note: Allele Ag^x is dominant to allele Ag.

Table 84. Beta lipoprotein, Ld system: LD

Place/Population	Ld^a	Ld	n	H	PE	Source
EUROPE						
NORWAY						
Norwegians	.242	.758	162	36.7	8.0	Berg 1965

Note: Allele Ld^a is dominant to allele Ld.

Table 85. Beta lipoprotein, Lp system: LP

Place/Population	Lp^a	Lp	n	H	PE	Source
EUROPE						
AUSTRIA	.171	.829	310	28.4	8.1	Mourant et al. 1976(c)
BULGARIA	.179	.821	609	29.4	8.1	Mourant et al. 1976(c)
GERMANY (Gottingen)	.162	.838	1,030	27.2	8.0	Jørgensen et al. 1965
HUNGARY	.106	.894	150	19.0	6.8	Mourant et al. 1976(c)
ICELAND	.136	.864	340	23.5	7.6	Berg & Eriksson 1973c
IRELAND	.160	.840	431	26.9	8.0	Tills 1977
NORWAY						
Norwegians	.195	.805	1,109	31.4	8.2	Berg 1974(c)
Lapps/Saami (Finnmark)	.193	.807	252	31.2	8.2	Monn et al. 1971
UNITED KINGDOM						
Northern Ireland	.245	.755	300	37.0	8.0	Mourant et al. 1976(c)
U.S.S.R.						
Svani (Alkhazskaya)	.067	.933	263	12.5	5.1	Geserick et al. 1972
ASIA						
ISRAEL						
Yemenite Jews	.206	.794	84	32.7	8.2	Tills et al. 1977b
KOREA	.193	.807	115	31.2	8.2	Bajatzadeh & Walter 1969a
AFRICA						
TANZANIA						
Hadza	.191	.809	107	30.9	8.2	Berg 1968
NORTH AMERICA						
GREENLAND						
Eskimos (Augpilagtok Is.)	.164	.836	153	27.4	8.0	Berg & Eriksson 1971
CANADA						
Indians	.009	.991	234	1.8	.9	Berg 1968
UNITED STATES						
Whites	.218	.782	126	34.1	8.2	Berg 1968
Blacks	.189	.811	242	30.7	8.2	Berg 1968
SOUTH AMERICA						
BRAZIL	.168	.832	104	28.0	8.1	Berg 1968
VENEZUELA						
Makiritare (S. Venezuela &						
N. Brazil)	.144	.856	221	24.7	7.7	Arends et al. 1970
Yanomama (")	.089	.919	2,262	14.8	6.3	Weitkamp et al. 1972
OCEANIA						
POLYNESIA						
Easter Is.	.043	.957	106	8.2	3.6	Berg 1968

Note: Allele Lp^a is dominant to allele Lp.

Table 86. Beta-2-glycoprotein I, Bg system: BG

Place/Population	Bg^N	Bg^D	n	H	PE	Source
EUROPE						
GERMANY (Mainz)	.953	.047	210	9.0	4.3	Koppe et al. 1970
GREECE	.924	.076	157	14.0	6.5	"

Continued

Table 86. Beta-2-glycoprotein I, Bg system: BG (cont'd)

Place/Population	Bg^N	Bg^D	n	H	PE	Source
HUNGARY	.940	.060	151	11.6	5.5	"
ICELAND	.940	.060	97	11.6	5.5	"
IRELAND	.948	.052	107	9.9	4.7	"
UNITED KINGDOM						
England (London)	.941	.059	381	11.1	5.2	Atkin & Rundle 1974

ASIA

Place/Population	Bg^N	Bg^D	n	H	PE	Source
IRAN	.886	.114	141	20.2	9.1	Koppe et al. 1970
AFGHANISTAN						
Pushtoon (Kabul)	.876	.124	210	21.7	9.7	Rahimi et al. 1977
PAKISTAN	.949	.051	79	9.7	4.6	Koppe et al. 1970
INDIA						
Indians (N. India)	.954	.046	108	8.8	4.2	"
Bengali (E. India)	.954	.046	964	8.8	4.2	Walter et al. 1972b
PHILIPPINES						
Filipino	.937	.063	88	11.8	5.6	Walter et al. 1979
KOREA						
Koreans	.780	.220	105	34.3	14.2	Koppe et al. 1970

AFRICA

Place/Population	Bg^N	Bg^D	n	H	PE	Source
MOZAMBIQUE						
Negroes	.742	.258	151	38.3	15.5	Koppe et al. 1970
SOUTHERN AFRICA						
Negroes	.950	.050	250	9.5	4.5	Walter et al. 1979

SOUTH AMERICA

Place/Population	Bg^N	Bg^D	n	H	PE	Source
ECUADOR						
Shuara	.967	.033	90	6.4	3.1	Goedde et al. 1977a

Note: Alleles Bg^N and Bg^D are codominant.

Table 87. Ceruloplasmin: CP

Place/Population	A	B	C	x	n	H	PE	Source
EUROPE								
GERMANY	.013	.985	.002		224	3.0	1.5	Bajatzadeh & Walter 1969b
GREECE	.036	.960	.004		210	7.7	3.8	Kellermann & Walter 1972
ICELAND	.010	.988	.002		240	2.4	1.2	"
IRELAND	.012	.983	.004		240	3.4	1.6	Walter & Palsson 1973
SPAIN								
Spanish (Madrid)		1.000			164			Goedde et al. 1973
Basques								
(Berneo & Pamplona)	.005	.995			218	1.0	.5	"
U.S.S.R.								
EUROPEAN PART								
Abkhazia (Ochamchir Dist.)		1.000			112			Ferrell et al. 1985
ASIA								
SAUDI ARABIA								
Tribes (Western)		1.000			185			Saha et al. 1980
IRAN								
Turkoman (Gonbad, NE Iran)	.003	.994		$.003^{NH}$	155	1.2	.6	Kirk et al. 1977
AFGHANISTAN								
Pushtoon (Kabul)		1.000			210			Rahimi et al. 1977
PAKISTAN	.005	.995			96	1.0	.5	Bajatzadeh & Walter 1969b

Continued

Table 87. Ceruloplasmin: CP (cont'd)

Place/Population	A	B	C	x	n	H	PE	Source
INDIA								
Indians (N. India)[b]	.001[Ni]	.999			485	.2	.1	Blake et al. 1971a
Gujarati & Marathi (W. India)		1.000			287			McDermid 1971b
Irula Tr. (Nilgiri Hills, S. India)		1.000			168			Saha et al. 1976
Bengali (E. India)	.012	.986	.002		978	2.8	1.4	Kellermann & Walter 1972
Moslems (")		.996	.004		123	.8	.4	Mukherjee et al. 1974
INDONESIA								
Batak (Samosir Is.)		.991		.009[V]	167	1.8	.9	McDermid et al. 1973
CHINA								
Chinese (Beijing)		.999		.001[CZ]	1,042	.2	.1	Zhang et al. 1985
JAPAN								
Japanese (Hiroshima & Nagasaki)		.990		.010[a]	14,964	2.0	1.0	Fujita et al. 1985a
KOREA	.009	.978	.013		115	4.3	2.2	Bajatzadeh & Walter 1969b

AFRICA

Place/Population	A	B	C	x	n	H	PE	Source
EGYPT								
Nubians (S. Egypt)		1.000			155			Bertin et al. 1978
SUDAN								
Sudanese (Khartoum)		1.000			200			Saha et al. 1978b
NIGERIA								
Negroes (Northern)	.149	.837	.003	.011[NH]	520	27.7	12.8	Shokeir & Shreffler 1970
ANGOLA								
Negroes	.052	.935	.004	.009[NH]	909	12.3	6.1	Kellermann & Walter 1972
MOZAMBIQUE	.041	.955	.004		580	8.6	4.2	"
SOUTHERN AFRICA								
Bantu (Natal)	.035	.960		.005[NH]	302	7.7	3.8	McDermid & Vos 1971a

NORTH AMERICA

Place/Population	A	B	C	x	n	H	PE	Source
ALASKA								
Eskimos (St. Lawrence Is.)		1.000			222			Ferrell et al. 1981b
CANADA								
Eskimos (Igloolik)		1.000			356			McAlpine et al. 1974
Ojibwa Indians		1.000			205			Szathmary et al. 1974
Dogrib " (NW Territories)	.003	.997			153	.6	.3	Szathmary et al. 1983
UNITED STATES								
Whites	.003	.997			1,270	.6	.3	Shokeir & Shreffler 1970
Blacks	.052	.939	.003	.006[NH]	1,126	11.6	5.7	"

MIDDLE AMERICA

Place/Population	A	B	C	x	n	H	PE	Source
GUATEMALA								
Black Caribs (Livingston)	.022	.978			205			Crawford et al. 1981
COSTA RICA								
Guaymi		1.000			286			Barrantes et al. 1982
PANAMA								
Blacks (Costa Arriba)	.047	.952	.001		415	9.1	4.4	Ferrell et al. 1978b
WEST INDIES (Haiti)	.113	.873	.002	.012[NH]	323	22.5	10.6	Shokeir & Shreffler 1970

SOUTH AMERICA

Place/Population	A	B	C	x	n	H	PE	Source
ARGENTINA								
Mapuche		1.000			71			Haas et al. 1985
BOLIVIA								
Aymara (Western)		1.000			316			Ferrell et al. 1978a
BRAZIL								
Parakana		.928		.072	125			Blake et al. 1980
Baniwa		1.000			377			Salzano et al. 1986
Pacaás Novos		1.000			222			Salzano et al. 1985
Caingang		1.000			214			Salzano, et al. 1980
Cayapo (Para & Mato Grosso)	.043	.957			184	8.2	3.9	Salzano et al. 1972b
Macushi (N. Brazil & S. Guyana)		.996		.004[V]	509	.8	.4	Neel et al. 1977b

Continued

Table 87. Ceruloplasmin: CP (cont'd)

Place/Population	A	B	C	x	n	H	PE	Source
Wapishana (N. Brazil & S. Guyana)		.992		.008[V]	621	1.6	.8	"
Ticuna (C. Amazonas)		1.000			758			Neel et al. 1980
COLOMBIA								
Noanama		1.000			155			Kirk et al. 1974
VENEZUELA								
Makiritare								
(S. Venezuela & N. Brazil)		1.000			463			Arends et al. 1970
Yanomama (")		1.000			2,416			Weitkamp et al. 1972

OCEANIA

Place/Population	A	B	C	x	n	H	PE	Source
AUSTRALIA								
Aborigines (Elcho Is., N. Territory)	.002	.998			251	.4	.2	McDermid 1971b
MELANESIA								
New Guinea								
Asmat (S. Coastal Plain)		1.000			114			Gajdusek et al. 1978
Papua New Guinea								
Gainj & Kalam (N. Central Highlands)		1.000			281			Long et al. 1986
Fuyuge (C. Dist. Highlands)		1.000			145			Woodfield et al. 1974
Manus Is.		1.000			183			Malcolm et al. 1972
MICRONESIA								
Marshall Is.		1.000			370			Neel et al. 1976
POLYNESIA								
Samoans (Christchurch, New Zealand)		1.000			77			Booth et al. 1977

Note: [a] Pooled frequency of seven different kinds of variants.
 [b] An individual was a heterozygote ($A_{Nigeria}$/B)

Table 88. Coagulation factor; prothrombin: F2

Place/Population	1	2	3	4	n	H	PE	Source
OCEANIA								
AUSTRALIA								
Aborigines	1.000				42			Board et al. 1982
Caucasians (Canberra)	.995	.001	.004		415	1.0	.5	"
MELANESIA								
Karkar Is.	1.000				131			"
Fiji Is.	.980			.020	126	4.0	.2	"
POLYNESIA								
Samoans (Samoa Is.)	.990			.010	102	2.0	.1	"
Rarotonga	1.000				110			"

Table 89. Coagulation factor XIII-A: F13A

Place/Population	1	2	4	n	H	PE	Source
EUROPE							
GERMANY (Hessen)	.797	.203		239	32.4	13.6	Kreckel et al. 1982b
ASIA							
IRAN (Caspian Sea Region)	.774	.226		78	35.0	14.4	Castle & Board 1985
INDONESIA (Yogyakarta)	.952	.045	.003	157	9.2	4.4	"
JAPAN							
Japanese (Tokyo)	.887	.113		561	20.0	9.0	Nishigaki et al. 1981
Japanese (W. Honshu Is.)	.891	.107	.002	266	19.5	8.9	Suzuki et al. 1986
NORTH AMERICA							
UNITED STATES							
Whites (Minneapolis)	.774	.225	.001	585	35.0	14.5	Dykes & Polesky 1985
Blacks (")	.771	.229		105	35.3	14.5	"
Pima Indians	.841	.159		154	26.7	11.6	Castle & Board 1985
Mexican-Americans (Texas)	.764	.232	.004	140	36.2	15.2	Hewett-Emmett et al. 1986a
OCEANIA							
AUSTRALIA							
Whites	.79	.21		179	33.2	13.8	Board 1979
Aborigines	.784	.216		97	33.9	14.1	Castle & Board 1985
MELANESIA							
Melanesians	.783	.209	.008	127	34.3	14.8	Board & Coggan 1981
Papua New Guinea							
Eastern Highlands	.850	.150		150	25.5	11.1	Castle & Board 1985
Western Highlands	.759	.241		56	36.6	14.9	"
Buka Is.	.829	.171		76	28.4	12.2	"
Fiji Is.							
Melanesians	.783	.209	.008	127	34.3	14.8	Board & Coggan 1981
Indians	.768	.232		127	35.6	14.6	"
POLYNESIA							
Samoa Is.	.901	.099		147	17.8	8.1	Castle & Board 1985
Cook Is.	.980	.020		100	3.9	1.9	Board & Coggan 1981

Table 90. Coagulation factor XIII-B: F13B

Place/Population	1	2	3	x	n	H	PE	Source
EUROPE								
GERMANY (Hessen)	.735	.095	.170		428	42.2	21.8	Kreckel et al. 1982a
NORWAY	.690	.150	.160		283	47.6	25.2	Olaisen et al. 1983
ASIA								
JAPAN								
Japanese (Tokyo)	.252	.013	.735		304	39.6	16.9	Nakamura & Abe 1982a
Japanese (")	.298	.018	.680	.003[a]	435	44.8	19.3	Nakamura et al. 1986b
NORTH AMERICA								
ALASKA								
Eskimos (Kodiak Is.)	.238	.058	.704		225	44.4	21.5	Kamboh & Ferrell 1986a
Eskimos (St. Lawrence Is.)	.080		.920		143	14.7	6.8	"
CANADA								
Dogrib Indians (NW Canada)	.082		.918		201	15.1	7.0	Kamboh & Ferrell 1986a
UNITED STATES								
Whites (Minneapolis)	.776	.088	.136		328	37.2	19.3	Miller et al. 1985

Continued

Table 90. Coagulation factor XIII-B: F13B(cont'd)

Place/Population	1	2	3	x	n	H	PE	Source
Blacks (Minneapolis)	.286	.635	.079		178	50.9	25.0	Miller et al. 1985
Whites (Pittsburgh)	.693	.105	.202		171	46.8	24.2	Kamboh & Ferrell 1986a
Blacks (")	.388	.561	.051		98	53.2	24.2	"

MIDDLE AMERICA

MEXICO

Mayan Indians	.320	.008	.672		61	44.6	18.1	"

OCEANIA

AUSTRALIA

Whites	.747	.084	.169		245	40.6	20.8	Board 1980b

MELANESIA

Melanesians	.610	.020	.370		127	49.1	20.4	Board & Castle 1982

Note: [a] Pooled frequency of three variant alleles.

Table 91. Complement component C1r, subcomponent: C1R

Place/Population	1	2	n	H	PE	Source

NORTH AMERICA

UNITED STATES

	1	2	n	H	PE	Source
Whites (Pittsburgh)	.934	.066	175	12.3	5.8	Kamboh & Ferrell 1986b
Blacks (")	.899	.101	109	18.2	8.3	"

Table 92. Complement component 2: C2

Place/Population	C	B	AT	x	n	H	PE	Source

EUROPE

	C	B	AT	x	n	H	PE	Source
GERMANY (Bonn)	.965	.035			289	6.8	3.3	Dewald & Rittner 1979
NORWAY (Oslo)	.967	.033			122	6.4	3.1	Olaisen et al. 1978
UNITED KINGDOM								
England	.942	.058			52	10.9	5.2	Baur et al. 1984

ASIA

	C	B	AT	x	n	H	PE	Source
THAILAND (Northern)	.951	.049			184	9.3	4.4	Greiner et al. 1980
JAPAN	.939	.022	.034	.006[BH]	521	11.4	6.1	Tokunaga et al. 1981
KOREA (Seoul)	.961	.018	.009	.011[BH]	220	7.5	3.7	Park et al. 1985

NORTH AMERICA

	C	B	AT	x	n	H	PE	Source
CANADA (Newfoundland)	.950	.050			800	9.5	4.5	Noel et al. 1980
UNITED STATES								
Whites	.973	.020		.007	75	5.3	2.6	Alper 1976
Blacks	.967	.033			30	6.4	3.1	"

OCEANIA

MELANESIA

	C	B	AT	x	n	H	PE	Source
Fiji	.994	.006			178	1.2	.6	Ranford et al. 1982

POLYNESIA

	C	B	AT	x	n	H	PE	Source
W. Samoa	.987	.013			116	2.6	1.3	"

Table 93. Complement component 3: C3

Place/Population	S	F	Others	n	H	PE	Source
EUROPE							
AUSTRIA	.793	.207		600	32.8	13.7	Glahs 1974
BELGIUM (Liége)	.811	.186	.003	818	30.8	13.2	Brocteur et al. 1980
BULGARIA	.815	.185		127	30.2	12.8	Farhud & Walter 1973
DENMARK (Copenhagen)	.799	.201		1,554	32.1	13.5	Stoffersen & Jørgensen 1980
FINLAND							
Finns	.830	.170		1,034	28.2	12.1	Arvilommi 1972
Lapps/Saami	.934	.066		909	12.3	5.8	Arvilommi et al. 1973
GERMANY (Hessen)	.779	.215	.006	1,322	34.7	14.8	Kühnl & Spielmann 1972
GREECE	.786	.211	.003	1,055	33.8	14.3	Germenis et al. 1985
HUNGARY	.802	.198		134	31.8	13.4	Goedde et al. 1986b
ICELAND	.807	.193		246	31.2	13.1	Berg & Eriksson 1973b
ITALY							
Italians (Tuscany)	.805	.193	.002	1,000	31.5	13.4	Domenici et al. 1986
NETHERLANDS (Leiden)	.787	.213		790	33.5	14.0	Fraser et al. 1974
NORWAY							
Norwegians	.787	.208	.005	2,454	33.7	14.4	Teisberg 1971a
Lapps/Saami	.937	.063		198	11.8	5.6	Teisberg 1971b
POLAND (Wroclaw)	.823	.175	.005	4,741	29.2	12.9	Manczak 1984a
SPAIN							
Spanish (Madrid)	.772	.224	.004	219	35.4	14.9	Goedde et al. 1973
Basques (Bermeo/Pamplona)	.779	.214	.007	284	34.7	14.9	"
SWEDEN							
Swedes	.797	.194	.008	1,196	32.7	14.2	Brönnestam 1973
Lapps/Saami	.973	.027		148	5.3	2.6	Brönnestam et al. 1971
SWITZERLAND (Bern)	.792	.203	.005	2,961	33.2	14.2	Pflugshaupt et al. 1975
UNITED KINGDOM							
England (Northeast)	.778	.218	.004	268	34.7	14.6	MacDonald 1975
U.S.S.R.							
EUROPEAN PART							
Georgians	.762	.228	.010	232	36.6	15.7	Geserick et al. 1971
ASIA							
SAUDI ARABIA							
Shiites & Sunnites (al-Hasa)	.900	.092	.008	359	18.1	8.6	Goedde et al. 1979b
IRAQ							
Arabs	.788	.212		574	33.4	13.9	Roberts & Al-Agidi 1979
Kurds	.743	.257		235	38.2	15.4	"
Turks	.756	.244		209	36.9	15.0	"
IRAN							
Iranians	.792	.208		101	32.9	13.8	Farhud & Walter 1973
Turkoman (Gonbad, NE Iran)	.882	.118		85	20.8	9.3	Akbari et al. 1984
AFGHANISTAN							
Pushtoon (Kabul)	.881	.119		210	21.0	9.4	Rahimi et al. 1977
INDIA							
Punjabi (N. India)	.859	.141		99	24.2	10.6	Papiha 1981
Hindu (Madhya Pradesh)	.929	.071		91	13.2	6.2	Papiha et al. 1979
Telugu (Andhra Pradesh)	.923	.077		195	14.2	6.6	"
THAILAND	.996	.004		113	.8	.4	Goedde et al. 1987
PHILIPPINES							
Negrito (Angeles, C. Luzon)	.986	.014		106	2.8	1.4	Omoto et al. 1978
CHINA							
Mongolians (N. China)	.983	.012	.005	210	3.4	1.7	Goedde et al. 1984b
Han (Shanghai)	.996	.004		388	.8	.4	Zhao 1983
Zhuang (S. China)	.995	.005		208	1.0	.5	Goedde et al. 1984b
Tibetans (Switzerland)	1.000			110			Jeannet et al. 1972
TAIWAN							
Toroko-Aborigines	1.000			120			Chen et al. 1985
JAPAN							
Japanese	.994	.001	.005[a]	1,692	1.2	.6	Nishimukai et al. 1985
KOREA							
Koreans (Seoul)	.993	.007		150	1.4	.7	Goedde et al. 1986c
Koreans (NE China)	.995	.002	.003	213	1.0	.5	Goedde et al. 1984b

Continued

145

Table 93. Complement component 3: C3 (cont'd)

Place/Population	S	F	Others	n	H	PE	Source
AFRICA							
EGYPT							
Egyptians (Cairo & Anshas)	.792	.208		130	32.9	13.8	Goedde et al. 1980
TUNISIA	.833	.152	.015	233	28.3	13.0	Davrinche et al. 1981b
CAMEROON (Bamenda)	.951	.045	.004	275	9.4	4.5	Goedde et al. 1979a
ETHIOPIA	.888	.112		218	19.9	9.0	Agarwal et al. 1974
ANGOLA	.953	.041	.006	439	9.0	4.4	Farhud & Walter 1973
NORTH AMERICA							
GREENLAND							
Eskimos (Umanak Dist.)	.944	.056		125	10.6	5.0	Stoffersen et al. 1982
UNITED STATES							
Whites	.773	.219	.008	472	35.4	15.2	Alper & Rosen 1976
Blacks	.948	.049	.003	154	9.9	4.8	"
SOUTH AMERICA							
ARGENTINA							
Mapuche	.949	.051		107	9.7	4.6	Haas et al. 1985
CHILE							
Atacameño (Toconao, N. Chile)	.995	.005		180	1.0	.5	Goedde et al. 1985b
COLOMBIA							
Colombians (Bogota)	.794	.202		403	32.9	13.6	Bernal et al. 1985
ECUADOR							
Shuara	1.000			90			Goedde et al. 1977a
FRENCH GUIANA							
Wayampi (Trois-Sauts & Camopi)	1.000			226			Tchen et al. 1978
OCEANIA							
POLYNESIA							
Easter Is.	.960	.040		48	7.7	3.7	Thorsby et al. 1972

Note: [a] Pooled frequency of five rare alleles.

Table 94. Complement component 4A: C4A

Place/Population	1	2	3	4	5	6	Qo	Others	n	H	PE	Source
EUROPE												
FINLAND		.081	.754	.036			.016	.113	254	41.1	23.4	Partanen & Koskimies 1986
GERMANY	.005	.043	.672	.056	.004		.031	.188	382	50.7	29.5	Bertrams et al. 1984
ASIA												
JAPAN		.106	.686	.132			.067	.009	341	49.6	29.0	Tokunaga et al. 1985
NORTH AMERICA												
UNITED STATES												
Whites		.080	.695	.055	.005	.055	.110		100	49.2	30.2	Awdeh & Alper 1980
SOUTH AMERICA												
COLOMBIA												
Colombians	.026	.069	.707	.138	.017	.017		.026	58	47.4	28.5	Bernal et al. 1985

Table 95. Complement component 4B: C4B

Place/Population	1	2	3	6	Qo	Others	n	H	PE	Source
EUROPE										
FINLAND	.657	.153	.016		.175		254	51.4	28.3	Partanen & Koskimies 1986
GERMANY	.749	.077	.035	.012	.127		382	41.6	23.5	Bertrams et al. 1984
ASIA										
JAPAN	.587	.167			.158	.088[5]	341	59.5	36.1	Tokunaga et al. 1985
NORTH AMERICA										
UNITED STATES										
Whites	.760	.105			.135		100	39.3	20.5	Awdeh & Alper 1980
SOUTH AMERICA										
COLOMBIA										
Colombians(Bogota)	.845	.095	.034		.009	.017	58	27.5	14.6	Bernal et al. 1985

Table 96. Complement component 5: C5

Place/Population	1	2	n	H	PE	Source
EUROPE						
PORTUGAL	1.000		66			Vaz-Guedes et al. 1978
UNITED KINGDOM						
England	1.000		1,585			"
ASIA						
BURMA	1.000		229			"
AFRICA						
GAMBIA	1.000		296			"
OCEANIA						
AUSTRALIA						
Aborigines (Elcho Is.,						
N. Territory)	.962	.038	52	7.3	3.5	Hobart et al. 1981
MELANESIA						
Papua New Guinea (N. Coast)	.922	.078	45	14.4	6.7	"
Solomon Is.						
Santa Cruz	.928	.072	97	13.4	6.2	"

Table 97. Complement component 6: C6

Place/Population	A	B	B2	Others	n	H	PE	Source
EUROPE								
GERMANY	.601	.388	.003	.008	709	48.7	22.2	Knustmann et al. 1980
NORWAY								
Norwegians (Oslo)	.587	.409		.004	1,623	48.8	18.9	Olving et al. 1980
Lapps/Saami (C. Finnmark)	.533	.467			167	49.8	18.7	"
SWEDEN (Lund)	.647	.353			218	45.7	17.6	Rudduck et al. 1985

Continued

147

Table 97. Complement component 6: C6 (cont'd)

Place/Population	A	B	B2	Others	n	H	PE	Source
UNITED KINGDOM								
England	.683	.307		.010	202	43.9	18.0	Whitehouse & Putt 1983

ASIA

CHINA								
Chinese (Beijing)	.416	.532	.042	.010	155	54.2	24.8	Zeng et al. 1986
Chinese (Guangzhou)	.445	.518	.033	.004	255	53.1	26.6	"
JAPAN								
Japanese (Tokyo)	.432	.503	.060	.005	565	55.7	26.1	Nakamura et al. 1984b
Japanese	.446	.466	.081	.007	278	57.7	28.4	Tokunaga et al. 1986

NORTH AMERICA

UNITED STATES								
Blacks	.559	.381		.059	59	53.9	25.0	Hobart et al. 1974

OCEANIA

MELANESIA								
Fiji	.693	.307			186	42.6	16.7	Ranford et al. 1982
MICRONESIA								
Nauru	.446	.452	.067	.035	186	59.1	30.5	"
POLYNESIA								
W. Samoa	.629	.359	.006	.006	245	47.5	19.3	"

Table 98. Complement component 7: C7

Place/Population	1	2	4	5	6	n	H	PE	Source
EUROPE									
UNITED KINGDOM									
England (Cambridge)	.995	.002	.004			1,228	1.0	.6	Hobart et al. 1978
ASIA									
CHINA									
Chinese (Beijing)	.865	.069	.043	.020	.003	152	24.5	13.1	Zeng et al. 1986
Chinese (Guangzhou)	.884	.075	.010	.031		255	21.2	10.8	"
Chinese (Shanghai)	.850	.150				30	25.5	11.1	York et al. 1986
JAPAN									
Japanese (Tokyo)	.858	.096	.046			494	25.3	12.8	Nakamura et al. 1984a
Japanese	.875	.087	.038			278	22.5	11.3	Tokunaga et al. 1986
Japanese (W. Japan)	.809	.104	.038	.049		183	33.1	18.0	Nishimukai & Tamaki 1986

Table 99. Complement component 8, alpha-gamma polypeptide: C8A

Place/Population	A	B	Others	n	H	PE	Source
EUROPE							
GERMANY (Bonn)	.553	.429	.018	196	50.9	20.8	Rittner et al. 1984
NORWAY	.586	.386	.029	105	50.7	21.7	Rogde et al. 1985

Continued

Table 99. Complement component 8, alpha-gamma polypeptide: C8A (cont'd)

Place/Population	A	B	Others	n	H	PE	Source
ASIA							
JAPAN (Tokyo)	.623	.367	.010	448	47.7	19.1	Nakamura et al. 1986a
NORTH AMERICA							
UNITED STATES							
Whites	.649	.348	.003	165	45.8	17.9	Raum et al. 1979
Blacks	.700	.246	.054	130	44.7	21.4	"

Table 100. Complement component 8, beta polypeptide: C8B

Place/Population	B	A	A_1	n	H	PE	Source
EUROPE							
NORWAY	.943	.052	.004	105	10.8	5.1	Rogde et al. 1985
NORTH AMERICA							
UNITED STATES							
Whites	.952	.044	.004	125	9.2	4.5	Alper et al. 1983

Table 101. Double-band parotid salivary protein: DB

Place/Population	+	-	n	H	PE	Source
EUROPE						
NETHERLANDS	.190	.810	100	30.8	8.2	Pronk et al. 1984(c)
ASIA						
JAPAN	.053	.947	N.A.	10.0	4.3	Ikemoto 1983(c)
AFRICA						
KENYA	.550	.450	200	49.5	2.3	Pronk et al. 1984
NORTH AMERICA						
UNITED STATES						
Whites	.123	.877	100	21.6	7.3	Azen & Denniston 1974
Blacks	.564	.436	100	49.2	2.0	"
Chinese	.067	.933	54	12.5	5.1	"

Note: Allele Db^+ is considered to be dominant to Db^-.

Table 102. Factor H: HF, CFH

Place/Population	1	2	3	n	H	PE	Source
NORTH AMERICA							
UNITED STATES							
Whites (New York area)	.691	.302	.006	61	43.1	7.5	de Córdoba & Rubinstein 1984

Table 103. Factor I: FI

Place/Population	A	B	n	H	PE	Source
ASIA						
JAPAN (Tokyo)	.107	.893	435	19.1	8.6	Nakamura & Abe 1985

Table 104. Group specific component; vitamin D binding protein: GC, DBP

Place/Population	1	2	x	n	H	PE	Source
EUROPE							
AUSTRIA	.721	.279		1,000	40.2	16.1	Wing 1974 (c)
BELGIUM (Liége)	.750	.250		1,000	37.5	15.2	Brocteur et al. 1980
BULGARIA	.768	.232		138	35.6	14.6	Walter et al. 1972a
DENMARK	.725	.275		1,312	39.9	16.0	Nerstrøm 1965
FINLAND	.795	.205		5,536	32.6	13.6	Nevanlinna 1972
FRANCE (Paris)	.728	.272		221	39.6	15.9	Moullec 1963
GERMANY (Freiburg)	.715	.285		2,786	40.8	16.2	Hummel et al. 1970
GREECE	.752	.248		600	37.3	15.2	Germenis et al. 1983
ICELAND	.710	.290		369	41.2	16.4	Karlsson et al. 1980b
IRELAND	.710	.290		1,765	41.2	16.4	Tills 1977
ITALY (Ferrara)	.734	.266		160	39.0	15.7	Cleve & Vierucci 1965
NETHERLANDS (Leiden)	.706	.294		792	41.5	16.4	Fraser et al. 1974
NORWAY							
Norwegians	.737	.263		6,472	38.8	15.6	Reinskou 1974
Lapps/Saami (Finnmark)	.801	.199		301	31.9	13.4	Monn et al. 1971
POLAND (S. Poland)	.623	.377		3,624	47.0	18.0	Turowska et al. 1977
PORTUGAL							
Portuguese	.712	.288		1,500	41.0	16.3	Torrinha 1969
Portuguese (Terras de Bouro)	.772	.228		112	35.2	14.5	Cruz et al. 1973
SPAIN							
Spanish (Madrid)	.775	.225		187	34.9	14.4	Goedde et al. 1973
Basques	.687	.313		278	43.0	16.9	"
SWEDEN							
Swedes	.743	.257		1,744	38.2	15.4	Monn et al. 1971(c)
Lapps/Saami	.845	.155		190	26.2	11.4	"
SWITZERLAND	.712	.288		200	41.0	16.3	Hess & Bütler 1962
UNITED KINGDOM							
England	.714	.286		49	40.8	16.3	Hirschfeld 1962
Orkney Is. (Westray)	.780	.220		307	34.3	14.2	Welch et al. 1973
YUGOSLAVIA	.702	.298		205	41.8	16.5	Fraser et al. 1969a

Continued

Table 104. Group-specific component; vitamin D binding protein: GC, DBP (cont'd)

Place/Population	1	2	x	n	H	PE	Source
U.S.S.R.							
EUROPEAN PART							
Russians (Moscow)	.651	.347		1,553	45.4	17.5	Altukhov et al. 1981
Abkhazia (Ochamchir Dist.)	.773	.227		293	35.1	14.5	Ferrell et al. 1985
ASIA							
TURKEY (Asia Minor)	.743	.257		274	38.2	15.4	Hummel et al. 1970
ISRAEL							
Ashkenazi Jews	.662	.338		99	44.8	17.4	Cleve et al. 1962
Yemenite Jews	.806	.194		49	31.3	13.2	"
SAUDI ARABIA							
Shiites & Sunnites (Eastern)	.854	.146		352	24.9	10.9	Goedde et al. 1979b
IRAN	.761	.239		178	36.4	14.9	Ohkura et al. 1984
AFGHANISTAN							
Pushtoon (Kabul)	.769	.231		210	35.5	14.6	Rahimi et al. 1977
PAKISTAN							
Punjabi	.706	.294		90	41.5	16.4	Kirk et al. 1963
INDIA							
Punjabi (Northern)	.734	.266		500	39.0	15.7	Singh et al. 1974a
Irula tr. (Nilgiri Hills,							
S. India)	.902	.098		61	17.7	8.1	Kirk et al. 1963
Bengali (E. India)	.760	.240		983	36.5	14.9	Walter et al. 1972b
BANGLADESH							
Moslems (Dacca)	.752	.248		121	37.3	15.2	Papiha et al. 1975
MALAYSIA							
Malays (Singapore)	.843	.157		51	26.5	11.5	Kenrick & Douglas 1967
THAILAND							
Thais (Bangkok)	.770	.230		163	35.4	14.6	Kirk et al. 1963
VIETNAM	.872	.128		153	22.3	9.9	Herzog et al. 1976
PHILIPPINES							
Filipino (Germany)	.892	.108		143	19.3	8.7	Windhof & Walter 1983
Negrito (Angeles, C. Luzon)	.627	.167	$.206^N$	126	53.7	28.7	Omoto et al. 1978
CHINA							
Mongolians (N. China)	.765	.235		832	36.0	14.7	Herzog et al. 1978
Tibetans (Switzerland)	.718	.282		110	40.5	16.1	Jeannet et al. 1972
Chinese (Singapore)	.756	.244		201	36.9	15.0	Kenrick & Douglas 1967
TAIWAN							
Toroko-Aborigines	.828	.172		128	28.5	12.2	Chen et al. 1985
JAPAN							
Japanese (Tokyo)	.756	.244		1,962	36.9	15.0	Yuasa et al. 1983a(c)
Ainu (Hokkaido)	.777	.223		141	34.7	14.3	Harvey et al. 1978
KOREA							
Koreans (Germany)	.696	.304		115	42.3	16.7	Bajatzadeh & Walter 1969a
AFRICA							
LIBYA (Tripoli & Benghasi)	.845	.155		168	26.2	11.4	Walter et al. 1975
EGYPT							
Egyptians (Cairo & Anshas)	.834	.166		154	27.7	11.9	Goedde et al. 1980
SUDAN							
Beja-Amarar (NE Africa)	.832	.168		95	28.0	12.0	el Hassan et al. 1968
LIBERIA	.918	.082		220	15.1	7.0	Willcox et al. 1986
MALI	.775	.225		138	34.9	14.4	Goedde et al. 1975
NIGERIA							
Yoruba (Southwestern)	.963	.037		428	7.1	3.4	Heiken et al. 1974
CAMEROON (Bamenda)	.907	.093		274	16.9	7.7	Goedde et al. 1979a
CENTRAL AFRICAN REPUBLIC							
Babinga Pygmies (Bagandou)	.964	.036		70	6.9	3.3	Cavalli-Sforza et al. 1969
ETHIOPIA	.862	.138		80	23.8	10.5	Goedde et al. 1975
UGANDA (Karamojo)	.686	.314		105	43.1	16.9	Spitsyn et al. 1978
SOUTHERN AFRICA							
San/Bushmen							
Sekele	.773	.077	$.150^{Ab}$	117	37.4	19.2	Nurse et al. 1985
Negroes/Dama (Okambahe &							
Sesfontein)	.880	.120		445	21.1	9.4	Knussmann & Knussmann 1976
Bantu (Natal)	.846	.118	$.036^{Ab}$	169	26.9	13.3	McDermid & Vos 1971b

151

Place/Population	1	2	x	n	H	PE	Source
NORTH AMERICA							
GREENLAND							
Eskimos (Thule)	.704	.296		179	41.7	16.5	Gilberg & Persson 1967
ALASKA							
Eskimos (N. Alaska)	.689	.311		103	42.9	16.8	Scott et al. 1966
Athabaskan Indians							
(Interior valley)	.847	.153		108	25.9	11.3	"
CANADA							
Eskimos (Igloolik)	.652	.337	$.010^{Igl}$	338	46.1	18.7	Cox et al. 1978
Ojibwa Indians (Pikangikum)	.821	.168	$.011^{Chip}$	92	29.8	13.4	Szathmary et al. 1974
UNITED STATES							
Blackfeet Indians (Montana)	.805	.195		95	31.4	13.2	Rokala et al. 1977
Papago Indians (S. Arizona)	.832	.168		521	28.0	12.0	Niswander et al. 1970
Whites (Southern)	.699	.301		292	42.1	16.6	Blumberg et al. 1964
Negroes (Southern)	.913	.087		231	15.9	7.3	"
MIDDLE AMERICA							
MEXICO							
Indians (San Pablo del Monte)	.819	.181		254	29.6	12.6	Crawford et al. 1974
Mestizo (Tlaxcala)	.868	.132		129	22.9	10.1	"
GUATEMALA							
Black Caribs (Livingston)	.895	.105		204	18.8	8.5	Crawford et al. 1981
WEST INDIES (Haiti)	.966	.034		308	6.6	3.2	Basu et al. 1976
SOUTH AMERICA							
BRAZIL							
Baniwa (Jandu Cachoeira)	.824	.176		377	29.5	12.6	Salzano et al. 1986
Cayapo (Para & Mato Grosso)	.640	.360		531	46.1	17.7	Salzano et al. 1972b
Caingang (Rio Grande do Sul)	.602	.398		302	47.9	18.2	Salzano & Shreffler 1966
Parakana	.372	.628		113	46.7	17.9	Black et al. 1980
Macushi (N. Brazil & S. Guyana)	.936	.064		509	12.0	5.6	Neel et al. 1977a
Wapishana (")	.820	.180		612	29.5	12.6	"
Ticuna (C. Amazonas)	.838	.162		1,765	27.2	11.7	Neel et al. 1980
CHILE							
Atacameño (Tocanao)	.767	.233		176	35.7	14.7	Goedde et al. 1985b
ECUADOR							
Shuara	.772	.228		90	35.2	14.5	Goedde et al. 1977a
FRENCH GUIANA							
Wayampi (Trois-Sauts & Camopi)	.865	.135		111	23.4	10.3	Tchen et al. 1978
SURINAM							
Trio (Southern)	.824	.176		413	29.0	12.4	Geerdink et al. 1974b
VENEZUELA							
Makiritare (S. Venezuela &							
N. Brazil)	.821	.179		535	29.4	12.5	Arends et al. 1970
Yanomama (")	.883	.117		2,503	20.7	9.3	Weitkamp et al. 1972
OCEANIA							
AUSTRALIA							
Aborigines (N. Territory)	.900	.078	$.022^{Ab}$	295	18.3	9.1	Nicholls et al. 1965
MELANESIA							
W. New Guinea							
Aborigines (S. Coast)	.708	.224	$.068^{Ab}$	250	44.4	21.9	"
Papua New Guinea							
Gainj & Kalam (N. Central							
Highlands)	.824	.152	$.023^{Ab}$	522	29.7	13.9	Long et al. 1986
Aborigines (E. Highlands)	.493	.339	$.168^{Ab}$	1,011	61.4	32.7	Cleve 1974
Aborigines (W. Highlands)	.576	.389	$.035^{Ab}$	72	51.6	22.5	"
Aborigines (S. Highlands)	.520	.440	$.040^{Ab}$	150	53.4	23.4	"
MICRONESIA							
Caroline Is. (Truk)	.892	.108		37	19.3	8.7	Kirk et al. 1963
Marshall Is.	.732	.268		362	39.2	15.8	Neel et al. 1976
POLYNESIA							
Cook Is.	.753	.247		91	37.2	15.1	Kenrick & Douglas 1967

Table 104.1 Group specific component; vitamin D binding protein (subtypes): GC, DBP

Place/Population	1F	1S	2	x	y	n	H	PE	Source
EUROPE									
BELGIUM (Liége)	.167	.543	.290			267	59.3	31.7	Hoste 1979
DENMARK (Copenhagen)	.159	.572	.269			1,674	57.5	30.6	Thymann 1981
FINLAND	.139	.661	.200			574	50.4	26.6	Lukka et al. 1986
FRANCE									
French (Pyrenees)	.171	.470	.353	$.006^{1C1}$		167	62.5	33.9	Constans et al. 1985
Basques (Navarra)	.060	.548	.392			200	54.2	25.2	"
GERMANY									
Germans	.125	.603	.272			261	54.7	28.4	Kühnl et al. 1978
Germans (Munich)	.144	.592	.261	.002		440	56.1	29.7	Cleve et al. 1978
GREECE	.140	.648	.209	$.002^{2A2}$	$.002^{2C1}$	301	51.7	27.6	Kouvatsi & Triantaphyllidis 1987
HUNGARY	.085	.572	.343			118	54.8	26.9	Goedde et al. 1986b
ICELAND	.107	.631	.262			382	52.2	26.7	Karlsson et al. 1983
ITALY									
Italians (Rome)	.158	.591	.251			397	56.3	30.0	Petrucci & Congedo 1983
Italians (Bari Dist.)	.155	.629	.214		$.002^{1C3}$	271	53.5	28.6	Walter et al. 1986
SWEDEN (Umea)	.117	.619	.264			239	53.3	27.6	Fröhlander & Ljungberg 1986
UNITED KINGDOM									
England	.165	.575	.260			100	57.5	30.7	Papiha et al. 1982a
U.S.S.R.									
EUROPEAN PART									
Abkhazia (Ochamchir									
Dist.)	.192	.616	.192			112	54.7	29.4	Ferrell et al. 1985
ASIA									
ISRAEL									
Arabs	.212	.602	.186			342	55.8	30.0	Nevo & Cleve 1983
YEMEN									
Bedouins (Northern)	.270	.589	.137	$.004^{1A1}$		135	56.1	29.8	Constans et al. 1985
IRAQ									
Kurds	.224	.596	.172	$.008^{1A1}$		58	56.5	30.8	"
IRAN									
Zooastrians	.144	.629	.227		$.003^{1CJ}$	236	53.2	28.2	Papiha et al. 1982b
INDIA									
N. Indians (Delhi)	.135	.555	.309			488	57.8	30.2	Kamboh et al. 1984
Jat Sikhs (Punjab)	.154	.531	.312			146	59.7	31.8	Papiha et al. 1982a
Tamil (S. India)	.143	.625	.232			112	53.5	28.3	Constans et al. 1985
Koya Dora tr. (")	.133	.678	.189			177	48.7	25.7	Walter et al. 1984
Brahmins (E. India)	.180	.528	.292			89	60.4	32.4	Walter et al. 1986
Ahoms (")	.297	.444	.259			116	64.8	35.7	"
NEPAL									
Jirel Sherpa	.251	.518	.226	$.005^{1A9}$		195	61.8	34.0	Constans et al. 1985
Nepalese	.245	.482	.273			144	63.3	34.7	Yuasa et al. 1983b
MALAYSIA									
Malays	.795	.149	.056			134	34.3	17.2	Tan et al. 1981
INDONESIA									
Javanese	.534	.281	.176	.008		176	60.5	33.1	Matsumoto et al. 1980
Balinese (Bali Is.)	.710	.246	.044			136	43.3	20.3	Constans et al. 1986
THAILAND									
Thais (Bangkok)	.405	.354	.236	$.005^{1A3}$		199	65.5	36.4	Constans et al. 1985
Thais	.395	.342	.263			114	65.8	36.3	Goedde & Benkmann 1987
CHINA									
Chinese	.478	.257	.261	$.002^{1A2}$	$.002^{1C4}$	113	63.7	35.3	Kamboh et al. 1984
Tibetans (Dogpa)	.364	.376	.249	$.009^{1A9}$	$.002^{1C7}$	231	66.4	37.5	
Chinese (Hong Kong)	.494	.258	.247			362	62.8	34.4	Kwok & Lewis 1981
Chinese (Malaysia)	.678	.157	.165			121	48.8	26.0	Tan et al. 1981
JAPAN									
Japanese (Yamaguchi									
& Izumo)	.448	.245	.278	$.019^{1A2}$	$.010^{c}$	1,000	66.2	38.6	Yuasa et al. 1984a
Japanese (Osaka)	.421	.301	.258	.021		316	66.5	38.3	Matsumoto et al. 1980
Ainu (Hokkaido)	.579	.203	.208	.009		271	58.0	31.9	"
KOREA									
Koreans (Seoul)	.403	.307	.290			150	65.9	36.5	Goedde et al. 1986c

Continued

Table 104.1. Group specific component; vitamin D binding protein (subtypes): GC, DBP (cont'd)

Place/Population	1F	1S	2	x	y	n	H	PE	Source
Koreans	.434	.234	.304	.028		303	66.4	38.6	Matsumoto et al. 1980
AFRICA									
ALGERIA	.482	.446	.054	.005[A1]	.013[a]	161	56.6	27.3	Constans et al. 1985
SENEGAL	.780	.115	.053	.027[1A1]	.025[b]	357	37.4	21.0	"
CONGO & CAMEROON									
Bantu/Bamileke	.825	.077	.086	.004[1A1]	.008[1C10]	123	30.6	16.3	Constans et al. 1985
CENTRAL AFRICAN REPUBLIC									
Bi-Aka Pygmies	.608	.185	.083	.034[1A1]	.090[2A3]	914	58.0	35.8	Constans et al. 1981b
SOUTHERN AFRICA									
San/Bushmen	.629	.306	.065			31	50.6	24.2	Constans et al. 1985
Bantu (Transkei)	.841	.099	.048	.008[1AT]	.004[1C34]	126	28.1	14.6	Papiha et al. 1985
NORTH AMERICA									
ALASKA									
Eskimos	.267	.492	.189	.052		328	64.8	38.6	Matsumoto et al. 1980
CANADA									
Dogrib Indians									
(NW Territories)	.380	.551	.070			158	54.7	26.0	Szathmary et al. 1983
UNITED STATES									
Whites (Minnesota)	.155	.566	.279			7,247	57.8	30.7	Dykes et al. 1983a
Blacks (")	.678	.186	.106	.015[1A1]	.015[1C10]	540	49.4	27.5	"
Whites (Pennsylvania)	.149	.572	.279			110	57.3	30.3	Kueppers & Harpel 1979
Blacks (Baltimore)	.667	.167	.130	.032[1A1]	.004[2A3]	126	50.9	28.9	Constans et al. 1985
Blackfeet Indians	.189	.615	.196			74	54.8	29.4	Dykes et al. 1983a
Navaho "	.403	.529	.068			103	55.3	26.1	"
Pima "	.449	.410	.141			332	61.0	31.8	"
Apache "	.318	.596	.086			457	53.6	26.5	"
Mexican-Americans									
(Texas)	.195	.523	.282			692	60.9	32.9	Hewett-Emmett et al. 1986a
MIDDLE AMERICA									
GUATEMALA									
Black Caribs									
(Livingston)	.637	.256	.107			215	51.7	26.5	Dykes et al. 1983a
SOUTH AMERICA									
BOLIVIA									
Aymara (Altiplano)	.231	.637	.123	.009[1A9]		253	52.6	28.1	Constans et al. 1985
Quechua (Tarabulo)	.178	.714	.105	.003[1A9]		171	44.7	23.5	"
BRAZIL									
Caingang	.200	.480	.320			106	62.7	34.0	Constans et al. 1985
Pacaás Novos	.451	.370	.179			216	62.8	33.7	Salzano et al. 1985
Macushi	.448	.453	.099			211	58.4	29.0	Constans et al. 1985
CHILE									
Atacameño (Toconao,									
N. Chile)	.301	.466	.233			176	63.8	34.9	Goedde et al. 1985b
OCEANIA									
AUSTRALIA									
Aborigines	.311	.587	.034	.034[1A1]	.034[1C2]	29	55.5	29.6	Constans et al. 1985
MELANESIA									
Papua New Guinea									
Fore (Kuru area)	.244	.250	.343	.163[1A1]		80	73.4	48.4	"
New Hebrides	.288	.386	.284	.042[1A1]		141	68.6	41.0	"
Solomon Is.	.385	.303	.311			66	66.3	36.8	"
MICRONESIA									
W. Caroline Is.									
Fais Is.	.538	.187	.275			131	60.0	32.3	"
POLYNESIA									
Samoans (Western)	.352	.362	.286			192	66.3	36.8	Kamboh et al. 1984
Cook Is.	.488	.255	.257			200	63.1	34.6	"

Note: [a] Pooled frequency of alleles 1C3 and 2A3. [b] Pooled frequency of alleles 1C3 and 2A5. [c] Pooled frequency
of alleles 1A3, 1A9, 1C2, 1C35, and 2A4.

154

Table 105. Haptoglobin, alpha: HPA*

Place/Population	1	2	x	n	H	PE	Source
EUROPE							
AUSTRIA	.386	.614		973	47.4	18.1	Pausch et al. 1985
BELGIUM (Liége)	.416	.584		800	48.6	18.4	Brocteur et al. 1980
BULGARIA	.379	.621		136	47.1	18.0	Walter et al. 1972a
CZECHOSLOVAKIA (Ovcia)	.286	.714		316	40.8	16.3	Matousek & Seemanová 1973
DENMARK	.403	.597		2,408	48.1	18.3	Berg 1974(c)
FAROE IS. (N. Atlantic)	.436	.564		674	49.2	18.5	Tills et al. 1985
FINLAND	.381	.619		5,536	47.2	18.0	Nevanlinna 1972
FRANCE							
French (Paris)	.391	.609		1,750	47.6	18.1	Moullec et al. 1961
Basques	.381	.619		63	47.2	18.0	Levine et al. 1974
GERMANY	.398	.602		4,242	47.9	18.2	Hennig & Hoppe 1964
GREECE							
Greeks	.339	.661		2,026	44.8	17.4	Germenis et al. 1983
Greeks (Plati)	.368	.632		1,038[a]	46.5	17.8	Tills et al. 1983b
HUNGARY	.364	.636		125	46.3	17.8	Goedde et al. 1986b
ICELAND	.394	.605		1,634	47.9	18.2	Tills et al. 1982b
IRELAND	.380	.620		1,766	47.1	18.0	Tills 1977
ITALY (Roma)	.383	.617		355	47.3	18.0	Serafini et al. 1968
NETHERLANDS (Leiden)	.421	.579		801	48.8	18.4	Fraser et al. 1974
NORWAY							
Norwegians	.380	.620		6,484	47.1	18.0	Berg 1974(c)
Lapps/Saami (Finnmark)	.311	.689		296	42.9	16.8	Fleischer & Monn 1970
POLAND (S. Poland)	.362	.638		22,296	46.2	17.8	Turowska et al. 1977
PORTUGAL							
Portugese	.394	.606		1,000	47.8	18.2	Torrinha 1967
Portugese (Terras de Bouro)	.295	.705		117	41.6	16.5	Cruz et al. 1973
SPAIN							
Spanish (Madrid)	.394	.606		211	47.8	18.2	Goedde et al. 1973
Basques (Bermeo/Pamplona)	.441	.559		274	49.3	18.6	"
SWEDEN							
Swedes (Northern)	.368	.632		4,333	46.5	17.8	Beckman et al. 1983
Lapps/Saami	.316	.684		322	43.2	16.9	Beckman & Mellbin 1959
SWITZERLAND (Bern)	.446	.554		444	49.4	18.6	Bütler et al. 1959
UNITED KINGDOM							
England (Nottingham)	.394	.606		533	47.8	18.2	Cartwright et al. 1977
Scotland (South-west)	.399	.601		874	48.0	18.2	Mitchell et al. 1976
Isle of Lewis	.431	.569		94	49.0	18.5	Clegg et al. 1985
Isle of Man	.350	.650		354	45.5	17.6	Mitchell et al. 1982
Orkney Is. (Westray)	.380	.620		300	47.1	18.0	Welch et al. 1973
YUGOSLAVIA	.373	.627		459	46.8	17.9	Grünwald & Herman 1963
U.S.S.R.							
EUROPEAN PART							
Russians (Moscow)	.339	.661		1,540	44.8	17.4	Altukhov et al. 1981
Abkhazia (Ochamchir Dist.)	.364	.636		287	46.3	17.8	Ferrell et al. 1985
ASIAN PART							
Eskimos (New Chaplino)	.242	.758		99	36.7	15.0	Sukernik et al. 1981
Chukchi (Chukotka/N. Kamchatka)	.275	.725		1,057	39.9	16.0	"
Nganasan-Avam (Taimir Pen.)	.239	.761		270	36.4	14.9	Sukernik et al. 1978
Nentzi (Between Ob R. & Taj R.)	.390	.610		666	47.6	18.1	Sukernik et al. 1980
ASIA							
TURKEY (Asia Minor)	.313	.687		299	43.0	16.9	Erdem et al. 1966
ISRAEL							
Ashkenazi Jews	.290	.710		499	41.2	16.4	Goldschmidt et al. 1962
Yemenite Jews	.312	.678		155	44.3	17.2	Tills et al. 1977b
JORDAN							
Bedouins	.326	.674		109	43.9	17.1	Saha & Banerjee 1986
KUWAIT							
Arabs	.345	.655		161	45.2	17.5	Sawhney et al. 1984
SAUDI ARABIA							
Shiites & Sunnites (al-Hasa)	.432	.568		343	49.1	18.5	Goedde et al. 1979b

Continued

Table 105. Haptoglobin, alpha: HPA* (cont'd)

Place/Population	1	2	x	n	H	PE	Source
Tribes (Western)	.366	.634		292	46.4	17.8	Saha et al. 1980
Arabs (Southern)	.445	.555		261	49.4	18.6	Marengo-Rowe et al. 1974
IRAN							
Turkoman (Gonbad, NE Iran)	.217	.783		152	34.0	14.1	Kirk et al. 1977
AFGHANISTAN							
Pushtu & Dari (Kabul)	.271	.729		253	39.5	15.9	Papiha et al. 1977a
PAKISTAN							
Punjabi	.196	.804		201	31.5	13.3	Kirk & Lai 1961
INDIA							
Punjabi (Chandigarh)	.237	.763		112	36.2	14.8	Papiha 1973
Gujarati (W. India)	.143	.857		164	24.5	10.8	Papiha et al. 1981
Hindu & Moslems							
(Madhya Pradesh)	.226	.774		301	35.0	14.4	Roberts et al. 1974
Bhil tr. (Madhya Pradesh)	.103	.897		136	18.5	8.4	Papiha et al. 1978
Hindu (Andhra Pradesh)	.161	.839		84	27.0	11.7	Naidu et al. 1985
Irula tr. (Nilgiri Hills,							
S. India)	.059	.941		169	11.1	5.2	Saha et al. 1976
Bengali (E. India)	.154	.845	.001[J]	309	26.2	11.5	Mukherjee & Das 1970
SRI LANKA							
Sinhalese (Anuradhapura)	.180	.820		147	29.5	12.6	Papiha 1973
Veddah	.133	.867		60	23.1	10.2	Kirk et al. 1962
BANGLADESH							
Moslems (Dacca)	.185	.815		192	30.2	12.8	Papiha et al. 1975
NEPAL							
Nepalese	.258	.742		248	38.3	15.5	Sunderland et al. 1979
Nepalese	.177	.823		144	29.1	12.4	Yuasa et al. 1983b
BHUTAN	.130	.870		154	22.6	10.0	Mourant et al. 1969
MALAYSIA							
Malays	.240	.760		266	36.5	14.9	Lie-Injo et al. 1967
Negrito	.050	.950		N.A.	9.5	4.5	Lie-Injo 1976
INDONESIA							
Batak (Samosir Is.)	.298	.702		168	41.8	16.5	McDermid et al. 1973
Balinese (Bali Is.)	.279	.721		140	40.2	16.1	Breguet et al. 1982a
THAILAND (Bangkok)	.226	.774		274	35.0	14.4	Kirk & Lai 1961
VIETNAM (N. Vietnam)	.356	.644		150	45.9	17.7	Herzog et al. 1976
PHILIPPINES							
Filipino (Germany)	.433	.567		142	49.1	18.5	Windhof & Walter 1983
Negrito (Angeles, C. Luzon)	.219	.781		129	34.2	14.2	Omoto et al. 1978
CHINA							
Mongolians (N. China)	.280	.720		1,615	40.3	16.1	Herzog et al. 1978
Mongolians (")	.260	.740		186	38.5	15.5	Xu et al. 1986
Han (Beijing)	.270	.730		1,121[a]	39.4	15.8	Chih-chuan et al. 1983
Zhuang (S. China)	.357	.643		211	45.9	17.7	Xu et al. 1986
Hui (Ningxia)	.352	.648		219	45.6	17.6	Niu & Du 1984
Dong (Guangxi)	.348	.652		201	45.4	17.5	"
Tibetans (Switzerland)	.202	.798		109	32.2	13.5	Jeannet et al. 1972
Chinese (SE Asia)	.280	.720		165	40.3	16.1	Kirk & Lai 1961
TAIWAN							
Toroko- Aborigines	.256	.744		129	38.1	15.4	Chen et al. 1985
JAPAN							
Japanese (Tokyo)	.266	.734		300	39.0	15.7	Ishimoto et al. 1967
Ainu (Hokkaido)	.113	.887		244	20.0	9.0	Harvey et al. 1978
KOREA							
Koreans (Seoul)	.243	.757		148	36.8	15.0	Goedde et al. 1986c
Koreans (NE. China)	.315	.685		214	43.2	16.9	Xu et al. 1986

AFRICA

LIBYA (Tripoli & Benghasi)	.438	.562		167	49.2	18.6	Walter et al. 1975
EGYPT							
Egyptians (Cairo & Anshas)	.390	.610		123	47.6	18.1	Goedde et al. 1980
Egyptians	.341	.659		505	44.9	17.4	Habib 1983
Nubians (S. Egypt)	.384	.616		146	47.3	18.1	Bertin et al. 1978
SUDAN							
Sudanese (Khartoum)	.358	.642		264	46.0	17.7	Saha et al. 1978b
Beja-Amarar (NE Sudan)	.495	.505		98	50.0	18.7	el Hassan et al. 1968
CHAD							
Sara Majingay	.489	.511		193	50.0	18.7	Hiernaux 1976

Continued

Table 105. Haptoglobin, alpha: HPA* (cont'd)

Place/Population	1	2	x	n	H	PE	Source
MALI							
Kel Kummer Twareg (Menaka)	.444	.556		285	49.4	18.6	Constans et al. 1981a
SENEGAL							
Bedik (E. Senegal)	.725	.275		496	39.9	16.0	Bouloux et al. 1972
LIBERIA	.703	.297		229	41.8	16.5	Willcox et al. 1986
IVORY COAST							
N'Da tr.	.616	.384		253	47.3	18.1	Constans et al. 1981a
NIGERIA							
Yoruba (Southwestern)	.686	.314		425	43.1	16.9	Heiken et al. 1974
CAMEROON (Bamenda)	.683	.317		275	43.3	17.0	Goedde et al. 1979a
CENTRAL AFRICAN REPUBLIC							
Babinga Pygmies	.386	.614		88	47.4	18.1	Cavalli-Sforza et al. 1969
Bi Aka Pygmies	.352	.648		606	45.6	17.6	Constans et al. 1981b
BURUNDI							
Hutu	.567	.433		91	49.1	18.5	Giblett et al. 1966
ETHIOPIA							
Amhara tr.	.447	.553		107	49.4	18.6	Barnicot et al. 1962
UGANDA							
Baganda (Kampala)	.634	.366		82	46.4	17.8	Roberts et al. 1977
KENYA	.568	.432		227	49.1	18.5	Herzog et al 1970
TANZANIA							
Sandawe (Kondoa-Irang Dt.)	.639	.356	$.005^{2M}$	184	46.6	18.4	Godber et al. 1976
Nyaturu (")	.569	.420	$.011^{2M}$	175	50.0	19.9	"
Hadza (W. Lake Eyasi)	.582	.413	$.005^{2M}$	104	49.1	19.0	Tills et al. 1982a
ANGOLA							
Njinga	.682	.318		74	43.4	17.0	Nurse et al. 1982
SOUTHERN AFRICA							
Baster (Namibia)	.523	.477		107	49.9	18.7	Nurse et al. 1982
San/Bushmen							
!Kung, Dobe	.318	.682		371	43.4	17.0	Nurse & Jenkins 1977
Khoi/Hottentot							
Keetmanshoop Nama	.594	.406		153	48.2	18.3	Jenkins 1972
Negroes/Dama (Okombahe &							
Sesfontein)	.644	.356		111	45.9	17.7	Nurse et al. 1976
Bantu (Natal)	.514	.486		295	50.0	18.7	McDermid & Vos 1971a
Griqua (Campbell, Cape Province)	.545	.455		77	49.6	18.6	Nurse & Jenkins 1975
TRISTAN DA CUNHA	.444	.556		160	49.4	18.6	Jenkins et al. 1985

NORTH AMERICA

Place/Population	1	2	x	n	H	PE	Source
GREENLAND							
Eskimos (Thule)	.341	.659		179	44.9	17.4	Gilberg & Persson 1967
ALASKA							
Eskimos (Northern)	.320	.680		104	43.5	17.0	Scott et al. 1966
Eskimos (St. Lawrence Is.)	.322	.678		219	43.7	17.1	Ferrell et al. 1981b
Athabaskan Indians							
(Interior valley)	.365	.635		104	46.4	17.8	Scott et al. 1966
CANADA							
Eskimos (Igloolik)	.344	.656		356	45.1	17.5	McAlpine et al. 1974
Cree Indians (Plains)	.412	.588		605	48.5	18.4	Bowen et al. 1971
Ojibwa Indians (Pikangikum)	.260	.740		98	38.5	15.5	Szathmary et al. 1974
Dogrib Indians (NW Territory)	.361	.639		158	46.1	17.7	Szathmary et al. 1983
UNITED STATES							
Blackfeet Indians (Montana)	.458	.542		95	49.6	18.7	Rokala et al. 1977
Papago Indians (S. Arizona)	.452	.548		693	49.5	18.6	Niswander et al. 1970
Whites (Southern)	.410	.590		143	48.4	18.3	Cooper et al. 1963
Negroes (")	.520	.480		164	49.9	18.7	"
Whites (Louisiana)	.395	.605		369	47.8	18.2	Fox et al. 1981
Blacks (")	.488	.512		308	50.0	18.7	"
Mexican-Americans (Texas)	.426	.574		679	48.9	18.5	Hewett-Emmett et al. 1986b

MIDDLE AMERICA

Place/Population	1	2	x	n	H	PE	Source
MEXICO							
Indians (San Pablo del Monte)	.548	.452		252	49.5	18.6	Crawford et al. 1974
Mestizo (Tlaxcala)	.559	.441		135	49.3	18.6	"
GUATEMALA							
Black Caribs (Livingston)	.596	.395	$.010^{2M}$	204	48.9	19.5	Crawford et al. 1981
COSTA RICA							

Continued

Table 105. Haptoglobin, alpha: HPA* (cont'd)

Place/Population	1	2	x	n	H	PE	Source
Guaymi	.570	.430		286	49.0	18.5	Barrantes et al. 1982
PANAMA							
Blacks (Costa Arriba)	.656	.344		411	45.1	17.5	Ferrell et al. 1978b
WEST INDIES (Haiti)	.589	.411		451	48.4	18.3	Basu et al. 1976

SOUTH AMERICA

Place/Population	1	2	x	n	H	PE	Source
ARGENTINA							
Mapuchi	.807	.193		70	31.2	13.1	Haas et al. 1985
BOLIVIA							
Aymara (W. Bolivia)	.698	.302		286	42.2	16.6	Ferrell et al. 1978a
BRAZIL							
Baniwa (Jandu Cachoeira)	.516	.484		363	49.9	18.7	Salzano et al. 1986
Cayapo (Para & Mato Grosso)	.580	.420		504	48.7	18.4	Salzano et al. 1972b
Caingang	.716	.284		214	40.7	16.2	Salzano et al. 1980
Parakana	.247	.753		89	37.2	15.1	Black et al. 1980
Pacaás Novos	.817	.183		208	29.9	12.7	Salzano et al. 1985
Macushi (N. Brazil & S. Guyana)	.538	.462		509	49.7	18.7	Neel et al. 1977b
Wapishana (")	.475	.525		623	49.9	18.7	"
Ticuna (C. Amazonas)	.662	.338		1,765	44.8	17.4	Neel et al. 1980
CHILE							
Atacameño (Toconao, N. Chile)	.700	.300		175	42.0	16.6	Goedde et al. 1985b
COLOMBIA							
Noanama	.432	.568		155	49.1	18.5	Kirk et al. 1974
ECUADOR							
Shuara	.645	.355		90	45.8	17.7	Goedde et al. 1977a
SURINAM							
Trio (S. Surinam)	.528	.472		483	49.8	18.7	Geerdink et al. 1974b
VENEZUELA							
Makiritare (S. Venezuela &							
N. Brazil)	.420	.580		525	48.7	18.4	Arends et al. 1970
Yanomama (")	.830	.170		2,224	28.2	12.1	Weitkamp et al. 1972

OCEANIA

Place/Population	1	2	x	n	H	PE	Source
AUSTRALIA							
Aborigines/ Malag (Elcho Is.,							
N. Territory)	.162	.838		643	27.2	11.7	Kirk et al. 1969
Aborigines (Amoonguna,							
C. Australia)	.207	.793		104	32.8	13.7	Kirk et al. 1971a
Aborigines/Ngarinjin							
(Kimberley, W. Australia)	.212	.788		104	33.4	13.9	Blake & Spargo 1986
MELANESIA							
W. New Guinea							
Asmat (S. Coastal Plain)	.797	.203		113	32.4	13.6	Gajdusek et al. 1978
Papua New Guinea							
Gainj & Kalam (N. Central							
Highlands)	.739	.261		470	38.6	15.6	Long et al. 1986
Jimi Valley (W. Highlands)	.678	.322		325	43.7	17.1	Mourant et al. 1981
Yagaria (E. Highlands)	.613	.387		397	47.4	18.1	Mourant et al. 1982
Fuyuge (C. Highlands)	.690	.310		147	42.8	16.8	Woodfield et al. 1974
Daga (Bay Province)	.760	.240		151	36.5	14.9	Jenkins et al. 1983
Karkar Is.							
Takia	.736	.264		265	38.9	15.7	Booth et al. 1982
Manus Is.	.632	.368		133	46.5	17.8	Malcolm et al. 1972
Buka Is.	.506	.494		79	50.0	18.7	McLoughlin et al. 1982a
Solomon Is.	.554	.446		2,544	49.4	18.6	Lai & Bloom 1982
MICRONESIA							
E. Caroline Is.							
(Pingelap)	.750	.250		414	37.5	15.2	Morton & Yamamoto 1973
Marshall Is.	.521	.479		364	49.9	18.7	Neel et al. 1976
POLYNESIA							
Samoans (Christchurch,							
New Zealand)	.552	.448		77	49.5	18.6	Booth et al. 1977

Note: [a] Variant "Johnson" was found in one individual.

Table 105.1. Haptoglobin (subtypes): HPA*

Place/Population	1S	1F	2	n	H	PE	Source
EUROPE							
FINLAND							
Finns	.327	.241	.431	58	64.9	35.7	Ehnholm 1969
Lapps/Saami	.525	.067	.408	60	55.3	26.0	Ehnholm & Eriksson 1969
GERMANY	.261	.144	.595	232	55.7	29.4	Cleve 1966
GREECE	.203	.136	.660	2,026	50.5	26.6	Angelopoulos et al. 1966
HUNGARY	.118	.221	.661	675	50.0	26.1	Hevér & Hajpal 1978
IRELAND	.098	.229	.673	188	48.5	24.8	Kehr-Löke et al. 1966
ITALY	.213	.136	.650	441	51.4	27.1	Santoro et al. 1983
NORWAY							
Norwegians	.236	.140	.624	388	53.5	28.4	Fleischer & Monn 1970
Lapps/Saami	.218	.093	.689	124	46.9	23.8	"
POLAND	.218	.155	.626	222	53.7	28.5	Schlesinger 1971
SPAIN (Barcelona)	.238	.148	.620	317	53.7	28.6	Moral & Panadero 1983
ASIA							
IRAN	.162	.108	.729	360	43.1	22.5	Farhud & Walter 1972
INDIA							
Indians (Bombay)	.104	.046	.850	85	26.5	13.4	Shim & Bearn 1964
INDONESIA							
Balinese (Bali Is.)	.288		.712	210	41.0	16.3	Constans et al. 1986
THAILAND	.236		.764	68	36.1	14.8	Shim & Bearn 1964
PHILIPPINES							
Filipino (Honolulu)	.250	.190	.560	42	58.8	31.7	Kehr-Löke et al. 1966
CHINA							
Chinese (New York, USA)	.341		.659	113	44.9	17.4	Shim & Bearn 1964
JAPAN							
Japanese	.227	.003	.771	170	35.4	14.8	"
KOREA	.321		.679	120	43.6	17.0	"
AFRICA							
NIGERIA	.258	.473	.269	78	63.7	35.0	"
CONGO & CAMEROON							
Bi-Aka Pygmies	.171	.165	.664	464	50.3	26.8	Constans et al. 1981b
NORTH AMERICA							
CANADA							
Eskimos	.239		.761	67	36.4	14.9	"
UNITED STATES							
Whites	.251	.133	.616	66	53.6	28.1	Giblett & Brooks 1963
Blacks	.293	.259	.448	222	64.6	35.5	"
Am. Indians (Alabama)	.374		.626	143	46.8	17.9	Shim & Bearn 1964
SOUTH AMERICA							
CHILE							
Araucanian	.774		.226	31	35.0	14.4	"
OCEANIA							
AUSTRALIA							
Aborigines (Arnhem Land)	.290		.710	50	41.2	16.4	"
Aborigines (W. Desert)	.169	.025	.806	101	32.1	15.0	"

Table 106. Hemoglobin, beta: HBB

Place/Population	A	S	E	C	D	n	H	PE	Source
EUROPE									
FAROE IS. (N. Atlantic)	1.000					668			Tills et al. 1985
FINLAND									
Finns	1.000					1,769			Nilsson & Eriksson 1972
Lapps/Saami	1.000					1,776			"
GREECE	.988	.012				215	2.4	1.2	Schneider et al. 1975
ICELAND	1.000					499			Bjarnason et al. 1973
IRELAND	1.000					1,800			Tills 1977
ITALY	1.000					14,536			Silvestroni et al. 1978
NETHERLANDS (Leiden)	1.000					804			Fraser et al. 1974
PORTUGAL (Lisbon)	.9995	.0005				3,042	.2	.1	Trincao & Cordeiro 1962
SWEDEN	1.000					3,171			Nilsson & Eriksson 1972
UNITED KINGDOM									
England	1.000					1,971			Liddell et al. 1964
Isle of Lewis	1.000					95			Clegg et al. 1985
U.S.S.R.									
EUROPEAN PART									
Russians (Moscow)	1.000					1,695			Altukhov et al. 1981
Abkhazia (Ochamchir									
Dist.)	1.000					262			Ferrell et al. 1985
Maris (Mari Republic)	1.000					317			Nilsson & Ericksson 1972
ASIA									
KUWAIT									
Arabs	.989	.011				89	2.2	1.1	Al-Nassar et al. 1981
SAUDI ARABIA									
Shiites & Sunnites									
(al-Hasa)	.919	.081				357	14.9	6.9	Goedde et al. 1979b
Tribes (Western)	.984	.016				334	3.1	1.5	Bayoumi et al. 1979
UNITED ARAB EMIRATES									
Abu Dhabians	.988	.012	'			500	2.4	1.2	Kamel et al. 1980
IRAN									
Turkoman (Gonbad,									
NE Iran)	.997				.003	155	.6	.3	Kirk et al. 1977
INDIA									
Punjabi (Chandigarh,									
N. India)	.989	.011				140	2.2	1.1	Papiha 1973
Gujaratis (W. India)	.988	.012				281	2.4	1.2	Papiha et al. 1981
Hindu & Moslems									
(Madhya Pradesh)	.985	.015				339	3.0	1.5	Roberts et al. 1974
Bhil tr. (Madhya Pradesh)	.914	.086				145	15.7	7.2	Papiha et al. 1978
Hindu (Andhra Pradesh)	1.000					85			Naidu et al. 1985
Irula tr. (Nilgiri Hills,									
S. India)	.857	.143				175	24.5	10.8	Saha et al. 1976
Bengali (E. India)	.980		.012		.008	490	3.9	1.2	Ghosh et al. 1981
SRI LANKA									
Sinhalese (Anuradhapura)	.991	.003	.006			157	1.8	.9	Papiha 1973
Veddah	.891		.109			184	19.4	8.8	Wickremasinghe et al. 1963
BANGLADESH									
Moslems (Dacca)	.975	.002	.023			200	4.9	2.4	Papiha et al. 1975
NEPAL									
Sherpas	.980		.020			74	3.9	1.9	Santachiara et al. 1976
BHUTAN	.981		.019			289	3.7	1.8	Mourant et al. 1969
MALAYSIA									
Malays (Singapore)	.971		.027		.002	259	5.6	2.8	Blake et al. 1973a
INDONESIA									
Batak (Samosir Is.)	1.000					188			McDermid et al. 1973
Balinese (Bali Is.)	.997		.003			148	.6	.3	Breguet et al. 1982b
THAILAND (Bangkok)	.947		.054			149	10.0	4.8	Flatz et al. 1965
VIETNAM	.975		.025			259	4.9	2.4	Bowman et al. 1971
CHINA									
Tibetans	1.000					13			Santachiara et al. 1976
Chinese (Malaysia)	.997		.003			378	.6	.3	Blake et al. 1973a

Continued

Table 106. Hemoglobin, beta: HBB (cont'd)

Place/Population	A	S	E	C	D	n	H	PE	Source
JAPAN									
Japanese	1.000					2,200			Ohta 1963
AFRICA									
LIBYA (Tripoli)	.996	.002		.002		680	.8	.4	Kamel et al. 1975
EGYPT									
Nubians (Southern)	1.000					155			Bertin et al. 1978
SUDAN									
Sudanese (Khartoum)	.954	.046				294	8.8	4.2	Saha et al. 1978b
Beja-Amarar (NE Sudan)	1.000					100			el Hassan et al. 1968
CHAD									
Sara Majingay	.979	.021				257	4.1	2.0	Hiernaux 1976
SENEGAL									
Bedik (Eastern)	.858	.142				780	24.4	10.7	Bouloux et al. 1972
Bedik	.858	.142				875	24.4	10.7	Mauran-Sendrall et al. 1975
Niokholonko	.893	.096		.004	.007	612	19.3	8.5	"
LIBERIA	.950	.045	.006			616	9.5	4.2	Willcox & Beckman 1981
UPPER VOLTA	.846	.070		.084		1,059	27.2	14.0	Labie et al. 1984
NIGERIA									
Yoruba	.820	.133		.047		347	30.8	15.3	Roberts & Boyo 1962
CAMEROON	.905	.095				1,183	17.2	7.9	Bernstein et al. 1980
CENTRAL AFRICAN REPUBLIC									
Babinga Pygmies	.917	.083				96	15.2	7.0	Cavalli-Sforza et al. 1969
BURUNDI									
Hutu	.936	.064				1,404	12.0	5.6	Gall et al. 1982
ETHIOPIA									
Amhara tr.	1.000					115			Lehmann et al. 1962
KENYA	.881	.104	.014[F]		.001	473	21.3	9.7	Herzog et al. 1970
TANZANIA									
Sandawe	.986	.014				218	2.8	1.4	Godber et al. 1976
Nyaturu	.963	.037				216	7.1	3.4	"
ANGOLA									
Njinga	.794	.202			.004	104	32.9	13.6	Nurse et al. 1979
SOUTHERN AFRICA									
San/Bushmen									
Dobe	1.000					424			Jenkins 1972
Khoi/Hottentot									
Keetmanshoop Nama	1.000					146			"
Negroes/Dama (Okombahe)	1.000					92			Nurse et al. 1976
Bantu	.9994	.0006				1,741	.2	.1	Griffiths 1954
Griqua (Campbell, Cape									
Province)	1.000					100			Nurse & Jenkins 1975
TRISTAN DA CUNHA	1.000					66			Jenkins et al. 1985
NORTH AMERICA									
GREENLAND									
Eskimos	1.000					153			Nilsson & Eriksson 1972
ALASKA									
Eskimos	1.000					N.A.			Scott 1973
Eskimos									
(St. Lawrence Is.)	1.000					222			Ferrell et al. 1981b
Athabaskan Indians	1.000					N.A.			Scott 1973
CANADA									
Eskimos (Igloolik)	1.000					361			McAlpine et al. 1974
Dogrib Indians (NW									
Territories)	1.000					158			Szathmary et al. 1983
UNITED STATES									
Whites (Louisiana)	1.000					376			Fox et al. 1981
Blacks (")	.954	.038		.008		373	8.8	4.4	"
Blacks (Southern)	.946	.042	.004[F]	.008		247	10.2	5.0	Cooper et al. 1963
MIDDLE AMERICA									
GUATEMALA									
Black Caribs									
(Livingston)	.923	.077				202	14.2	6.6	Crawford et al. 1981

Continued

Table 106. Hemoglobin, beta: HBB (cont'd)

Place/Population	A	S	E	C	D	n	H	PE	Source
COSTA RICA									
Guaymi	1.000					286			Barrantes et al. 1982
PANAMA									
Blacks (Costa Arriba)	.910	.081		.010		421	16.5	8.0	Ferrell et al. 1978b
WEST INDIES (Haiti)	.934	.058		.008		463	12.4	6.0	Basu et al. 1976

SOUTH AMERICA

Place/Population	A	S	E	C	D	n	H	PE	Source
ARGENTINA									
Mapuchi	1.000					96			Haas et al. 1985
BOLIVIA									
Aymara (Western)	1.000					2,720			Ferrell et al. 1978a
BRAZIL									
Baniwa (Jandu Cachoeira)	1.000					377			Salzano et al. 1986
Cayapo (Pará & Mato Grossa)	1.000					493			Salzano et al. 1972b
Caingang	1.000					214			Salzano et al. 1980
Parakana	1.000					97			Black et al. 1980
Pacaás Novos	1.000					217			Salzano et al. 1985
Macushi (N. Brazil & S. Guyana)	1.000					500			Neel et al. 1977b
Wapishana (")	1.000					615			"
Ticuna (C. Amazonas)	1.000					1,765			Neel et al. 1980
Tiriyo	1.000					127			Salzano et al. 1974
FRENCH GUIANA									
Wayampi (Trois-Sauts & Camopi)	1.000					213			Tchen et al. 1978
VENEZUELA									
Makiritare (S. Venezuela & N. Brazil)	1.000					484			Tanis et al. 1973
Yanomama (")	1.000					2,506			"

OCEANIA

Place/Population	A	S	E	C	D	n	H	PE	Source
AUSTRALIA									
Aborigines /Malag (Elcho Is., N. Territory)	1.000					272			Blake 1979
Aborigines (Hooker Creek, N. Territory)	1.000					285			"
Aborigines (Doomadgee, C. Australia)	1.000					106			"
Aborigines/Walmadjari (Kimberly, W. Australia)	1.000					171			Blake & Spargo 1986
MELANESIA									
W. New Guinea									
Asmat (S. Coastal Plain)	1.000					653			Gajdusek et al. 1978
Papua New Guinea									
Gainj & Kalam (N. Central Highlands)	1.000					588			Long et al. 1986
Fuyuge (C. Highlands)	1.000					173			Woodfield et al. 1974
Daga (Bay Province)	1.000					139			Jenkins et al. 1983
Manus Is.	1.000					134			Malcolm et al. 1972
MICRONESIA									
Marshall Is.	1.000					375			Neel et al. 1976
POLYNESIA									
Samoans (Christchurch, New Zealand)	1.000					77			Booth et al. 1977

Note: There are many variant alleles at this locus. See Lehmann and Kynoch (1976) and Livingstone (1985) for detail.

Table 107. Hemoglobin, delta: HBD

Place/Population	A2	B2	FLB[a]	BAB[b]	n	H	PE	Source
AFRICA								
CAMEROON								
Pygmies	.979	.013	.008		234	4.1	2.1	Bernini 1986
CENTRAL AFRICAN REPUBLIC								
Pygmies (Western)	.966	.033	.023	.008	1,039	6.5	3.0	"
Bagandu	.998	.002			236	.4	.2	"
RWANDA								
Pygmies	1.000				130			"
Hutu	.991	.009			116	1.8	.9	"
ZAIRE								
Pygmies (Eastern)	1.000				132			"
SOUTHERN AFRICA								
San/Bushmen								
!Kung, Dobe	.998	.002			424[c]	.4	.2	Jenkins 1972
Khoi/Hottentot								
Keetmanshoop Nama	1.000				146			"
Zulu/Bantu	.985	.015			102	3.0	1.5	Nurse et al. 1985
NORTH AMERICA								
UNITED STATES								
Blacks (Maryland)	.993	.007			1,151	1.4	.7	Boyer et al. 1963b

Note: [a] Rare Hb Flatbush allele. [b] Rare Hb Babinga allele. [c] One individual with an unidentified variant was included.

Table 108. Hemoglobin, gamma: HBG

Place/population	$A_\gamma I$	$A_\gamma T$	n	H	PE	Source
EUROPE						
ITALY	.763	.237	150	36.2	14.8	Huisman et al. 1983
YUGOSLAVIA	.762	.238	189	36.3	14.8	"
ASIA						
INDIA						
Indians (Bombay)	.827	.173	208	28.6	12.3	"
VIETNAM						
Vietnamese (USA)	.865	.135	26	23.4	10.3	"
CHINA						
Chinese (Shanghai)	.921	.079	841	14.6	6.7	"
JAPAN						
Japanese (Tokyo)	.822	.178	275	29.3	12.5	"
NORTH AMERICA						
UNITED STATES						
Whites (Georgia)	.776	.224	78	34.8	14.4	"
Blacks (")	.901	.099	343	17.8	8.1	"

Note: There are many variant alleles at this locus. See Lehmann and Kynoch (1976) and Livingstone (1985) for detail.

Table 109. Hemopexin: HPX

Place/Population	1	2	3	n	H	PE	Source
NORTH AMERICA							
ALASKA							
Eskimos (S. Alaska)	1.000			64			Stewart & Lovrien 1971
UNITED STATES							
Papago Indians	1.000			50			"
Whites (Pittsburgh)	1.000			175			Kamboh & Ferrell 1986c
Blacks (Pittsburgh)	.931	.018	.051	109	13.0	6.5	"
" (Houston)	.941	.018	.041	85	11.3	5.6	"

Table 110. Lymphocyte cytosol 40k polypeptide: LC40P

Place/Population	1	2	n	H	PE	Source
ASIA						
JAPAN	.543	.457	69	49.6	18.7	Hamaguchi et al. 1982c

Table 111. Lymphocyte cytosol 49k polypeptide: LC49P

Place/Population	1	2	n	H	PE	Source
ASIA						
JAPAN	.683	.317	93	43.3	17.0	Hamaguchi et al. 1982c

Table 112. Lymphocyte cytosol 64k polypeptide: LC64P, LCP1

Place/Population	1	2	n	H	PE	Source
ASIA						
JAPAN	.936	.064	110	12.0	5.6	Hamaguchi et al. 1982a

Table 113. Lymphocyte cytosol 100k polypeptide: LC100P

Place/Population	1	2	n	H	PE	Source
ASIA						
JAPAN	.907	.093	113	16.9	7.7	Hamaguchi et al. 1982b

Table 114. Orosomucoid 1; alpha-1-acid glycoprotein (locus 1): ORM1

Place/Population	1	2	3	n	H	PE	Source
EUROPE							
FINLAND	.500	.500		49	50.0	18.8	Johnson et al. 1969
FRANCE							
French	.563	.388	.049	112	53.0	24.0	Yuasa et al. 1986b
SWEDEN	.670	.330		52	44.2	17.2	Johnson et al. 1969
ASIA							
INDIA (Calcutta)	.440	.560		50	49.3	18.6	"
NEPAL							
Newars (Katmandu)	.674	.312	.014	141	44.8	18.7	Yuasa et al. 1986b
CHINA	.470	.530		73	49.8	18.7	Johnson et al. 1969
JAPAN							
Japanese	.270	.730		64	39.4	15.8	"
Japanese (Yamagata)	.221	.779		500	34.4	14.3	Umetsu et al. 1985
AFRICA							
NIGERIA							
Nigerians	.410	.590		51	48.4	18.3	Johnson et al. 1969
Nigerians	.618	.382		187	47.2	18.0	Escallon 1987
SOUTHERN AFRICA							
Zulu	.370	.630		67	46.6	17.9	Johnson et al. 1969
NORTH AMERICA							
ALASKA							
Eskimos (Kodiak & St. Lawrence Is.)	.573	.427		220	48.9	18.5	Escallon et al. 1987
CANADA							
Dogrib Indians	.547	.453		169	49.6	18.6	"
UNITED STATES							
Whites (Pittsburgh & Iowa)	.559	.441		228	49.3	18.6	"
Blacks (")	.616	.384		181	47.3	18.1	"
MIDDLE AMERICA							
MEXICO							
Indians	.540	.460		24	49.7	18.7	Johnson et al. 1969
Mayan Indians (Yucatan)	.621	.379		62	47.1	18.0	Escallon 1987
SOUTH AMERICA							
COLOMBIA							
Huitoto Indians	.556	.444		62	49.4	18.6	Escallon et al. 1987
OCEANIA							
MELANESIA							
Papua New Guinea							
Papuans (Highlands)	.841	.159		110	26.7	11.6	"

Table 115.   Orosomucoid 2; alpha-1-acid glycoprotein (locus 2): ORM2

Place/Population	1	2	3	4	n	H	PE	Source
EUROPE								
FRANCE	1.000				112			Yuasa et al. 1986b
ASIA								
NEPAL								
Newars	1.000				141			"
AFRICA								
NIGERIA	.987	.005	.005	.003	220	2.6	1.3	Escallon 1987
NORTH AMERICA								
ALASKA								
Eskimos (Kodiak &								
St. Lawrence Is.)	1.000				220			Escallon et al. 1987
CANADA								
Dogrib Indians	1.000				169			"
UNITED STATES								
Whites (Pittsburgh								
& Iowa)	1.000				228			"
Blacks (")	.958	.025	.006	.011	181	8.1	4.1	"
MIDDLE AMERICA								
MEXICO								
Mayan Indians (Yucatan)	1.000				62			Escallon 1987
SOUTH AMERICA								
COLOMBIA								
Huitoto Indians	1.000				62			Escallon et al. 1987
OCEANIA								
MELANESIA								
Papua New Guinea								
Papuans (Highlands)	1.000				110			"

Table 116.   Parotid acidic protein Pa: PA

Place/Population	+	−	n	H	PE	Source
EUROPE						
NETHERLANDS	.120	.880	100	21.1	7.2	Pronk et al. 1984(c)
ASIA						
JAPAN	.225	.775	145	34.9	8.1	Ikemoto et al. 1977
AFRICA						
KENYA	.180	.820	200	29.5	8.1	Pronk et al. 1984

Continued

Table 116. Parotid acidic protein Pa: PA (cont'd)

Place/Population	+	−	n	H	PE	Source
NORTH AMERICA						
UNITED STATES						
Whites	.214	.786	330	33.6	8.2	Friedman et al. 1975
Blacks	.136	.864	122	23.5	7.6	"

Note: Allele Pa$^+$ is dominant to allele Pa$^-$. The phenotypes at this locus are highly correlated with those at the PXS locus (see Azen 1977; Table 60)

Table 117. Parotid basic protein Pb: PB

Place/Population	1	2	n	H	PE	Source
EUROPE						
NETHERLANDS	1.000		110			Pals & Pronk 1979
ASIA						
JAPAN	1.000		100			Minaguchi et al. 1976
AFRICA						
MALI						
Bozo tr.	.800	.200	71	32.0	13.4	Pals & Pronk 1979
KENYA	.880	.120	200	21.1	9.4	Pronk et al. 1984
NORTH AMERICA						
UNITED STATES						
Whites	.995	.005	101	1.0	.5	Azen 1972
Blacks	.840	.160	90	26.9	11.6	"

Note: Alleles Pb^1 and Pb^2 are codominant.

Table 118. Parotid isoelectric focusing protein: PIF

Place/Population	+	−	n	H	PE	Source
NORTH AMERICA						
UNITED STATES						
Whites	.66	.34	148	44.9	.9	Azen & Denniston 1981
Blacks	.35	.65	90	45.5	6.2	"
Chinese	.56	.44	78	49.3	2.1	"

Note: Allele PIF$^+$ is dominant to allele PIF$^-$.

Table 119. Parotid middle-band protein (fast): PMF*

Place/Population	+	−	n	H	PE	Source
ASIA						
JAPAN	.380	.620	195	47.1	5.6	Ikemoto et al. 1977
NORTH AMERICA						
UNITED STATES						
Whites	.150	.850	140	25.5	7.8	Azen & Denniston 1980

Note: Allele PMF$^+$ is dominant to allele PMF$^-$. This locus may be the same as the locus PMS (Goodman et al. 1985).

Table 120. Parotid middle-band protein (slow): PMS*

Place/Population	+	−	n	H	PE	Source
NORTH AMERICA						
UNITED STATES						
Whites	.12	.88	150	21.1	7.2	Azen & Denniston 1980
Blacks	.24	.76	101	36.5	8.0	"

Note: Allele PMS$^+$ is dominant to allele PMS$^-$. This locus may be the same as the locus PMF (Goodman et al. 1985).

Table 121. Parotid salivary glycoprotein: G1, PRB3

Place/Population	1	2	3	4	0	n	H	PE	Source
NORTH AMERICA									
UNITED STATES									
Whites	.742	.040	.155	.017	.046	143	42.1	15.2	Azen et al. 1979
Blacks	.459	.050	.337	.044	.110	82	65.9	24.1	"

Note: Alleles GL^1, GL^2, GL^3, and GL^4 are codominant to one another, and they are all dominant to allele GL^0.

Table 122. Parotid-size variant: PS

Place/Population	1	2	0	n	H	PE	Source
NORTH AMERICA							
UNITED STATES							
Whites	.598	.101	.301	150	54.2	9.6	Azen & Denniston 1980
Blacks	.185	.126	.689	101	47.5	18.1	"

Note: Alleles PS^1 and PS^2 are codominant to each other and dominant to allele PS^0.

Table 123. Pepsinogen: PGA

Place/Population	a	b	c	d	n	H	PE	Source
ASIA								
PHILIPPINES								
Filipino (USA)	1.000				61			Samloff et al. 1973
CHINA								
Chinese (Los Angeles)	1.000				80			"
JAPAN								
Japanese (Hawaii)	1.000				229			"
NORTH AMERICA								
UNITED STATES								
Whites	.540	.370	.063	.028	215	56.7	28.7	Taggart et al. 1979

Table 124. Plasminogen: PLG

Place/Population	1	2	3	4	Others	n	H	PE	Source
EUROPE									
SWITZERLAND	.688	.281			.031	308	44.7	19.9	Dimo-Simonin et al. 1985
UNITED KINGDOM									
England	.708	.292				327	41.3	16.4	Hobart 1979
ASIA									
THAILAND	.987	.014				111	2.6	1.4	Goedde & Benkmann 1987
JAPAN (Tokyo)	.956	.011	.023	.009		750	8.5	4.3	Nakamura & Abe 1982b
KOREA (Seoul)	.980	.020				149	3.9	1.9	Goedde et al. 1986c
AFRICA									
GAMBIA	.860	.140				89	24.1	10.6	Hobart 1979
SOUTHERN AFRICA									
Negroes/Bantu	.698	.274			.029	1,252	43.7	19.5	Hitzeroth et al. 1986
NORTH AMERICA									
ALASKA									
Eskimos (St. Lawrence Is. & Bering Strait)	.989	.011				273	2.2	1.1	Dykes et al. 1983b
UNITED STATES									
Whites	.686	.299			.015	102	44.0	18.5	Raum et al. 1980
Blacks	.795	.193			.012	127	33.1	14.6	"
Orientals	.964	.029			.007	69	7.0	3.5	"
Whites (Minneapolis)	.665	.304	.018		.013[B,D,M]	1,501	46.5	20.7	Dykes et al. 1983b
Blacks (")	.741	.214	.003		.042[B,D]	203	40.3	19.3	"
Pima Indians (Arizona)	.993	.007				200	1.4	.7	"
Mexican-Americans (Texas)	.854	.143	.003			295	25.0	11.1	"
MIDDLE AMERICA									
GUATEMALA									
Black Caribs (Livingston)	.677	.262	.003	.058		189	47.0	22.7	"
SOUTH AMERICA									
CHILE									
Atacameño (N. Chile)	.901	.089	.010			151	18.0	8.6	Goedde et al. 1985b

Table 125. Proline-rich salivary protein Pr: PR

Place/Population	PR^1	$PR^{1'}$	PR^2	$PR^{2'}$	n	H	PE	Source
EUROPE								
NETHERLANDS	.810		.190		100	30.8	13.0	Pronk et al. 1984(c)
ASIA								
CHINA								
Chinese (USA)	.770			.230	54	35.4	14.6	Azen & Denniston 1974
JAPAN	.759			.241		36.6	14.9	Ikemoto 1983
AFRICA								
KENYA	.660		.330		200	45.6	17.6	Pronk et al. 1984
NORTH AMERICA								
UNITED STATES								
Whites	.640	.005	.080	.275	100	50.8	25.5	Azen & Denniston 1974
Blacks	.700	.050	.080	.170	100	47.2	26.5	"

Note: All four alleles are codominant.

Table 126. Proline-rich salivary protein Pc: PCS

Place/Population	1	2	n	H	PE	Source
NORTH AMERICA						
UNITED STATES						
Whites	.461	.539	178	49.7	18.7	Karn et al. 1985
Blacks	.670	.330	47	44.2	17.2	"

Note: Alleles PCS^1 and PCS^2 are codominant.

Table 127. Properdin factor B; glycine-rich beta-glycoprotein: BF, GBG

Place/Population	S	F	F1	S0.7	Others	n	H	PE	Source
EUROPE									
DENMARK	.780	.203	.008	.009		318	35.0	15.7	Mortensen & Lamm 1981
FINLAND	.73	.26	.01			70	39.9	16.8	Partanen & Koskimies 1986
FRANCE									
French	.821	.157	.008	.013		469	30.1	14.0	Hauptmann et al. 1977
Basques	.550	.296	.139	.015		201	59.0	32.1	Ohayon et al. 1980
GERMANY									
Germans (Hamburg)	.791	.196	.007	.006		1,112	33.6	14.9	Püschel et al. 1979
HUNGARY	.818	.165			.017	N.A.	30.3	13.9	Goedde et al. 1986b
ITALY									
Italians (Northern)	.768	.205	.014	.013		431	36.8	17.0	Scherz et al. 1982b
Italians (Tuscany)	.713	.250	.013	.024		1,000	42.8	19.9	Domenici et al. 1986
NORWAY									
Norwegians (Oslo)	.817	.172	.005	.007		300	30.3	13.7	Teisberg & Olaisen 1977
Lapps/Saami (Finnmark)	.888	.112				197	19.9	9.0	"
POLAND (Wroclaw)	.834	.151	.004	.011		890	28.2	13.0	Manczak 1984b
PORTUGAL (Lisbon/Faro)	.654	.278	.043	.026		456	49.2	24.5	Weissmann & Reuter 1981
SPAIN									
Spanish (Madrid)	.650	.270	.053	.027		330	50.1	25.5	Rodriguez-Córdoba et al. 1981
Basques (Bilbao)	.562	.270	.145	.023		24	59.0	32.8	"
SWEDEN	.814	.174	.008	.004		1,660	30.7	13.8	Rudduck et al. 1984
SWITZERLAND	.805	.176	.010	.009		654	32.1	14.7	Scherz et al. 1977
UNITED KINGDOM									
England (Northeast)	.821	.167	.002		.010	210	29.8	13.4	Papiha & Rodger 1986
U.S.S.R.									
ASIAN PART									
Tuvinsti (C. Siberia)	.738	.246		.014	.002	258	39.5	17.1	Dykes et al. 1981
ASIA									
JORDAN	.546	.374	.004	.076		426	55.6	27.0	Saleh et al. 1986
SAUDI ARABIA									
Shiites & Sunnites (al-Hasa)	.474	.447	.020	.059		246	57.2	28.0	Goedde et al. 1979b
Arabs (Riyadh)	.517	.321	.007	.151	.003	918	60.7	32.8	Klouda et al. 1984
AFGHANISTAN									
Pushtoon (Kabul)	.696	.268		.036		207	44.2	20.1	Benkmann et al. 1980
INDIA									
Punjabi (N. India)	.606	.394				99	47.8	18.2	Papiha 1981
THAILAND									
Thais (N. Thailand)	.902	.098				184	17.7	8.1	Greiner et al. 1980
Thais	.933	.067				112	12.5	5.9	Goedde & Benkmann 1987
PHILIPPINES									
Filipino (Manila)	.717	.283				46	40.6	16.2	Miyano et al. 1986
CHINA									
Han (Shanghai)	.870	.128		.002		200	22.7	10.2	Zhao 1983
Chinese (Singapore)	.900	.100				55	18.0	8.2	Miyano et al. 1986
TAIWAN									
Toroko-Aborigines	.885	.115				122	20.4	9.1	Chen et al. 1985
JAPAN									
Japanese (Nara)	.824	.176				360	29.0	12.4	Horai 1976
KOREA									
Koreans (Seoul)	.723	.273		.003		150	40.3	16.3	Goedde et al. 1986c
AFRICA									
TUNISIA	.617	.281	.019	.083		375	53.3	27.8	Davrinche et al. 1981b
ALGERIA									
Kabyle	.532	.321	.083		.064	78	60.3	33.8	Bouali et al. 1981
NIGER	.393	.510	.077		.019	103	57.9	28.5	Davrinche et al. 1981a
NIGERIA	.209	.724	.056		.010	98	42.9	21.5	Larsen et al. 1981
CAMEROON (Bamenda)	.373	.595	.029		.003	275	50.6	22.0	Goedde et al. 1979a
SOUTH AFRICA									
Negroes	.282	.655	.034	.025	.004	944	49.0	24.0	Mauff et al. 1976

Continued

Table 127. Properdin factor B; glycine-rich beta-glycoprotein: BF, GBG (cont'd)

Place/Population	S	F	F1	S0.7	Others	n	H	PE	Source
NORTH AMERICA									
ALASKA									
Eskimos (St. Lawrence									
Is. & Bering Strait)	.989	.010	.001			368	2.2	1.1	Dykes et al. 1981
UNITED STATES									
Whites	.709	.278			.013	158	42.0	17.7	Alper et al. 1972
Blacks	.437	.512	.051			127	54.4	24.6	"
Whites (Minnesota)	.786	.194	.011	.009		6,454	34.4	15.7	Dykes et al. 1981
Blacks (")	.478	.501	.018	.003		357	52.0	21.4	"
Blackfeet Indians	.974	.019	.006			78	5.1	2.4	"
Apache " (San									
Carlos)	.986	.004		.005	.005	510	2.8	1.4	"
Pima Indians	.972	.019	.006	.002		231	5.5	2.7	"
Navajo "	.967	.010		.024		105	6.4	3.3	"
MIDDLE AMERICA									
GUATEMALA									
Black Caribs (Livingston)	.393	.519	.086	.002		215	56.9	27.9	Dykes et al. 1981
SOUTH AMERICA									
ARGENTINA									
Mapuche	.907	.077			.016	91	17.1	8.4	Haas et al. 1985
CHILE									
Atacameño (Toconao, N.									
Chile)	.915	.063		.005	.017	172	15.9	8.0	Goedde et al. 1985b
COLOMBIA									
Colombians	.771	.179	.029	.019		309	37.2	18.4	Bernal et al. 1985
ECUADOR									
Shuara	.978	.011	.005		.005	90	4.3	2.1	Goedde et al. 1977a
OCEANIA									
MELANESIA									
Fiji	.894	.106				179	19.0	8.6	Ranford et al. 1982
MICRONESIA									
Nauru	.916	.084				184	15.4	7.1	"
POLYNESIA									
W. Samoa	.711	.283	.004		.002	247	41.4	17.0	"

Table 127.1 Properdin factor B ; glycine-rich beta-glycoprotein (subtypes): BF

Place/Population	S	Sa	Sb	F	Fa	Fb	F1	S1	n	H	PE	Source
EUROPE												
FRANCE		.566	.434						121	49.1	18.5	David et al. 1983
GERMANY												
Germans	.802		.007	.176			.005	.010	516	32.6	15.1	Weidinger et al. 1984b
OCEANIA												
AUSTRALIA												
Caucasians	.808				.038	.128	.019	.005	183	32.9	17.1	Teng & Tan 1982
(Canberra)												

Table 128. Salivary protein 1: CON1

Place/Population	+	−	n	H	PE	Source
NORTH AMERICA						
UNITED STATES						
Whites	.396	.604	134	47.8	5.3	Azen & Yu 1984a
Blacks	.581	.419	74	48.7	1.8	"
Chinese	.580	.420	79	48.7	1.8	"

Note: Allele CON1$^+$ is dominant to allele CON1$^-$.

Table 129. Salivary protein 2: CON2

Place/Population	+	−	n	H	PE	Source
NORTH AMERICA						
UNITED STATES						
Whites	.034	.966	134	6.6	3.0	Azen & Yu 1984a
Blacks	.007	.993	74	1.4	.7	"
Chinese		1.000	79			"

Note: Allele CON2$^+$ is dominant to allele CON2$^-$.

Table 130. Salivary protein Pe: PE

Place/Population	+	−	n	H	PE	Source
NORTH AMERICA						
UNITED STATES						
Whites	.760	.240	317	36.5	.3	Azen & Yu 1984b
Blacks	.760	.240	51	36.5	.3	"

Note: Allele Pe$^+$ is dominant to allele Pe$^-$.

Table 131. Salivary protein Po: PO

Place/Population	+	−	n	H	PE	Source
NORTH AMERICA						
UNITED STATES						
Whites	.750	.250	408	37.5	.3	Azen & Yu 1984b
Blacks	.770	.230	59	35.4	.2	"

Note: Allele Po$^+$ is dominant to allele Po$^-$.

Table 132. Salivary protein SAL I: SAL1

Place/Population	F	f	n	H	PE	Source
ASIA						
JAPAN						
Japanese (Hawaii)	.364	.636	42	46.3	6.0	Balakrishnan & Ashton 1974
NORTH AMERICA						
UNITED STATES						
Whites (Hawaii)	.403	.597	154	48.1	5.1	"

Note: Allele F is dominant to allele f.

Table 133. Salivary protein SAL II: SAL2

Place/Population	S	s	n	H	PE	Source
ASIA						
JAPAN						
Japanese (Hawaii)	.591	.409	42	48.3	1.7	Balakrishnan & Ashton 1974
NORTH AMERICA						
UNITED STATES						
Whites (Hawaii)	.573	.427	154	48.9	1.9	"

Note: Allele S is dominant to allele s.

Table 134. Thyroxine-binding globulin: TBG

Place/Population	C	S	n	H	Source
EUROPE					
Europeans	1.000		50		Whitehouse et al. 1985
ASIA					
INDIA					
Indians (Delhi)	1.000		151		Kamboh & Kirwood 1984
Indians (Madras)	1.000		120		"
Soliga tr. (S. India)	1.000		80		"
INDONESIA					
Lesser Sunda Is.	.997	.003	251	.6	"
THAILAND	1.000		59		"
CHINA	1.000		114		"
JAPAN	1.000		104		"
AFRICA					
CAMEROON	.917	.083	45		Daiger et al. 1981
CENTRAL AFRICAN REPUBLIC					
Pygmies	.951	.049	704	9.3	"
ZAIRE	1.000		62		"
SOUTHERN AFRICA					
San/Bushmen (Botswana)	1.000		52		Wang & Cavalli-Sforza 1986
Bantu (Durban)	.941	.059	163	11.1	Kamboh & Kirwood 1984
NORTH AMERICA					

Continued

Table 134. Thyroxine-binding globulin: TBG (cont'd)

Place/Population	C	S	n	H	Source
ALASKA					
Eskimos	.993	.007	97	1.4	Daiger et al. 1981
UNITED STATES					
Whites	1.000		404		Daiger et al. 1981
Blacks	.892	.108	580	19.3	"

MIDDLE AMERICA

PANAMA					
Blacks	.908	.092	103	16.7	"

OCEANIA

AUSTRALIA					
Aborigines/Malag (Elcho Is.,					
N. Territory)	1.000		181		Kamboh & Kirwood 1984
Aborigines (Mowanjum,					
W. Australia)	1.000		140		"
MELANESIA					
Papua New Guinea					
E. Highlands	.891	.109	131	19.4	"
Port Moresby	.933	.067	129	12.5	"
MICRONESIA					
Kiribati	.991	.009	142	1.8	"
POLYNESIA					
Cook Is.	.967	.033	143	6.4	"
Western Samoa	.951	.049	148	9.3	"

Table 135. Transcobalamin II: TC2

Place/Population	X	S	M	F	Others	n	H	PE	Source
EUROPE									
SWITZERLAND	.424	.012	.558	.003	.002	659	50.9	20.7	Scherz et al. 1982a
ASIA									
TAIWAN									
Toroko-Aborigines	.657		.343			127	45.1	17.5	Chen et al. 1985
AFRICA									
NIGERIA									
Yoruba (West & Midwest)	.070	.105	.547	.279		44	60.7	35.3	Porck et al. 1984
RWANDA									
Twa Pygmies	.091	.607	.282	.020		126	54.3	28.6	Wang & Cavalli-Sforza 1986
Hutu	.075	.678	.215	.033		107	48.7	26.3	"
NORTH AMERICA									
UNITED STATES									
Whites	.450		.531	.020		131	51.5	21.1	Daiger et al. 1978
Blacks	.178		.635	.187		163	53.0	28.4	"
Chinese & Japanese	.486		.514			34	50.0	18.7	"
Whites	.406	.010	.578	.004	.002	249	50.1	20.4	Yang et al. 1981
Whites	.380	.017	.590	.013	.002	510	50.7	22.1	Fráter-Schröder et al. 1979
MIDDLE AMERICA									
GUATEMALA									
Amerindians	.236		.764			236	36.1	14.8	"

Table 136. Transferrin: TF

Place/Population	C	B	D	Others	n	H	PE	Source
EUROPE								
AUSTRIA (Wien)	.995	.005			1,030	1.0	.5	Glahs 1974
BELGIUM (Liége)	.990	.007	.003		576	2.0	1.0	Brocteur et al. 1980
BULGARIA	1.000				138			Walter et al. 1972a
CZECHOSLOVAKIA (Praha)	.963	.016[1,2]	.021[D,1,3]		283	7.2	3.6	Geserick et al. 1969
FAROE IS. (N. Atlantic)	.986	.014			681	2.8	1.4	Tills et al. 1985
FINLAND	.978	.012[0-1]	.009[Chi]	.001	5,536	4.3	2.2	Nevanlinna 1972
FRANCE	.993	.006	.001		2,400	1.4	.7	Moullec 1967
GERMANY	.995	.002	.002		2,000	1.0	.4	Ritter 1969
GREECE	.998	.001[1]	.001[1]		2,050	.4	.2	Angelopoulos et al. 1967
ICELAND	.999	.001[2]	.001[1]		1,692	.2	.1	Tills et al. 1982b
IRELAND	.990	.009	.001		1,795	2.0	1.0	Tills 1977
ITALY	.993	.007			599	1.4	.7	Santachiara & Modiano 1964
NETHERLANDS (Leiden)	.988	.010[1-2,2]	.002		790	2.4	1.2	Fraser et al. 1974
NORWAY								
Norwegians	.995		.005		950	1.0	.5	Braend et al. 1965
Lapps/Saami	.996	.002[0-1]	.002[1]		301	.8	.4	Monn et al. 1971
POLAND (Southern)	.984	.016			1,119	3.1	1.5	Próchnicka 1968
SPAIN								
Spanish (Madrid)	.995	.005[2]			217	1.0	.5	Goedde et al. 1973
Basques	.985	.015			259	3.0	1.5	"
SWEDEN								
Swedes	.995		.005		2,395	1.0	.5	Beckman et al. 1961
Lapps/Saami	.991		.009		329	1.8	.9	Beckman & Holmgren 1961
UNITED KINGDOM								
England	.990	.009	.001		570	2.0	1.0	Cartwright et al. 1977
Scotland (South-west)	.996	.003	.001		876	.8	.4	Mitchell et al. 1976
Orkney Is. (Westray)	1.000				300			Welch et al. 1973
Isle of Lewis	.984	.016			94	3.1	1.5	Clegg et al. 1982
Isle of Man	.990	.008	.001		356	2.0	1.0	Mitchell et al. 1982
U.S.S.R.								
EUROPEAN PART								
Russians (Moscow)	.986	.008	.006		1,560	2.8	1.4	Altukhov et al. 1981
Abkhazia (Ochamchir								
Dist.)	1.000				293			Ferrell et al. 1985
ASIAN PART								
Eskimos (New Chaplino)	1.000				99			Sukernik et al. 1981
Chukchi (Chukotka/								
N. Kamchatka)	1.000				1,057			"
Nganasan-Avam (Taimir								
Penn.)	.952	.048[0-1]			270	9.1	4.4	Sukernik et al. 1978
ASIA								
TURKEY (Asia Minor)	1.000				300			Dinçol et al. 1976
ISRAEL								
Yemenite Jews	1.000				157			Tills et al. 1977b
KUWAIT								
Arabs	1.000				88			Al-Nassar et al. 1981
SAUDI ARABIA								
Shiites/Sunnites								
(al-Hasa)	.994		.006[Chi]		359	1.2	.6	Goedde et al. 1979b
Tribes (Western)	.966		.034		191	6.6	3.2	Saha et al. 1980
Arabs (Southern)	.998		.002		261	.4	.2	Marengo-Rowe et al. 1974
IRAN								
Turkoman (Gonbad,								
NE Iran)	1.000				153			Kirk et al. 1977
AFGHANISTAN								
Pushtu & Dari (Kabul)	.994	.006			255	1.2	.6	Papiha et al. 1977a
PAKISTAN								
Punjabi	1.000				201			Kirk & Lai 1961
INDIA								
Punjabi (N.India)	1.000				161			Tiwari 1961
Gujarati (W. India)	.997	.003			170	.6	.3	Papiha et al. 1981

Continued

176

Table 136. Transferrin: TF (cont'd)

Place/Population	C	B	D	Others	n	H	PE	Source
Bhil tr. (Madhya Pradesh)	.989	.007	.004		142	2.2	1.1	Papiha et al. 1978
Hindu (Andhra Pradesh)	1.000				85			Naidu et al. 1985
Irula (Nilgiri Hills, S. India)	1.000				106			Saha et al. 1976
Bengali (E. India)	1.000				288			Mukherjee & Das 1970
SRI LANKA								
Sinhalese	1.000				220			Papiha & Wastell 1974
Veddah	.921		.079[1]		64	14.6	6.7	Kirk et al. 1962
BANGLADESH	.995	.002	.003		198	1.0	.5	Papiha & Wastell 1974
NEPAL	.995	.005			206	1.0	.5	Sunderland et al. 1979
BHUTAN	.990		.010[Chi]		152	2.0	1.0	Mourant et al. 1969
BURMA	.987		.013[Chi]		150	2.6	1.3	Than-Than-Sint & Mya-Tu 1973
MALAYSIA								
Malays	.977		.023[1]		236	4.5	2.2	Kirk & Lai 1961
Aborigines	.985		.015[Chi]		202	3.0	1.5	Lie-Injo et al. 1967
INDONESIA								
Batak (Samosir Is.)	.946	.009	.045[Chi]		167	10.3	5.1	McDermid et al. 1973
Balinese (Bali Is.)	.829		.171[Chi]		140	28.4	12.2	Breguet et al. 1982a
THAILAND								
Thais (Bangkok)	.969		.031		274	6.0	2.9	Kirk & Lai 1961
Thais	.990		.010		113	2.0	1.0	Goedde & Benkmann 1987
VIETNAM (N. Vietnam)	.967		.033[1]		153	6.4	3.1	Herzog et al. 1976
PHILIPPINES								
Filipino (Germany)	.977		.023		107	4.5	2.2	Windhof & Walter 1983
Negrito (Angeles, C. Luzon)	.996		.004		128	.8	.4	Omoto et al. 1978
CHINA								
Mongolians (N. China)	.983	.003[2]	.014[1]		210	3.4	1.7	Goedde et al. 1984b
Zhauang (S. China)	.959		.034[1]	.007[Chi]	207	7.9	3.2	"
Tibetans/Dogpas	1.000				132			Sunderland et al. 1979
Chinese (Taiwan)	.945		.055		100	10.4	4.9	Shih & Hsia 1969
JAPAN								
Japanese (Hiroshima & Nagasaki)	.989		.006[Chi]	.005[a]	13,314	2.2	1.1	Fujita et al. 1985b
Ainu (Hokkaido)	.990	.010			246	2.0	1.0	Harvey et al. 1978
KOREA								
Koreans (NE. China)	.988	:002[2]	.005[1]	.005[Chi]	210	2.4	1.2	Goedde et al. 1984b

AFRICA

Place/Population	C	B	D	Others	n	H	PE	Source
EGYPT								
Egyptians (Cairo & Anshas)	.994		.006[1]		239	1.2	.6	Goedde et al. 1980
Nubians (Southern)	1.000				155			Bertin et al. 1978
SUDAN								
Sudanese (Khartoum)	.934		.066		113	12.3	5.8	Saha et al. 1978b
Beja-Amarar (NE Sudan)	.990		.010		99	2.0	1.0	el Hassan et al. 1968
CHAD								
Sara Majingay	1.000				247			Hiernaux 1976
SENEGAL								
Bedik (E. Senegal)	1.000				791			Bouloux et al. 1972
MALI	.982		.018[1]		142	3.5	1.7	Goedde et al. 1975
NIGERIA								
Yoruba (Southwestern)	.929		.071		421	13.2	6.2	Heiken et al. 1974
CAMEROON (Bamenda)	.960		.040[1]		275	7.7	3.7	Goedde et al. 1979a
CENTRAL AFRICAN REPUBLIC								
Babinga Pygmies	.862		.138[1]		160	23.8	10.5	Cavalli-Sforza et al. 1969
BURUNDI								
Hutu	.989		.011[1]		91	2.2	1.1	Giblett et al. 1966
ETHIOPIA								
Ethiopians	.994		.006[1]		78	1.2	.6	Goedde et al. 1975
Amhara tr.	1.000				107			Barnicot et al. 1962
UGANDA								
Baganda (Kampala)	.971	.004	.025		122	5.7	2.8	Roberts et al. 1977
KENYA	.971	.004[B2]	.025[1]		357	5.7	2.8	Herzog et al. 1970
TANZANIA								

Continued

Table 136. Transferrin: TF (cont'd)

Place/Population	C	B	D	Others	n	H	PE	Source
Sandawe	.953		.047		212	9.0	4.3	Godber et al. 1976
Nyaturu	.984		.016		214	3.1	1.5	"
Hadza (W. Lake Eyasi)	.968		.032		110	6.2	3.0	Tills et al. 1982a
ANGOLA								
Njinga	.995		.005[1]		109	1.0	.5	Nurse et al. 1979
SOUTHERN AFRICA								
Basters (Namibia)	.988	.004	.008		119	2.4	1.2	Nurse et al. 1982
San/Bushmen								
!Kung, Dobe	.848		.152		418	25.8	11.2	Nurse & Jenkins 1977
Khoi/Hottentot								
Keetmanshoop Nama	.990		.010		153	2.0	1.0	Jenkins 1972
Negroes/Dama (Okombahe)	1.000				117			Nurse et al. 1976
Bantu (Natal)	.980		.020		304	3.9	1.9	McDermid & Vos 1971a
Griqua (Cape Province)	.970		.030[1]		84	5.8	2.8	Nurse & Jenkins 1975
TRISTAN DA CUNHA	1.000				160			Jenkins et al. 1985

NORTH AMERICA

Place/Population	C	B	D	Others	n	H	PE	Source
GREENLAND								
Eskimos	.998	.001[1]	.001[1]		1,277	.4	.2	Persson 1968
ALASKA								
Eskimos								
(St. Lawrence Is.)	1.000				222			Ferrell et al. 1981b
CANADA								
Eskimos (Igloolik)	1.000				356			McAlpine et al. 1974
Cree Indians (Plains)	.971		.029[Chi]		605	5.6	2.7	Bowen et al. 1971
Ojibwa Indians								
(Pikangikum)	.960		.040[Chi]		100	7.7	3.7	Szathmary et al. 1974
Dogrib Indians (NW								
Territory)	1.000				158			Szathmary et al. 1983
UNITED STATES								
Blackfeet Indians								
(Montana)	.983	.017[0-1]			180	3.3	1.6	Rokala et al. 1977
Papago Indians (Arizona)	.980		.020		541	3.9	1.9	Brown & Johnson 1970
Whites (Southern)	.980	.010[2]	.010[1]		107	3.9	2.0	Cooper et al. 1963
Blacks (")	.950		.050[1]		133	9.5	4.5	"
Whites (Louisiana)	.997	.001	.001		376	.6	.2	Fox et al. 1981
Blacks (")	.969		.031		373	6.0	2.9	"

MIDDLE AMERICA

Place/Population	C	B	D	Others	n	H	PE	Source
MEXICO								
Indians (San Pablo del								
Monte)	.976	.006	.018		254	4.7	2.3	Crawford et al. 1974
Mestizo (Tlaxcala)	.926	.023	.051		128	13.9	7.0	"
GUATEMALA								
Black Caribs								
(Livingston)	.973		.027[1]		206	5.3	2.6	Crawford et al. 1981
COSTA RICA								
Guaymi	.970		.030[Chi]		286	5.8	2.8	Barrantes et al. 1982
PANAMA								
Blacks (Costa Arriba)	.942		.058		415	10.9	5.2	Ferrell et al. 1978b
WEST INDIES (Haiti)	.946	.004	.050		451	10.3	5.0	Basu et al. 1976

SOUTH AMERICA

Place/Population	C	B	D	Others	n	H	PE	Source
ARGENTINA								
Mapuche	1.000				71			Haas et al. 1985
BOLIVIA								
Aymara (Western)	1.000				316			Ferrell et al. 1978a
BRAZIL								
Baniwa (Jandu								
Cachoeira)	.940		.016[Chi]		377	3.9	1.9	Salzano et al. 1986
Cayapo (Para & Mato								
Grosso)	1.000				551			Salzano et al. 1972b
Caingang	1.000				214			Salzano et al. 1980
Parakana	.996		.004[1]		129	.8	.4	Black et al. 1980
Pacaás Novos	1.000				222			Salzano et al. 1985

Continued

Table 136. Transferrin: TF (cont'd)

Place/Population	C	B	D	Others	n	H	PE	Source
Macushi (N. Brazil & S. Guyana)	1.000				508			Neel et al. 1977b
Wapishana (")	1.000				623			"
Ticuna (C. Amazonas)	.988		.012[Chi]		1,765	2.4	1.2	Neel et al. 1980
CHILE								
Atacameño (Toconao, N. Chile)	1.000				180			Goedde et al. 1985b
COLOMBIA								
Noanama	1.000				155			Kirk et al. 1974
ECUADOR								
Shuara	1.000				90			Goedde et al. 1977a
FRENCH GUIANA								
Wayampi (Trois-Sauts & Camopi)	.967		.033		226	6.4	3.1	Tchen et al. 1978
SURINAM								
Trio (Southern)	1.000				413			Geerdink et al. 1974b
VENEZUELA								
Makiritare (S. Venezuela & N. Brazil)	1.000				536			Arends et al. 1970
Yanomama (")	1.000				2,425			Weitkamp et al. 1972

OCEANIA

Place/Population	C	B	D	Others	n	H	PE	Source
AUSTRALIA								
Aborigines/ Malag (Elcho Is., N. Territory)	.830		.170		641	28.2	12.1	Kirk et al. 1969
Aborigines (Amoonguna, C. Australia)	.957		.043		104	8.2	4.0	Kirk et al. 1971a
Aborigines (Kimberley, W. Australia)	.885		.115[1]		104	20.4	9.1	Blake & Spargo 1986
MELANESIA								
W. New Guinea								
Asmat (S. Coastal Plain)	.966		.034[1]		118	6.6	3.2	Gajdusek et al. 1978
Papua New Guinea								
Gainj & Kalam (N. Central Highlands)	.946		.054		575	10.2	4.8	Long et al. 1986
Jimi Valley (W. Highlands)	.950		.050		343	9.5	4.5	Mourant et al. 1981
Yagaria (E. Highlands)	.909		.091		407	16.5	7.6	Mourant et al. 1982
Fuyuge (C. Highlands)	.970		.030		148	5.8	2.8	Woodfield et al. 1974
Daga (Bay Province)	.975		.025[1]		142	4.9	2.4	Jenkins et al. 1983
Karkar Is.								
Takia	.978		.022		293	4.3	2.1	Booth et al. 1982
Manus Is.	.982		.018		138	3.5	1.7	Malcolm et al. 1972
Buka Is.	.987		.013[1]		80	2.6	1.3	McLoughlin et al. 1982
Solomon Is.	.992		.008		2,034	1.6	.8	Lai & Bloom 1982
MICRONESIA								
Caroline Is.	.991		.009[1]		438	1.8	.9	Hainline et al. 1969
Marshall Is.	.996		.004		372	.8	.4	Neel et al. 1976
POLYNESIA								
Samoans (Christchurch, New Zealand)	1.000				77			Booth et al. 1977

Note: [a] Pooled frequency of 15 different kinds of rare alleles.

Table 136.1 Transferrin (subtypes): TF

Place/Population	C_1	C_2	C_3	C_x	D	D_{Chi}	B	n	H	PE	Source
EUROPE											
BELGIUM	.784	.206			.003		.007	253	34.3	14.9	Hoste 1979
DENMARK (Copenhagen)	.814	.186						132	30.3	12.8	Thymann 1978
FINLAND (Uleaborg)	.869	.106					.024	306	23.3	11.3	Beckman et al. 1980b
FRANCE (Pyrenees)	.788	.132	.053				.027[2]	250	35.8	19.1	Constans et al. 1980a
GERMANY											
Germans (Hessen)	.820	.172			.001		.008	942	29.8	13.3	Kühnl & Spielmann 1978
Germans (Southern)	.787	.136	.068	.001[V]	.001[V]		.006[V]	1,125	35.7	18.6	Weidinger et al. 1984a
GREECE	.739	.185	.059	.002[6]	.015			295	41.6	21.5	Kouvatsi & Triantaphyllidis 1987
HUNGARY	.693	.276	.031					145	44.3	19.8	Goedde et al. 1986b
ICELAND	.841	.158						227	26.8	11.6	Beckman et al. 1980b(c)
ITALY (Rome)	.761	.180	.053	.001[6]	.001		.003	1,352	38.6	19.2	Pascali & Auconi 1983
SWEDEN											
Swedes	.828	.167					.004	1,084	28.7	12.5	Beckman et al. 1980b(c)
Lapps/Saami	.838	.153					.007	222	27.4	12.2	"
SWITZERLAND (Bern)	.773	.163	.057		.003		.003	759	37.3	18.9	Scherz et al. 1985
ASIA											
JORDAN											
Bedouin	.721	.275	.004					111	40.5	16.5	Saha & Banerjee 1986
INDIA											
Indians (Delhi)	.736	.216	.048					498	40.9	19.6	Kamboh & Kirk 1983
Indian s(Madras)	.774	.184	.042					95	36.5	17.6	"
Bengali-Bagdi (E. India)	.726	.268	.003	.003[4]				164	40.1	16.6	Reddy et al. 1984
NEPAL	.722	.250	.017	.010[9]				144	41.6	18.7	Yuasa et al. 1983b
MALAYSIA											
Malays (Kuala Lumpur)	.793	.178	.019			.011		135	33.9	16.1	Tan et al. 1982
INDONESIA											
Batak (Sumatra)	.774	.162	.019			.043	.002	232	37.4	19.2	"
Balinese (Bali Is.)	.493	.344			.004[1]	.169		136	61.0	33.0	Constans et al. 1986
THAILAND											
Thais	.700	.283			.008	.008		60	43.0	18.3	Kamboh & Kirk 1983
Thais	.728	.238	.024		.010			103	41.3	19.1	Goedde et al. 1987
PHILIPPINES											
Filipino	.813	.164			.023			107	31.2	14.5	Windhof & Walter 1983
Filipino (Hawaii)	.896	.067			.037			82	19.1	9.7	Beckman et al. 1980b
CHINA											
Chinese (Malaysia)	.754	.194	.030			.022		116	39.2	19.5	Tan et al. 1982
JAPAN											
Japanese (Osaka)	.765	.227			.006	.002[2]		342	36.3	15.5	Shibata 1983
KOREA											
Koreans (Seoul)	.733	.250				.017		148	40.0	17.4	Goedde et al. 1986c
AFRICA											
TUNISIA	.770[a]	.215			.015			404	36.1	15.9	Lefranc et al. 1981
EGYPT	.764	.193	.031		.006		.006	161	37.8	18.4	Sebetan et al. 1985
LIBERIA	.871	.057		.013	.051[1,2]		.008	236			Willcox et al. 1986
NIGERIA	.863	.053			.084			131	24.5	12.5	Beckman et al. 1980b
CONGO & CAMEROON											
Bi-Aka Pygmies	.820	.027			.153			337	30.3	14.4	Constans et al. 1981b
NORTH AMERICA											
UNITED STATES											
Whites (Minnesota)	.775	.163	.056	.001[6]			.005[2,1-2]	947	37.0	18.7	Dykes et al. 1982
Blacks (")	.842	.119	.008	.003[8]	.028			194	27.6	13.9	"
Whites (Pennsylvania)	.802	.188			.003[2]		.007[2]	149	32.1	14.2	Kueppers & Harpel 1980
Blacks (")	.843	.111			.045[1]			166	27.5	13.8	"
Blackfeet Indians	.715	.108		.177[4]				79	44.6	23.2	Dykes et al. 1982

Continued

Table 136.1 Transferrin (subtypes): TF (cont'd)

Place/Population	C_1	C_2	C_3	C_x	D	D_{Chi}	B	n	H	PE	Source
Pima Indians	.839	.054	.006	.081[4]			.020[0-1]	398	28.6	15.5	"
Apache Indians	.807	.014		.179[4]				140	31.7	14.2	"

SOUTH AMERICA

BRAZIL
Caingang	.830	.150		.020[4]				102	28.8	13.5	Constans & Salzano 1980

CHILE
Atacameño (Toconao, N. Chile)	.937	.063						174	11.8	5.6	Goedde et al. 1985b

OCEANIA

AUSTRALIA
Aborigines
(C. Australia)	.902	.034		.015[4,6]	.049			366	18.3	9.6	Kamboh & Kirk 1983

MELANESIA
Papua New Guinea
E. Highlands	.811	.086			.103			185	32.4	16.8	"
Port Moresby	.883	.074			.043			196	21.3	10.8	"
Sepik River	.831	.163			.006			242	28.3	12.5	"

MICRONESIA
Kiribati	.953	.047						211	9.0	4.3	"
Nauru	.955	.037	.008					200	8.7	4.3	"

POLYNESIA
Samoa Is.	.907	.087	.003					161	17.0	7.7	"
Cook Is.	.771	.224	.005					216	35.5	15.0	"

Note: C_x indicates an allele of C other than C_1, C_2, and C_3. [a] Tf^{C1} is the sum of Tf^{C1} and Tf^{C3}. [b] Other Tf alleles.

Table 137. Vitamin B_{12} binding (R) protein: VBRP*

Place/Population	1	2	n	H	PE	Source

EUROPE

UNITED STATES
Whites	.885	.115	143	20.4	9.1	Azen & Denniston 1979
Blacks	.942	.058	104	10.9	5.2	"
Chinese	1.000		75			"

Note: Alleles 1 and 2 are codominant.

Table 138. Xm system: XM

Place/Population	Xm^a	Xm	n^b	H	Source
EUROPE					
NORWAY	.230	.770	100	35.4	Berg & Bearn 1966
NORTH AMERICA					
UNITED STATES					
Whites	.263	.737	57	38.8	"
Blacks	.309	.691	81	42.7	"
OCEANIA					
POLYNESIA					"
Easter Is.	.242	.758	66	36.7	

Note: [b] All individuals studied were males.

Table 139. Proteins with no variants or few variants

Protein: Symbol	Populations studied (n) & remarks	Source
Haptoglobin, beta: HPB	Germans, Marburg. Two different variants (Hp1-1Mb, Hp2-1Mb, were found in a family.	Cleve & Deicher 1965
	US Blacks. A variant (Hp2-1 Bellevue) was found in one individual.	Javid 1967
	England. Four variants (Hp1-P, Hp2-P, Hp1-H, Hp2-H) in two British families and one variant (Hp2-L) in one West Indian blood donor were found.	Robson et al. 1964 "
Hemoglobin, alpha: HBA[a]	Russians, Moscow, USSR (1695)	Altukhov et al. 1981
	Abhkazia, USSR (262)	Ferrell et al. 1985
	Eskimos, St Lawrence Is., Alaska (222)	Ferrell et al. 1981b
	Aymara Indians, Bolivia (2,720)	Ferrell et al. 1978a
	Cayapo Indians, Brazil (894)	Neel 1978
	Macushi " " (500)	"
	Wapishana ' " (615)	"
	Makiritare " " (480)	"
	Yanomama " " (2,603)	"
Thyroxine-binding prealbumin: TBPA, TTR, PALB	Europeans (51)	Whitehouse et al. 1985

Note: [a] There are many variant alleles at this locus. See Lehmann and Kynoch (1976) and Livingstone (1985) for detail.

6 Blood Groups and Platelet Antigen Systems

Table 140. ABH secretion: SE

Place/Population	Se	se	n	H	PE	Source
EUROPE						
AUSTRIA	.570	.430	1,659	49.0	1.9	Mayr et al. 1970
BULGARIA	.548	.452	666	49.5	2.3	Popivanov et al. 1965
DENMARK	.559	.441	4,469	49.3	2.1	Jordal 1958
FINLAND						
Finns	.624	.376	241	46.9	1.2	Eriksson et al. 1986
Lapps/Saami	.786	.214	306	33.6	.2	"
GERMANY	.530	.470	363	49.8	2.6	Schiff 1940
ICELAND	.358	.642	228	46.0	6.1	Bjarnason et al. 1973
IRELAND	.444	.556	1,973	49.4	4.2	Mitchell 1976(c)
ITALY	.574	.426	518	48.9	1.9	Ceppellini 1954
NETHERLANDS	.535	.465	1,437	49.8	2.5	Haverkorn & Goslings 1969
NORWAY						
Norwegians	.594	.406	109	48.2	1.6	Brendemoen 1950
Lapps/Saami	.686	.314	183	43.1	.7	Allison et al. 1952
POLAND (Southern)	.599	.401	21,088	48.0	1.5	Turowska et al. 1977
SPAIN						
Spanish	.506	.494	336	50.0	3.0	Goti Iturriaga & Velasco Alonso 1965
Basques	.568	.432	386	49.1	2.0	"
SWEDEN	.530	.470	2,093	49.8	2.6	Nerell 1963
SWITZERLAND	.504	.496	1,000	50.0	3.1	Hässig et al. 1955
UNITED KINGDOM						
England (London)	.518	.482	1,560	49.9	2.8	Mitchell 1976(c)
Scotland (Aberdeen)	.454	.546	510	49.6	2.3	"
Isle of Man	.463	.537	163	49.7	3.9	"
YUGOSLAVIA	.572	.428	459	49.0	1.9	Grünwald & Herman 1963
U.S.S.R.						
ASIAN PART						
Chukchi (Chukotka & NE Kamchatka)	1.000		212			Sukernik et al. 1981
ASIA						
ISRAEL						
Ashkenazi Jews (E. Europe)	.521	.479	288	49.9	2.7	Kobyliansky et al. 1982
Yemenite Jews	.534	.466	23	49.8	2.5	"
IRAN						
Turkoman (Gonbad, NE Iran)	.324	.676	153	43.8	6.8	Kirk et al. 1977
INDIA						
Punjabi (N. India)	.548	.452	500	49.5	2.3	Seth 1968
Bhil tr. (Madhya Pradesh)	.633	.367	89	46.5	1.1	Vyas et al. 1962
Kurumba tr. (S. India)	.527	.473	116	49.9	2.6	Büchi 1959
Bengali (E. India)	.483	.517	438	49.9	3.5	Roy & Chatterjea 1965
NEPAL						
Shrestha	.491	.509	282	50.0	3.3	Bhasin 1970
BURMA	.493	.507	101	50.0	3.3	Mya-Tu & Ma Than Saw 1970
MALAYSIA						
Negrito	.505	.495	49	50.0	3.0	Polunin & Sneath 1953
INDONESIA						
Balinese (Bali Is.)	.271	.729	65	39.5	7.7	Breguet et al. 1982a
THAILAND (Bangkok)	.557	.443	515	49.4	2.1	Chandanayingyong et al. 1979
CHINA						
Han (N. China)	.476	.524	153	49.9	3.6	Yuan et al. 1984b
Dong (Guangxi)	.486	.514	201	50.0	3.4	Yuan et al. 1984c
Zhuang (S. China)	.462	.538	211	49.7	3.9	Yuan et al. 1984a
Tibetans	.508	.492	128	50.0	3.0	Tiwari 1966
Chinese (Calcutta)	.482	.518	555	49.9	3.5	Chaudhuri et al. 1967
JAPAN						
Japanese (Tokyo & Chiba)	.510	.490	960	50.0	2.9	Nakajima 1958

Continued

Table 140. ABH secretion: SE (cont'd)

Place/Population	Se	se	n	H	PE	Source
Ainu (Hokkaido)	.339	.661	110	44.8	6.5	Harvey et al. 1978
KOREA						
Koreans (NE China)	.491	.509	216	50.0	3.3	Yuan et al. 1984a

AFRICA

EGYPT	.529	.471	500	49.8	2.6	Moharram 1943
NIGERIA						
Yoruba	.497	.503	300	50.0	3.2	Ball 1962
CENTRAL AFRICAN REPUBLIC						
Babinga Pygmies (Bagandou)	.500	.500	64	50.0	3.1	Cavalli-Sforza et al. 1969
UGANDA						
Karamojo	.506	.494	45	50.0	3.0	Allbrook et al. 1965
SOUTHERN AFRICA						
Basters (Namibia)	.507	.493	115	50.0	3.0	Nurse et al. 1982
San/Bushmen						
!Kung, Dobe	.582	.418	308	48.7	1.8	Nurse & Jenkins 1977
Khoi/Hottentot						
Keetmanshoop Nama	.578	.422	129	48.8	1.8	Jenkins 1972
Negroes/Dama (Okombahe)	.602	.398	82	47.9	1.5	Nurse et al. 1976
Bantu (Natal)	.480	.520	181	49.9	3.5	Moores & Brain 1968
Griqua (Campbell, Cape Province)	.515	.485	51	50.0	2.8	Nurse & Jenkins 1975
TRISTAN DA CUNHA	.570	.430	54	49.0	1.9	Jenkins et al. 1985

NORTH AMERICA

GREENLAND						
Eskimos (Augpilagtok)	.884	.116	75	20.5	.0	Eriksson et al. 1986
ALASKA						
Eskimos-Inupiak (Wainwright)	1.000		111			Corcoran et al. 1959
Athabaskan Indians (Fort Yukon)	1.000		99			"
CANADA						
Eskimos (Victoria Is.)	1.000		200			Chown & Lewis 1959
UNITED STATES						
Blackfeet Indians (Montana)	.779	.221	822	34.4	.2	Kaklamani & Holborow 1963

MIDDLE AMERICA

WEST INDIES (Haiti)	.384	.616	103	47.3	5.5	Basu et al. 1976

SOUTH AMERICA

BRAZIL						
Baniwa	.879	.121	136	21.3	.0	Salzano et al. 1986
Cayapo (Para & Mato Grosso)	.707	.293	406	41.4	.5	Salzano et al. 1972a
Caingang	1.000		58			Salzano et al. 1980
Pacaás Novos	1.000		173			Salzano et al. 1985
Macushi (N. Brazil & S. Guyana)	.710	.290	230	41.2	.5	Salzano et al. 1984
FRENCH GUIANA						
Wayampi (Trois-Sauts & Camopi)	1.000		176			Tchen et al. 1981
VENEZUELA						
Makiritare (S. Venezuela &						
N. Brazil)	1.000		464			Gershowitz et al. 1970
Yanomama (")	.907	.093	1,726	16.9	.0	Gershowitz et al. 1972

OCEANIA

AUSTRALIA						
Aborigines (Western)	.988	.012	355	2.4	.0	Vos & Comley 1967
MELANESIA						
Papua New Guinea						
Papuans	.687	.313	255	43.0	.7	McConnell 1969
Kuni (Goilala)	.837	.163	300	27.3	.1	Booth et al. 1981
Karkar Is.						
Waskia	.695	.305	247	42.4	.6	Booth et al. 1982
Solomon Is.	.505	.495	2,034	50.0	3.0	Lai & Bloom 1982

Continued

Table 140. ABH secretion: SE (cont'd)

Place/Population	Se	se	n	H	PE	Source
MICRONESIA						
Marshall Is.	.467	.533	211	49.8	3.8	Simmons et al. 1952
POLYNESIA						
Cook Is.	.437	.563	312	49.2	4.4	Douglas et al. 1966

Note: Allele Se is dominant to se.

Table 141. ABO system: ABO

Place/Population	A_1	A	A_2	B	O	n	H	PE	Source
EUROPE									
AUSTRIA (Wien)		.294		.107	.599	1,000	54.3	16.8	Speiser 1958
CZECHOSLOVAKIA (Ovcie)		.205		.222	.573	321	58.0	19.9	Matousek & Seemanová 1973
DENMARK	.206		.074	.074	.646	12,123	52.9	18.0	Gürtler 1970
FAROE IS. (N. Atlantic)	.130	.001[a]	.060	.044	.765	682	39.2		Tills et al. 1985
FINLAND	.213		.096	.132	.559	5,536	61.6	21.8	Nevanlinna 1972
FRANCE	.228		.062	.059	.651	7,090	51.7	16.2	Khérumian et al. 1958
GERMANY	.198		.068	.080	.654	100,000	52.2	18.4	Rasch 1960
GREECE									
Greeks (Plati)	.279	.001[a]	.041	.072	.607	1,038	54.7		Tills et al. 1983b
HUNGARY		.297		.141	.562	7,770	57.6	18.2	Rex-Kiss et al. 1973
ICELAND	.124		.062	.056	.758	1,684	55.9	15.9	Tills et al. 1982b
IRELAND	.130		.040	.070	.760	1,699	39.9	16.2	Tills et al. 1977a
ITALY	.167		.078	.096	.659	1,042	52.3	19.8	Angelini Rota et al. 1961
NETHERLANDS (Leiden)	.192		.060	.055	.693	795	47.6	16.2	Fraser et al. 1974
NORWAY									
Norwegians	.230		.082	.068	.620	6,568	55.1	17.4	Berg 1974(c)
Lapps/Saami (Finnmark)	.174		.256	.107	.463	423	67.8	21.4	Kornstad 1972
POLAND (Olkusz)	.223		.047	.145	.585	213	58.5	20.9	Wolanski et al. 1983
PORTUGAL (Terras de Bouro)	.263		.097	.052	.588	118	57.3	15.5	Cruz et al. 1973
SPAIN									
Spanish (Madrid)		.292		.065	.642	1,000	49.8	14.1	Goedde et al. 1973
Basques (Bermeo/									
Pamplona)		.225		.032	.744	146	39.5	11.8	"
SWEDEN									
Swedes	.222		.085	.077	.616	10,457	55.8	18.2	Beckman 1959
Lapps/Saami	.063		.372	.031	.534	419	57.2	12.5	Beckman et al. 1959
SWITZERLAND	.279		.048	.070	.602	104	55.3	16.0	Greuter et al. 1963
UNITED KINGDOM									
England (S. England)	.209		.070	.061	.660	3,469	51.2	16.8	Ikin et al. 1939
Scotland (South-west)		.214		.070	.716	786	43.7	15.2	Mitchell et al. 1976
Isle of Lewis	.122	.014[a]	.040	.073	.752	284	41.2		Clegg et al. 1985
Isle of Man		.222		.049	.729	219	41.7	13.5	Mitchell et al. 1982
Orkney Is. (Westray)		.220		.130	.650	345	51.2	18.4	Welch et al. 1973
YUGOSLAVIA	.213		.052	.143	.591	459	58.2	21.1	Grünwald & Herman 1963
U.S.S.R.									
EUROPEAN PART									
Russians	.215		.040	.157	.589	1549	58.1	21.3	Budyakov 1966
Abkhazia (Ochamchir Dist.)		.189		.070	.741	759	41.0	15.1	Ferrell et al. 1985
ASIAN PART									
Eskimos (New Chaplino)	.143			.184	.672	102	49.4	18.6	Sukernik et al. 1981
Chukchi (Chukotka &									
NE Kamchatka)	.195			.149	.656	1,066	50.9	18.9	"
Nganasan-Avam (Taimir									
Pen.)	.253			.097	.650	430	49.3	16.4	Sukernik et al. 1978

Continued

Table 141. ABO system: ABO (cont'd)

Place/Population	A$_1$	A	A$_2$	B	O	n	H	PE	Source
Nentzi (Between Ob R. & Taj R.)	.119		.021	.230	.630	753	53.6	19.6	Sukernik et al. 1980

ASIA

Place/Population	A$_1$	A	A$_2$	B	O	n	H	PE	Source
TURKEY (Asia Minor)	.218		.068	.149	.565	108	60.6	21.7	Aksoy et al. 1958
ISRAEL									
Ashkenazi Jews (E. Europe)	.227		.075	.122	.576	306	59.6	20.8	Kobyliansky et al. 1982
Yemenite Jews	.078		.093	.058	.771	157	38.7	16.0	Tills et al. 1977b
IRAQ		.230		.156	.605	2,156	55.7	19.1	Al-Khafaji & Al-Rubeai 1976
JORDAN									
Bedouin		.220		.146	.634	108	52.8	19.0	Saha & Banerjee 1986
KUWAIT									
Arabs	.152		.021	.127	.700	162	47.0	18.8	Sawhney et al. 1984
SAUDI ARABIA									
Shiah (Eastern)	.084		.036	.150	.730	465	43.6	18.2	Maranjian et al. 1966
Tribes (Western)	.107		.055	.126	.714	210	46.0	19.5	Saha et al. 1980
Arabs (Southern)	.066	.024[a]	.089	.064	.757	261	41.0		Marengo-Rowe et al. 1974
UNITED ARAB EMIRATES									
Abu Dhabians		.196		.082	.722	624	43.4	15.9	Kamel et al. 1980
IRAN									
Turkoman (Gonbad, NE Iran)	.247		.036	.173	.544	156	61.2	21.0	Kirk et al. 1977
AFGHANISTAN									
Pushtu & Dari (Kabul)	.178		.054	.186	.582	283	59.2	22.3	Papiha et al. 1977a
PAKISTAN									
Punjabi (Lahore)	.151		.033	.260	.556	228	59.9	21.1	Boyd & Boyd 1954
INDIA									
Punjabi (Chandigarh)	.128		.025	.281	.566	137	58.4	19.9	Papiha et al. 1972
Gujarati (W. India)	.184		.020	.248	.548	289	60.4	20.9	Papiha et al. 1981
Hindu & Moslems (Madhya Pradesh)	.176		.031	.254	.538	343	61.4	21.4	Roberts et al. 1974
Bhil tr. (Madhya Pradesh)	.219		.025	.221	.535	145	61.6	21.3	Papiha et al. 1978
Hindu (Andhra Pradesh)	.119		.039	.193	.648	211	52.7	20.6	Roberts et al. 1980
Irula tr. (Nilgiri Hills, S. India)	.172		.033	.337	.458	61	64.6	20.2	Saha et al. 1976
Bengali (E. India)	.173		.012	.221	.594	372	56.8	20.4	Bhattacharjee 1956
SRI LANKA									
Sinhalese (Anuradhapura)	.122		.011	.167	.700	157	46.7	18.5	Roberts et al. 1972a
Veddah	.040		.010	.277	.673	254	46.9	13.3	Wickremasinghe et al. 1963
BANGLADESH									
Moslems	.140		.033	.231	.596	236	57.1	21.0	Boyd & Boyd 1954
NEPAL									
Shrestha	.211		.070	.198	.521	150	64.0	22.9	Bhasin 1970
BHUTAN	.132	.112[a]		.190	.565	154	61.5		Mourant et al. 1969
BURMA (Central)	.165		.010	.249	.576	120	57.9	20.0	Mya-Tu et al. 1971
MALAYSIA									
Malays (Singapore)		.178		.203	.619	5,461	54.4	19.6	Chan 1962
INDONESIA									
Toba Batak (Samosir Is.)	.077		.007	.213	.703	150	45.4	16.3	Hawkins et al. 1973
Balinese (Bali Is.)		.111		.225	.664	148	49.6	17.6	Breguet et al. 1982a
THAILAND (Bangkok)	.137		.007	.190	.666	227	50.2	19.0	Chandanayingyong et al. 1967
PHILIPPINES									
Filipino		.174		.156	.671	403	49.5	18.9	Fraser et al. 1964
Negrito/Aeta (Zambales)	.269			.094	.637	353	51.3	16.4	Misawa & Omoto 1985
CHINA									
Han (N. China)		.208		.213	.579	640	57.6	20.0	Yuan et al. 1984b
Mongolians (")		.202		.231	.567	200	58.4	19.9	Yuan et al. 1984a
Uygur (NW China)		.215		.221	.564	221	58.7	20.0	Yuan et al. 1984d
Zhaung (S. China)		.161		.208	.631	211	53.3	19.3	Yuan et al. 1984a
Hui (Ningxia)		.223		.253	.524	219	61.2	19.8	Yuan et al. 1985
Dong (Guangxi)		.204		.167	.629	201	53.5	19.4	Yuan et al. 1984c
Tibetans (Switzerland)		.181		.218	.601	115	55.9	19.7	Jeannet et al. 1972
Cantonese (Singapore)		.193		.171	.636	1,007	52.9	19.4	Hawkins & Simons 1976
Chinese (New York)	.174			.164	.662	400	50.5	19.0	Wiener 1969
TAIWAN									
Toroko-Aborigines	.184			.124	.692	127	47.2	18.0	Chen et al. 1985
JAPAN									
Japanese (Tokyo)	.271			.170	.559	65,743	58.5	19.2	Nei & Imaizumi 1966(c)

Continued

186

Table 141. ABO system: ABO (cont'd)

Place/Population	A_1	A	A_2	B	O	n	H	PE	Source
Japanese (Colorado, USA)	.277		.004	.183	.536	180	60.2	19.5	Mourant et al. 1976(c)
Ainu (Hokkaido)	.243			.199	.558	144	59.0	19.8	Harvey & Giblett 1968
KOREA									
Koreans (Seoul)	.221			.207	.572	322	58.1	20.0	Won et al. 1960
Koreans (NE. China)		.237		.219	.543	216	60.1	19.9	Yuan et al. 1984a

AFRICA

Place/Population	A_1	A	A_2	B	O	n	H	PE	Source
LIBYA (Tripoli & Benghasi)	.147		.078	.132	.643	168	54.1	21.5	Walter et al. 1975
EGYPT									
Nubians (S.Egypt)		.222		.104	.674	208	48.6	17.3	Azim et al. 1974
SUDAN									
Sudanese (Khartoum)	.128		.047	.126	.699	300	47.7	19.7	Saha et al. 1978b
Beja-Amarar (NE Sudan)	.078	.028[a]	.088	.083	.723	100	45.6		el Hassan et al. 1968
CHAD									
Sara Majingay		.113		.144	.744	258	41.3	16.8	Hiernaux 1976
SENEGAL									
Bedik (E. Senegal)	.165		.038	.230	.567	771	59.7	21.8	Bouloux et al. 1972
NIGERIA									
Yoruba (Western)	.094		.055	.188	.664	135	51.2	20.6	Blumberg et al. 1961
CENTRAL AFRICAN REPUBLIC									
Babinga Pygmies (Bagandou)	.130		.032	.102	.737	163	42.9	17.8	Cavalli-Sforza et al. 1969
BURUNDI									
Hutu		.192		.119	.689	99	47.4	17.9	Fraser et al. 1966a
ZAIRE									
Bantu	.136		.080	.096	.688	93	49.3	19.6	Govaerts et al. 1972
ETHIOPIA									
Amhara tr.	.141		.110	.146	.603	107	58.3	22.8	Ikin & Mourant 1962
TANZANIA									
Sandawe (Kondoa-Irang Dist.)	.100	.026[a]	.044	.129	.700	215	48.1		Godber et al. 1976
Nyaturu (")	.089	.038[a]	.046	.112	.715	217	46.5		"
Hadza (W. Lake Eyasi)	.238	.170[a]	.016	.019	.557	162	60.4		Tills et al. 1982a
ANGOLA									
Njinga	.131	.012[a]	.013	.127	.717	109	45.2		Nurse et al. 1979
SOUTHERN AFRICA									
Basters (Namibia)	.190	.038[c]	.052	.115	.605	120	58.0		Nurse et al. 1982
San/Bushmen									
Dobe	.239	.040[b]	.018	.020	.684	436	47.3		Nurse & Jenkins 1977
Khoi/Hottentot									
Keetmanshoop Nama	.120	.022[b]	.126	.171	.561	153	62.5		Jenkins 1972
Negroes/Dama (Okombahe & Sesfontein)	.086	.013[a]		.272	.628	119	52.4		Nurse et al. 1976
Bantu/Zulu	.106		.069	.086	.739	91	43.0	18.0	Nurse et al. 1985
Griqua (Cape Province)	.185	.026[b]	.072	.152	.566	100	61.6		Nurse & Jenkins 1975
TRISTAN DA CUNHA	.375			.071	.552	160	55.0		Jenkins et al. 1985

NORTH AMERICA

Place/Population	A_1	A	A_2	B	O	n	H	PE	Source
GREENLAND									
Eskimos (W. Greenland)	.246	.012[a]	.018	.067	.658	2,082	50.2		Gürtler 1971
ALASKA									
Eskimos-Inupiak (Wainwright)	.350		.007	.090	.553	111	56.4	15.0	Corcoran et al. 1959
Eskimos (St. Lawrence Is.)	.228			.134	.638	313	52.3	18.6	Ferrell et al. 1981b
Athabaskan Indians (Fort Yukon)	.076		.010		.914	110	15.9	6.2	Corcoran et al. 1959
CANADA									
Eskimos (NW Territories)	.293			.033	.674	320	45.9	11.2	Chown & Lewis 1959
Ojibwa Indians (Pikangikum)	.219		.007		.774	95	35.3	8.4	Szathmary et al. 1975
Cree Indians (Manitoba)	.157		.005	.012	.826	200	29.3	9.6	Lucciola et al. 1974
Dogrib Indians (NW Territories)	.177				.823	158	29.1	8.1	Szathmary 1983
UNITED STATES									
Blackfeet Indians (Montana)	.508			.010	.482	148	51.0	4.2	Rokala et al. 1977
Papago Indians									

Continued

187

Table 141. ABO system: ABO (cont'd)

Place/Population	A_1	A	A_2	B	O	n	H	PE	Source
(S. Arizona)	.061		.002	.003	.934	709	12.4	5.2	Niswander et al. 1970
Whites (Southern)	.169		.077	.050	.704	333	46.7	16.2	Cooper et al. 1963
Negroes (")	.097		.061	.129	.713	300	46.2	19.6	"
Mexican-Americans	.152		.029	.074	.745	991	41.6	16.4	Hewett-Emmett et al. 1986b

MIDDLE AMERICA

Place/Population	A_1	A	A_2	B	O	n	H	PE	Source
MEXICO									
Indians (San Pablo del Monte)	.048		.002	.004	.946	257	10.3	4.5	Crawford et al. 1974
Mestizo (Tlaxcala)	.095		.020	.044	.841	138	28.1	12.1	"
GUATEMALA									
Black Caribs (Livingston)	.064		.032	.160	.744	204	41.6	17.0	Crawford et al. 1981
COSTA RICA									
Guaymi		.001		.001	.998	286	.4	.2	Barrantes et al. 1982
WEST INDIES (Haiti)		.125		.180	.695	470	46.9	18.0	Basu et al. 1976

SOUTH AMERICA

Place/Population	A_1	A	A_2	B	O	n	H	PE	Source
ARGENTINA									
Mapuche		.024		.015	.961	105	7.6	3.7	Haas et al. 1985
BOLIVIA									
Aymara (Western)		.028		.004	.967	315	6.4	2.9	Ferrell et al. 1978a
BRAZIL									
Baniwa (Jandu Cachoeira)					1.000	363			Salzano et al. 1986
Cayapo (Para & Mata Grosso)					1.000	522			Salzano et al. 1972a
Caingang					1.000	214			Salzano et al. 1980
Parakana					1.000	100			Black et al .1980
Pacaás Novos					1.000	222			Salzano et al. 1985
Macushi (N. Brazil & S. Guyana)	.001				.999	503	.2	.1	Neel et al. 1977a
Wapishana (")	.011	.004	.005		.980	618	3.9	1.9	"
Ticuna (C. Amazonas)					1.000	1,760			Neel et al. 1980
CHILE									
Atacameño (Toconao, N. Chile)		.056		.003	.941	182	11.1	4.8	Rothhammer et al. 1984
COLOMBIA									
Noanama					1.000	168			Kirk et al. 1974
FRENCH GUIANA									
Wayampi (Trois-Sauts & Camopi)					1.000	240			Tchen et al. 1981
SURINAM									
Trio (Southern)					1.000	497			Geerdink et al. 1974c
VENEZUELA									
Makiritare (S. Venezuela & N. Brazil)					1.000	539			Gershowitz et al. 1970
Yanomama (")					1.000	2,516			Gershowitz et al. 1972

OCEANIA

Place/Population	A_1	A	A_2	B	O	n	H	PE	Source
AUSTRALIA									
Aborigines/Malag (Elcho Is., N. Territory)	.121			.014	.865	352	23.7	8.8	Simmons & Cooke 1969
Aborigines (Haast Bluff, N. Territory)		.33			.67	126	44.2	6.6	Kirk 1965(c)
Aborigines (Yuendumu, C. Australia)		.31			.69	93	42.8	7.0	"
Aborigines (W. Australia)		.25			.75	1,698	37.5	7.9	"
Aborigines (Queensland)		.21		.02	.77	447	36.3	10.6	"
MELANESIA									
W. New Guinea									
Asmat (S. Coastal Plain)	.217			.138	.645	402	51.8	18.7	Gajdusek et al. 1978
Papua New Guinea									
Gainj & Kalam (N. Central Highlands)	.347			.126	.527	415	58.6	16.7	Long et al. 1986
Jimi Valley (W. Highlands)	.200			.196	.604	347	55.7	19.8	Mourant et al. 1981

Continued

Table 141. ABO system: ABO (cont'd)

Place/Population	A_1	A	A_2	B	O	n	H	PE	Source
Yagaria (E. Highlands)	.211			.109	.680	374	48.1	17.5	Mourant et al. 1982
Fuyuge (C. Highlands)		.107		.072	.821	202	30.9	13.3	Woodfield et al. 1974
Daga (Bay Province)	.213			.058	.729	134	42.0	14.2	Jenkins et al. 1983
Karkar Is.									
Takia		.262		.106	.632	661	52.1	17.2	Booth et al. 1982
Manus Is.	.116			.087	.797	138	34.4	14.5	Malcolm et al. 1972
Buka Is.	.139			.093	.768	79	38.2	15.6	McLoughlin et al. 1982a
Solomon Is.		.274		.058	.668	2,926	47.5	13.8	Lai & Bloom 1982
MICRONESIA									
E. Caroline Is.									
(Pingelap)	.300			.056	.644	1,035	49.2	13.3	Morton & Yamamoto 1973
Marshall Is.	.127		.002	.103	.768	310	38.3	15.9	Sussman et al. 1960
POLYNESIA									
Samoans (Christchurch,									
New Zealand)		.258		.130	.610	77	54.4	18.2	Booth et al. 1977

Note: Alleles A and B are codominant to each other and dominant to allele O. A_1 and A_2 are suballeles of A, and A_1 is dominant to A_2. [a] and [b] refer to alleles A_{int} and A_{Bantu}, respectively. [c] Frequency of .007 of allele A_{Bantu} is added to that of allele A_{int}.

Table 142. Auberger system: AU

Place/Population	Au^a	Au	n	H	PE	Source
EUROPE						
FRANCE (Paris)	.564	.436	389	49.2	2.0	Salmon et al. 1961
UNITED KINGDOM						
England (London)	.607	.393	155	47.7	1.4	"
AFRICA						
SENEGAL						
Negroes	.692	.308	21	42.6	.6	"

Note: Allele Au^a is dominant to Au.

Table 143. Colton system: CO

Place/Population	Co^a	Co^b	n	H	PE	Source
EUROPE						
NETHERLANDS (Amsterdam)	.954	.046	1,430	8.8	.0	Heistö et al. 1967
NORWAY (Oslo)	.922	.078	500	14.4	.0	"
UNITED KINGDOM						
England	.958	.042	1,100	8.0	.0	"
NORTH AMERICA						
CANADA						
Cree Indians (Manitoba)	.990	.010	100	2.0	.0	Lucciola et al. 1974

Continued

Table 143. Colton system: CO (cont'd)

Place/Population	Co^a	Co^b	n	H	PE	Source
UNITED STATES						
Whites	.970	.030	1,083	5.8	.0	Crawford 1967
Blacks	1.000		1,706			"
Papago & Pima Indians	1.000		573			Matson et al. 1968

Note: Co^a and Co^b are codominant. However, since antigen Co^b is rarely tested, PE is computed under the assumption that Co^a is dominant.

Table 144. Cs system: CS

Place/Population	Cs^a	Cs	n	H	PE	Source
Europeans	.842	.158	363	26.6	.1	Giles et al. 1965a
Negroes	.806	.194	53	31.3	.1	"

Note: Allele Cs^a is dominant to Cs.

Table 145. Diego system: DI

Place/Population	Di^a	Di^b	n	H	PE	Source
EUROPE						
FAROE IS. (N. Atlantic)		1.000	484			Tills et al. 1985
FINLAND		1.000	500			Tiilikainen et al. 1969
GREECE						
Greeks	.008	.992	61	1.6	.8	Fraser et al. 1969b
Greeks (Plati)		1.000	391			Tills et al. 1983b
ICELAND		1.000	90			Tills et al. 1982b
IRELAND		1.000	100			Tills et al. 1977a
ITALY		1.000	285			Spedini 1960
NETHERLANDS		1.000	200			Layrisse 1958
NORWAY						
Lapps/Saami		1.000	433			Kornstad 1960
SWEDEN						
Lapps/Saami		1.000	220			Beckman et al. 1959
U.S.S.R.						
EUROPEAN PART						
Russians (Moscow)		1.000	323			Umnova et al. 1964
ASIAN PART						
Eskimos (New Chaplino)	.020	.980	102	3.9	1.8	Sukernik et al. 1981
Chukchi (Chukotka & NE Kamchatka)	.030	.970	1,066	5.8	2.7	"
Nganasan-Avam (Taimir Pen.)	.006	.994	348	1.2	.6	Sukernik et al. 1978
ASIA						
ISRAEL						
Yemenite Jews		1.000	157			Tills et al. 1977b
Habbanite Jews		1.000	101			Bonné et al. 1970
SAUDI ARABIA						

Continued

Table 145. Diego system: DI (cont'd)

Place/Population	Di^a	Di^b	n	H	PE	Source
Arabs (Southern)	.015	.985	103	3.0	1.4	Marengo-Rowe et al. 1974
IRAN						
Mazandaranians &						
Guilanians	.001	.999	361	.2	.1	Ohkura et al. 1984
PAKISTAN						
Pathans		1.000	139			Vos & Kirk 1961
INDIA						
Irula tr. (Nilgiri Hills,						
S. India)		1.000	72			Kirk et al. 1962
Bengali (E. India)	.009	.991	110	1.8	.9	Das et al. 1967
SRI LANKA						
Sinhalese		1.000	72			Kirk et al. 1962
Veddah		1.000	39			"
BHUTAN	.024	.976	128	4.7	2.2	Mourant et al. 1969
MALAYSIA						
Malays	.004	.996	40	.8	.4	Vos & Kirk 1961
Aborigines		1.000	270			Chin 1964
INDONESIA						
Indonesians (Java)		1.000	91			Nijenhuis 1961
THAILAND						
Thais (Bangkok)	.012	.988	429	2.4	1.1	Chandanayingyong et al. 1967
PHILPPINES						
Filipino (Manila)	.018	.982	403	3.5	1.7	Fraser et al. 1964
Negrito/Aeta (Zambales)	.012	.988	335	2.4	1.1	Misawa & Omoto 1985
CHINA						
Han (Northern)	.057	.943	290	10.8	4.5	Yuan et al. 1982
Uygur (NW China)	.039	.961	211	7.5	3.3	Yuan et al. 1984d
Hui (Ningxia)	.035	.965	219	6.8	3.0	Yuan et al. 1985
Dong (Guangxi)	.023	.977	201	4.5	2.1	Yuan et al. 1984c
Zhuang (S. China)	.044	.956	211	8.4	3.7	Yuan et al. 1984a
Tibetans	.012	.988	42	2.4	1.1	Ørjasaeter et al. 1966
Chinese (Taiwan)	.030	.970	300	5.8	2.7	Fraser et al. 1965
TAIWAN						
Toroko-Aborigines		1.000	129			Chen et al. 1985
JAPAN						
Japanese (Tokyo)	.049	.951	163	9.3	4.0	Nakajima 1973b
Ainu (Hokkaido)	.024	.976	248	4.7	2.2	Harvey et al. 1978
KOREA						
Koreans (Seoul)	.054	.946	212	10.2	4.3	Nakajima 1973a
Koreans (SE China)	.040	.960	216	7.7	3.4	Yuan et al. 1984a

AFRICA

Place/Population	Di^a	Di^b	n	H	PE	Source
NIGERIA						
Yoruba (Western)		1.000	59			Blumberg et al. 1961
CENTRAL AFRICAN REPUBLIC						
Babinga Pygmies (Bagandou)		1.000	95			Cavalli-Sforza et al. 1969
ZAIRE						
Bantu		1.000	93			Govaerts et al. 1972
TANZANIA						
Sandawe (Kondoa-Irang Dist.)		1.000	215			Godber et al. 1976
Nyaturu (")		1.000	215			"
Hadza (Munguli)		1.000	44			Tills et al. 1982a
SOUTHERN AFRICA						
San/Bushmen		1.000	114			Weiner & Zoutendyk 1959

NORTH AMERICA

Place/Population	Di^a	Di^b	n	H	PE	Source
ALASKA						
Eskimos-Inupiak (Wainwright)	.009	.991	111	1.8	.9	Corcoran et al. 1959
Eskimos (St. Lawrence Is.)	.042	.958	217	8.0	3.5	Ferrell et al. 1981b
Athabaskan Indians (Fort Yukon)	.005	.995	110	1.0	.5	Corcoran et al. 1959
CANADA						
Eskimos (Victoria Is.)		1.000	320			Chown & Lewis 1959
Ojibwa Indians (Pikangikum)	.182	.818	96	29.8	8.1	Szathmary et al. 1975
Cree " (Manitoba)	.050	.950	200	9.5	4.1	Lucciola et al. 1974
Cree " (Alberta)	.015	.985	1,598	3.0	1.4	Buchanan et al. 1983
Dogrib " (Yellowknife)	.008	.992	315	1.6	.8	"

Continued

Table 145. Diego system: DI (cont'd)

Place/Population	Di^a	Di^b	n	H	PE	Source
Dogrib " (NW Territories)		1.000	158			Szathmary 1983
UNITED STATES						
Blackfeet Indians (Montana)	.024	.976	148	4.7	2.2	Rokala et al. 1977
Papago Indians (S. Arizona)	.028	.972	709	5.4	2.5	Niswander et al. 1970
Blacks (Southern)	.003	.997	188	.6	.3	Cooper et al. 1963
Mexican-Americans (Texas)	.052	.948	1,685	9.9	4.2	Edwards-Moulds & Alperin 1986

MIDDLE AMERICA

Place/Population	Di^a	Di^b	n	H	PE	Source
MEXICO						
Indians (San Pablo del Monte)	.056	.944	257	10.6	4.4	Crawford et al. 1974
Mestizo (Tlaxcala)	.076	.924	138	14.0	5.5	"
GUATEMALA						
Black Caribs (Livingston)	.020	.980	201	3.9	1.8	Crawford et al. 1981
COSTA RICA						
Guaymi		1.000	286			Barrantes et al. 1982
WEST INDIES (Haiti)	.034	.966	222	6.6	3.0	Basu et al. 1976

SOUTH AMERICA

Place/Population	Di^a	Di^b	n	H	PE	Source
BOLIVIA						
Aymara	.050	.950	491	9.5	4.1	Matson et al. 1966
BRAZIL						
Baniwa (Jandu Cachoeira)	.24	.76	363	36.5	8.0	Salzano et al. 1986
Cayapo (Pará & Mato Grosso)	.215	.785	515	33.8	8.2	Salzano et al. 1972a
Caingang	.236	.764	214	36.1	8.0	Salzano et al. 1980
Parakana	.39	.61	49	47.6	5.4	Black et al. 1980
Macushi (N. Brazil & S. Guyana)	.117	.883	499	20.7	7.1	Neel et al. 1977a
Wapishana (")	.162	.838	617	27.2	8.0	"
Ticuna (C. Amazonas)	.159	.841	1,747	26.7	8.0	Neel et al. 1980
CHILE						
Atacameño (Toconao, N. Chile)	.008	.992	182	1.6	.8	Rothhammer et al. 1984
COLOMBIA						
Noanama		1.000	168			Kirk et al. 1974
FRENCH GUIANA						
Wayampi (Trois-Sauts & Camopi)	.319	.681	95	43.4	6.9	Tchen et al. 1981
SURINAM						
Trio (Southern)	.171	.829	112	28.4	8.1	Geerdink et al. 1974c
VENEZUELA						
Makiritare (S. Venezuela & N. Brazil)	.190	.810	539	30.8	8.2	Gershowitz et al. 1970
Yanomama (")		1.000	2,504			Gershowitz et al. 1972

OCEANIA

Place/Population	Di^a	Di^b	n	H	PE	Source
AUSTRALIA						
Aborigines/Malag (Elcho Is., N. Territory)		1.000	179			Simmons & Cooke 1969
MELANESIA						
W. New Guinea						
Asmat (S. Coastal Plain)		1.000	125			Gajdusek et al. 1978
Papua New Guinea						
Gainj & Kalam (N. Central Highlands)		1.000	54			Long et al. 1986
Yagaria (E. Highlands)		1.000	379			Mourant et al. 1982
Karkar Is.						
Takia		1.000	11			Booth et al. 1982
MICRONESIA						
Marshall Is.		1.000	310			Sussman et al. 1960
POLYNESIA						
Cook Is.		1.000	214			Douglas & Staveley 1959

Note: Alleles Di^a and Di^b are codominant. However, since antigen Di^b is rarely tested, PE is computed under the assumption that Di^a is dominant to Di^b.

Table 146. Dombrock system: DO

Place/Population	Do^a	Do	n	H	PE	Source
EUROPE						
Europeans (Northern)	.420	.580	755	48.7	4.8	Tippett 1967
ASIA						
ISRAEL						
Jews	.407	.593	128	48.3	5.0	"
THAILAND						
Thais (Bangkok)	.070	.930	423	13.0	5.2	Chandanayingyong et al. 1967
JAPAN						
Japanese (Chiba)	.126	.874	760	22.0	7.4	Nakajima & Moulds 1980
NORTH AMERICA						
CANADA						
Cree Indians (Manitoba)	.219	.781	200	34.2	8.1	Lucciola et al. 1974
UNITED STATES						
Blacks	.331	.669	161	44.3	6.6	Polesky & Swanson 1966
Papago Indians (Arizona)	.423	.577	21	48.8	4.7	Matson et al. 1968

Note: Allele Do^a is dominant to allele Do.

Table 147. Duffy system (Fy^a; see note): FY

Place/Population	Fy^a	Fy^b+Fy	n	H	PE	Source
EUROPE						
AUSTRIA (Wein)	.436	.564	2,242	49.2	4.4	Mayr et al. 1970
FINLAND	.471	.529	5,536	49.8	3.7	Nevanlinna 1972
FRANCE						
French	.429	.571	533	49.0	4.6	André et al. 1956
Basques	.358	.642	63	46.0	6.1	Nijenhuis 1956
GERMANY	.407	.593	2,000	48.3	5.0	Ritter 1969
GREECE						
Greeks (Sphakiá, Crete)	.440	.560	115	49.3	4.3	Barnicot et al. 1965
HUNGARY (Budapest)	.472	.528	480	49.8	3.7	Joó-Szabados 1970
ITALY (Southern)	.437	.563	167	49.2	4.4	Sangiorgi et al. 1982
NETHERLANDS (Leiden)	.444	.556	795	49.4	4.2	Fraser et al. 1974
NORWAY						
Norwegians	.430	.570	447	49.0	4.5	Berg 1974(c)
Lapps/Saami (Finnmark)	.541	.459	423	49.7	2.4	Kornstad 1972
PORTUGAL (Terras de Bouro)	.415	.585	114	48.6	4.9	Cruz et al. 1973
SPAIN						
Basques	.295	.705	161	41.6	7.3	Alberdi et al. 1957
SWEDEN						
Swedes (Northern)	.411	.589	3,151	48.4	4.9	Cedergren et al. 1983
Lapps/Saami	.498	.502	222	50.0	3.2	"
SWITZERLAND	.416	.584	417	48.6	4.8	Holländer 1951
UNITED KINGDOM						
England	.413	.587	1,166	48.5	4.9	Mourant et al. 1976(c)
Orkney Is. (Westray)	.443	.557	354	49.4	4.3	Welch et al. 1973
U.S.S.R.						
EUROPEAN PART						
Russians (Moscow)	.445	.505	1.445	50.0	3.2	Umnova et al. 1964

Continued

Table 147. Duffy system (Fy^a): FY (cont'd)

Place/Population	Fy^a	Fy^b+Fy	n	H	PE	Source
ASIA						
TURKEY (Asia Minor)	.488	.512	107	50.0	3.4	Aksoy et al. 1958
ISRAEL						
Yemenite Jews	.176	.824	157	29.0	8.1	Tills et al. 1977b
KUWAIT						
Arabs	.288	.712	140	41.0	7.4	Sawhney et al. 1984
SAUDI ARABIA						
Shiah (Eastern)	.045	.955	465	8.6	3.7	Maranjian et al. 1966
Tribes (Western)	.137	.863	208	23.6	7.6	Saha et al. 1980
IRAN						
Mazandaranians & Guilanians	.463	.537	537	49.7	3.9	Ohkura et al. 1984
PAKISTAN						
Punjabi (Lahore)	.515	.485	68	50.0	2.8	Boyd & Boyd 1954
INDIA						
Punjabi (Chandigarh)	.466	.534	137	49.8	3.8	Papiha et al. 1972
Irula tr. (Nilgiri Hills, S. India)	.735	.265	100	39.0	.4	Lehmann & Cutbush 1952
Bengali (E. India)	.358	.642	102	46.0	6.1	Das et al. 1967
SRI LANKA						
Sinhalese (Anuradhapura)	.595	.405	134	48.2	1.6	Roberts et al. 1972a
Veddah	.693	.307	106	42.6	.6	Wickremasinghe et al. 1963
NEPAL	.813	.187	200	30.4	.1	Bird et al. 1957
BURMA	.752	.248	81	37.3	.3	Ikin et al. 1969
MALAYSIA						
Negrito	.553	.447	35	49.4	2.2	Polunin & Sneath 1953
INDONESIA						
Toba Batak (Samosir Is.)	.884	.116	150	20.5	.0	Hawkins et al. 1973
PHILIPPINES						
Filipino	.750	.250	399	37.5	.3	Fraser et al. 1964
Negrito/Aeta (Zambales)	.886	.114	335	20.2	.0	Misawa & Omoto 1985
CHINA						
Chinese (Hong Kong)	1.000		500			Grimmo & Lee 1964b
AFRICA						
SUDAN						
Sudanese (Khartoum)	.017	.983	116	3.3	1.6	Saha et al. 1978b
CHAD						
Sara Majingay	.002	.998	257	.4	.2	Hiernaux 1976
ETHIOPIA						
Amhara tr.	.146	.854	107	24.9	7.8	Ikin & Mourant 1962
SOUTHERN AFRICA						
San/Bushmen						
!Kung, Dobe	.260	.740	385	38.5	7.8	Nurse & Jenkins 1977
Khoi/Hottentot						
Keetmanshoop Nama	.300	.700	153	42.0	7.2	Jenkins 1972
Bantu (Natal)	.060	.940	365	11.3	4.7	Shapiro 1953
Griqua (Campbell, Cape Province)	.251	.749	98	37.6	7.9	Nurse & Jenkins 1975
NORTH AMERICA						
GREENLAND						
Eskimos (Holsteinsborg)	.808	.192	27	31.0	.1	Laughlin 1957
ALASKA						
Eskimos-Inupiat (Wainwright)	.790	.210	111	33.2	.2	Corcoran et al. 1959
Athabaskan Indians (Fort Yukon)	1.000		110			"
CANADA						
Eskimos (Victoria Is.)	.750	.250	320	37.5	.3	Chown & Lewis 1959
UNITED STATES						
Blackfeet Indians (Montana)	.768	.232	148	35.6	.2	Rokala et al. 1977
Papago Indians (S. Arizona)	.820	.180	709	29.5	.1	Niswander et al. 1970
Whites (Southern)	.422	.578	332	48.8	4.7	Cooper et al. 1963
Blacks (")	.046	.954	304	8.8	3.8	"
MIDDLE AMERICA						
MEXICO						

Continued

Table 147. Duffy system (Fy^a): FY (cont'd)

Place/Population	Fy^a	Fy^b+Fy	n	H	PE	Source
Indians (San Pablo del Monte)	.784	.216	257	33.9	.2	Crawford et al. 1974
Mestizo (Tlaxcala)	.526	.474	138	49.9	2.7	"
WEST INDIES (Haiti)	.023	.977	447	4.5	2.1	Basu et al. 1976

SOUTH AMERICA

BRAZIL						
Caingang	.710	.290	214	41.2	.5	Salzano et al. 1980
Parakana	1.000		100			Black et al. 1980
Pacaás Novos	.720	.280	213	40.3	.4	Salzano et al. 1985
COLOMBIA						
Noanama	.733	.267	168	39.1	.4	Kirk et al. 1974
SURINAM						
Trio (Southern)	.794	.206	496	32.7	.1	Geerdink et al. 1974c

OCEANIA

AUSTRALIA						
Aborigines (Yuendumu, N. Territory)	1.000		140			Kirk 1965 (c)
Aborigines (W. Australia)	1.000		903			"
Aborigines (Cape York, Queensland)	1.000		50			"
MELANESIA						
W. New Guinea						
Asmat (S. Coastal Plain)	.733	.267	364	39.1	.4	Gajdusek et al. 1978
Papua New Guinea						
Fuyuge (C. Highlands)	1.000		103			Woodfield et al. 1974
Daga (Bay Province)	1.000		134			Jenkins et al. 1983
Buka Is.	1.000		79			McLoughlin et al. 1982a
MICRONESIA						
Caroline Is. (Truk)	.941	.059	284	11.1	.0	Plato & Cruz 1966
Marshall Is.	.671	.329	129	44.2	.8	Sussman et al. 1959

Note: Data obtained by tests with anti-Fy^a only are given in this table. Data obtained by tests with anti-Fy^a and anti-Fy^b are presented in Table 147.1. Alleles Fy^a and Fy^b are codominant to each other and dominant to Fy. PE was computed under the assumption that Fy^a is dominant to Fy^b and Fy.

Table 147.1. Duffy system (Fy^a, Fy^b; see note): FY

Place/Population	Fy^a	Fy^b	Fy	n	H	PE	Source
EUROPE							
CZECHOSLOVAKIA (Ovcie)	.213	.787		247	33.5	14.0	Matousek & Seemanová 1973
DENMARK	.419	.581		2,736	48.7	18.4	Gürtler 1970
FAROE IS. (N. Atlantic)	.432	.568		675	49.1	18.5	Tills et al. 1985
GREECE							
Greeks (Plati)	.443	.557		1,027	49.4	18.6	Tills et al. 1983b
ICELAND	.473	.527		1,682	49.9	18.7	Tills et al. 1982b
IRELAND	.420	.580		1,694	48.7	18.4	Tills et al. 1977a
POLAND (Olkusz)	.498	.502		213	50.0	18.7	Wolanski et al. 1983
SPAIN							
Spanish	.365	.635		483	46.4	17.8	Valls 1975
UNITED KINGDOM							
Isle of Lewis	.396	.574	.030	282	51.3	7.2*	Clegg et al. 1985
Isle of Man	.428	.572		293	49.0	18.5	Mitchell et al. 1982
U.S.S.R.							

Continued

Table 147.1. Duffy system (Fy^a, Fy^b; see note): FY (cont'd)

Place/Population	Fy^a	Fy^b	Fy	n	H	PE	Source
EUROPEAN PART							
Abkhazia (Ochamchir Dist.)	.483	.517		355	49.9	18.7	Ferrell et al. 1985
ASIAN PART							
Eskimos (New Chaplino)	.916	.084		101	15.4	7.1	Sukernik et al. 1981
Chukchi (Chukotka							
& NE Kamchatka)	.944	.056		931	10.6	5.0	"
Nganasan-Avam (Taimir Pen.)	.905	.095		429	17.2	7.9	Sukernik et al. 1978
Nentzi (Between Ob R. and							
Taj R.)	.809	.191		719	30.9	13.1	Sukernik et al. 1980

ASIA

Place/Population	Fy^a	Fy^b	Fy	n	H	PE	Source
ISRAEL							
Ashkenaji Jews	.436	.531	.033	936	52.7	7.1*	Sandler et al. 1979
SAUDI ARABIA EMIRATES							
Arabs (Southern)	.104	.122	.774	243	37.5	15.6*	Marengo-Rowe et al. 1974
UNITED ARAB							
Abu Dhabians	.187	.133	.680	100	48.5	18.4*	Kamel et al. 1980
AFGHANISTAN							
Pushtu & Dari (Kabul)	.576	.377	.047	283	52.4	7.7*	Papiha et al. 1977a
INDIA							
Gujarati (W. India)	.574	.426		287	48.9	18.5	Papiha et al. 1981
Hindu & Moslems							
(Madhya Pradesh)	.536	.391	.073	277	55.4	8.2*	Roberts et al. 1974
Bhil tr. (")	.724	.276		145	40.0	16.0	Papiha et al. 1978
Hindu (Andhra Pradesh)	.685	.315		211	43.2	16.9	Roberts et al. 1980
BHUTAN	.815	.079	.106	128	31.8	6.0*	Mourant et al. 1969
INDONESIA							
Balinese (Bali Is.)	.918	.082		148	15.1	7.0	Breguet et al. 1982a
THAILAND (Bangkok)	.888	.112		1,000	19.9	9.0	Chandanayingyong et al. 1979
PHILIPPINES							
Filipino (Germany)	.913	.087		143	15.9	7.3	Windhof & Walter 1983
Negrito/Aeta (Zambales)	.886	.114		335	20.2	9.1	Misawa & Omoto 1985
CHINA							
Han (Northern)	.775	.140	.035	276	37.3	8.1*	Yuan et al. 1982
Mongolians (Northern)	.922	.078		129	14.4	6.7	Yuan et al. 1984a
Uygur (NW China)	.898	.102		221	18.3	8.3	Yuan et al. 1984d
Hui (Ningxia)	.927	.073		219	13.5	6.3	Yuan et al. 1985
Dong (Guangxi)	.965	.035		201	6.8	3.2	Yuan et al. 1984c
Zhuang (S. China)	.981	.019		211	3.7	1.8	Yuan et al. 1984a
Tibetans (Switzerland)	.913	.087		115	15.9	7.3	Jeannet et al. 1972
TAIWAN							
Toroko-Aborigines	.799	.201		127	32.1	13.5	Chen et al. 1985
JAPAN							
Japanese (Tokyo)	.871	.129		167	22.5	10.0	Nakajima 1973b
Ainu (Hokkaido)	.858	.142		248	24.4	10.7	Harvey et al. 1978
KOREA							
Koreans (Seoul)	.930	.070		179	13.0	6.1	Nakajima 1973a
Koreans (NE China)	.940	.060		216	11.3	5.3	Yuan et al. 1984a

AFRICA

Place/Population	Fy^a	Fy^b	Fy	n	H	PE	Source
LIBYA (Tripoli & Benghasi)	.390	.299	.311	169	66.2	15.7*	Walter et al. 1975
SUDAN							
Beja -Amarar (NE Sudan)	.127	.086	.786	97	35.9	14.9*	el Hassan et al. 1968
NIGERIA							
Nigerians (Northern)			1.000	141			Tills et al. 1979
CENTRAL AFRICAN REPUBLIC							
Babinga Pygmies (Bagandou)			1.000	163			Cavalli-Sforza et al. 1969
ZAIRE							
Bantu		.029	.971	93	5.6	2.7	Govaerts et al. 1972
UGANDA (Kampala)		.043	.957	108	8.2	3.6	Ssebabi & Roberts 1975
TANZANIA							
Sandawe (Konda-Irang Dist.)	.007	.002	.991	214	1.8	.9*	Godber et al. 1976
Nyaturu (")	.002	.007	.991	217	1.8	.9*	"
Hadza (W. Lake Eyasi)			1.000	104			Tills et al. 1982a

Continued

Table 147.1. Duffy system (Fy^a, Fy^b; see note): FY (cont'd)

Place/Population	Fy^a	Fy^b	Fy	n	H	PE	Source
ANGOLA							
Njinga	.005		.995	109	1.0	.5*	Nurse et al. 1979
SOUTHERN AFRICA							
Basters (Namibia)	.521	.431	.048	116	54.0	7.4*	Nurse et al. 1982
Negroes/Dama (Okombahe)	.046	.065	.889	90	20.3	9.3*	Nurse et al. 1976

NORTH AMERICA

Place/Population	Fy^a	Fy^b	Fy	n	H	PE	Source
ALASKA							
Eskimos (St. Lawrence Is.)	.886	.114		211	20.2	9.1	Ferrell et al. 1981b
CANADA							
Ojibwa Indians (Pikangikum,							
NW Ontario)	.729	.271		96	39.5	15.9	Szathmary et al. 1975
Cree Indians (Manitoba)	.810	.190		100	30.8	13.0	Lucciola et al. 1974
Dogrib Indians							
(NW Territories)	.952	.048		158			Szathmary 1983
UNITED STATES							
Mexican Americans (Texas)	.494	.506		494	50.0	18.7	Hewett-Emmett et al. 1986b

MIDDLE AMERICA

Place/Population	Fy^a	Fy^b	Fy	n	H	PE	Source
COSTA RICA							
Guaymi	.480	.520		286	49.9	18.7	Barrantes et al. 1982

SOUTH AMERICA

Place/Population	Fy^a	Fy^b	Fy	n	H	PE	Source
BOLIVIA							
Aymara (Western)	.821	.174	.005	309	29.6	8.2*	Ferrell et al. 1978a
BRAZIL							
Baniwa (Jandu Cachoeira)	.738	.262			38.7	15.6	Salzano et al. 1986
Cayapo (Para & Mato Grosso)	.703	.281	.016	363	42.7	8.1*	Salzano et al. 1972a
Macushi (N. Brazil							
& S. Guyana)	.667	.343		503	43.7	17.1	Neel et al. 1977a
Wapishana (")	.699	.301		617	42.1	16.6	"
Ticuna (C. Amazonas)	.644	.356		1,751	45.9	17.7	Neel et al. 1980
CHILE							
Atacameño (Toconao, N. Chile)	.761	.239		180	36.4	14.9	Rothhammer et al. 1984
FRENCH GUIANA							
Wayampi (Trois-Sauts & Camopi)	.687	.313		158	43.0	16.9	Tchen et al. 1981
VENEZUELA							
Makiritare (S. Venezuela &							
N. Brazil)	.727	.273		392	39.7	15.9	Gershowitz et al. 1970
Yanomama (")	.575	.425		1,789	48.9	18.5	Gershowitz et al. 1972

OCEANIA

Place/Population	Fy^a	Fy^b	Fy	n	H	PE	Source
AUSTRALIA							
Aborigines/Malag (Elcho Is.,							
N. Territory)	.913	.087		179	15.9	7.3	Simmons & Cooke 1969
MELANESIA							
Papua New Guinea							
Gainj & Kalam (N. Central							
Highlands)	.994	.006		390	1.2	.6	Long et al. 1986
Jimi Valley (W. Highlands)	1.000			344			Mourant et al. 1981
Yagaria (E. Highlands)	1.000			371			Mourant et al. 1982
Karkar Is.							
Takia	.958	.042		106	8.0	3.9	Booth et al. 1982
Manus Is.	.846	.154		84	26.1	11.3	Malcolm et al. 1972
POLYNESIA							
Samoans (Christchurch,							
New Zealand)	.930	.070		64	13.0	6.1	Booth et al. 1977

Note: Data obtained by tests with anti-Fy^a and anti-Fy^b are given in this table. Data obtained by tests with anti-Fy^a only are presented in Table 147. Alleles Fy^a and Fy^b are codominant to each other and dominant to Fy. The PE values with the * mark were computed by using equation (21) in chapter 1, whereas those without the mark were computed by using equation (17).

Table 148. Gerbich system: GE

Place/Population	Ge^a	Ge	n	H	PE	Source
EUROPE						
DENMARK	1.000		<500			Rosenfield et al. 1960
FAROE IS. (N. Atlantic)	1.000		653			Tills et al. 1985
GREECE						
Greeks (Plati)	1.000		728			Tills et al. 1983b
UNITED KINGDOM						
England	1.000		22,331			Cleghorn 1961
Isle of Lewis	1.000		283			Clegg et al. 1985
ASIA						
THAILAND	.985	.015	4,253	3.0	.0	Chandanayingyong et al. 1979
JAPAN						
Japanese	1.000		22,000			Okubo et al. 1984b
NORTH AMERICA						
CANADA						
Cree Indians	1.000		200			Lucciola et al. 1974
UNITED STATES						
Whites, Blacks & Chinese	1.000		>11,000			Rosenfield et al. 1960
OCEANIA						
MELANESIA						
Papua New Guinea						
Gainj & Kalam (N. Central						
Highlands	.623	.377	300	47.0	1.3	Long et al. 1986
Jimi Valley (W. Highlands)	1.000		316			Mourant et al. 1981
Yagaria (E. Highlands)	1.000		20			Mourant et al. 1982
Fuyuge (C. Highlands)	1.000		103			Woodfield et al. 1974
Karkar Is.						
Takia	.710	.290	427	41.0	.5	Booth et al. 1982
Manus Is.	.799	.201	74	32.1	.1	Malcolm et al. 1972
Buka Is.	1.000		79			McLoughlin et al. 1982
Solomon Is.	1.000		28			Booth & McLoughlin 1972
POLYNESIA						
Samoans (New Zealand)	1.000		77			Booth et al. 1977

Note: Allele Ge^a is dominant to Ge.

Table 149. Henshaw antigen: HE

Place/Population	He$^+$	He$^-$	n	H	PE	Source
EUROPE						
FAROE IS. (N. Atlantic)	1.000	680				Tills et al. 1985
GREECE						
Greeks (Plati)	1.000	1,038				Tills et al. 1983b
IRELAND	1.000	100				Tills et al. 1977a
UNITED KINGDOM						
Isle of Lewis	1.000	283				Clegg et al. 1985
ASIA						
ISRAEL						

Continued

Table 149. Henshaw antigen: HE (cont'd)

Place/Population	He$^+$	He$^-$	n	H	PE	Source
Yemenite Jews	.045	.955	155	8.6	3.7	Tills et al. 1977b
Habbanite Jews		1.000	596			Bonné et al. 1970
SAUDI ARABIA						
Arabs (Southern)		1.000	261			Marengo-Rowe et al. 1974
Shiah (Eastern)	.005	.995	412	1.0	.5	Maranjian et al. 1966
NEPAL						
Gorkhas		1.000	200			Bird et al. 1957
THAILAND (Bangkok)		1.000	439			Chandanayingyong et al. 1967
JAPAN						
Ainu (Hokkaido)		1.000	104			Harvey et al. 1978

AFRICA

Place/Population	He$^+$	He$^-$	n	H	PE	Source
ETHIOPIA						
Amhara tr.	.019	.981	107	3.7	1.8	Ikin & Mourant 1962
TANZANIA						
Sandawe (Konda-Irag Dist.)	.016	.984	214	3.1	1.5	Godber et al. 1976
Nyaturu	.026	.974	217	5.1	2.3	"
Hadza	.002	.998	264	.4	.2	Tills et al. 1985
ANGOLA						
Njinga	.014(NS)[a]	.986	109	2.8	1.3	Nurse et al. 1979
NAMIBIA						
Basters	.061(MS & NS)[a]	.939	120	11.5	4.7	Nurse et al. 1982
NIGERIA (Northern)	.064	.936	141	12.0	4.9	Tills et al. 1979
SOUTHERN AFRICA						
San/Bushmen (Botswana)	.023(Ms)[a]	.977	94	4.5	2.1	Jenkins et al. 1975
Negroes/Dama	.023	.977	92	4.5	2.1	Nurse et al. 1976
Griqua	.069(Ms)[a]	.931	99	12.8	5.2	Nurse & Jenkins 1975
TRISTAN DA CUNHA		1.000	160			Jenkins et al. 1985

OCEANIA

Place/Population	He$^+$	He$^-$	n	H	PE	Source
MELANESIA						
Papua New Guinea						
Daga (Bay Province)		1.000	134			Jenkins et al. 1983
Yagaria (E. Highlands)		1.000	373			Mourant et al. 1982
Karkar Is.						
Takia		1.000	151			Booth et al. 1982

Note: Allele He$^+$ is dominant to allele He$^-$. [a] He$^+$ is associated with NS, MS & NS or Ms, depending on the population.

Table 150. Karhula antigen: UL

Place/Population	Ul^a	Ul	n	H	PE	Source

EUROPE

Place/Population	Ul^a	Ul	n	H	PE	Source
FINLAND						
Finnish (Helsinki)	.013	.987	2,620	2.6	1.2	Furuhjelm et al. 1968
Lapps/Saami		1.000	140			"
SWEDEN	.001	.999	501	.2	.1	"

ASIA

Place/Population	Ul^a	Ul	n	H	PE	Source
JAPAN	.002	.998	8,000[b]	.4	.2	Okubo et al. 1986b

NORTH AMERICA

Place/Population	Ul^a	Ul	n	H	PE	Source
UNITED STATES						
Blacks		1.000	66			Furuhjelm et al. 1968

Note: Allele Ul^a is dominant to allele Ul. [b] More than 66,000 donor serum samples were also tested, but no more Ul^a was found.

Table 151. Kell system: K, KEL

Place/Population	K	k	n	H	PE	Source
EUROPE						
AUSTRIA (Wien)	.043	.957	2,247	8.2	3.9	Mayr et al. 1970
CZECHOSLOVAKIA (Ovcie)	.065	.935	316	12.2	5.7	Matousek & Seemanová 1973
DENMARK	.039	.961	2,763	7.5	3.6	Gürtler 1970
FAROE IS. (N. Atlantic)	.043	.952	656	9.2	3.9	Tills et al. 1985
FINLAND	.020	.980	5,536	3.9	1.9	Nevanlinna 1972
FRANCE	.042	.958	1,582	8.0	3.9	André et al. 1954
GERMANY	.039	.961	10,000	7.5	3.6	Hoppe 1957
GREECE						
Greeks	.037	.963	418	7.1	3.4	Fraser et al. 1969b
Greeks (Plati)	.129	.871	1.032	22.5	10.0	Tills et al. 1983b
ICELAND	.056	.944	1,683	10.6	5.0	Tills et al. 1982b
IRELAND	.040	.960	1,701	7.7	3.7	Tills et al. 1977a
ITALY (Southern)	.035	.965	117	6.8	3.3	Sangiorgi et al. 1982
NETHERLANDS (Leiden)	.037	.963	795	7.1	3.4	Fraser et al. 1974
NORWAY						
Norwegians	.036	.964	2,793	6.9	3.3	Berg 1974(c)
Lapps/Saami	.006	.994	423	1.2	.6	"
POLAND (Olkusz)	.038	.962	213	7.3	3.5	Wolanski et al. 1983
PORTUGAL (Terras de Bouro)	.041	.959	113	7.9	3.9	Cruz et al. 1973
SPAIN						
Spanish	.038	.962	4,300[a]	7.3	3.5	Campillo et al. 1973
Basques	.047	.953	120	9.0	4.3	Moya 1971
SWEDEN						
Swedes	.036	.964	4,527	6.9	3.3	Heiken 1962
Lapps/Saami	.006	.994	423	1.2	.6	Beckman et al. 1959
SWITZERLAND	.037	.963	616	7.1	3.4	Hässig 1952
UNITED KINGDOM						
England	.046	.954	8,769	8.8	4.2	Race & Sanger 1968
Isle of Lewis	.053	.947	283	10.0	4.8	Clegg et al. 1985
Isle of Man	.048	.952	351	9.1	4.4	Mitchell et al. 1982
Orkney Is.	.030	.970	349	5.8	2.8	Welch et al. 1973
U.S.S.R.						
EUROPEAN PART						
Abkhazia (Ochamchir Dist.)	.045	.955	574	8.6	4.1	Ferrell et al. 1985
ASIAN PART						
Eskimos (New Chaplino)		1.000	102			Sukernik et al. 1981
Chukchi (Chukotka/N. Kamchatka)		1.000	1,066			"
Nganasan-Avam (Taimir Pen.)		1.000	430			Karaphet et al. 1981
Nentzi (Between Ob R. & Taj R.)		1.000	753			Sukernik et al. 1980
ASIA						
TURKEY (Asia Minor)	.019	.981	107	3.7	1.8	Aksoy et al. 1958
ISRAEL						
Ashkenazi Jews (E. Europe)	.041	.959	171	7.9	3.8	Kobyliansky et al. 1982
Yemenite Jews	.010	.990	157	2.0	1.0	Tills et al. 1977b
KUWAIT						
Arabs	.019	.981	162	3.7	1.8	Sawhney et al. 1984
SAUDI ARABIA						
Shiah (Eastern)	.032	.968	465	6.2	3.0	Maranjian et al. 1966
Tribes (Western)	.061	.939	205	11.5	5.4	Saha et al. 1980
Arabs (Southern)	.047	.953	243	9.0	4.3	Marengo-Rowe et al. 1974
IRAN	.029	.971	459	5.6	2.7	Ohkura et al. 1984
AFGHANISTAN						
Pushtu & Dari (Kabul)	.018	.982	283	3.5	1.7	Papiha et al. 1977a
PAKISTAN						
Punjabi (Lahore)	.084	.916	87	15.4	7.1	Boyd & Boyd 1954
INDIA						
Punjabi (Chandigarh)	.004	.996	137	.8	.4	Papiha et al. 1972
Gujarati (W. India)	.012	.988	289	2.4	1.2	Papiha et al. 1981
Hindu & Moslems (Madhya Pradesh)	.006	.994	333	1.2	.6	Roberts et al. 1974

Continued

200

Table 151. Kell system: K, KEL (cont'd)

Place/Population	K	k	n	H	PE	Source
Bhil tr. (Madhya Pradesh)		1.000	145			Papiha et al. 1978
Hindu (Andhra Pradesh)	.002	.998	211	.4	.2	Roberts et al. 1980
Bengali (E. India)	.040	.960	203	7.7	3.7	Sen 1960
SRI LANKA						
Sinhalese (Anuradhapura)	.003	.997	157	.6	.3	Roberts et al. 1972a
Veddah	.078	.922	67	14.4	6.7	Wickremasinghe et al. 1963
NEPAL						
Shrestha	.052	.948	162	9.9	4.7	Bhasin 1970
BHUTAN		1.000	154			Mourant et al. 1969
BURMA	.044	.956	82	8.4	4.0	Ikin et al. 1969
INDONESIA						
Toba Batak (Samosir Is.)		1.000	150			Hawkins et al. 1973
Balinese (Bali Is.)		1.000	148			Breguet et al. 1982a
THAILAND (Bangkok)	.001	.999	1,000	.2	.1	Chandanayingyong et al. 1979
PHILIPPINES						
Filipino	.009	.991	403	1.8	.9	Fraser et al. 1964
Negrito/Aeta (Zambales)		1.000	335			Misawa & Omoto 1985
CHINA						
Han (N. China)	.002	.998	290	.4	.2	Yuan et al. 1982
Mongolians (N. China)	.004	.996	129	.8	.4	Yuan et al. 1984a
Uygur (NW China)	.036	.964	156	6.9	3.3	Yuan et al. 1984d
Hui (Ningxia)		1.000	219			Yuan et al. 1985
Zhuang (S. China)		1.000	211			Yuan et al. 1984a
Tibetans (Switzerland)		1.000	115			Jeannet et al. 1972
Chinese (Hong Kong)		1.000	500			Grimmo & Lee 1964a
TAIWAN						
Toroko-Aborigines		1.000	127			Chen et al. 1985
JAPAN						
Japanese (Tokyo)		1.000	150			Nakajima 1973b
Ainu (Hokkaido)		1.000	248			Harvey et al. 1978
KOREA						
Koreans (Seoul)	.002	.998	255	.4	.2	Nakajima 1973a
Koreans (NE China)		1.000	216			Yuan et al. 1984a

AFRICA

Place/Population	K	k	n	H	PE	Source
LIBYA (Tripoli and Benghasi)	.055	.945	169	10.4	4.9	Walter et al. 1975
SUDAN						
Sudanese (Khartoum)	.030	.970	116	5.8	2.8	Saha et al. 1978b
Beja-Amarar (NE Sudan)	.005	.995	97	1.0	.5	el Hassan et al. 1968
CHAD						
Sara Majingay	.002	.998	258	.4	.2	Hiernaux 1976
SENEGAL						
Bedik (E. Senegal)	.001	.999	792	.2	.1	Bouloux et al. 1972
NIGERIA						
Yoruba (Western)	.015	.985	135	3.0	1.5	Blumberg et al. 1961
CENTRAL AFRICAN REPUBLIC						
Babinga Pygmies (Bagandou)	.006	.994	163	1.2	.6	Cavalli-Sforza et al. 1969
ZAIRE						
Bantu		1.000	93			Govaerts et al. 1972
ETHIOPIA						
Amhara tr.	.028	.972	107	5.4	2.6	Ikin & Mourant 1962
UGANDA (Kampala)[a]	.083	.917	108	15.2	7.0	Ssebabi & Roberts 1975
TANZANIA						
Sandawe (Kondoa-Irang Dist.)	.003	.997	163	.6	.3	Godber et al. 1976
Nyaturu (")	.005	.995	217	1.0	.5	"
Hadza (W. Lake Eyasi)		1.000	105			Tills et al. 1982a
ANGOLA						
Njimba		1.000	109			Nurse et al. 1979
SOUTHERN AFRICA						
Basters (Nambia)	.117	.883	120	20.7	9.3	Nurse et al. 1982
San/Bushmen						
!Kung, Dobe		1.000	385			Nurse & Jenkins 1975
Khoi/Hottentot						
Keetmanshoop Nama	.007	.993	153	1.4	.7	Jenkins 1972
Negroes/Dama (Okomhabe)		1.000	92			Nurse et al. 1976
Bantu (Natal)	.003	.997	705	.6	.3	Shapiro 1953
Griqua (Campbell, Cape Province)	.042	.958	98	8.0	3.9	Nurse & Jenkins 1975

Continued

Table 151. Kell system: K, KEL (cont'd)

Place/Population	K	k	n	H	PE	Source
NORTH AMERICA						
GREENLAND						
Eskimos (Angmagssalik)		1.000	739			Gürtler 1971
ALASKA						
Eskimos-Inupiak (Wainwright)		1.000	111			Corcoran et al. 1959
Eskimos (St. Lawrence Is.)		1.000	222			Ferrell et al. 1981b
Athabaskan Indians (Fort Yukon)	.005	.995	110	1.0	.5	Corcoran et al. 1959
CANADA						
Eskimos (Manitoba)		1.000	67			Chown & Lewis 1952
Ojibwa Indians (Pikangikum)		1.000	96			Szathmary et al. 1975
Cree Indians (Manitoba)	.005	.995	100	1.0	.5	Lucciola et al. 1974
Dogrib Indians (NW Territories)		1.000	158			Szathmary 1983
UNITED STATES						
Blackfeet Indians (Montana)		1.000	148			Rokala et al. 1977
Papago Indians (S. Arizona)		1.000	709			Niswander et al. 1970
Whites (Southern)	.042	.958	333	8.0	3.9	Cooper et al. 1963
Blacks (")	.005	.995	303	1.0	.5	"
Mexican-Americans (Texas)	.011	.989	668	2.2	1.1	Hewett-Emmett et al. 1986b
MIDDLE AMERICA						
MEXICO						
Indians (San Pablo del Monte)		1.000	257			Crawford et al. 1974
Mestizo (Tlaxcala)	.004	.996	138	.8	.4	"
GUATEMALA						
Black Caribs (Livingston)		1.000	202			Crawford et al. 1981
WEST INDIES (Haiti)	.014	.986	473	2.8	1.4	Basu et al. 1976
SOUTH AMERICA						
BOLIVIA						
Aymara (W. Bolivia)	.018	.982	315			Ferrell et al. 1978a
BRAZIL						
Baniwa (Jandu Cachoeira)		1.000	363			Salzano et al. 1986
Cayapo (Para & Mato Grosso)		1.000	526			Salzano et al. 1972a
Caingang		1.000	214			Salzano et al. 1980
Parakana		1.000	100			Black et al. 1980
Pacaás Novos		1.000	215			Salzano et al. 1985
Macushi (N. Brazil & S. Guyana)		1.000	503			Neel et al. 1977a
Wapishana (")	.003	.997	617	.6	.3	"
Ticuna (C. Amazonas)		1.000	1,760			Neel et al. 1980
CHILE						
Atacameños (Toconao, N. Chile)		1.000	182			Rothhammer et al. 1984
COLOMBIA						
Noanama		1.000	168			Kirk et al. 1974
FRENCH GUIANA						
Wayampi (Trois-Sauts & Camopi)		1.000	240			Tchen et al. 1981
SURINAM						
Trio (Southern)	.006	.994	494	1.2	.6	Geerdink et al. 1974c
VENEZUELA						
Makiritare (S. Venezuela &						
N. Brazil)		1.000	538			Gershowitz et al. 1970
Yanomama (")		1.000	1,812			Gershowitz et al. 1972
OCEANIA						
AUSTRALIA						
Aborigines (N. Territory)	.008	.992	332	1.6	.8	Nicholls et al. 1965
Aborigines (W. Australia)		1.000	1,444			Kirk 1965(c)
Aborigines (Morrington,						
& Forsyth, Queensland)		1.000	67			"
MELANESIA						
W. New Guinea						
Asmat (S. Coastal Plain)		1.000	402			Gajdusek et al. 1978
Papua New Guinea						
Gainj & Kalam (N. Central						

Continued

Table 151. Kell system: K, KEL (cont'd)

Place/Population	K	k	n	H	PE	Source
Highlands)		1.000	86			Long et al. 1986
Jimi Valley (W. Highlands)		1.000	344			Mourant et al. 1981
Yagaria (E. Highlands)		1.000	372			Mourant et al. 1982
Fuyuge (C. Highlands)		1.000	34			Woodfield et al. 1974
Daga (Bay Province)		1.000	134			Jenkins et al. 1983
Karkar Is.						
Takia		1.000	197			Booth et al. 1982
Manus Is.		1.000	40			Malcolm et al. 1972
Buka Is.		1.000	79			McLoughlin et al. 1982a
MICRONESIA						
Caroline Is. (Truk)		1.000	357			Plato & Cruz 1966
Marshall Is.		1.000	310			Sussman et al. 1960
POLYNESIA						
Samoans (Christchurch, New Zealand)		1.000	77			Booth et al. 1977

Note: [a] Tested with anti-Js[a] and anti-Js[b]. Alleles K and k are codominant.

Table 152. Kidd system (Jk[a]; see note): JK

Place/Population	Jk[a]	Jk[b]+Jk	n	H	PE	Source
EUROPE						
DENMARK	.517	.483	399	49.9	2.8	Gürtler 1970
FRANCE						
French	.490	.510	300	50.0	3.3	André et al. 1956
Basques	.529	.471	63	49.8	2.6	Levine et al. 1974
ICELAND	.458	.542	929	49.6	4.0	Tills et al. 1982b
NETHERLANDS	.584	.416	156	48.6	1.7	Nijenhuis 1961
SWEDEN	.498	.502	119	50.0	3.2	Heiken 1967
UNITED KINGDOM						
England	.488	.512	343	50.0	3.4	Race & Sanger 1968(c)
U.S.S.R.						
EUROPEAN PART						
Russians	.554	.446	728	49.4	2.2	Arzhelas & Reznikova 1968
ASIAN PART						
Nentzi (Between Ob R. and Taj R.)	.406	.594	519	48.2	5.1	Sukernik et al. 1980
Chukchi (Chukotka)	.540	.460	352	49.9	2.5	Sukernik et al. 1981
ASIA						
ISRAEL						
Ashkenazi Jews (E. Europe)	.584	.416	3,108	48.6	1.7	Kobyliansky et al. 1982
Yemenite Jews	.649	.351	73	45.6	1.0	Tills et al. 1977b
KUWAIT						
Arabs	.580	.420	74	48.7	1.8	Al-Nassar et al. 1981
SAUDI ARABIA						
Arabs (Southern)	.534	.466	221	49.8	2.5	Marengo-Rowe et al. 1974
IRAN						
Mazandarians & Guilanians	.510	.490	234[a]	50.0	2.9	Ohkura et al. 1984
PAKISTAN						
Punjabi (Lahore)	.482	.518	67	49.9	3.5	Boyd & Boyd 1954
SRI LANKA						
Sinhalese (Anuradhapura)	.418	.582	56	48.7	4.8	Roberts et al. 1972a
BHUTAN	.366	.634	97	46.4	5.9	Mourant et al. 1969

Continued

Table 152. Kidd system (Jk^a): JK (cont'd)

Place/Population	Jk^a	Jk^b+Jk	n	H	PE	Source
INDONESIA						
Toba Batak (Samosir Is.)	.493	.507	70	50.0	3.3	Hawkins et al. 1973
CHINA						
Hui (Ningxia)	.443	.557	219	49.4	4.3	Yuan et al. 1985
Dong (Guangxi)	.463	.537	201	49.7	3.9	Yuan et al. 1984c
TAIWAN						
Toroko-Aborigines	.330	.670	127	44.2	6.6	Chen et al. 1985
JAPAN						
Ainu (Hokkaido)	.276	.724	242	40.0	7.6	Harvey et al. 1978

AFRICA

Place/Population	Jk^a	Jk^b+Jk	n	H	PE	Source
SUDAN						
Beja-Amarar (NE Sudan)	.546	.454	97	49.6	2.3	El Hassan et al. 1968
SENEGAL						
Bedik (E. Senegal)	.396	.604	167	47.8	5.3	Bouloux et al. 1972
NIGERIA & KENYA						
Yoruba, Ibo, Jal. etc.	.782	.218	105	34.1	.2	Ikin & Mourant 1952
CENTRAL AFRICAN REPUBLIC						
Babinga Pygmies (Bagandou)	.922	.078	163	14.4	.0	Cavalli-Sforza et al. 1969
TANZANIA						
Sandawe (Kondoa-Irang Dist.)	.717	.283	175	40.6	.5	Godber et al. 1976
Nyaturu (")	.614	.386	67	47.4	1.4	"
Hadza (W. Lake Eyasi)	.582	.418	103	48.7	1.8	Tills et al. 1982a

NORTH AMERICA

Place/Population	Jk^a	Jk^b+Jk	n	H	PE	Source
GREENLAND						
Eskimos	.570	.430	27	49.0	1.9	Laughlin 1957
ALASKA						
Eskimos-Inupiak (Wainwright)	.455	.545	111	49.6	4.0	Corcoran et al. 1959
Athabaskan Indians (Fort Yukon)	.469	.531	110	49.8	18.7	"
CANADA						
Eskimos (Southampton Is.)	.629	.371	87	46.7	1.2	Chown & Lewis 1960
UNITED STATES						
Blackfeet Indians (Montana)	.715	.285	148			Rokala et al. 1977
Papago Indians (S. Arizona)	.360	.640	557	46.1	6.0	Niswander et al. 1970

MIDDLE AMERICA

Place/Population	Jk^a	Jk^b+Jk	n	H	PE	Source
MEXICO						
Indians (San Pablo del Monte)	.358	.642	257	46.0	6.1	Crawford et al. 1974
Mestizo (Tlaxcala)	.369	.631	138	46.6	5.8	"
COSTA RICA						
Guaymi	.135	.865	286	23.4	7.6	Barrantes et al. 1982
WEST INDIES (Haiti)	.706	.294	473	41.5	.5	Basu et al. 1976

SOUTH AMERICA

Place/Population	Jk^a	Jk^b+Jk	n	H	PE	Source
BOLIVIA						
Aymara (W. Bolivia)	.303	.697	313	42.2	7.2	Ferrell et al. 1978a
BRAZIL						
Baniwa (Jandu Cachoeira)	.552	.448	363	49.5	2.2	Salzano et al. 1986
Cayapo (Pará & Mato Grosso)	.448	.552	518	49.5	4.2	Salzano et al. 1972a
Caingang	.629	.371	80	46.7	1.2	Salzano et al. 1980
Parakana	.549	.451	64	49.5	2.3	Black et al. 1980
Macushi (N. Brazil & S. Guyana)	.368	.632	503	46.5	5.9	Neel et al. 1977a
Wapishana (")	.429	.571	617	49.0	4.6	"
FRENCH GUIANA						
Wayampi (Trois-Sauts & Camopi)	.383	.617	63	47.3	5.6	Tchen et al. 1981
SURINAM						
Trio (Southern)	.491	.509	247	50.0	3.3	Geerdink et al. 1974c
VENEZUELA						
Makiritare (S. Venezuela & N. Brazil)	.318	.682	537	43.4	6.9	Gershowitz et al. 1970
Yanomama (")	.538	.462	2,512	49.7	2.5	Gershowitz et al. 1972

OCEANIA

Continued

Table 152. Kidd system (Jk[a]): JK (cont'd)

Place/Population	Jk[a]	Jk[b]+Jk	n	H	PE	Source
AUSTRALIA						
Aborigines (N. Territory)	.374	.626	23	46.8	5.7	Simmons et al. 1962
MELANESIA						
Papua New Guinea						
Asaro (E. Highlands)	.446	.554	88	49.4	4.2	McLoughlin et al. 1982b
MICRONESIA						
Gilbert Is.	.509	.491	236	50.0	3.0	Douglas et al. 1961
POLYNESIA						
Samoans (Christchurch, New Zealand)	.376	.624	77	46.9	5.7	Booth et al. 1977

Note: Data obtained by tests with anti-Jk[a] only are given in this table. Data obtained by tests with anti-Jk[a] and anti-Jk[b] are presented in Table 152.1. PE was computed under the assumption that Jk[a] is dominant to Jk[b] and Jk.

Table 152.1 Kidd system (Jk[a], Jk[b]; see note): JK

Place/Population	Jk[a]	Jk[b]	Jk	n	H	PE	Source
EUROPE							
AUSTRIA	.521	.479		628	49.9	18.7	Mayr et al. 1970
DENMARK	.517	.483		399	49.9	18.7	Gürtler 1970
FAROE IS. (N. Atlantic)	.532	.468		673	49.8	18.7	Tills et al. 1985
FINLAND	.536	.464		100	49.7	18.7	Kaarsalo et al. 1962
GERMANY (Hessen)	.505	.495		1,010	50.0	18.7	Spielmann 1966
GREECE							
Greeks (Plati)	.452	.548		673	49.5	18.6	Tills et al. 1983b
ITALY (Southern)	.444	.556		117	49.4	18.6	Sangiorgi et al. 1982
NORWAY							
Norwegians	.506	.494		1,816	50.0	18.7	Berg 1974(c)
Lapps/Saami (Finnmark)	.577	.423		423	48.8	18.5	Kornstad 1972
POLAND (Olkusz)	.439	.561		213	49.3	18.6	Wolanski et al. 1983
SPAIN	.537	.463		244	49.7	18.7	Valls 1974
UNITED KINGDOM							
England (Holy Is.)	.484	.516		122	49.9	18.7	Cartwright 1976
Isle of Lewis	.498	.502		282	50.0	18.7	Tills et al. 1985
U.S.S.R.							
EUROPEAN PART							
Abkhazia (Ochamchir Dist.)	.516	.484		224	49.9	18.7	Ferrell et al. 1985
ASIAN PART							
Nganasan-Avam (Taimir Pen.)	.833	.167		78	27.8	12.0	Sukernik et al. 1978
ASIA							
AFGHANISTAN							
Pushtu & Dari (Kabul)	.558	.442		60	49.3	18.6	Papiha et al. 1977a
INDIA							
Jat Sikhs (Punjab)	.520	.480		148	49.9	18.7	Chahal & Papiha 1981
Gujarati (W.India)	.581	.419		87	48.7	18.4	Papiha et al. 1981
Hindu & Moslems (Madhya Pradesh)	.543	.316	.141	175	58.5	10.3*	Roberts et al. 1974
Hindu (Andhra Pradesh)	.427	.339	.234	211	64.8	13.3*	Roberts et al. 1980
PHILIPPINES							
Filipino (Germany)	.629	.371		93	46.7	17.9	Windhof & Walter 1983
Negrito/Aeta (Zambales)	.576	.424		335	48.8	18.5	Misawa & Omoto 1985
CHINA							
Mongolians (N. China)	.409	.591		129	48.3	18.3	Yuan et al. 1984a
Han (Beijing)	.349	.651		290	45.4	17.6	Yuan et al. 1982

Continued

Table 152.1 Kidd system (Jk^a, Jk^b): JK (cont'd)

Place/Population	Jk^a	Jk^b	Jk	n	H	PE	Source
Uygur (NW China)	.314	.686		221	43.1	16.9	Yuan et al. 1984d
Hui (Ningxia)	.443	.557		219	49.4	18.6	Yuan et al. 1985
Zhuang (S. China)	.365	.635		211	46.4	17.8	Yuan et al. 1984a
JAPAN							
Japanese (Tokyo)	.472	.528		628	49.8	18.7	Nakajima 1973b
KOREA							
Koreans (Seoul)	.500	.500		169	50.0	18.8	Nakajima 1973a
Koreans (NE China)	.431	.569		216	49.0	18.5	Yuan et al. 1984a

AFRICA

Place/Population	Jk^a	Jk^b	Jk	n	H	PE	Source
LIBYA (Tripoli & Benghasi)	.532	.468		169	49.8	18.7	Walter et al. 1975
ZAIRE							
Bantu	.859	.141		93	24.2	10.6	Govaerts et al. 1972
SOUTHERN AFRICA							
San/Bushmen							
Dobe	.888	.112		340	19.9	9.0	Jenkins 1972
Khoi/Hottentot							
Keetmanshoop Nama	.946	.054		153	10.2	4.8	"

NORTH AMERICA

Place/Population	Jk^a	Jk^b	Jk	n	H	PE	Source
ALASKA							
Eskimos (St. Lawrence Is.)	.736	.264		216	38.9	15.7	Ferrell et al. 1981b
CANADA							
Ojibwa (Pikangikum)	.438	.562		96	49.2	18.6	Szathmary et al. 1975
Cree Indians (Manitoba)	.500	.500		100	50.0	18.8	Lucciola et al. 1974
Dogrib Indians							
(NW Territories)	.672	.328		158	44.1	17.2	Szathmary 1983
UNITED STATES							
Whites (Southern)	.536	.464		333	49.7	6.3	Cooper et al. 1963
Blacks (")	.743	.257		303	38.2	15.4	"
Mexican-Americans (Texas)	.457	.543		620	49.6	18.7	Hewett-Emmett et al. 1986b

SOUTH AMERICA

Place/Population	Jk^a	Jk^b	Jk	n	H	PE	Source
BRAZIL							
Ticuna (C. Amazonas)	.407	.593		1,747	48.3	18.3	Neel et al. 1980

OCEANIA

Place/Population	Jk^a	Jk^b	Jk	n	H	PE	Source
MELANESIA							
Papua New Guinea							
Gainj & Kalam (N. Central							
Highlands)	.246	.754		240	37.1	15.1	Long et al. 1986
Karkar Is.							
Takia	.549	.451		92	49.5	18.6	Booth et al. 1982

Note: Data obtained by tests with anti-Jk^a and anti-Jk^b are given in this table. Data obtained by tests with anti-Jk^a only are given in Table 152. Alleles Jk^a and Jk^b are codominant to each other and dominant to allele Jk. The PE values with the * mark were computed by using equation (21) in chapter 1, whereas those without the mark were computed by using equation (17).

Table 153. Lewis system: LE, LES

Place/Population	Le	le	n	H	PE	Source

EUROPE

Place/Population	Le	le	n	H	PE	Source
DENMARK	.758	.242	238	36.7	.3	Andersen 1948
FINLAND	.801	.199	76	31.9	.1	Kaarsalo et al. 1962
ICELAND	.773	.227	58	35.1	.2	Bjarnason et al. 1973

Continued

Table 153. Lewis system: LE, LES (cont'd)

Place/Population	Le	le	n	H	PE	Source
ITALY						
Italians (Bologna)	.759	.241	400	36.6	.3	Facchini et al. 1973
NETHERLANDS	.718	.282	676	40.5	.5	Nijenhuis 1961
NORWAY						
Lapps/Saami	.684	.316	183	43.2	.7	Allison et al. 1952
SPAIN						
Basques	.698	.302	193	42.2	.6	Goti Iturriaga 1966
SWITZERLAND	.631	.369	1,000	46.6	1.2	Hässig et al. 1955

ASIA

Place/Population	Le	le	n	H	PE	Source
INDONESIA						
Balinese (Bali Is.)	.876	.124	65	21.7	.0	Breguet et al. 1982a
THAILAND						
Thais (Bangkok)	.504	.496	1,668	50.0	3.1	Chandanayingyong et al. 1979
CHINA						
Han (N. China)	.492	.508	120	50.0	3.3	Yuan et al. 1984b
Uygur (NW China)	.431	.569	221	49.0	4.5	Yuan et al. 1984d
Hui (Ningxia)	.388	.612	219	47.5	5.4	Yuan et al. 1985
Dong (Guangxi)	.323	.677	201	43.7	6.8	Yuan et al. 1984c
TAIWAN						
Toroko-Aborigines	.414	.586	64	48.5	4.9	Chen et al. 1985
JAPAN						
Japanese	.812	.188	485	30.5	.1	Iseki et al. 1957
Ainu (Hokkaido)	.684	.316	40	43.2	.7	Harvey et al. 1978
KOREA	1.000		80			Won et al. 1960

AFRICA

Place/Population	Le	le	n	H	PE	Source
LIBYA (Tripoli & Benghasi)	.614	.386	167	47.4	1.4	Walter et al. 1975
ZAIRE						
Bantu	.308	.692	93	42.6	7.1	Govaerts et al. 1972
SOUTHERN AFRICA						
Basters (Namibia)	.651	.349	115	45.4	1.0	Nurse et al. 1982
San/Bushmen						
Dobe	.437	.563	218	49.2	4.4	Jenkins 1972
Khoi/Hottentot						
Keetmanshoop Nama	.309	.691	115	42.7	7.0	"
Negroes/Dama	.380	.620	52	47.1	18.0	"

NORTH AMERICA

Place/Population	Le	le	n	H	PE	Source
ALASKA						
Eskimos	1.000		84			Chown & Lewis 1962
UNITED STATES						
Blackfeet Indians (Montana)	.400	.600	148	48.0	5.2	Rokala et al. 1977

MIDDLE AMERICA

Place/Population	Le	le	n	H	PE	Source
MEXICO						
Mestizo	1.000		138			Crawford et al. 1974
COSTA RICA						
Guaymi	.184	.816	286	30.0	8.2	Barrantes et al. 1982
WEST INDIES (Haiti)	.392	.608	92	47.7	5.4	Basu et al. 1976

SOUTH AMERICA

Place/Population	Le	le	n	H	PE	Source
BRAZIL						
Baniwa (Jandu Cachoeira)	.50	.50	363	50.0	3.1	Salzano et al. 1986
Cayapo (Para & Mato Grosso)	.512	.488	474	50.0	2.9	Salzano et al. 1972a
Macushi	.397	.603	503	47.9	5.2	Neel et al. 1977a
Wapishana	.440	.560	618	49.3	4.3	"
Ticuna (C. Amazonas)	.368	.632	1,753	46.5	5.9	Neel et al. 1980
COLOMBIA						
Noanama	.202	.798	168	32.2	8.2	Kirk et al. 1974
FRENCH GUIANA						
Wayampi (Trois-Sauts & Camopi)	.465	.535	63	40.2	7.5	Tchen et al. 1981
SURINAM						

Continued

Table 153. Lewis system: LE, LES

Place/Population	Le	le	n	H	PE	Source
Trio (S. Surinam)	.524	.476	131	49.9	2.7	Geerdink et al. 1974c
VENEZUELA						
Makiritare (S. Venezuela &						
N. Brazil)	.429	.571	539	49.0	4.6	Gershowitz et al. 1970
Yanomama (")	.492	.508	2,510	50.0	3.3	Gershowitz et al. 1972

OCEANIA

AUSTRALIA						
Aborigines (N. Territory)	.566	.434	331	49.1	2.0	Nicholls et al. 1965
MELANESIA						
Papua New Guinea						
Gainj & Kalam (N.Central						
Highlands)	.133	.867	53	23.1	7.5	Long et al. 1986

Note: Allele Le is dominant to allele le.

Table 154. Lutheran system: LU

Place/Population	a	b	n	H	PE	Source

EUROPE

DENMARK	.042	.958	245	8.0	3.9	Linnet-Jepsen et al. 1958
FAROE IS. (N. Atlantic)	.011	.989	680	2.2	1.1	Tills et al. 1985
FINLAND	.019	.981	129	3.7	1.8	Kaarsalo et al. 1962
FRANCE		1.000	132			Ikin 1963
GERMANY						
Jews	.038	.962	67	7.3	3.5	Bonné-Tamir 1975
GREECE						
Greeks (Plati)	.020	.980	1,037	3.9	1.9	Tills et al. 1983b
ICELAND	.015	.985	1,684	3.0	1.5	Tills et al. 1982b
IRELAND	.020	.980	1,690	3.9	1.9	Tills et al. 1977a
ITALY (Southern)	.020	.980	147	3.9	1.9	Sangiorgi et al. 1982
NETHERLANDS (Leiden)	.021	.979	795	4.1	2.0	Fraser et al. 1974
NORWAY						
Norwegians	.042	.958	2,582	8.0	3.9	Hartmann et al. 1965
Lapps/Saami	.022	.978	183	4.3	2.1	Allison et al. 1952
POLAND (Olkusz)		1.000	98			Wolanski et al. 1983
SPAIN						
Basques	.041	.959	161	7.9	3.8	Alberdi et al. 1957
SWITZERLAND	.027	.973	894	5.3	2.6	Gonzenbach et al. 1955
UNITED KINGDOM						
England	.039	.961	1,373	7.5	3.6	Race & Sanger 1968
Isle of Lewis	.048	.952	283	9.1	4.4	Clegg et al. 1985
Isle of Man	.057	.943	332	10.8	5.1	Mitchell et al. 1982

U.S.S.R.

ASIAN PART						
Nganasan-Avam (Taimir Pen.)	.001	.999	555	.2	.1	Sukernik et al. 1978

ASIA

TURKEY (Asia Minor)	.014	.986	108	2.8	1.4	Aksoy et al. 1958
ISRAEL						
Yemenite Jews	.006	.994	157	1.2	.6	Tills et al. 1977b
SAUDI ARABIA						
Shiah (Eastern)	.010	.990	465	2.0	1.0	Maranjian et al. 1966

Continued

Table 154. Lutheran system: LU (cont'd)

Place/Population	a	b	n	H	PE	Source
Tribes (Western)	.122	.878	185	21.4	9.6	Saha et al. 1980
Arabs (Southern)	.006	.994	258	1.2	.6	Marengo-Rowe et al. 1974
IRAN						
Guilanians	.005	.995	224	1.0	.5	Ohkura et al. 1984
AFGHANISTAN						
Pushtu & Dari (Kabul)	.025	.975	60	4.9	2.4	Papiha et al. 1977a
INDIA						
Jat Sikhs (Punjab)		1.000	148			Chahal & Papiha 1981
Gujarati (W. India)	.017	.983	176	3.3	1.6	Papiha et al. 1981
Hindu (Andhra Pradesh)	.002	.998	211	.4	.2	Roberts et al. 1980
Irula tr. (Nilgiri Hills, S. India)		1.000	100			Lehmann & Cutbush 1952
SRI LANKA						
Sinhalese (Anuradhapura)		1.000	36			Roberts et al. 1972a
Veddah	.008	.992	63	1.6	.8	Wickremasinghe et al. 1963
NEPAL		1.000	200			Bird et al. 1957
BHUTAN		1.000	154			Mourant et al. 1969
BURMA	.018	.982	83	3.5	1.7	Ikin et al. 1969
MALAY						
Negrito		1.000	105			Polunin & Sneath 1953
INDONESIA						
Toba Batak (Samosir Is.)	.009	.991	57	1.8	.9	Hawkins et al. 1973
THAILAND (Bangkok)	.001	.999	455	.2	.1	Chandanayingyong et al. 1967
PHILIPPINES						
Negrito/Aeta (Zambales)		1.000	335			Misawa & Omoto 1985
CHINA						
Han (N. China)	.004	.996	290	.8	.4	Yuan et al. 1982
Uygur (NW China)	.028	.972	221	5.4	2.6	Yuan et al. 1984d
Hui (Ningxia)		1.000	35			Yuan et al. 1985
Tibetans (Switzerland)		1.000	115			Jeannet et al. 1972
JAPAN						
Japanese (Tokyo)		1.000	157			Nakajima 1973b
Ainu (Hokkaido)		1.000	248			Harvey et al. 1978
KOREA (Seoul)		1.000	255			Nakajima 1973a

AFRICA

Place/Population	a	b	n	H	PE	Source
LIBYA (Tripoli & Benghasi)	.006	.994	169	1.2	.6	Walter et al. 1975
SUDAN						
Beja-Amarar (NE Sudan)	.025	.975	100	4.9	2.4	el Hassan et al. 1968
NIGERIA						
Yoruba	.041	.959	200	7.9	3.8	Garlick & Barnicot 1957
CENTRAL AFRICAN REPUBLIC						
Babinga Pygmies (Bagandou)	.003	.997	163	.6	.3	Cavalli-Sforza et al. 1969
ZAIRE						
Bantu	.023	.977	93	4.5	2.2	Govaerts et al. 1972
ETHIOPIA						
Amhara tr.	.043	.957	107	8.2	3.9	Ikin & Mourant 1962
TANZANIA						
Sandawe (Kondoa-Irang Dist.)	.036	.964	215	6.9	3.3	Godber et al. 1976
Nyaturu (")	.026	.974	217	5.1	2.5	"
Hadza (W. Lake Eyasi)	.038	.962	162	7.3	3.5	Tills et al. 1982a
SOUTHERN AFRICA						
Negroes/Dama (Okombahe)	.030	.970	86	5.8	2.8	Nurse et al. 1976
Bantu (Natal)	.027	.973	205	5.3	2.6	Shapiro 1953

NORTH AMERICA

Place/Population	a	b	n	H	PE	Source
ALASKA						
Eskimos-Inupiak (Wainwright)		1.000	241			Corcoran et al. 1959
Athabaskan Indians		1.000	255			"
CANADA						
Eskimos (Victoria Is.)		1.000	146			Chown & Lewis 1959
Cree Indians (Manitoba)	.010	.990	100	2.0	1.0	Lucciola et al. 1974
Ojibwa Indians (Pikangikum)		1.000	96			Szathmary et al. 1975
UNITED STATES						
Blackfeet Indians (Montana)		1.000	98			Rokala et al. 1977
Whites (Southern)	.036	.964	333	6.9	3.3	Cooper et al. 1963

Continued

Table 154. Lutheran system: LU (cont'd)

Place/Population	a	b	n	H	PE	Source
Blacks (Southern)	.044	.956	304	8.4	4.0	"

MIDDLE AMERICA

MEXICO
Maya (Chiapas & Yucatan)		1.000	173			Matson & Swanson 1959

COSTA RICA
Guaymi		1.000	286			Barrantes et al. 1982

SOUTH AMERICA

BOLIVIA
Ayamara	.004	.996	120	.8	.4	Matson et al. 1966

BRAZIL
Cayapo (Parà & Mato Grosso)	.002	.998	270	.4	.2	Salzano et al. 1972a
Ticuna (C. Amazonas)	.002	.998	1,744	.4	.2	Neel et al. 1980

SURINAM
Trio (Southern)		1.000	497			Geerdink et al. 1974c

VENEZUELA
Makiritare (S. Venezuela & N. Brazil)		1.000	392			Gershowitz et al. 1970
Yanomama (")		1.000	1,794			Gershowitz et al. 1972

OCEANIA

AUSTRALIA
Aborigines (N. Territory)		1.000	178			Sanger et al. 1951
Aborigines (W. Australia)		1.000	94			Kirk 1965(c)

MELANESIA
Papua New Guinea
Weri (Middle Waria)		1.000	90			Booth et al. 1981
Yagaria (E. Highlands)		1.000	373			Mourant et al. 1982

Karkar Is.
Takia	.008	.992	202	1.6	.8	Booth et al. 1982
Solomon Is.		1.000	186			Douglas et al. 1962

POLYNESIA
Easter Is.		1.000	51			Simmons 1966

Note: Alleles Lu^a and Lu^b are codominant.

Table 155. McCoy antigen: MCC*

Place/Population	a	-	n	H	PE	Source

NORTH AMERICA

UNITED STATES
Whites	.879	.121	962	21.3	.0	Molthan & Moulds 1978
Blacks	.819	.181	645	29.6	.1	"

Note: Allele MCC^a is dominant to allele MCC$^-$

Table 156. MNS system: MNS

Place/Population	M		N					Source
	MS	Ms	NS	Ns	n	H	PE	
EUROPE								
AUSTRIA	.254	.297	.080	.369	2,455	70.5	31.8	Mayr et al. 1970
CZECHOSLOVAKIA (Ovcie)	.687		.313		321	43.0	16.9	Matousek & Seemanová 1973
DENMARK	.229	.335	.061	.376	366	69.0	31.2	Mourant et al. 1976(c)
FINLAND	.247	.396	.077	.280	5,536	69.8	31.7	Nevanlinna 1972
FRANCE								
French	.273	.243	.078	.405	132	69.6	31.2	Ikin 1963
Basques	.323	.310	.110	.256	75	72.2	32.6	Levine et al. 1974
GERMANY	.236	.304	.066	.394	2,000	69.2	31.2	Ritter 1969
GREECE								
Greeks (Plati)	.319	.231	.143	.308	1,035	73.0	33.0	Tills et al. 1983b
HUNGARY	.239	.311	.084	.366	300	70.5	31.8	Rex-Kiss 1967
ICELAND	.153	.423	.064	.360	1,684	66.4	29.9	Tills et al. 1982b
IRELAND	.270	.300	.060	.370	1,698	69.7	31.3	Tills et al. 1977a
ITALY (Southern)	.281	.381	.037	.301	147	68.4	30.8	Sangiorgi et al. 1982
NETHERLANDS (Leiden)	.238	.274	.051	.437	795	67.5	30.3	Fraser et al. 1974
NORWAY								
Norwegians	.233	.317	.071	.379	526	69.7	31.4	Reinskou 1967
Lapps/Saami (Finnmark)	.268	.256	.154	.322	359	73.5	33.3	Kornstad 1972
POLAND (Olkusz)	.213	.360	.113	.314	213	71.4	32.4	Wolanski et al. 1983
PORTUGAL (Terras de Bouro)	.654		.346		114	45.3	17.5	Cruz et al. 1973
SPAIN	.243	.311	.058	.389	76	69.0	31.1	Agosti Romero et al. 1950
SWEDEN								
Swedes	.241	.322	.082	.354	4,116	70.6	31.8	Heiken 1965
Lapps/Saami	.237	.150	.210	.403	193	71.5	32.3	Allison et al. 1956
SWITZERLAND	.218	.282	.088	.412	104	69.5	31.4	Greuter et al. 1963
UNITED KINGDOM								
England	.240	.298	.056	.406	1,166	68.6	30.8	Ikin et al. 1952
Isle of Lewis	.283	.412	.017	.288	95	66.7	30.0	Clegg et al. 1985
Isle of Man	.212	.305	.055	.428	593	67.6	30.4	Mitchell et al. 1982
YUGOSLAVIA	.560		.440		459	49.3	18.6	Grünwald & Herman 1963
U.S.S.R.								
EUROPEAN PART								
Russians (Moscow)	.253	.343	.083	.321	563	70.8	31.7	Umnova et al. 1964
Abkhazia (Ochamchir Dist.)	.134	.482	.062	.322	159	64.2	29.0	Ferrell et al. 1985
ASIAN PART								
Eskimos (New Chaplino)	.088	.456	.010	.446	102	58.5	25.1	Sukernik et al. 1981
Chukchi (Chukotka &								
NE Kamchatka)	.051	.367	.003	.579	1,066	52.7	21.9	"
Nganasan-Avam (Taimir								
Pen.)		.274	.007	.719	430	40.8	16.6	Sukernik et al. 1978
Nentzi (Between Ob R.								
Taj R.)	.081	.424	.008	.487	753	57.6	24.5	Sukernik et al. 1980
ASIA								
TURKEY (Asia Minor)	.190	.301	.091	.418	108	69.0	31.2	Aksoy et al. 1958
ISRAEL								
Ashkenazi Jews	.518		.482		168	49.9	18.7	Kobyliansky et al. 1982
Yemenite Jews	.351	.466	.039	.144	157	63.7	28.6	Tills et al. 1977b
KUWAIT								
Arabs	.222	.381	.051	.345	159	68.4	30.8	Sawhney et al. 1984
SAUDI ARABIA EMIRATES								
Shiah (Eastern)	.256	.375	.081	.289	463	70.4	32.0	Maranjian et al. 1966
Tribes (Western)	.304	.571	.034	.091	176	57.2	25.5	Saha et al. 1980
Arabs (Southern)	.366	.366	.088	.181	261	69.2	31.4	Marengo-Rowe et al. 1974
UNITED ARAB								
Abu Dhabians	.247	.363	.105	.285	100	71.5	32.5	Kamel et al. 1980
IRAN								
Mazandaranians &								
Guilanians	.298	.368	.085	.249	522	70.7	32.0	Ohkura et al. 1984
AFGHANISTAN								
Pushtu & Dari (Kabul)	.266	.370	.089	.275	283	70.9	32.2	Papiha et al. 1977a

Continued

Table 156. MNS system: MNS (cont'd)

Place/Population	M		N					
	MS	Ms	NS	Ns	n	H	PE	Source
PAKISTAN								
Punjabi (Lahore)	.187	.386	.118	.309	226	70.7	32.1	Boyd & Boyd 1954
INDIA								
Punjabi (Chandigarh)	.331	.318	.143	.208	137	72.6	32.9	Papiha et al. 1972
Gujarati (W. India)	.224	.393	.085	.298	286	69.9	31.8	Papiha et al. 1981
Hindu & Moslems								
(Madhya Pradesh)	.247	.371	.127	.255	343	72.0	32.7	Roberts et al. 1974
Bhil tr. (Madhya Pradesh)	.167	.399	.128	.306	145	70.3	32.0	Papiha et al. 1978
Hindu (Andhra Pradesh)	.253	.391	.067	.289	211	69.5	31.5	Roberts et al. 1980
Irula tr. (Nilgiri Hills,								
S. India)	.17	.55	.03	.25	72	60.5	27.7	Kirk et al. 1962
Bengali (E. India)	.594		.406		201	48.2	18.3	Bhattacharjee 1956
SRI LANKA								
Sinhalese (Anuradhapura)	.203	.297	.096	.404	92	69.8	31.6	Roberts et al. 1972a
Veddah	.216	.350	.091	.343	106	70.5	31.9	Wickremasinghe et al. 1963
BANGLADESH								
Moslems	.201	.380	.130	.289	230	71.5	32.5	Boyd & Boyd 1954
NEPAL								
Shrestha	.233	.460	.117	.189	153	68.5	31.2	Bhasin 1970
BHUTAN	.141	.545	.034	.280	97	60.4	27.4	Mourant et al. 1969
BURMA (Central)	.074	.667		.259	120	48.2	21.4	Mya-Tu et al. 1971
MALAYSIA								
Malays (Singapore)	.099	.568		.333	42	55.7	24.4	Gibson-Hill 1953
Aboriginal Malays	.029	.766		.206	107	37.0	16.3	Polunin & Sneath 1953
Negrito	.087	.649	.051	.213	102	52.3	24.1	"
INDONESIA								
Toba Batak (Samosir Is.)	.022	.378	.071	.529	105	57.2	24.7	Hawkins et al. 1973
Balinese (Bali Is.)	.048	.652	.085	.216	148	51.9	23.7	Breguet et al. 1982a
THAILAND (Bangkok)	.062	.617	.022	.299	456	52.6	23.0	Chandanayingyong et al. 1967
PHILIPPINES								
Filipino	.103	.473	.062	.362	403	63.1	28.2	Fraser et al. 1964
Negrito/Aeta (Zambales)	.037	.354	.110	.499	282	61.2	27.2	Misawa & Omoto 1985
CHINA								
Han (N. China)	.498		.502		640	50.0	18.7	Yuan et al. 1984b
Mongolians (N. China)	.055	.496	.011	.438	129	55.9	23.4	Yuan et al. 1984a
Uygur (NW China)	.158	.396	.039	.407	221	65.1	29.1	Yuan et al. 1984d
Hui (Ningxia)	.028	.442	.031	.498	219	55.5	22.9	Yuan et al. 1985
Dong (Guangxi)		.629	.012	.358	201	47.6	18.7	Yuan et al. 1984c
Zhuang (S. China)	.012	.727		.261	211	40.3	16.7	Yuan et al. 1984a
Tibetans (Switzerland)	.142	.458	.032	.368	115	63.4	28.3	Jeannet et al. 1972
Chinese (Hong Kong)	.017	.565	.002	.417	406	50.7	20.0	Grimmo & Lee 1964a
Chinese (Singapore)	.030	.496	.012	.462	213	53.9	21.9	Hawkins & Simons 1976
TAIWAN								
Toroko-Aborigines		.500	.063	.437	127	55.5	23.0	Chen et al. 1985
JAPAN								
Japanese (Tokyo)	.064	.466	.025	.444	230	58.1	24.7	Nakajima 1973b
Ainu (Hokkaido)	.020	.393	.231	.355	248	66.6	29.7	Harvey et al. 1978
KOREA								
Koreans (Seoul)	.038	.496	.009	.457	255	54.4	22.2	Nakajima 1973a
Koreans (NE China)	.019	.490	.004	.487	216	52.2	20.5	Yuan et al. 1984a
AFRICA								
LIBYA (Tripoli & Benghasi)	.276	.310	.052	.362	168	69.4	31.2	Walter et al. 1975
EGYPT	.231	.284	.068	.418	144	68.7	31.0	Donegani et al. 1950
SUDAN								
Sudanese (Khartoum)	.075	.425	.082	.418	115	63.2	28.1	Saha et al. 1978b
Beja-Amarar (SE Sudan)	.202	.278	.070	.450	100	67.5	30.4	el Hassan et al. 1968
CHAD								
Sara Majingay	.578		.422		255	48.8	18.4	Hiernaux 1976
SENEGAL								
Bedik (E. Senegal)	.099	.429	.178	.294	459	68.8	31.2	Bouloux et al. 1972
NIGERIA								
Yoruba (Western)	.053	.410	.052	.485	135	59.1	25.6	Blumberg et al. 1961
CENTRAL AFRICAN REPUBLIC								
Babinga Pygmies (Bagandou)	.226	.322	.184	.268	158	74.0	33.6	Cavalli-Sforza et al. 1969
BURUNDI								

Continued

Table 156. MNS system: MNS (cont'd)

Place/Population	M MS	M Ms	N NS	N Ns	n	H	PE	Source
Hutu (Burundi)	.245	.281	.116	.358	99	71.9	32.5	Fraser et al. 1966a
ZAIRE								
Bantu	.118	.353	.157	.372	93	69.8	31.7	Govaerts et al. 1972
ETHIOPIA								
Amhara tr.	.219	.412	.039	.330	107	67.2	30.3	Ikin & Mourant 1962
UGANDA (Kampala)	.241	.335	.137	.287	138	72.9	33.1	Ssebabi & Roberts 1975
TANZANIA								
Sandawe (Kondoa-Irang								
Dist.)	.298	.298	.138	.267	215	73.2	33.2	Godber et al. 1976
Nyaturu (")	.183	.350	.136	.331	215	71.6	32.5	Godber et al. 1976
Hadza (W. Lake Eyasi)	.107	.443	.040	.410	162	62.3	27.5	Tills et al. 1982a
ANGOLA								
Njinga	.083	.472	.074	.371	109	62.7	27.9	Nurse et al. 1979
SOUTHERN AFRICA								
Basters (Namibia)	.117	.416	.161	.305	120	69.4	31.4	Nurse et al. 1982
San/Bushmen								
!Kung, Dobe	.121	.432	.027	.420	258	62.2	27.4	Nurse & Jenkins 1977
Khoi/Hotentot								
Keetmanshoop Nama	.205	.455	.050	.291	391	66.4	30.2	Jenkins 1972
Negroes/Dama (Okombahe &								
Sesfontein)	.051	.392	.173	.384	117	66.6	30.0	Nurse et al. 1976
Bantu (Natal)	.093	.488	.044	.375	205	61.1	27.0	Shapiro 1953
Griqua (Campbell, Cape								
Province)	.170	.460	.042	.328	99	65.0	29.4	Nurse & Jenkins 1975
TRISTAN DA CUNHA	.215	.233	.099	.453	67	68.4	30.0	Jenkins et al. 1985

NORTH AMERICA

Place/Population	M MS	M Ms	N NS	N Ns	n	H	PE	Source
GREENLAND								
Eskimos (Angmagssalik)	.202	.723		.075	180	43.1	19.6	Skeller 1954
ALASKA								
Eskimos-Inupiak								
(Wainwright)	.156	.668		.176	111	49.8	23.1	Corcoran et al. 1959
Eskimos								
(St. Lawrence Is.)	.103	.463	.031	.404	221	61.1	26.9	Ferrell et al. 1981b
Athabaskan Indians								
(Fort Yukon)	.221	.597		.182	110	56.2	25.8	Corcoran et al. 1959
CANADA								
Eskimos (Victoria Is.								
NW Territories)	.183	.659		.158	320	50.7	23.5	Chown & Lewis 1959
Ojibwa Indians (Pikan-								
gikum, N. Ontario)	.341	.200	.174	.284	96	73.3	33.0	Szathmary et al. 1975
Cree Indians								
(Manitoba)	.255	.435	.035	.275	100	66.9	30.3	Lucciola et al. 1974
Dogrib Indians								
(NW Territories)	.095	.820	.022	.063	158	31.4	15.1	Szathmary 1983
UNITED STATES								
Blackfeet Indians								
(Montana)	.315	.481	.073	.131	148	64.7	29.3	Rokala et al. 1977
Papago Indians								
(S. Arizona)	.346	.444	.080	.130	709	66.0	29.8	Niswander et al. 1970
Whites (Southern)	.205	.303	.074	.418	333	68.6	30.9	Cooper et al. 1963
Blacks (")	.137	.347	.155	.361	304	70.6	32.1	"
Mexican Americans	.299	.361	.071	.269	965	70.3	31.0	Hewett-Emmet et al. 1986b

MIDDLE AMERICA

Place/Population	M MS	M Ms	N NS	N Ns	n	H	PE	Source
MEXICO								
Indians (San Pablo del								
Monte)	.376	.406	.071	.147	257	66.7	30.1	Crawford et al. 1974
Mestizo (Tlaxcala)	.358	.334	.055	.253	138	69.3	31.2	"
GUATEMALA								
Black Caribs								
(Livingston)	.110	.314	.108	.468	202	65.9	29.8	Crawford et al. 1981
COSTA RICA								
Guaymi	.311	.398	.078	.213	286	69.3	31.5	Barrantes et al. 1982
WEST INDIES (Haiti)	.079	.419	.118	.384	470	65.7	29.5	Basu et al. 1976

Continued

Table 156. MNS system: MNS (cont'd)

Place/Population	M		N		n	H	PE	Source
	MS	Ms	NS	Ns				

SOUTH AMERICA

Place/Population	MS	Ms	NS	Ns	n	H	PE	Source
BOLIVIA								
Aymara (W. Bolivia)	.188	.444	.040	.326	311	66.0	29.7	Ferrell et al. 1978a
BRAZIL								
Baniwa (Jandu Cachoeira)	.20	.63	.08	.09	363	54.9	25.3	Salzano et al. 1986
Cayapo (Para & Mato Grosso)	.220	.541	.036	.203	480	61.6	28.4	Salzano et al. 1972a
Caingang	.465	.325	.126	.084	214	65.5	29.6	Salzano et al. 1980
Parakana	.19	.81			100	30.8	13.0	Black et al. 1980
Macushi (N. Brazil & S. Guyana)	.132	.549	.052	.267	503	60.7	27.7	Neel et al. 1977a
Wapishana (")	.288	.479	.045	.188	617	65.0	29.6	Neel et al. 1977a
Ticuna (C. Amazonas)	.089	:807	.019	.086	1,747	33.3	16.1	Neel et al. 1980
CHILE								
Atacameño (Toconao, N. Chile)	.628		.372		180	46.7	17.9	Rothhammer et al. 1984
COLOMBIA								
Noanama	.150	.546	.020	.284	168	59.8	27.1	Kirk et al. 1974
FRENCH GUIANA								
Wayampi	.281	.338	.196	.185	240	73.4	33.3	Tchen et al. 1981
SURINAM								
Trio (S. Surinam)	.136	.566	.043	.255	497	59.4	27.2	Geerdink et al. 1974c
VENEZUELA								
Makiritare (S. Venezuela & N. Brazil)	.320	.400	.150	.130	538	69.8	31.7	Gershowitz et al. 1970
Yanomama (")	.147	.489	.028	.337	2,516	62.5	28.1	Gershowitz et al. 1972

OCEANIA

Place/Population	MS	Ms	NS	Ns	n	H	PE	Source
AUSTRALIA								
Aborigines/Malag (Elcho Is., N. Territory)		.263		.737	179	38.8	15.6	Simmons & Cooke 1969
Aborigines (Haast's Bluff, N. Territory)	.24		.76		125	36.5	14.9	Kirk 1965(c)
Aborigines (Yuendumu, N. Territory, C. Australia)	.28		.72		91	40.3	16.1	"
Aborigines (Queensland)	.18		.82		372	29.5	12.6	"
Aborigines (W. Australia)	.26		.74		1,698	38.5	15.5	"
MELANESIA								
W. New Guinea								
Asmat (S. Coastal Plain)	.020	.080	.097	.803	401	33.9	16.3	Gajdusek et al. 1978
Papua New Guinea								
Gainj & Kalam (N. Central Highlands)	.019		.981		415	3.7	1.8	Long et al. 1986
Jimi Valley (W. Highlands)		.014	.142	.843	344	26.9	12.0	Mourant et al. 1981
Yagaria (E. Highlands)		.064	.106	.830	371	29.6	14.1	Mourant et al. 1982
Fuyuge (C. Dist. Highlands)	.002	.113	.144	.741	100	41.7	19.6	Woodfield et al. 1974
Daga (Bay Province)		.015	.198	.787	134	34.1	14.8	Jenkins et al. 1983
Karkar Is.								
Takia	.012	.173	.146	.669	322	50.1	23.3	Booth et al. 1982
Manus Is.	.429		.571		91	49.0	18.5	Malcolm et al. 1972
Buka Is.	.025	.211	.011	.753	79	38.8	16.9	McLoughlin et al. 1982
Solomon Is.	.065	.246	.128	.561	2,774	60.4	27.7	Lai & Bloom 1982
MICRONESIA								
Caroline Is. (Pingelap)	.005	.177	.009	.809	1,035	31.4	13.7	Morton & Yamamoto 1973
Marshall Is.		.235	.120	.645	305	51.4	23.4	Simmons et al. 1952
POLYNESIA								
Samoans (New Zealand)	.013	.662		.325	77	45.6	18.3	Booth et al. 1977

Note: Alleles M and N are codominant to each other, as are S and s. The M and S loci are closely linked. The gene frequencies under the headings of M and N represent the sum of the frequencies of MS and Ms and that of NS and Ns, respectively.

Table 157. Ne[a] antigen: NE*

Place/Population	a	-	n	H	PE	Source
EUROPE						
FINLAND	.025	.975	3,370	4.9	2.3	Sistonen et al. 1981
SWEDEN	.003	.997	395	.6	.3	"
SWITZERLAND	.001	.999	502	.2	.1	"

Note: Allele Ne[a] is dominant to allele Ne[-].

Table 158. P system: P

Place/Population	P_1	P_2+p	n	H	PE	Source
EUROPE						
AUSTRIA (Wien)	.527	.473	2,100	49.9	2.6	Mayr et al. 1970
BULGARIA	.418	.582	1,200	48.7	4.8	Baev & Popwassilew 1969
DENMARK	.531	.469	12,123	49.8	2.6	Gürtler 1970
FAROE IS. (N. Atlantic)	.463	.537	679	49.7	3.9	Tills et al. 1985
FINLAND	.451	.549	5,536	49.5	4.1	Nevanlinna 1972
FRANCE						
French	.534	.466	2,305	49.8	2.5	Khérumian & Moullec 1956
Basques	.496	.504	63	50.0	3.2	Levine et al. 1974
GERMANY	.555	.445	10,000	49.4	2.2	Hoppe 1957
GREECE						
Greeks (Plati)	.429	.571	1,038	49.0	4.6	Tills et al. 1983b
ICELAND	.412	.588	1,684	48.5	4.9	Tills et al. 1982b
IRELAND	.490	.510	1,540	50.0	3.3	Tills et al. 1977a
ITALY	.489	.511	1,585	50.0	3.3	Marras et al. 1964
NETHERLANDS (Leiden)	.519	.481	795	49.9	2.8	Fraser et al. 1974
NORWAY						
Norwegians	.499	.501	1,162	50.0	3.1	Berg 1974(c)
Lapps/Saami (Finnmark)	.306	.694	336	42.5	7.1	Kornstad 1972
POLAND (Olkusz)	.506	.494	213	50.0	3.0	Wolanski et al. 1983
SPAIN						
Spanish	.586	.414	532	48.5	1.7	Casado 1975
Basques	.529	.471	36	49.8	2.6	"
SWEDEN						
Swedes	.540	.460	10,457	49.7	2.4	Beckman 1959
Lapps/Saami	.386	.614	419	47.4	5.5	Beckman et al. 1959
SWITZERLAND	.579	.421	1,038	48.8	1.8	Beringer 1967
UNITED KINGDOM						
England	.516	.484	1,166	49.9	2.8	Mourant et al. 1976(c)
Isle of Lewis	.543	.457	283	49.6	2.4	Clegg et al. 1985
Isle of Man	.426	.574	182	48.9	4.6	Mitchell et al. 1982
Orkney Is. (Westray)	.618	.382	349	47.2	1.3	Welch et al. 1973
U.S.S.R.						
EUROPEAN PART						
Russians (Moscow)	.485	.515	1444	49.9	3.2	Umnova 1959
Abkhazia (Ochamchir Dist.)	.278	.722	290	40.1	7.6	Ferrell et al. 1985
ASIAN PART						
Eskimos (New Chaplino)	.204	.796	101	32.5	8.2	Sukernik et al. 1981
Chukchi (Chukotka & NE Kamchatka)	.422	.578	981	48.8	4.7	"
Nganasan-Avam (Taimir Pen.)	.235	.765	424	36.0	8.0	Sukernik et al. 1978
Nentzi (Between Ob R. & Taj R.)	.233	.767	753	35.7	8.1	Sukernik et al. 1980

Continued

Table 158. P system: P (cont'd)

Place/Population	P_1	P_2+p	n	H	PE	Source
ASIA						
ISRAEL						
Ashkenazi Jews (E. Europe)	.353	.647	177	45.7	6.2	Kobyliansky et al. 1982
Yemenite Jews	.338	.662	157	44.8	6.5	Tills et al. 1977b
KUWAIT						
Arabs	.520	.480	74	49.9	2.8	Al-Nassar et al. 1981
SAUDI ARABIA EMIRATES						
Shiah (Eastern)	.442	.558	465	49.3	4.3	Maranjian et al. 1966
Arabs (Southern)	.509	.491	261	50.0	3.0	Marengo-Rowe et al. 1974
UNITED ARAB						
Abu Dhabians	.461	.539	100	49.7	3.9	Kamel et al. 1980
IRAN						
Mazandaranians & Guilanians	.418	.582	786	48.7	4.8	Ohkura et al. 1984
AFGHANISTAN						
Pushtu & Dari (Kabul)	.462	.538	283	49.7	3.9	Papiha et al. 1977a
INDIA						
Jat Sikhs (Punjab)	.442	.558	148	49.3	4.3	Chahal & Papiha 1981
Gujarati (W. India)	.490	.510	200	50.0	3.3	Papiha et al. 1981
Bhil tr. (Madhya Pradesh)	.488	.512	145	50.0	3.4	Papiha et al. 1978
Hindu (Hyderabad)	.400	.600	211	48.0	5.2	Roberts et al. 1980
Irula tr. (S. India)	.559	.441	72	49.3	2.1	Kirk et al. 1962
Bengali (E. India)	.300	.700	481	42.0	7.2	Sen 1960
SRI LANKA						
Sinhalese (Anuradhapura)	.330	.670	156	44.2	6.6	Roberts et al. 1972a
Veddah	.459	.541	106	49.7	3.9	Wickremasinghe et al. 1963
NEPAL	.252	.748	200	37.7	7.9	Bird et al. 1957
BHUTAN	.236	.764	154	36.1	8.0	Mourant et al. 1969
BURMA	.201	.799	83	32.1	8.2	Ikin et al. 1969
MALAY						
Negrito	.506	.494	86	50.0	3.0	Polunin & Sneath 1953
INDONESIA						
Toba Batak (Samosir Is.)	.312	.688	150	42.9	7.0	Hawkins et al. 1973
Balinese (Bali Is.)	.126	.874	148	22.0	7.4	Breguet et al. 1982a
THAILAND (Bangkok)	.154	.846	2,500	26.1	7.9	Chandanayingyong et al. 1979
PHILIPPINES						
Filipino (Germany)	.147	.853	143	25.1	7.8	Windhof & Walter 1983
Negrito/Aeta (Zambales)	.234	.766	335	35.8	8.1	Misawa & Omoto 1985
CHINA						
Han (N. China)	.213	.787	357	33.5	8.2	Yuan et al. 1984b
Mongolians (N. China)	.193	.807	129	31.2	8.2	Yuan et al. 1984a
Uygur (NW China)	.417	.583	221	48.6	4.8	Yuan et al. 1984d
Hui (Ningxia)	.132	.868	179	22.9	7.5	Yuan et al. 1985
Dong (Guangxi)	.133	.867	201	23.1	7.5	Yuan et al. 1984c
Zhuang (S. China)	.102	.898	211	18.3	6.6	Yuan et al. 1984a
Tibetans (Switzerland)	.130	.870	115	22.6	7.4	Jeannet et al. 1972
Chinese (Singapore)	.159	.841	185	26.7	8.0	Hawkins & Simons 1976
JAPAN						
Japanese (Tokyo)	.198	.802	412	31.8	8.2	Nakajima 1973b
Ainu (Hokkaido)	.098	.902	248	17.7	6.5	Harvey et al. 1978
KOREA						
Koreans (Seoul)	.208	.792	255	32.9	8.2	Nakajima 1973a
Koreans (NE China)	.142	.858	216	24.4	7.7	Yuan et al. 1984a
AFRICA						
LIBYA (Tripoli & Benghasi)	.578	.422	169	48.8	1.8	Walter et al. 1975
EGYPT	.526	.474	1,000	49.9	2.7	Moharram 1942
SUDAN						
Beja-Amarar (NE Africa)	.588	.412	100	48.5	1.7	el Hassan et al. 1968
CHAD						
Sara Majingay	.823	.177	256	29.1	.1	Hiernaux 1976
SENEGAL						
Bedik (E. Senegal)	.705	.295	793	41.6	.5	Bouloux et al. 1972
Yoruba (Western)	.690	.310	135	42.8	.6	Blumberg et al. 1961
CENTRAL AFRICAN REPUBLIC						
NIGERIA						

Continued

Table 158. P system: P (cont'd)

Place/Population	P_1	P_2+p	n	H	PE	Source
Babinga Pygmies (Bagandou)	.889	.111	163	19.7	.0	Cavalli-Sforza et al. 1969
ZAIRE						
Bantu	.806	.194	93	31.3	.1	Govaerts et al. 1972
ETHIOPIA						
Amhara tr.	.651	.349	107	45.4	1.0	Ikin & Mourant 1962
UGANDA (Kampala)	.669	.331	137	44.3	.8	Ssebabi & Roberts 1975
TANZANIA						
Sandawe (Kondoa-Irang Dist.)	.626	.374	215	46.8	1.2	Godber et al. 1976
Nyaturu (")	.634	.366	217	46.4	1.1	"
Hadza (W. Lake Eyasi)	.728	.272	162	39.6	.4	Tills et al. 1982a
SOUTHERN AFRICA						
Basters (Namibia)	.533	.467	119	49.8	2.5	Nurse et al. 1982
San/Bushmen						
!Kung, Tsumkwe	.639	.361	277	46.1	1.1	Nurse & Jenkins 1977
Khoi/Hottentot						
Topnaar, Kuiseb	.491	.509	58	50.0	3.3	"
Negroes/Dama (Okomhabe)	.791	.209	92	33.1	.2	Nurse et al. 1976
Bantu (Natal)	.717	.283	400	40.6	.5	Shapiro 1953
Griqua (Campbell, Cape Province)	.733	.267	98	39.1	.4	Nurse & Jenkins 1975

NORTH AMERICA

GREENLAND						
Eskimos (Thule)	.197	.803	2,067	31.6	8.2	Gürtler 1971
ALASKA						
Eskimos (Wainwright)	.090	.910	111	16.4	6.2	Corcoran et al. 1959
Eskimos (St. Lawrence Is.)	.211	.789	209	33.3	8.2	Ferrell et al. 1981b
Athabaskan Indians (Fort Yukon)	.184	.816	99	30.0	8.2	Corcoran et al. 1959
CANADA						
Eskimos (Victoria Is.,						
NW Territories)	.184	.816	320	30.0	8.2	Chown & Lewis 1959
Ojibwa Indians (Pikangikum)	.750	.250	96	37.5	0.3	Szathmary et al. 1975
Dogrib Indians (NW Territories)	.177	.823	158	29.1	8.1	Szathmary 1983
UNITED STATES						
Blackfeet Indians (Montana)	.415	.585	148	48.6	4.9	Rokala et al. 1977
Papago Indians (S. Arizona)	.499	.501	709	50.0	3.1	Niswander et al. 1970
Whites (Southern)	.526	.474	333	49.9	2.7	Cooper et al. 1963
Blacks (")	.757	.243	304	36.8	.3	"

MIDDLE AMERICA

MEXICO						
Indians (San Pablo del Monte)	.367	.633	257	46.5	5.9	Crawford et al. 1974
Mestizo (Tlaxcala)	.330	.670	138	44.2	6.6	"
COSTA RICA						
Guaymi	.566	.434	286	49.1	2.0	Barrantes et al. 1982

SOUTH AMERICA

BOLIVIA						
Aymara (W. Bolivia)	.427	.573	314	48.9	4.6	Ferrell et al. 1978a
BRAZIL						
Baniwa (Jandu Cachoeira)	.53	.47	363	49.8	2.6	Salzano et al. 1986
Cayapo (Pará & Mato Grosso)	.570	.430	480	49.0	1.9	Salzano et al. 1972a
Caingang	.404	.596	214	48.2	5.1	Salzano et al. 1980
Parakana	.68	.32	100	43.5	.7	Black et al. 1980
Pacaás Novos	.497	.503	79	50.0	3.2	Salzano et al. 1985
Macushi (N. Brazil & S. Guyana)	.594	.406	503	48.2	1.6	Neel et al. 1977a
Wapishana (")	.491	.509	618	50.0	3.3	"
Ticuna (C. Amazonas)	.479	.521	1,754	49.9	3.5	Neel et al. 1980
COLOMBIA						
Noanama	.433	.567	168	49.1	4.5	Kirk et al. 1974
FRENCH GUIANA						
Wayampi (Trois-Sauts & Camopi)	.337	.663	239	44.7	6.5	Tchen et al. 1981
SURINAM						
Trio (Southern)	.418	.582	499	48.7	4.8	Geerdink et al. 1974c
VENEZUELA						
Makiritare (S. Venezuela &						

Continued

217

Table 158. P system: P (cont'd)

Place/Population	P_1	P_2+p	n	H	PE	Source
N. Brazil)	.440	.560	538	49.3	4.3	Gershowitz et al. 1970
Yanomama (")	.578	.422	2,506	48.8	1.8	Gershowitz et al. 1972

OCEANIA

AUSTRALIA						
Aborigines/Malag (Elcho Is., NT)	.148	.852	179	25.2	7.8	Simmons & Cooke 1969
Aborigines (Haast Bluff, NT)	.123	.877	100	21.6	7.3	Kirk 1965(c)
Aborigines (W. Australia)	.177	.823	1,444	29.1	8.1	"
MELANESIA						
W. New Guinea						
Asmat (S. Coastal Plain)	.208	.792	327	32.9	8.2	Gajdusek et al. 1978
Papua New Guinea						
Gainj & Kalam (N. Central Highlands)	.046	.954	157	8.8	3.8	Long et al. 1986
Jimi Valley (W. Highlands)	.308	.692	98	42.6	7.1	Mourant et al. 1981
Yagaria (E. Highlands)	.095	.905	447	17.2	6.4	Mourant et al. 1982
Fuyuge (C. Dist. Highlands)	.599	.401	56	48.0	1.5	Woodfield et al. 1974
Daga (Bay Province)	.011	.989	134	2.2	1.1	Jenkins et al. 1983
Karkar Is.						
Takia	.183	.817	262	29.9	8.2	Booth et al. 1982
Manus Is.	.274	.726	74	39.8	7.6	Malcolm et al. 1972
MICRONESIA						
Caroline Is. (Truk)	.261	.739	284	38.6	7.8	Plato & Cruz 1966
Marshall Is.	.553	.447	100	49.4	2.2	Simmons et al. 1952
POLYNESIA						
Samoans (New Zealand)	.270	.730	77	39.4	7.7	Booth et al. 1977

Note: Allele P_1 is dominant to alleles P_2 and p, and P_2 is dominant to p. The frequency of allele p is exceedingly low. PE was computed under the assumption that P_1 is dominant to P_2 and p.

Table 159. Rhesus system: RH

Place/Population	CDE	CDe	CdE	Cde	cDE	cDe	cdE	cde	n	H	PE	Source
EUROPE												
AUSTRIA[d]	.001	.433		.012	.140	.019	.007	.388	2,407	64.2	27.2	Mayr et al. 1970
CZECHOSLOVAKIA (Ovcie)[c]		.331		.054	.103	.120		.392	316	70.9	26.1	Matousek & Seema- nová 1973
DENMARK[d]		.417		.017	.153	.018	.007	.387	12,123	65.2	27.6	Gürtler 1970
FAROE IS. (N. Atlantic)[e]	.004	.346		.010	.261	.083		.296	678	71.8	29.8	Tills et al. 1985
FINLAND[d]		.428		.011	.184	.037	.001	.338	5,536	66.7	28.1	Nevanlinna 1972
FRANCE												
French[b]		.410		.009	.131	.039	.004	.408	1,672	64.7	26.9	Cazal et al. 1951
Basques[c]		.405			.079	.016		.499	63	58.0	24.0	Levine et al. 1974
GERMANY[b]	.001	.418		.012	.138	.018	.004	.409	10,000	63.8	27.1	Hoppe 1957
GREECE												
Greeks (Plati)[e]	.001	.552		.020	.113	.036		.278	1,038	60.4	24.6	Tills et al. 1983b
HUNGARY[d]	.002	.424		.011	.143	.024	.006	.391	5,000	64.6	27.4	Rex-Kiss 1970
ICELAND[e]		.452		.004	.168	.013	.008	.355	1,683	64.1	27.7	Tills et al. 1982b
IRELAND[e]	.001	.400		.004	.157	.014	.004	.421	1,692	63.8	27.6	Tills et al. 1977a
ITALY (Southern)[c]		.627			.108	.045	.007	.212	368	54.8	23.1	Sangiorgi et al. 1
NETHERLANDS (Leiden)[e]		.434		.011	.150	.013	.011	.380	795	64.4	27.5	Fraser et al. 1974
NORWAY												
Norwegians[c]		.443		.003	.154	.016	.006	.378	6,588	63.7	27.4	Berg 1974(c)
Lapps/Saami (Finnmark)[d]		.584		.001	.252			.163	423	56.9	24.9	Kornstad 1972
POLAND (Olkusz)[c]		.411		.026	.150	.028		.385	213	65.9	27.4	Wolanski et al. 19
PORTUGAL (Terras de Bouro)[b]		.328			.094	.036		.542	116	58.9	24.4	Cruz et al. 1973

Continued

Table 159. Rhesus system: RH (cont'd)

Place/Population	CDE	CDe	CdE	Cde	cDE	cDe	cdE	cde	n	H	PE	Source
SPAIN												
Spanish[b]	.070	.410		.017	.071	.054	.002	.377	859	67.7	27.9	Guasch 1950
Basques[b]	.014	.403		.027	.043	.073	.004	.436	626	63.9	23.8	"
SWEDEN												
Swedes[c]		.415		.013	.170	.018	.004	.380	8,488	65.4	27.9	Berg 1974
Lapps/Saami[d]		.608			.106	.060		.226	419	56.4	23.4	Beckman et al. 1959
SWITZERLAND (Bern)[b]	.004	.444		.011	.151	.023	.005	.362	1,095	64.8	27.5	Hässig 1952
UNITED KINGDOM												
England[b]	.001	.430		.008	.138	.028	.008	.388	1,038	64.5	27.2	Murray 1946
Isle of Lewis[e]	.054	.337		.026	.226	.015	.012	.329	284	72.3	31.3	Clegg et al. 1985
Isle of Man[c]	.003	.369		.005	.162	.018	.008	.434	625	64.9	28.0	Mitchell et al. 1982
Orkney Is. (Westray)[c]	.003	.445			.135	.003		.413	337	61.3	26.6	Welch et al. 1973

U.S.S.R.

Place/Population	CDE	CDe	CdE	Cde	cDE	cDe	cdE	cde	n	H	PE	Source
EUROPEAN PART												
Russians (Moscow)	.002	.411		.020	.164	.003	.005	.367	1,173	66.9	27.3	Umnova et al. 1964
Abkhazia (Ochamchir Dist.)[c]	.045	.405		.033	.141	.048		.328	534	70.3	29.2	Ferrell et al. 1985
ASIAN PART												
Eskimos (New Chaplino)[d]		.436			.549	.015[i]			102	50.8	19.6	Sukernik et al. 1981
Chukchi (Chukotka & NE Kamchatka)[d]		.710			.216	.074[i]			1,066	44.4	19.4	"
Nganasan-Avam (Taimir Pen.)[d]	.006	.371			.604	.019[i]			430	49.7	19.7	Sukernik et al. 1978
Nentzi (Between Ob R. & Taj R.)[d]		.470			.462			.068	753	56.1	22.8	Sukernik et al. 1980

ASIA

Place/Population	CDE	CDe	CdE	Cde	cDE	cDe	cdE	cde	n	H	PE	Source
TURKEY (Asia Minor)												
Eti-Turks[g]		.547			.131	.014		.308	118	58.9	25.3	Aksoy et al. 1958
ISRAEL												
Sephardic Jews[c]	.009	.458		.056	.092	.110	.009	.266	200	69.6	25.8	Margolis et al. 1960
Yemenite Jews[e]	.020	.611			.072	.016		.281	65	54.2	23.2	Tills et al. 1977b
JORDAN												
Bedouins[c]		.405		.044	.133	.162		.256	109	72.5	26.9	Saha & Banerjee 1986
KUWAIT												
Arabs[c]	.022	.492			.050	.088	.023	.325	110	64.1	25.1	Sawhney et al. 1984
SAUDI ARABIA												
Shiah (Eastern)[e]		.456		.011	.118	.224		.190	465	69.2	25.9	Maranjian et al. 1966
Tribes (Western)[c]		.390	.004		.100	.226	.026	.255	178	72.1	26.4	Saha et al. 1980
Arabs (Hadhramaut)[e]	.010	.388			.174	.109		.319	103	70.5	28.5	Marengo-Rowe et al. 1974
IRAN												
Turkoman (Gonbad, NE Iran)[d]		.417		.133	.227	.106		.116	110	73.2	27.1	Akbari et al. 1984
AFGHANISTAN												
Pushtu & Dari (Kabul)[c]	.003	.501		.005	.160	.049		.283	283	64.1	26.9	Papiha et al. 1977a
PAKISTAN												
Punjabi (Lahore)[b]	.014	.587		.019	.096	.079		.206	227	59.7	24.0	Boyd & Boyd 1954
INDIA												
Punjabi (Chandigarh)[c]		.610			.095	.054		.241	137	55.8	23.1	Papiha et al. 1972
Gujarati (W. India)[c]		.638		.007	.076	.033		.246	289	52.6	21.6	Papiha et al. 1981
Hindu & Moslems (Madhya Pradesh)[c]	.002	.650		.036	.085	.045		.182	343	53.4	20.9	Roberts et al. 1974
Bhil tr. (Madhya Pradesh)[c]	.009	.718			.099	.046	.016	.111	145	46.0	20.2	Papiha et al. 1978
Hindu (Andhra Pr.)[c]	.008	.608		.043	.075	.055		.211	211	57.5	22.0	Roberts et al. 1980
Irula tr. (Nilgiri Hills, S. India)[c]	.057	.499			.186			.258	72	64.7	29.1	Kirk et al. 1962
Bengali (E. India)[b]	.006	.637			.118	.082		.157	140	54.9	23.2	Bhattacharjee 1956
SRI LANKA												
Sinhalese (Anuradhapura)[c]	.004	.674		.032	.082	.038		.170	157	50.8	20.2	Roberts et al. 1972a
Veddah[f]	.002	.774		.041	.059	.013		.112	254	38.3	14.8	Wickremasinghe et al. 1963
BANGLADESH												
Moslems[b]	.017	.631		.066	.077	.038		.171	236	56.1	21.1	Boyd & Boyd 1954

Continued

Table 159. Rhesus system: RH (cont'd)

Place/Population	CDE	CDe	CdE	Cde	cDE	cDe	cdE	cde	n	H	PE	Source
NEPAL												
Shrestha[c]	.052	.590		.053	.120	.086		.099	183	61.5	24.8	Bhasin 1970
BHUTAN[e]	.006	.666			.289			.039	154	47.1	19.6	Mourant et al. 1969
BURMA (Central)[c]	.028	.780			.088			.103	120	37.2	17.6	Mya-Tu et al. 1971
MALAY (Singapore)[b]		.857			.063	.080			42	25.5	12.0	Gibson-Hill 1953
INDONESIA												
Toba Batak (Samosir Is.)[d]	.007	.846			.106	.040			150	27.1	12.5	Hawkins et al. 1973
Balinese (Bali Is.)[c]		.990			.007			.003	148	2.0	1.0	Breguet et al. 1982a
THAILAND (Bangkok)[c]	.017	.730	.016	.178	.037			.022	1,000	43.3	18.4	Chandanayingyong et al. 1979
PHILIPPINES												
Filipino (Germany)[c]	.009	.802			.141	.047			143	33.5	15.2	Windhof & Walter 1983
Negrito (Zambales)[c]	.022	.726			.189	.063			123	43.3	19.3	Misawa & Omoto 1985
CHINA												
Han (N. China)[b]	.013	.648			.221	.064		.054	336	52.4	21.6	Yuan et al. 1984b
Uygur (NW China)[b]	.016	.529	.026		.180	.051	.026	.174	221	65.3	27.1	Yuan et al. 1984d
Hui (Ningxia)[b]	.046	.622			.220	.056		.056	219	55.6	24.7	Yuan et al. 1985
Dong (Guangxi)[b]	.023	.753			.140	.084			201	40.6	18.6	Yuan et al. 1984c
Tibetans (Switzerland)[d]		.604			.304	.091			115	53.4	22.6	Jeannet et al. 1972
Chinese (Singapore)[d]	.006	.643	.028	.239	.012			.072	631	52.3	21.4	Hawkins & Simons 1976
TAIWAN												
Toroko-Aborigines[c]		.882			.098	.019			127	21.2	9.7	Chen et al. 1985
JAPAN												
Japanese (Tokyo)	.003	.646	.011	.262	.017	.029		.033	2,455	51.2	20.4	Nakajima 1973b
Ainu (Hokkaido)[c]		.539	.011	.268			.158	.024	249	61.2	21.3	Harvey et al. 1978
KOREA												
Koreans (Seoul)	.004	.639			.306			.051	255	49.5	20.6	Nakajima 1973a
Koreans (NE China)[b]	.073	.642			.121	.095		.068	216	55.4	24.5	Yuan et al. 1984a
AFRICA												
LIBYA (Tripoli & Benghasi)[c]	.412	.008			.133	.110	.008	.329	168	69.2	27.1	Walter et al. 1975
EGYPT[a]	.463	.005			.140	.234		.157	720	68.7	26.7	el-Dewi 1951
SUDAN												
Sudanese (Khartoum)[g]	.171				.070	.522		.237	115	63.7	18.2	Saha et al. 1978b
Beja-Amarar (NE Sudan)[e]	.270				.085	.428		.217	100	69.0	22.6	el Hassan et al. 1968
CHAD												
Sara Majingay[g]	.035		.019		.143	.532	.001	.270	257	62.2	15.9	Hiernaux 1976
SENEGAL												
Bedik (E. Senegal)[c]	.003	.016			.098	.620		.264	791	53.6	10.2	Bouloux et al. 1972
NIGERIA												
Yoruba (Western)[7]	.050		.031		.011	.680		.228	135	48.2	8.3	Blumberg et al. 1961
CENTRAL AFRICAN REPUBLIC												
Babinga Pygmies (Bagandou)[e]				.044	.006	.829	.006	.114	163	29.8	5.2	Cavalli-Sforza et al. 1969
Hutu (Burundi)[e]	.079				.031	.750		.140	99	41.1	9.7	Fraser et al. 1966a
ZAIRE												
Bantu	.101				.015	.678		.206	93	48.7	9.8	Govaerts et al. 1972
ETHIOPIA												
Amhara tr.[g]	.182				.047	.535		.236	107	62.3	17.0	Ikin & Mourant 1962
UGANDA (Kampala)[g]	.018				.065	.749		.168	138	40.6	7.5	Ssebabi & Roberts 1975
TANZANIA												
Sandawe (Kondoa-Irang Dist.)[e]	.035				.051	.707		.207	215	45.3	8.0	Godber et al. 1976
Nyaturu (")[e]	.075				.047	.760		.118	214	40.1	10.7	"
Hadza (W. Lake Eyasi)[e]	.009				.068	.749		.174	162	40.4	6.9	Tills et al. 1982a
ANGOLA												
Njinga[c]	.089		.025		.077	.714		.095	109	46.7	15.6	Nurse et al. 1979
SOUTHERN AFRICA												
Basters (Namibia)[g]	.278		.030		.075	.390	.032	.195	120	72.5	24.8	Nurse et al. 1982
San/Bushmen												
!KUNG, Dobe[c]	.046				.003	.798		.154	383	33.7	4.5	Nurse & Jenkins 1977
Khoi/Hottentot												
Keetmanshoop Nama[c]	.118				.056	.744		.081	152	42.3	14.2	Jenkins 1972
Negroes/Dama (Okombahe &												

Continued

Table 159. Rhesus system: RH (cont'd)

Place/Population	CDE	CDe	CdE	Cde	cDE	cDe	cdE	cde	n	H	PE	Source
Sesfontein)[g]		.035			.071	.673	.069	.152	119	51.3	13.9	Nurse et al. 1976
Bantu (Natal)[e]		.041		.059	.043	.758		.100	644	40.8	12.1	Shapiro 1953
Griqua (Campbell, Cape Province)[c]		.061		.059	.058	.720		.101	99	46.1	14.5	Nurse & Jenkins 1975
TRISTAN DA CUNHA[e]		.256		.013	.406	.148		.177	160	71.6	29.4	Jenkins et al. 1985

NORTH AMERICA

GREENLAND

Place/Population	CDE	CDe	CdE	Cde	cDE	cDe	cdE	cde	n	H	PE	Source
Eskimos (Western)[b]	.001	.561		.021	.308	.005		.104	2,074	57.9	23.9	Gürtler 1971
ALASKA												
Eskimos-Inupiak (Wainwright)[g]		.478			.514	.009			111	50.7	19.3	Corcoran et al. 1959
Eskimos (St. Lawrence Is.)[c]		.454			.444	.102			216	58.6	24.4	Ferrell et al. 1981b
Athabaskan Indians (Fort Yukon)[g]	.005	.291			.622	.082			110	52.2	22.2	Corcoran et al. 1959
CANADA												
Eskimos (NW Territories)[c]		.498			.491	.011			320	51.1	19.5	Chown & Lewis 1959
Ojibwa Indians (Pikangikum)[c]	.203	.323			.474				96	63.0	27.4	Szathmary et al. 1975
Cree Indians (Manitoba)[c]	.082	.354			.469			.095	164	63.9	27.9	Lucciola et al. 1974
Dogrib Indians (NW Territories)[c]	.014	.221			.695	.023		.047	158	46.5	20.4	Szathmary 1983
UNITED STATES												
Blackfeet Indians (Montana)[c]	.004	.564			.344		.001	.087	148	55.6	23.3	Rokala et al. 1977
Papago Indians (S. Arizona)[c]	.022	.627			.329	.022			709	49.8	20.4	Niswander et al. 1970
Whites (Southern)[e]	.003	.431		.006	.148	.037	.017	.358	333	66.2	27.8	Cooper et al. 1963
Blacks (")[e]		.103		.025	.108	.535		.230	304	63.8	18.7	"
Mexican Americans	.017	.438			.175	.066	.003	.301	990	68.2		Hewett-Emmett et al. 1986b

MIDDLE AMERICA

MEXICO

Place/Population	CDE	CDe	CdE	Cde	cDE	cDe	cdE	cde	n	H	PE	Source
Indians (San Pablo del Monte)[c]	.040	.602			.323	.035			256	53.0	22.3	Crawford et al. 1974
Mestizo (Tlaxcala)[c]	.025	.577			.265	.126		.007	138	58.0	25.4	Crawford et al. 1974
GUATEMALA												
Black Caribs (Livingstone)[c]	.034	.091		.121	.145	.508		.101	205	68.7	26.2	Crawford et al. 1981
COSTA RICA												
Guaymi[c]		.844			.122	.034			286	27.2	12.4	Barrantes et al. 1982
WEST INDIES (Haiti)[c]		.056		.019	.080	.581		.264	471	58.3	13.6	Basu et al. 1976

SOUTH AMERICA

ARGENTINA

Place/Population	CDE	CDe	CdE	Cde	cDE	cDe	cdE	cde	n	H	PE	Source
Mapuchi	.031	.383			.352	.028	.021	.185	104	69.3	29.6	Haas et al. 1985
BOLIVIA												
Aymara (Western)[c]	.044	.462			.468	.026			315	56.5	23.1	Ferrell et al. 1978a
BRAZIL												
Baniwa (Jandu Cachoeira)[c]	.01	.59			.38			.02[j]	363	50.7	20.1	Salzano et al. 1986
Cayapo (Pará & Mato Grosso)[d]	.039	.475			.429			.057	526	58.6	24.4	Salzano et al. 1972a
Caingang[c]	.098	.516			.318	.034		.034	214	62.1	27.1	Salzano et al. 1980
Pacaás Novos[c]	.060	.500			.210	.150		.080	215	67.3	29.3	Salzano et al. 1985
Macushi (N. Brazil & S. Guyana)[c]	.020	.601			.374	.005			503	49.8	19.7	Neel et al. 1977a
Wapishana (")[c]	.046	.537			.359	.058			618	57.7	24.4	"
Ticuna (C. Amazonas)[c]	.023	.651			.315	.011			1,755	47.6	19.4	Neel et al. 1980
CHILE												
Atacameño (Toconao, N. Chile)[c]	.046	.473			.434	.027		.020	181	58.5	24.3	Rothhammer et al. 1984

Continued

Table 159. Rhesus system: RH (cont'd)

Place/Population	CDE	CDe	CdE	Cde	cDE	cDe	cdE	cde	n	H	PE	Source
COLOMBIA												
Noanama[d]		.755			.234	.011			168	37.5	15.6	Kirk et al. 1974
FRENCH GUIANA												
Wayampi (Trois-Sauts & Camopi)[d]	.155	.617			.220	.008			239	54.7	24.2	Tchen et al. 1981
SURINAM												
Trio (S. Surinam)[g]	.123[h]	.480			.338			.059[i]	505	63.7	27.9	Geerdink et al. 1974c
VENEZUELA												
Makiritare (N. Brazil & S. Venezuela)[c]	.020	.410			.540	.030			539	53.9	21.7	Gershowitz et al. 1970
Yanomama (")[c]	.094	.823			.070	.014			2,516	30.9	14.6	Gershowitz et al. 1972

OCEANIA

Place/Population	CDE	CDe	CdE	Cde	cDE	cDe	cdE	cde	n	H	PE	Source
AUSTRALIA												
Aborigines/Malag (Elcho Is., N. Territory)[c]	.098	.777			.087	.038			352	37.8	17.8	Simmons & Cooke 1969
Aborigines (Yuendumu, N. Territory)		.57			.41	.02			93	50.7	19.7	Kirk 1965(c)
Aborigines (W. Australia)	.05	.65		.01	.21	.08			1,698	52.4	23.0	"
MELANESIA												
W. New Guinea												
Asmat (S. Coastal Plain)[c]	.003	.928			.046	.023			402	13.6	6.6	Gajdusek et al. 1978
Papua New Guinea												
Gainj & Kalam (N. Central Highlands)[c]	.010	.933			.021	.036			412	12.8	6.2	Long et al. 1986
Jimi Valley (W. Highlands)[d]		.878			.102	.020			344	21.8	10.1	Mourant et al. 1981
Yagaria (E. Highlands)[g]		.868			.093	.039			459	23.6	11.0	Mourant et al. 1982
Fuyuge (C. Highlands)[c]		.912			.073	.015			200	16.3	7.7	Woodfield et al. 1974
Daga (Bay Province)[c]	.044	.870			.019	.067			134	23.6	11.5	Jenkins et al. 1983
Karkar Is.												
Takia[g]		.845		.065	.081	.009			476	27.5	8.1	Booth et al. 1982
Manus Is.[g]		.884			.036	.080			138	21.1	9.9	Malcolm et al. 1972
Buka Is.[c]	.062	.755			.140	.043			79	40.5	18.7	McLoughlin et al. 1982
Solomon Is.[c]	.037	.820			.125	.018			2,430	31.0	14.3	Lai & Bloom 1982
MICRONESIA												
Caroline Is. (Pingelap)[c]		.951			.022			.027[i]	1,035	9.4	4.6	Morton & Yamamoto 1973
Marshall Is.[c]		.949			.044	.007			678	9.7	4.7	Simmons et al. 1952
POLYNESIA												
Samoans (Christchurch, New Zealand)[c]	.016	.574			.327	.082			77	55.7	23.6	Booth et al. 1977

Note: [a] Tested with anti-C, -D, and -E. [b] Tested with anti-C, -c, -D, and -E. [c] Tested with anti-C, -c, -D, -E, and -e. [d] Tested with anti-C, C^W, -c, -D, -E, and -e; C-- and C^W-- combined. [e] Tested with anti-C, -C^W, -c, -D, -D^u, -E, and -e; C-- and C^W-- combined; -D- and -D^u- combined. [f] Tested with anti-C, -c, -D, -D^u, and -E. [g] Tested with anti-C, -c, -D, -D^u, -E, and -e. [h] Sum of the frequencies of CDE and CdE. [i] Sum of the frequencies of cDe and cde. [j] cDe or cde. The loci C, D, and E are considered to be closely linked. Allele D is dominant to d, whereas alleles C and c and alleles E and e are codominant to each other.

Table 160. Rodgers system: RG

Place/Population	Rg[a]	Rg	n	H	PE	Source
EUROPE						
UNITED KINGDOM						
England (London & Leeds)	.825	.175	1,038	.1	12.4	Longster & Giles 1976

Table 161. Sd system: SD

Place/Population	Sd[a]	Sd	n	H	PE	Source
EUROPE						
UNITED KINGDOM						
England	.706	.194	290	46.4	.1	MacVie et al. 1967

Note: Allele Sd[a] is dominant to Sd.

Table 162. Stoltzfus: SF

Place/Population	Sf[a]	Sf	n	H	PE	Source
NORTH AMERICA						
UNITED STATES						
Whites	.190	.810	1,130	30.8	8.2	Bias et al. 1969
Blacks	.101	.899	78	18.2	6.6	"
Orientals	.203	.797	11	32.4	8.2	"

Note: Allele Sf[a] is dominant to Sf.

Table 163. Sutter antigen: JS

Place/Population	Js[a]	Js[b]	n	H	PE	Source
EUROPE						
GREECE						
Greeks (Plati)	.002	.998	399	.4	.2	Tills et al. 1983b
ICELAND		1.000	100			Bjarnson et al. 1973
ASIA						
ISRAEL						
Yemenite Jews	.006	.994	157	1.2	.6	Tills et al. 1977b
PAKISTAN						
Pathans		1.000	139			Vos & Kirk 1961
Punjabi		1.000	168			"
INDIA						
Irula tr. (S. India)		1.000	72			"
SRI LANKA						
Tamil		1.000	44			"
Sinhalese		1.000	93			"
Veddah		1.000	39			"
MALAYSIA						
Malays		1.000	40			"
THAILAND (Bangkok)		1.000	188			"
PHILIPPINES						
Filipino		1.000	297			Fraser et al. 1964
Negrito	.057	.943	45			Pascasio et al. 1974
JAPAN						
Ainu (Hokkaido)		1.000	248			Harvey et al. 1978

Continued

Table 163. Sutter antigen: JS (cont'd)

Place/Population	Js[a]	Js[b]	n	H	PE	Source
AFRICA						
NIGERIA (N. Nigeria)	.089	.911	141	16.2	7.5	Tills et al. 1979
BURUNDI						
Hutu	.096	.904	99	17.4	7.9	Fraser et al. 1966a
ZAIRE						
Bantu		1.000	93			Govaerts et al. 1972
ETHIOPIA						
Amhara tr.	.053	.947	107	10.0	4.8	Ikin & Mourant 1962
UGANDA (Kampala)	.083	.917	108	15.2	7.0	Ssebabi & Roberts 1975
TANZANIA						
Sandawe	.074	.926	163	13.7	6.4	Godber et al. 1976
Nyaturu	.089	.911	217	16.2	7.5	"
Hadza (Western)	.019	.981	104	3.7	1.8	Tills et al. 1982a
NORTH AMERICA						
ALASKA						
Eskimos-Inupiak						
(Wainwright)		1.000	111			Corcoran et al. 1959
Athabaskan Indians						
(Fort Yukon)		1.000	110			"
UNITED STATES						
Whites (Southern)	.002	.998	324	.4	.2	Cooper et al. 1963
Blacks (")	.003	.997	303	.6	.3	"
SOUTH AMERICA						
BOLIVIA						
Aymara	.011	.989	315	2.0	1.0	Ferrell et al. 1978a
BRAZIL						
Cayapo (Pará & Mato Grosso)		1.000	221			Salzano et al. 1972a
VENEZUELA						
Makiritare (S. Venezuela & N. Brazil)		1.000	538			Gershowitz et al. 1970
Yanomama		1.000	1851			Gershowitz et al. 1972
OCEANIA						
MELANESIA						
Papua New Guinea						
Yagaria (E. Highlands)		1.000	372			Mourant et al. 1982
Karkar Is. (Waskia)		1.000	136			Booth et al. 1982

Note: Alleles Js[a] and Js[b] are codominant.

Table 164. V factor: V

Place/Population	V	v	n	H	PE	Source
EUROPE						
UNITED KINGDOM						
England		1.000	407[a]			de Natle et al. 1955
ASIA						
SAUDI ARABIA						
Shiah (Eastern)	.056	.944	379	10.6	4.4	Maranjian et al. 1966

Continued

Table 164. V factor: V (cont'd)

Place/Population	V	v	n	H	PE	Source
PAKISTAN						
Pathans		1.000	139			Vos & Kirk 1961
Punjabi		1.000	168			"
INDIA						
Tamils/S. Indians						
(Sri Lanka)		1.000	44			"
Irula tr. (Nilgiri						
(S. India)		1.000	72			"
Indians (Malaysia)		1.000	131			"
SRI LANKA						
Sinhalese		1.000	93			"
Veddah		1.000	39			"
MALAYSIA						
Malays		1.000	40			"
THAILAND						
Thais (Bangkok)		1.000	188			"

AFRICA

NIGERIA						
Yoruba (Ibadan)	.376	.624	162	46.9	5.7	Garlick 1958
ETHIOPIA						
Amhara tr.	.283	.717	107	40.6	7.5	Ikin & Mourant 1962
UGANDA (Kampala)	.187	.813	112	30.4	8.2	Ssebabi et al. 1975
TANZANIA						
Sandawe	.328	.672	215	44.1	6.7	Godber et al. 1976
Nyaturu	.303	.697	214	42.2	7.2	"
Hadza (Western)	.104	.896	162	18.6	6.7	Tills et al. 1982a
Hadza (Eastern)		1.000	102			"
SOUTHERN AFRICA						
San/Bushmen						
!Kung, Dobe	.178	.822	383	29.3	8.1	Nurse & Jenkins 1977
Khoi/Hottentot						
Topnaar, Sesfontein	.024	.976	42	4.7	2.2	"
Negroes/Dama	.091	.909	52	16.5	6.2	"
Bantu	.183	.817	511	29.9	8.2	Shapiro 1964

NORTH AMERICA

UNITED STATES						
Whites (New York)		1.000	444[a]			de Natle et al. 1955
Blacks (")	.144	.856	168	24.7	7.7	"
Whites (Seattle)	.001	.999	514	.2	.1	Giblett et al. 1957
Blacks (")	.156	.844	327	26.3	7.9	"
Orientals	.002	.998	272	.4	.2	"
Am. Indians	.006	.994	174	1.2	.6	"

Note: Allele V is dominant to v. The V factor is generally associated with Ce and ce of the Rh system. [a] Two individuals with V positve were found.

Table 165. XG system: XG

Place/Population	Xg[a]	Xg	n	H	Source
EUROPE					
DENMARK	.548	.452	115	49.5	Noades et al. 1966
FINLAND/NORWAY	.503	.497	91	50.0	Noades et al. 1966
FRANCE	.601	.399	148	48.0	"
GERMANY/HOLLAND/POLAND	.673	.327	103	44.0	"
GREECE	.570	.430	240	49.0	Fraser et al. 1969

Continued

Table 165. XG system: XG (cont'd)

Place/Population	Xg[a]	Xg	n	H	Source
ITALY	.761	.239	322	36.4	Siniscalco et al. 1966
SPAIN					
Spanish	.570	.430	451	49.0	Vega 1975
Basques	.550	.450	239	49.5	"
SWEDEN	.668	.332	155	44.4	Noades et al. 1966
SWITZERLAND	.667	.333	558	44.4	Metaxas & Metaxas-Bühler 1970
UNITED KINGDOM					
England	.670	.330	1,947	44.2	Noades et al. 1966

ASIA

Place/Population	Xg[a]	Xg	n	H	Source
BAHRAIN					
Arabs	.408	.592	45	48.3	Dewey & Mann 1967
INDIA					
Indians (W. India)	.652	.348	100	45.4	Bhatia 1963
Indians (Singapore)	.567	.433	91	49.1	Saha & Banerjee 1973
MALAYSIA					
Malays (Singapore)	.495	.505	72	50.0	"
THAILAND					
Thais (Bangkok)	.57	.43	181	49.0	Ratanaubol & Ratanasiri-vanich 1971
PHILIPPINES					
Negrito	.311	.689	45	43.0	Pascasio et al. 1974
CHINA					
Han (N. China)	.327	.673	290	44.0	Yuan et al. 1982
Uygur (NW China)	.341	.659	221	44.9	Yuan et al. 1984d
Hui (Ningxia)	.443	.557	219	49.4	Yuan et al. 1985
Dong (Guangxi)	.375	.625	201	46.9	Yuan et al. 1984c
Tibetans (Switzerland)	.766	.234	110	35.8	Jeannet et al. 1972
Chinese (Taiwan)	.599	.401	171	48.0	Dewey & Mann 1967
Chinese (Singapore)	.446	.554	101	49.4	Saha & Banerjee 1973
JAPAN					
Japanese (Tokyo)	.650	.350	131	45.5	Nakajima 1973b
Ainu (Hokkaido)	.434	.566	144	49.1	Harvey et al. 1978

AFRICA

Place/Population	Xg[a]	Xg	n	H	Source
ZAIRE					
Bantu	.505	.495	93	50.0	Govaertz et al. 1972
SOUTHERN AFRICA					
San/Bushmen (!Xo)	.300	.700	37	42.0	Nurse & Jenkins 1977
Khoi/Hottentots					
Hei//Om	.414	.586	94	48.5	Nurse & Jenkins 1977
Negroes/Dama (Okombahe)	.325	.675	69	43.9	Nurse et al. 1976

NORTH AMERICA

Place/Population	Xg[a]	Xg	n	H	Source
ALASKA					
Eskimos	1.000				Scott 1973
Athabaskan Indians	1.000				"
CANADA					
Cree Indians (Manitoba)	.564	.436	100	49.2	Lucciola et al. 1974
UNITED STATES					
Zuni Indians (New Mexico)	.912	.088	57		Dewey & Mann 1967

SOUTH AMERICA

Place/Population	Xg[a]	Xg	n	H	Source
FR. GUIANA					
Wayampi (Trois-Sauts & Camopi)	.338	.662	54	44.8	Tchen et al. 1981
VENEZUELA					
Yanomama (S. Venezuela & N. Brazil)	.600	.400	99	48.0	Gershowitz et al. 1972

OCEANIA

AUSTRALIA

Continued

Table 165. XG system: XG (cont'd)

Place/Population	Xg^a	Xg	n	H	Source
Aborigines	.793	.207	352	32.8	Simmons 1970
MELANESIA					
New Guinea					
Aborigines	.845	.155	263	26.2	Simmons 1970
MICRONESIA					
Aborigines (Mariana Is.)	.651	.349	109	45.4	Plato et al. 1964

Note: Xg^a is located on the X chromosome and dominant to Xg.

Table 166. Yt (Cartwright) system: YT

Place/Population	Yt^a	Yt^b	n	H	PE	Source
EUROPE						
SWITZERLAND (Zurich)	.960	.040	369	7.7	3.7	Giles et al. 1967
UNITED KINGDOM						
England (London)	.958	.042	1,030	8.0	3.9	"
ASIA						
ISRAEL						
Jews	.862	.132	264	23.8	10.5	Levine et al. 1985
THAILAND						
Thais (Bangkok)	1.000		327			Chandanayingyong et al. 1967
NORTH AMERICA						
ALASKA						
Eskimos	1.000		146			Chown & Lewis 1959
UNITED STATES						
Blacks	.957	.043	714	8.2	3.9	Wurzel & Haesler 1968
MIDDLE AMERICA						
GUATEMALA						
Black Caribs						
(Livingston)	1.000		202			Crawford et al. 1981

Note: Allele Yt^a and Yt^b are codominant.

Table 167. Blood group loci with rare variants

Locus: Name of antigen	Populations studied	No. individuals tested n	No. variant individuals	Source
Ahonen[b]: AN	Finns, Finland	10,000	6	Furuhjelm et al. 1972
	Swedes, Sweden	3,266	2	"

Continued

Table 167. Blood group loci with rare variants (cont'd)

Locus: Name of antigen	Populations studied	No. individuals tested n	No. variant individuals	Source
August[a]: AUG*	New Yorkers including Blacks, USA	>6,600		Applewhaite et al. 1967
Batty[b]: BY	English, UK	31,522	2	Race & Sanger 1975(c)
	Negroes, UK	157		"
	Australian Aborigines	24		Simmons & Were 1955
Becker[b]: BEC*	Caucasians	122		Race & Sanger 1975(c)
Berrens[b]: BE	Eskimos, Alaska	241		Corcoran et al. 1959
	Am. Indians, "	255		"
	Whites, USA	323		Cooper et al. 1963
Biles[b]: BI	Whites, USA	179		Race & Sanger 1975(c)
	Blacks, USA	140		"
	Cherokee Indians, USA	181		"
Bishop[b]: BP	Norwegians, Norway	7,000		"
	English, UK	75,000	1	"
	Tibetans	42		Ørjasaeter et al. 1966
Box[b]: BX	English, UK	4,445		Race & Sanger 1975(c)
	Blood donors, London, UK	17,661	1	Contreras et al. 1980
Caldwell[b]: CL	Scotts, UK	11,000		Race & Sanger 1975(c)
	Swiss, Switzerland	1,541		"
	Tibetans in Switzerland	110		Jeannet et al. 1972
Chido: CH*	Blood donors, New York, USA	>2,000		Harris et al. 1967
Chr[b]: CHR	Danish, Denmark	500	1	Race & Sanger 1975(c)
Duch: DH	Danish, Denmark	2,493		Jørgensen et al. 1982
Dupuy[a]: DP	Blood donors, Miami, USA	600		Frank et al. 1970
Eldridge[a]: EL	"	3,000		"
Envelope[a]: EN	Finns, Finland	8,800		Furuhjelm et al. 1969
	Estonians	200		"
	English, UK	12,509		Darnborough et al. 1969
Er[a]: ER*	Blood donors, Quebec, Canada	5,000		Daniels et al. 1982
Evans[b]: EV*	English, UK	480	1	Race & Sanger 1975(c)
Fr[b]: FR	Unspecified individuals	415		Lewis et al. 1978
Gonsowski[a]: GN	Blood donors, Minnesota, USA	2,600		Fox & Taswell 1969
Gonzales[b]: GO	English, UK	2,750		Alter et al. 1967
	Indians, UK	34		"
	Nepalese	242		"
	Blacks, Africa	35	1	"
	Blacks, USA	1,131	20	"
	Chinese, USA	100		"
Good[b]: GD*	English, UK	397		Race & Sanger 1975(c)
Gregory[a]: GY	Faroe Is. (N. Atlantic)	438		Tills et al. 1985
	Isle of Lewis, UK	195		Clegg et al. 1985
	Greeks, Plati, Greece	289		Tills et al. 1983b
	Japanese, Japan	9,350		Okubo et al. 1986a
	Cree Indians, Canada	100		Lucciola et al. 1974
	Whites, Minneapolis, USA	9,459		Swanson et al. 1967
	Blacks, Minneapolis, USA	75		Swanson et al. 1967
	Am. Indians, USA	611		"

Continued

Table 167. Blood group loci with rare variants (cont'd)

Locus: Name of antigen	Populations studied	No. individuals tested n	No. variant individuals	Source
	Black Caribs, Guatemala	202		Crawford et al. 1981
Griffiths[b]: GF	English, UK	6,886		Race & Sanger 1975(c)
Heibel[b]: HEI*	Germans, Germany	>500		Ballowitz et al. 1968
Hey[b]: HEY	French, France	8,127	2	Yvart et al. 1974
Hov[b]: HOV	Dutch, Netherlands	1,155	2	Szaloky et al. 1973
Hughes[b]: HG	Wales, UK	5,434	2	Rowe & Hammond 1983
Hunt[b]: HT	English, UK	453		Race & Sanger 1975(c)
I: I	French, France	10,090	1	Ducos et al. 1965
	Thais, Bangkok, Thailand	389		Chandanayingyong et al. 1967
	Japanese, Japan	145		Furuhata et al. 1961
	Sara Majingay, Chad, Africa	256		Hiernaux 1976
	Bedik, Senegal	793		Bouloux et al. 1972
	Babinga Pygmies, C. Africa	59		Cavalli-Sforza et al. 1969
	Aymara, Bolivia	1,697		Quilici 1968
	Wayampi, French Guiana	99		Tchen et al. 1981
	Takia, Karkar Is., Melanesia	180	30[c]	Booth et al. 1982
	Samoans, Polynesia	76		Booth et al. 1977
Indian[b]: IN	Arabs	246	29	Badakere et al. 1980
	Iranians, Iran	557	59	"
	Parsi, Bombay, India	595		"
	Gujarati, "	132	6	Badakere et al. 1974
	Maharastrians, "	608	14	"
	Indonesians	201		Badakere et al. 1980
	Thais, Thailand	441		"
	Chinese, China	300		"
Jensen[b]: JEN	Danish, Denmark	>1,000		Skov 1972
	Faroe Is., N. Atlantic	622		Tills et al. 1985
	Greeks, Plati, Greece	690		Tills et al. 1983b
Jn[b]: JN	Norwegians, Norway	4,767		Kornstad et al. 1967
	Tibetans	42		Ørjasaeter et al. 1966
Jobbins[b]: JO	Caucasians	120		Race & Sanger 1975(c)
Jr[a]: JR	Japanese, Japan	298	5	Nakajima & Ito 1978
	Blood donors, Pittsburgh, USA	9,145		Race & Sanger 1975(c)
	Orientals, USA	1,041		"
	Eskimos	75		"
Kamhuber[b]: KA	Viennese, Austria	1,100		"
Knops-Helgeson[a]: KN	Blood donors, Minneapolis, USA	2,091	4	Helgeson et al. 1970
Lan[a]: LAN	Dutch, Netherlands	4,000	1	Hart et al. 1961
	English, UK	2,268		Smith et al. 1969
	Japanese, Japan	15,000		Okubo et al. 1984a
	Babinga Pygmies, C. Africa	53		Cavalli-Sforza et al. 1969
Levay[b]: LEV*	English, UK	350		Race & Sanger 1975(c)
Lewis II[b]: LS	English, UK	5,887		"
	Negroes, W. Africa	81	2	"
	Blacks, USA	110	1	"
	Negroes, W. Indies	878	9	"
LW: LW	Blood donors, Canada	10,552		de Veber et al. 1971

Continued

Table 167. Blood group loci with rare variants (cont'd)

Locus: Name of antigen	Populations studied	No. individuals tested n	No. variant individuals	Source
Marriott: ZT	Greeks, Plati, Greece	178		Tills et al. 1983b
	Takia, Karkar Is., Melanesia	132		Booth et al 1982
Martin: MT	Black Caribs, Guatemala	202		Crawford et al. 1981
Mg: MG	Faroe Is., N. Atlantic	679		Tills et al. 1985
	English, UK	61,128		Race & Sanger 1975(c)
	Isle of Lewis, UK	283		Clegg et al. 1985
	Swiss, Switzerland	6,530	10	Metaxas et al. 1966
	Greeks, Plati, Greece	1,038		Tills et al. 1983b
	Liberia, W. Africa	454	80	Neppert 1980
	Japanese, Japan	157		Furuhata et al. 1961
	Tibetans in Switzerland	110		Jeannet et al. 1972
	Eskimos, Alaska	239		Corcoran et al. 1959
	Am. Indians, "	255		"
	Cree Indians, Canada	100		Lucciola et al. 1974
	Whites, USA	333		Cooper et al. 1963
	Blacks, USA	303		"
	Yanomama, Venezuela	> 2,516		Gershowitz et al. 1972
Milne[b]: MIL	Blood donors of Caucasian and Polynesian origin	2,643		Pinder et al. 1984
Miltenberger: MI	Faroe Is., N. Atlantic	665		Tills et al. 1985
	Negrito, Phillipines	39		Misawa et al. 1981
	Eskimos, Alaska	241		Corcoran et al. 1959
	Am. Indians, "	255		"
	Ojibwa Indians, Canada	102		Szathmary et al. 1975
	Tibetans in Switzerland	110		Jeannet et al. 1972
Mit[b]: MIT	Caucasians, Canada	1,370	1	Battista et al. 1980
	Cree Indians, Manitoba, Canada	555		"
	Blacks, Natal, Africa	662		"
	Indians, Africa	500		"
Moen[b]: MO	Norwegians, Norway	9,000		Race & Sanger 1975(c)
	Belgians, Belgium	9,793		"
Mur: MUR	Faroe Is., N. Atlantic	665		Tills et al. 1985
	Greeks, Plati, Greece	1,023		Tills et al. 1983b
	Isle of Lewis, UK	282		Clegg et al. 1985
Newfoundland: NFLD	Japanese, Canada	30		Lewis et al. 1984
	Caucasians, "	1,125		"
	Cree Indians, "	111		"
	Hutterite, "	500		"
Nyberg: NY	Faroe Is., N. Atlantic	679		Tills et al. 1985
	Norwegians, Norway	3,746	8	Kornstad et al. 1971
	Lapps/Saami, Norway	305		"
	Germans, Germany	20,000		Race & Sanger 1975(c)
	Swiss, Zürich, Switzerland	7,994		"
	Greeks, Plati, Greece	1,038		Tills et al. 1983b
	Mazandaranian/Guilanian, Iran	522		Ohkura et al. 1984
	Japanese, Tokyo	1,347		Nakajima et al. 1967b
	Tibetans in Switzerland	110		Jeannet et al. 1972
	Cree Indians, Canada	100		Lucciola et al. 1974
Ok[a]: OK	Whites, USA	9,053		Morel & Hamilton 1979
	Blacks, USA	911		"
	Orientals, USA	261		"
	Mexican-Americans, USA	1,378		"
	Japanese, Hiroshima, Japan	800		"
Ol (Oldeide): OL	Norwegians, Norway	7,151	1	Kornstad 1986

Continued

Table 167. Blood group loci with rare variants (cont'd)

Locus: Name of antigen	Populations studied	No. individuals tested n	No. variant individuals	Source
Orriss[b]: ORR*	English, UK	887		Race & Sanger 1975(c)
	Blacks, USA	163	1	"
Os[b]: OS	Japanese, Gifu, Japan	50,000		Seno et al. 1983
Pe[b]: PEA*	Caucasians, USA	1,500		McPherson et al. 1979
Peters[b]: PT	Norwegians, Norway	14,674		Pinder et al. 1969
	Tibetans	43		"
	Chinese	83		"
Raddon: RDD*	English, UK	2,000	1	Race & Sanger 1968(c)
Radin[b]: RD	Danes, Denmark	4,933	24	Lundsgaard & Jensen 1968
	Faroe Is., N. Atlantic	551		Tills et al. 1985
	Icelanders, Iceland	365	1	Tills et al. 1982b
	Isle of Lewis, UK	282		Clegg et al. 1985
	Greeks, Plati, Greece	1,038	4	Tills et al. 1983b
	Ukrainians	170	1	Race & Sanger 1975(c)
	Jews, New York, USA	562	3	Rausen et al. 1967
	Yemenite Jews, Israel	157		Tills et al. 1977b
	Kurds, Iran	107		Lehmann et al. 1973
	Ainu, Hokkaido, Japan	104		Harvey et al. 1978
	Nigerians, Nigeria	61		Tills et al. 1979
	Sandawe, Tanzania	215		Godber et al. 1976
	Nyaturu, "	217		"
	Cree Indians, Canada	100		Lucciola et al. 1974
	Papua New Guineans, Jimi Valley	99		Mourant et al. 1981
	Icelanders, Iceland	365	1	Tills et al. 1982b
	Papua New Guinea, Yagaria	372		Mourant et al. 1982
	Waskia, Karkar Is., Melanesia	314		Booth et al. 1982
Reid[b]: RE	Canadians, Canada	>10,000		Race & Sanger 1975(c)
Ridley[b]: RI	English, UK	17,013	20	"
	English, UK	53,488		Contreras et al. 1984a
	Negro & Indian antenatal patients	1,650		"
Rm(Romunde)[b]: RM	Caucasians	200		Race & Sanger 1975(c)
Rosenlund[a]: RL	Norwegians, Norway	4,400	7	"
Scianna: SC	Polish, Warszawa, Poland	1,025	9	Seyfried et al. 1966
	Germans, Berlin	2,015	15	Fünfhausen & Gremplewski 1967
	English, UK	1,039	7	Seyfried et al. 1966
	Negroes, Canada	212		Race & Sanger 1975(c)
	Cree Indians, Canada	100		Lucciola et al. 1974
Skjelbred[b]: SKJ*	Norwegians, Norway	12,753		Kornstad et al. 1968
	Tibetans	42		"
	Chinese	58		"
Stobo[b]: ST*	Scots, Scotland, UK	3000	2	Race & Sanger 1975(c)
Swann[b]: SW	Faroe Is., N. Atlantic	523		Tills et al. 1985
	English, UK	23,888	4	Race & Sanger 1975(c)
	Isle of Lewis, UK	242		Clegg et al. 1985
	Greeks, Plati, Greece	881		Tills et al. 1983b
	Negroes	157		Cleghorn 1961
	Japanese, Japan	217		Furuhata et al. 1961
	Tibetans in Switzerland	110		Jeannet et al. 1972
Tc[a]: TCA*	Japanese, Tokyo	5,000		Laird-Fryer et al. 1983
	Whites, USA	5,000		"
	Blacks, USA	350		"

Continued

231

Table 167. Blood group loci with rare variants (cont'd)

Locus: Name of antigen	Populations studied	No. individuals tested n	No. variant individuals	Source
Torkildsen[b]: TO	Norwegians, Norway	6,461	1	Kornstad et al. 1968
Traversu[b]: TR	English, UK	38,069	2	Race & Sanger 1975(c)
	Norwegians, Norway	9,500		"
Tsunoi[b]: TSU*, TS	Japanese, Japan	2,000		Nakajima et al. 1967a
Vel: VEL	Faroe Is., N. Atlantic	594		Tills et al. 1985
	English, UK	16,548	4	Cleghorn 1961
	Isle of Lewis, UK	241	1	Clegg et al. 1985
	Finns, Finland	18,920		Nevanlinna 1966
	French, Paris, France	5,000	2	Battaglini et al. 1964
	Greeks, Plati, Greece	894		Tills et al. 1983b
	Abu Dhabians, United Arab Emirates	100	2	Kamel et al. 1980
	Thais, Bangkok, Thailand	328	4	Chandanayingyong et al. 1967
	Chinese, New York, USA	30		Race & Sanger 1975(c)
	Japanese, Tokyo, Japan	42		Furuhata et al. 1961
	Tibetans in Switzerland	110		Jeannet et al. 1972
	Eskimos-Inupik, Alaska	65		Corcoran et al. 1959
	Am. Indians, "	244		"
	Cree Indians, Canada	100		Lucciola et al. 1974
	Blacks, New York, USA	200		Race & Sanger 1975(c)
	Solomon Is., Melanesia	186		Douglas et al. 1962
	Gilbert Is., Micronesia	236		Douglas et al. 1961
	Cook Is., Polynesia	77		Douglas & Staveley 1959
Ven[b]: VEN*	Caucasians	170		Race & Sanger 1975(c)
Verweyst[b]: VW	Faroe Is., Atlantic	679		Tills et al. 1985
	Isle of Lewis, UK	282		Clegg et al. 1985
	Greeks, Plati, Greece	1,036		Tills et al. 1983b
	Thais, Bangkok, Thailand	321		Chardanayingyong et al. 1967
	Tibetans in Switzerland	110		Jeannet et al. 1972
	Japanese, Japan	217		Furuhata et al. 1961
	Bedik, Senegal	755		Bouloux et al. 1972
	Eskimos, Alaska	241		Corcoran et al. 1959
	Am. Indians, "	255		"
	Whites (Southern U.S.)	333		Cooper et al. 1963
	Blacks (")	302		"
	Yanomama Indians, S.Venezuela & Brazil	1770		Gershowitz et al. 1972
	Papua New Guineans, Yagaria	47		Mourant et al. 1982
	Takia, Karkar Is.,Melanesia	41		Booth et al. 1982
Waldner[b]: WAL*, WD	Caucasians, Canada	4,000		Lewis & Kaita 1981
Webb[b]: WB	English, UK	15,815	3	Race & Sanger 1975(c)
	Wales, "	10,117	8	Bloomfield et al. 1986
	Tibetans	42		Ørjasaeter et al. 1966
	Japanese, Japan	2,954		Nakajima et al. 1965
	Australian Aborigines	92		Simmons & Albrey 1963
	New Guineans	105		"
Wright[b]: WR	Czechoslovakians	1,500	1	Kout 1962
	Faroe Is., N. Atlantic	680		Tills et al. 1985
	Greeks, Plati, Greece	1038	3	Tills et al. 1983b
	Irish,Ireland	100		Tills et al. 1977a
	Icelanders,Iceland	365	4	Tills et al. 1982b
	Italians	6,350	7	Liotta et al. 1970
	Norwegians, Norway	3,140	2	Kornstad 1961
	Lapps, "	433		"
	Swiss	3,753	2	Metaxas & Metaxas-Bühler 1963
	English, UK	14,109	12	Race & Sanger 1975(c)
	Scotish, UK	1,000	1	"
	Isle of Lewis, UK	283		Clegg et al. 1985

Continued

Table 167. Blood group loci with rare variants (cont'd)

Locus: Name of antigen	Populations studied	No. individuals tested n	No. variant individuals	Source
	Yemenite Jews, Israel	157		Tills et al. 1977b
	Kurds, Iran	105	2	Lehmann et al. 1973
	Sinhalese, Sri Lanka	72		Kirk et al. 1962
	Tibetans	42		Ørjasaeter et al. 1966
	Bhutanese	154		Mourant et al. 1969
	Thais, Bangkok, Thailand	433		Chandanayingyong et al. 1967
	Japanese, Tokyo, Japan	217		Furuhata et al. 1961
	Ainu, Hokkaido, Japan	248		Harvey et al. 1978
	Negrito, Philippines	41		Misawa et al. 1981
	Nigerian, Nigerians	141		Tills et al. 1979
	Babinga Pygmies, C. Africa Rep.	163		Cavalli-Sforza et al. 1969
	Sandawe, Tanzania	215		Godber et al. 1976
	Nyaturu, "	217		"
	Hadza, "	308		Tills et al. 1982a
	Eskimos, Inupiak, Alaska	241		Corcoran et al. 1959
	Am. Indians, "	255		"
	Blackfeet Indians, Montana	97		Rokala et al. 1977
	Whites (Southern U.S.)	333		Cooper et al. 1963
	Blacks (")	304		"
	Cayapo, Brazil	435		Salzano et al. 1972a
	Makiritare, S. Venezuela & N. Brazil	538		Gershowitz et al. 1970
	Yanomama, Venezuela	>2,500		Gershowitz et al. 1972
	Papuans, Jimi Valley	344		Mourant et al. 1981
	Papua New Guineans, Yagaria	373	1	Mourant et al. 1982
	Takia, Karkar Is., Melanesia	46		Booth et al. 1982
Wulfsberg[b]: WU	Norwegians, Norway	7,000	1	Race & Sanger 1975(c)
Zd[b]: ZD	Czechoslovakians	270		Svanda et al. 1970

Note: [a] Public or very frequent antigens. [b] Private or very infrequent antigens. [c] Weak variant.

Table 168. Platelet antigen systems

Antigen/Locus	Population	Allele +	Allele -	n	H	PE	Source
Bak[a]: BAK	Dutch	.696	.304	119	42.3	16.7	von dem Borne et al. 1980
Duzo: DUZ*	French	.117	.883	82	20.7	9.3	Moulinier 1958
Ko[a]: KO	Dutch	.074	.926	1,696	13.7	6.4	van der Weerdt 1965
Lek[a]: LEK	French	.865	.135	165	23.4	10.3	Boizard & Wautier 1984
Pl[A1], Zw[a]: PLA*	Dutch	.844	.156	287	26.3	11.4	van Loghem et al. 1959
	British	.867	.133	1,983	23.1	10.2	Contreras et al. 1984b
	French	.844	.156	289	26.3	11.4	Soulier & Patereau 1976
	Germans	.853	.147	1,211	25.1	11.0	Mueller-Eckhardt et al. 1985
	Japanese	1.000		300			Shibata et al. 1986
	US Whites	.821	.179	250	29.4	12.5	Ramsey & Salamon 1986
	US Blacks	.936	.064	243	12.0	5.6	"
Pl[E1]: PLE1*	N. Americans	.968	.032	1,025	6.2	3.0	Shulman et al. 1965
Pl[E2]: PLE2*	"	.025	.975	200	4.9	2.4	"
Yuk[a]: YUK	Japanese	.008	.992	300	1.6	.8	Shibata et al. 1986

Note: The allele with the + sign is dominant to the allele with the - sign for all systems in this table.

233

Table 169. HLA-A histocompatibility type: HLAA

Allele	EUROPE								ASIA	
	England	France	Germany	Italy	Nether-lands[a]	Scandi-navia	Spain	Switzer-land[a]	Israel: Ash-kenazi Jews[a]	India
A1	.190	.125	.115	.107	.152	.170	.164	.136	.111	.176
A2	.319	.242	.292	.293	.264	.366	.284	.281	.227	.095
A3	.187	.172	.142	.104	.122	.116	.110	.123	.085	.108
A11	.048	.063	.069	.049	.059	.049	.041	.116	.056	.170
A23	.011	.008	.023	.020	.026	.013	.014	.041	.036	
A24	.032	.117	.092	.142	.081	.103	.116	.090	.102	.122
A25	.037	.031	.035	.020	.021	.009	.014		.008	
A26	.032	.039	.038	.020	.070	.018	.055	.018	.115	.027
A28	.021	.039	.038	.035	.048	.076	.021	.029	.036	.108
A29	.016	.023	.031	.032	.037	.022	.034	.023	.040	
A30	.027	.031	.023	.078	.021	.013	.048	.006	.028	
A31	.021	.055	.031	.015	.032	.018	.027	.041	.024	.054
A32	.059	.031	.050	.061	.026	.022	.041	.053	.024	.014
AW33		.023	.012	.023	.021		.027	.006	.032	.068
AW34					.001				.020	.014
AW36			.004							
AW43										
AW66			.004							
AX			.001		.019	.004	.004	.036	.059	.047
n	188	128	260	345	96	224	146	87	129	74
H	81.6	86.8	85.8	85.5	87.0	80.3	85.5	85.7	89.1	88.2
PE	65.1	74.1	73.5	72.2	74.1	62.8	72.2	69.9	76.2	73.7

Source: Baur et al. 1984; [a]Baur & Danilovs 1980   Continued

Table 169. HLA-A histocompatibility type: HLAA (cont'd)

Allele	ASIA		AFRICA	NORTH AMERICA			MID. AMERICA	SOUTH AMERICA
	China	Japan	Blacks[a]	Whites[b]	Blacks	Am. Indians	Mexicans[a]	Am. Indians[a,b]
A1	.023	.002	.025	.169	.083	.044	.099	.408
A2	.348	.262	.107	.263	.174	.369	.254	
A3	.041	.004	.061	.151	.136		.030	
A11	.198	.103	.004	.038	.046		.036	
A23			.092	.040	.091		.006	.242
A24	.111	.361	.050	.102	.046	.335	.139	
A25				.021			.112	
A26	.035	.104	.047	.027	.046	.044	.061	.076
A28	.041	.002	.115	.053	.046	.087	.030	
A29	.006		.069	.027	.023	.022	.054	
A30	.080	.002	.207	.029	.068		.018	.242
A31	.023	.062	.007	.038		.087	.067	
A32	.017		.021	.027	.046		.042	.032
AW33	.064	.071	.036	.011			.042	
AW34			.061		.046		.012	
AW36			.011		.105			
AW43			.021					
AW66		.009						
AX	.014	.019	.067	.003	.046	.013	.001	

Continued

Table 169. HLA-A histocompatibility type: HLAA (cont'd)

Allele	ASIA		AFRICA	NORTH AMERICA			MID. AMERICA	SOUTH AMERICA
	China	Japan	Blacks[a]	Whites[b]	Blacks	Am. Indians	Mexicans[a]	Am. Indians[a,b]
n	172	100	143	362	44	46	85	42
H	81.1	77.0	90.0	85.8	90.5	73.2	87.6	71.0
PE	63.4	53.1	76.7	72.5	79.1	46.4	76.3	46.2

Source: Baur et al. 1984; [a]Baur & Danilovs 1980
Note: [b] The original estimates of gene frequencies did not sum up to one. Therefore, they were proportionately adjusted to make the sum equal to one.

Table 170. HLA-B histocompatibility type: HLAB

Allele	EUROPE								ASIA	
	England	France	Germany	Italy	Nether-lands[a,b]	Scandi-navia[b]	Spain	Switzer-land[a]	Israel: Ash-kenazi Jews[a]	India
B7	.144	.133	.143	.066	.097	.141	.112	.129	.040	.056
B8	.137	.078	.084	.039	.074	.141	.112	.103	.044	.028
B13	.011	.023	.034	.042	.021	.025	.037	.023	.028	.014
B14					.041			.018	.124	
B18	.043	.047	.051	.099	.052	.013	.030	.035	.024	.028
B27	.060	.016	.029	.003	.047	.075	.015	.035	.008	.014
B35	.096	.109	.105	.140	.091	.075	.119	.041	.151	.139
B37	.005	.023	.029	.012	.025	.017	.022	.018	.024	.070
B38	.005	.016	.038	.018	.036	.017	.030	.029	.133	.014
B39	.021	.008	.013	.024	.016	.013	.030	.029	.020	
BW41	.005	.008	.008	.006	.016	.005	.015	.012	.044	
BW42						.005	.008	.006		
B44	.165	.102	.143	.081	.079	.104	.104	.149	.048	.042
B45	.005		.008	.003	.016	.005	.015	.006	.004	
BW46			.004							
BW47	.005	.008		.006	.011				.004	.014
BW48									.016	
B49	.021	.031	.017	.036	.011	.017	.030	.072	.004	
BW50	.016	.016	.013	.027	.011	.005		.006	.020	.014
B51	.032	.070	.055	.122	.079	.029	.075	.053	.032	.070
BW52		.016	.034	.012	.011	.008	.037	.018	.052	.111
BW53		.008		.009	.021		.015	.006		.014
BW54									.004	
BW55	.011	.031	.013	.024	.052	.005	.015	.006	.024	.014
BW56	.005	.016	.025	.006	.005	.017	.008	.006	.004	
BW57	.027	.016	.017	.045	.041	.021	.015	.041	.036	.070
BW58	.011	.008	.013	.049				.012	.032	.056
BW59					.005				.004	
BW60	.037	.023	.013	.012	.047	.083	.030	.041	.012	.014
BW61	.016	.031	.017	.018	.016	.083	.037	.012	.008	.111
BW62	.090	.094	.055	.048	.058	.075	.037	.065	.008	.056
BW63			.004	.015	.005			.035	.016	.028
BW64	.005		.017	.006		.005	.030			
BW65	.027	.063	.004	.021		.017	.015			
BW67										
BW71										.014
BW72			.004	.006		.005	.008			.014
BW73			.008	.003						
8W57										
8W58				.011						
8W66				.011						
BX		.008	.004	.005			.001		.041	
n	188	128	238	335	96	222	134	87	129	72

Continued

235

Table 170. HLA-B histocompatibility type: HLAB (cont'd)

Allele	EUROPE								ASIA	
	England	France	Germany	Italy	Nether-lands[a,b]	Scandi-navia[b]	Spain	Switzer-land[a]	Israel: Ash-kenazi Jews[a]	India
H	90.5	92.8	92.4	93.0	94.3	91.5	93.1	92.6	92.4	92.6
PE	81.0	85.5	85.4	86.2	89.1	83.5	87.0	85.8	84.9	85.6

Source: Baur et al. 1984; [a]Baur & Danilovs 1980
Note: [b] The original estimates of gene frequencies did not sum up to one. Therefore, they were proportionately adjusted to make the sum equal to one.

Table 170. HLA-B histocompatibility type: HLAB (cont'd)

Allele	ASIA		AFRICA	NORTH AMERICA			MID. AMERICA	SOUTH AMERICA
	China[b]	Japan[b]	Blacks[a]	Whites	Blacks	Am. Indians	Mexicans[a]	Am. Indians[a,b]
B7	.031	.060	.115	.122	.184		.030	
B8	.006		.043	.119		.021	.030	
B13	.124	.009	.004	.022	.026		.024	
B14			.039				.024	
B18			.043	.061	.026	.021		
B27	.024	.002	.014	.028	.026	.063	.024	
B35	.067	.109	.043	.080	.105	.288	.236	.069
B37	.031	.002		.011			.006	
B38	.018	.006		.033			.043	
B39	.044	.043	.036	.028		.083	.036	.105
BW41	.006		.021	.008			.006	
BW42	.013	.006	.103	.008	.026			
B44	.037	.081	.061	.133	.026	.021	.191	
B45			.004	.003	.053			
BW46	.007	.003						
BW47	.006	.002	.004			.021		
BW48	.006	.018	.014				.036	
B49			.007	.008	.026		.006	
BW50				.006		.063	.074	
B51	.037	.092	.007	.042	.026	.167	.068	.289
BW52	.068	.106	.007	.014		.021	.018	.011
BW53			.036	.003	.103	.021		
BW54	.031	.096		.003				
BW55	.018	.023	.014	.006				
BW56		.013						
BW57	.013		.028	.033	.053	.021	.006	
BW58	.062	.006	.164	.014	.132	.021	.012	
BW59		.018	.007					
BW60	.112	.045	.004	.055	.026	.021	.036	.094
BW61	.081	.142		.011	.026	.125	.018	.034
BW62	.145	.082	.004	.078			.055	.376
BW63				.011			.036	.022
BW64				.014				
BW65		.002		.042				
BW67		.002						
BW71		.006			.026			
BW72	.006	.004			.079	.021		
BW73								
8W57								
8W58			.111					
8W66			.004					
BX	.005	.019	.065	.006	.027	.004	.001	
n	172	544	143	361	38	48	84	42
H	92.2	91.6	91.7	92.5	90.8	85.4	88.3	74.9

Continued

Table 170. HLA-B histocompatibility type: HLAB (cont'd)

Allele	ASIA		AFRICA	NORTH AMERICA			MID. AMERICA	SOUTH AMERICA
	China[b]	Japan[b]	Blacks[a]	Whites	Blacks	Am. Indians	Mexicans[a]	Am. Indians[a]
PE	82.8	81.2	80.7	85.2	79.9	71.6	77.9	54.7

Source: Baur et al. 1984; [a]Baur & Danilovs 1980
Note: [b] The original estimates of gene frequencies did not sum up to one. Therefore, they were proportionately
adjusted to make the sum equal to one.

Table 171. HLA-C histocompatibility type: HLAC

Gene	EUROPE								ASIA	
	England	France	Germany	Italy	Nether-lands[a]	Scandi-navia	Spain	Switzer-land[a]	Israel: Ash-kenazi Jews[a]	India
Cw1	.011	.024	.051	.018	.037	.054	.051	.023	.036	.057
Cw2	.055	.032	.023	.032	.070	.076	.054	.059	.028	.028
Cw3	.160	.127	.114	.095	.196	.182	.080	.097	.040	.056
Cw4	.106	.111	.111	.144	.099	.097	.167	.059	.169	.113
Cw5	.104	.040	.063	.059	.065	.073	.058	.065	.032	
Cw6	.058	.087	.108	.117	.104	.071	.080	.097	.089	.153
Cw7	.301	.208	.256	.186	.026	.282	.268	.041	.020	.131
Cw8	.032	.064	.020	.027	.005	.022	.036		.048	
CX	.173	.308	.249	.323	.397	.142	.207	.558	.540	.464
n	189	126	254	340	96	224	138	87	129	72
H	82.4	81.9	82.8	81.2	77.2	83.8	83.4	65.6	66.5	72.4
PE	56.1	56.4	56.8	55.2	50.1	59.7	58.4	41.8	40.7	44.8

Source: Baur et al. 1984; [a]Baur & Danilovs 1980   Continued

Table 171. HLA-C histocompatibility type: HLAC (cont'd)

Gene	ASIA		AFRICA	NORTH AMERICA			MID. AMERICA	SOUTH AMERICA
	China	Japan	Blacks[a]	Whites	Blacks	Am. Indians	Mexicans[a]	Am. Indians[a]
Cw_1	.128	.198	.007	.022	.024		.006	.155
Cw_2	.012	.002	.147	.022	.158	.083	.036	.012
Cw_3	.283	.264	.084	.151	.047	.265	.101	.488
Cw_4	.047	.034	.143	.075	.257	.270	.221	
Cw_5	.006	.002	.018	.073		.021	.068	.024
Cw_6	.127		.123	.079	.095	.021	.055	
Cw_7	.182	.111	.011	.304	.142	.066	.012	
Cw_8		.004		.055	.024		.018	
CX	.216	.385	.467	.218	.254	.275	.484	.320
n	172	556	143	358	42	48	84	42
H	80.5	72.9	71.7	81.6	81.2	76.9	69.8	63.5
PE	49.8	35.9	42.4	54.1	52.1	41.4	40.2	18.6

Source: Baur et al. 1984; [a]Baur & Danilovs 1980

Table 172. HLA-DQ histocompatibility type: HLADQ

	EUROPE						ASIA			NORTH AMERICA		
Allele	England	France	Germany	Italy	Scandi-navia	Spain	India	China	Japan	Whites	Blacks	Am. Indians
DQw1	.340	.297	.357	.331	.284	.266	.308	.203	.332	.325	.484	.066
DQw2	.232	.102	.175	.194	.126	.149	.177	.168	.013	.168	.242	.088
DQw3	.246	.133	.240	.246	.189	.228	.135	.164	.344	.295	.156	.599
DQX	.182	.469	.228	.230	.401	.357	.381	.466	.312	.212	.118	.247
n	174	128	250	314	204	125	68	98	535	316	35	45
H	73.7	66.4	73.2	73.9	70.7	72.8	71.0	68.7	67.4	73.4	66.9	56.8
PE	32.1	27.9	31.1	32.1	30.9	32.3	30.6	32.1	18.1	31.2	26.4	15.9

Source: Baur et al. 1984

Table 173. HLA-DR histocompatibility type: HLADR

	EUROPE								ASIA	
Allele	England	France	Germany	Italy	Nether-lands[a]	Scandi-navia	Spain	Switzer-land[a]	Israel: Ash-kenazi Jews[a]	India
DR1	.145	.089	.097	.070	.109	.111	.099	.090	.082	.015
DR2	.173	.166	.189	.164	.164	.178	.126	.142	.091	.202
DR3	.133	.065	.080	.114	.116	.163	.174	.097	.070	.142
DR4	.199	.113	.157	.082	.132	.135	.104	.122	.129	.022
DR5					.109			.103	.220	
DRw6					.007			.041	.023	
DR7	.133	.097	.118	.122	.132	.075	.119	.149	.146	.165
DRw8	.036	.040	.022	.034	.042	.026	.026	.017	.027	.030
DRw9	.006	.008	.022			.011	.009	.023	.019	
DRw10			.009	.006		.016	.017	.012	.008	.060
DRw11	.048	.115	.116	.212		.095	.155			.045
DRw12	.006		.018	.021		.021	.017			.031
DRw13	.048	.113	.053	.055		.074	.044			.030
DRw14	.030	.040	.062	.055		.016	.061			.090
DRX	.043	.150	.057	.066	.189	.081	.050	.203	.187	.168
n	166	124	225	328	73	190	115	87		66
H	86.5	88.6	88.5	87.7	86.4	88.3	88.6	87.1	85.7	86.6
PE	69.7	70.4	73.7	71.2	64.4	72.4	73.9	66.6	63.6	66.0

Source: Baur et al. 1984; [a] Baur & Danilovs 1980.

Continued

Table 173. HLA-DR histocompatibility type: HLADR (cont'd)

Allele	ASIA China	ASIA Japan	AFRICA Blacks[a]	NORTH AMERICA Whites	NORTH AMERICA Blacks	NORTH AMERICA Am. Indians	MID. AMERICA Mexicans[a]	SOUTH AMERICA Am. Indians[a]
DR1	.011	.062	.026	.106	.086	.024	.041	
DR2	.158	.155	.107	.120	.196	.024	.077	.328
DR3	.046	.006	.206	.108	.128	.075	.083	.036
DR4	.087	.243	.064	.158	.057	.470	.154	.229
DR5			.095				.147	
DRw6			.064				.029	.061
DR7	.133		.072	.119	.115	.024	.121	
DRw8	.032	.091	.048	.016		.048	.041	.244
DRw9	.161	.127	.026	.012	.011		.005	.012
DRw10	.011	.002	.029	.003		.024	.005	.012
DRw11	.063	.023		.103	.128	.048		
DRw12	.053	.061		.037		.235		
DRw13		.039		.045	.057			
DRw14	.032	.079		.072	.172			
DRX	.214	.113	.263	.101	.049	.029	.297	.078
n	95	514	138	332	35	42	88	42
H	86.7	86.4	85.0	89.4	86.9	71.0	83.5	76.9
PE	66.3	66.9	63.2	73.9	69.7	44.7	60.0	46.1

Source: Baur et al. 1984; [a] Baur & Danilovs 1980.

239

8 Immunoglobulin Systems

Table 174. Immunoglobulin A2m system: IGA2

Place/Population	1	2	n	H	Source
EUROPE					
CZECHOSLOVAKIA[b]	.985	.015	341	3.0	Schanfield et al. 1975b
ITALY (Ferrara)[b]	.988	.012	86	2.4	Piazza et al. 1976
NETHERLANDS (Leiden)[a]	.939	.061	798	11.5	Fraser et al. 1974
ASIA					
THAILAND (Bangkok)[b]	.527	.473	200	49.9	van Loghem et al. 1975
CHINA					
Chinese[a]	.430	.570	121	49.0	Kunkel et al. 1969
TAIWAN					
Aborigines[b]	.248	.752	300	37.3	Matsumoto et al. 1973
JAPAN					
Japanese[a]	.530	.470	163	49.8	Kunkel et al. 1969
AFRICA					
CENTRAL AFRICAN REPUBLIC					
Pygmies[b]	.279	.721	77	40.2	van Loghem et al. 1973
SOUTH AFRICA					
Bantu[b]	.125	.875	24	21.9	"
NORTH AMERICA					
Eskimos[b]	.857	.143	21	24.5	"
Amerindians[b]	.988	.012	40	2.4	"
UNITED STATES					
Whites[a]	.990	.010	177	2.0	Kunkel et al. 1969
Blacks[a]	.170	.830	68	28.2	"
OCEANIA					
AUSTRALIA					
Aborigines[a] (W. Desert)	1.000		70		Curtain et al. 1972
MELANESIA					
Papua New Guinea					
Papuans[b]	.552	.448	48	49.5	van Loghem et al. 1973
Western Dist. Gogodara NAN[a]	.667	.333	99	44.4	Curtain et al. 1972
Central Dist. Motu MN[a]	.293	.707	38	41.4	"
POLYNESIA					
Easter Is.[a]	.460	.540	59	49.7	Kunkel et al. 1969

Note: [a] Tested for A2m(1); [b] Tested for A2m(1,2).

Table 175. Immunoglobulin Gm1 system: IGHG1

Place/Population	Haplotype									n	H	Source
	3,5 13,14	1,17 21	1,2 17,21	1,3,5 13,14	1,13 17	1,5,13 14,17	1,5 6,17	1,5,6 14,17	Others			
EUROPE												
DENMARK[a]	.662	.214	.124							1,000	50.1	Nielsen 1961
LAND												
Finns (Restiina)[1]	.580	.207	.208		.005					100	57.7	Steinberg et al. 1974

Continued

Table 175. Immunoglobulin Gm1 system: IGHG1 (cont'd)

	Haplotype											
Place/Population	3,5 13,14	1,17 21	1,2 17,21	1,3,5 13,14	1,13 17	1,5,13 14,17	1,5 6,17	1,5,6 14,17	Others	n	H	Source
Lapps/Saami[l]	.453	.497	.044	.006						254	54.6	"
FRANCE												
French[a]	.697	.213	.090							1,451	46.1	Ropartz et al. 1965
Basque[d]	.686	.237	.077							94	46.7	Steinberg & Cook 1981(c)
GERMANY[c]												
(Hessen)	.697	.188	.102						.013	2,000	46.8	Wiebecke et al. 1967
ICELAND[g]	.571	.240	.189							395	58.1	Bjarnason et al. 1973
IRELAND[a]	.639	.217	.144							294	52.4	Palsson et al. 1970
ITALY[a] (Ferrara)	.776	.188	.037							87	36.1	Vierucci 1965
NETHERLANDS[w]												
(Leiden)	.700	.187	.102		.004				.007	709	46.5	Fraser et al. 1974
NORWAY[f]	.611	.236	.142	.010						612	55.1	Gedde-Dahl et al. 1971
POLAND[a] (Kracow)	.806	.138	.056							600	32.8	Socha & Kaczera 1968
PORTUGAL[j]	.685	.210	.076	.030						419	48.0	Pereira & Manso 1975
SWEDEN[b]	.641	.354		.004						560	56.4	Màrtensson 1964
SWITZERLAND[a]	.628	.252	.120								52.8	
UNITED KINGDOM												
England[g]	.652	.229	.119							1,000	50.8	Brazier & Goldsmith 1968

ASIA

	Haplotype											
Place/Population	3,5 13,14	1,17 21	1,2 17,21	1,3,5 13,14	1,13 17	1,5,13 14,17	1,5 6,17	1,5,6 14,17	Others	n	H	Source
ISRAEL												
Ashkenazi Jews (USA)[t]	.798	.138	.046		.008	.010				248	34.2	Steinberg 1973
Yemenite Jews[g]	.633	.252	.020		.017	.078				75	52.9	Godber et al. 1973
IRAN[a] (Tehran)	.679	.183	.062	.076						374	49.6	Bajatzadeh & Walter 1969c
INDIA												
Punjabi[u] (N. India)	.396	.279	.149	.149	.022	.011				101	72.2	Daveau et al. 1980
Tamil[a] (S. India)	.223	.543	.234							132	60.1	Kirk et al. 1962
Irula tr.[d] (Nilgiri Hills, S. India)	.527	.361	.112							74	57.9	Vos et al. 1963
SRI LANKA												
Sinhalese[u]	.385	.327	.243	.045						119	68.4	Daveau et al. 1980
Veddah[d]	.596	.364	.040							52	51.1	Vos et al. 1963
BHUTAN[g]		.506	.189	.172	.125	.008				150	66.3	Mourant et al. 1969
MALAYSIA[o]												
Malays (Singapore)		.135	.087	.732	.023	.023				185	43.7	Daveau et al. 1980
THAILAND[d] (Bangkok)		.221	.073	.706						163	44.7	Vos et al. 1963
PHILIPPINES												
Filipino (Honolulu)[d]	.269	.080	.044	.606						138	55.2	Ropartz et al. 1964
Negritos (Zambales)[v]		.329	.097	.555	.020					127	57.4	Steinberg & Cook 1981(c)
CHINA												
Chinese[a] (N. China)		.420	.137	.444						161	60.8	Nakajima & Ohkura 1971
Chinese[a] (S. China)		.191	.096	.713						143	44.6	"
Chinese[o] (Singapore)		.213	.066	.640	.057	.024				181	53.7	Daveau et al. 1980
JAPAN												
Japanese[a] (Hokkaido)		.451	.221	.102	.226					87	68.6	Steinberg & Kageyama 1970
Ainu[a] (")		.511	.077	.026	.273				.113	159	64.5	"
KOREA[q]		.507	.213	.141	.138					195	65.9	Schanfield et al. 1972

AFRICA

	Haplotype											
Place/Population	3,5 13,14	1,17 21	1,2 17,21	1,3,5 13,14	1,13 17	1,5,13 14,17	1,5 6,17	1,5,6 14,17	Others	n	H	Source
LIBYA[a] (Tripoli & Benghasi)	.536	.292	.049	.124						168	61.0	Walter et al. 1975

Continued

Table 175. Immunoglobulin Gm1 system: IGHG1 (cont'd)

Place/Population	Haplotype									n	H	Source
	3,5 13,14	1,17 21	1,2 17,21	1,3,5 13,14	1,13 17	1,5,13 14,17	1,5 6,17	1,5,6 14,17	Others			
EGYPT[r] (Cairo)	.525	.100	.023			.293	.060			244	62.4	Steinberg & Cook 1981(c)
SUDAN												
Beja[g]	.235	.231	.021			.513				98	62.8	el Hassan et al. 1968
CENTRAL AFRICAN REPUBLIC												
Babinga Pygmies[i]						.852	.148			162	25.2	Cavalli-Sforza et al. 1969
NIGERIA												
Yoruba (Ibaden)[d]						.697	.303			35	42.2	Steinberg 1966
TANZANIA												
Sandawe[h]	.009		.009			.983				176	3.4	Godber et al. 1976
Nyaturu[h]	.015		.007			.978				67	4.3	"
SOUTHERN AFRICA												
San/Bushmen												
Dobe !Kung		.081			.603	.288	.004	.024		394	54.6	Steinberg et al. 1975
Khoi/Hottentot												
Sesfontein			.048		.333	.429	.027	.163		42	67.5	"
Negroes/Dama[a]												
(Okombahe)		.020			.014	.592	.272		.102	101	56.5	Nurse et al. 1985
Bantu/Zulu[k]					.191	.509	.159	.141		130	65.9	Jenkins et al. 1970

NORTH AMERICA

Place/Population	3,5 13,14	1,17 21	1,2 17,21	1,3,5 13,14	1,13 17	1,5,13 14,17	1,5 6,17	1,5,6 14,17	Others	n	H	Source
GREENLAND[a]												
Eskimo	.264	.656	.018		.062					144	49.6	Steinberg et al. 1974
ALASKA												
Eskimo[d]												
(Wainwright)		.864	.030	.105						50	24.2	Steinberg et al. 1961
Eskimo[p]												
(St.Lawrence Is.)	.007	.681	.007		.305					221	44.3	Ferrell et al. 1981b
Athabaskan Indians (Fort Yukon)		.808	.172	.020						51	31.7	Steinberg et al. 1961
CANADA												
Eskimo (Igloolik)[m]	.041	.782	.005		.171					365	35.8	McAlpine et al. 1974
Ojibwa Indians (Pikangikum)[q]		.860	.071		.069					102	25.1	Szathmary et al. 1974
Dogrib Indians (NW Territories)[p]	.010	.790	.066		.135					156	35.3	Szathmary et al. 1983

MIDDLE AMERICA

Place/Population	3,5 13,14	1,17 21	1,2 17,21	1,3,5 13,14	1,13 17	1,5,13 14,17	1,5 6,17	1,5,6 14,17	Others	n	H	Source
GUATEMALA												
Black Caribs[x] (Livingston)	.011	.134	.131			.495	.096	.118		187	69.7	Crawford et al. 1981

SOUTH AMERICA

Place/Population	3,5 13,14	1,17 21	1,2 17,21	1,3,5 13,14	1,13 17	1,5,13 14,17	1,5 6,17	1,5,6 14,17	Others	n	H	Source
BOLIVIA												
Aymara[p] (Altiplano)	.017	.902	.047			.033				528	18.3	Chakraborty et al. 1985
BRAZIL												
Cayapo[l]/ Txukahme		.709	.288	.003						154	41.4	Salzano et al. 1973
VENEZUELA												
Makiritare	.003	.563	.435							718	49.4	Gershowitz & Neel 1978
Yanomama[p]		.852	.148							3,447	25.2	"

OCEANIA

Place/Population	3,5 13,14	1,17 21	1,2 17,21	1,3,5 13,14	1,13 17	1,5,13 14,17	1,5 6,17	1,5,6 14,17	Others	n	H	Source
AUSTRALIA												
Aborigines[l] (N. Territory)		.636	.265		.099					96	51.5	Steinberg & Kirk 1970

Continued

Table 175. Immunoglobulin Gm1 system: IGHG1 (cont'd)

Place/Population	Haplotype									n	H	Source
	3,5 13,14	1,17 21	1,2 17,21	1,3,5 13,14	1,13 17	1,5,13 14,17	1,5 6,17	1,5,6 14,17	Others			
MELANESIA												
New Guinea[1]												
Huli (S. Highlands)	.507	.067	.426							149	55.7	Steinberg et al. 1972b
Solomon Is.[1]												
Aita Nan	.490	.091	.419							307	57.7	Steinberg et al. 1972a
MICRONESIA												
Caroline Is. (Pingelap)	.053	.011	.927			.009				409	13.8	Steinberg & Morton 1973

Note: Tested for GM
a. (1,2,5) b. (1,3,5) c. (1,2,3,5) d. (1,2,5,6) e. (1,2,3,5,10) f. (1,2,3,5,21) g. (1,2,3,5,10,11)
h. (1,2,3,5,11,13) i. (1,2,3,5,6,11,13) j. (1,2,3,5,10,13,21) k. (1,2,3,5,6,13,14) l. (1,2,3,5,6,13,14,21)
m. (1,2,3,5,6,13,16,21) n. (1,2,3,5,10,11,17,21) o. (1,2,3,5,10,11,13,21) p. (1,2,3,5,11,13,16,21)
q. (1,2,3,5,6,11,16,21,24) r. (1,2,3,5,6,13,14,17,21) s. (1,2,3,5,6,13,16,21,24) t. (1,2,3,5,6,13,14,15,16,21)
u. (1,2,3,5,10,11,13,14,21,23) v. (1,2,3,5,11,13,14,15,16,17,21,24) w. (1,2,3,5,6,10,11,13,15,16,17,21,23,24)
x. (1,2,3,5,6,10,11,13,15,16,17,24).

Table 176. Immunoglobulin Km (Inv) system: IGKC, KM

Place/Population	1 & 1,2[d]	3	n	H	Source
EUROPE					
DENMARK[b]	.095	.905	105	17.2	Ropartz et al. 1970
FINLAND					
Finns[a] (Ristinna)	.046	.954	100	8.8	Steinberg et al. 1974
Lapps[c]	.173	.827	142	28.6	"
FRANCE[b] (Normandy)	.107	.893	331	19.1	Ropartz et al. 1965
GERMANY[a] (Hessen)	.069	.931	2,000	12.8	Wiebecke et al. 1967
GREECE[a]	.065	.935	256	12.2	Archimandris et al. 1975
HUNGARY[a] (Budapest)	.073	.927	184	13.5	Schanfield et al. 1975a
ICELAND[a] (Dalasyla)	.092	.908	193	16.7	Pálsson & Walter 1967
IRELAND[a]	.076	.924	294	14.0	Pálsson et al. 1970
ITALY (Ferrara)[a]	.085	.915	86	15.6	Piazza et al. 1976
NETHERLANDS (Leiden)[b]	.094	.906	798	17.0	Fraser et al. 1974
NORWAY[a]	.072	.928	87	13.4	Gedde-Dahl & Berg 1965
POLAND[a] (Silesia)	.070	.930	1,051	13.0	Schlesinger 1968
SWITZERLAND[a]	.080	.920	98	14.7	Steinberg & Cook 1981(c)
UNITED KINGDOM					
England[b]	.092	.908	500	16.7	Brazier & Goldsmith 1968
ASIA					
TURKEY[a]	.074	.926	274	13.8	Hummel et al. 1970
ISRAEL					
Yemenite Jews[b]	.098	.902	75	17.7	Godber et al. 1973
Ashkenazi Jews[a] (USA)	.037	.963	248	7.1	Steinberg 1973
IRAN (Teheran)	.150	.850	354	25.5	Bajatzadeh & Walter 1969c
INDIA					
Brahmins[a] (Kumaon, N. India)	.103	.897	102	18.5	Chopra 1970
THAILAND[a] (Central)	.186	.814	887	38.3	Vogel et al. 1971
PHILIPPINES					
Filipino[b] (Honolulu)	.352	.649	126	45.5	Ropartz et al. 1964
Negritos[a] (Zambales, Luzon)	.163	.837	127	27.3	Steinberg & Cook 1981(c)
CHINA					
Chinese[a] (Northern)	.416	.584	161	48.6	Nakajima & Ohkura 1971
Chinese (Southern)	.342	.658	143	45.0	"

Continued

Table 176. Immunoglobulin Km (Inv) system: IGKC, KM (cont'd)

Place/Population	1 & 1,2[d]	3	n	H	Source
JAPAN					
Japanese[a] (Hokkaido)	.339	.661	87	44.8	Steinberg & Kageyama 1970
Ainu[a] (")	.168	.832	159	28.0	
KOREA					
Koreans[a] (Germany)	.290	.710	115	41.2	Bajatzadeh & Walter 1969a

AFRICA

EGYPT					
Egyptians[a] (Cairo)	.249	.751	245	37.4	Steinberg & Cook 1981(c)
CENTRAL AFRICAN REPUBLIC					
Babinga Pygmies[b]	.356	.644	164	45.9	Cavalli-Sforza et al. 1969
SOUTHERN AFRICA					
San[a]/Bushmen					
Dobe !Kung	.402	.598	394	48.1	Steinberg et al. 1975
Khoi[a]/Hottentot					
Keetmanshoop Nama	.562	.438	52	49.2	Jenkins 1972
Bantu[a]/Zulu	.358	.642	131	46.0	Jenkins et al. 1970

NORTH AMERICA

GREENLAND					
Eskimos[a] (Upernavik Dist.)	.138	.862	144	23.8	Steinberg et al. 1974
ALASKA					
Eskimos[c] (St. Lawrence Is.)	.330	.670	220	44.2	Ferrell et al. 1981b
CANADA					
Eskimos[a] (Igloolik)	.269	.731	365	39.3	McAlpine et al. 1974
Ojibwa Indians[a] (Pikangikum)	.286	.714	102	40.8	Szathmary et al. 1974
Dogrib Indians[c] (NW Territories)	.596	.404	156	48.2	Szathmary et al. 1983
UNITED STATES					
Whites[c] (Claxton, Georgia)	.069	.931	295	12.8	Blumberg et al. 1964
Blacks[c] (")	.393	.607	187	47.7	"

MIDDLE AMERICA

GUATEMALA					
Black Caribs (Livingston)	.401	.599	187	48.8	Crawford et al. 1981

SOUTH AMERICA

BOLIVIA					
Aymara[c] (Altiplano)	.268	.732	529	39.2	Chakraborty et al. 1985
BRAZIL					
Cayapo[a]/Txukahame	.497	.503	154	50.0	Salzano et al. 1973
Makiritare[c] (S. Brazil &					
N. Venezuela)	.574	.426	715	49.0	Gershowitz & Neel 1978
Yanomama[c] (")	.387	.613	3,447	47.4	"

OCEANIA

AUSTRALIA					
Aborigines[c] (N. Territory)	.208	.792	104	32.9	Steinberg & Kirk 1970
MELANESIA					
New Guinea					
Nan speakers[a]	.115	.885	721	28.4	Giles et al. 1965b
Solomon Is.					
Malaita	.221	.779	451	34.4	Steinberg et al. 1972a
MICRONESIA					
Caroline Is.[a] (Pingelap)	.253	.747	409	37.8	Steinberg & Morton 1973

Note: [a] Tested for Km(1); [b] Tested for Km(1,2); [c] Tested for Km(1,3); [d] Pooled frequencies of Km^1 and $Km^{1,2}$.

9 DNA Polymorphisms

Table 177. Albumin: ALB

	PstI(5'F) (14, 24/18)	MspI (13, 18)	PstI(5'I) (14/18, 24)	HaeIII(5'I) (3.6, 3.8)	HaeIII(M) (4.05, 4.1)	SacI (16, 20)	HaeIII(3'I) (.77, .9)	EcoRV (6.2, 9.0)	Cauca- sians	Blacks	Japanese & Chinese
			Haplotype (Fragment lengths in kb)								
1	-	+	-	+	+	+	+	-	.04	.12	
2	-	-	+	+	-	-	+	-	.49	.35	.68
3	+	-	+	-	+	+	-	-	.32	.46	.32
4	+	-	+	-	+	+	-	+	.14		
5	+	-	+	+	-	-	+	-	.01		
6	+	-	+	+	+	+	+	-	.08	.08	
7	-	-	+	-	+	+	-	-	.01		
n*									110	26	22
H									63.6	64.5	43.5
PE									38.0	38.5	17.0

Source: Murray et al. 1984b
Note: 5'F = 5' flanking, 5'I = 5' internal, M = midportion position, and 3'I = 3'internal. + and - denote the presence and absence of a restriction site, respectively.

Table 178. Alpha-1-antitrypsin; protease inhibitor: PI

Place/ Population	Enzyme	Alleles (kb)	Frequency +	Frequency -	n*	H	PE	Source
Caucasians	SstI	1.8, 1.9	.65	.35	64	45.5	17.6	Cox et al. 1985
	MspI	0.95, 0.98	.51	.49	64	50.0	18.7	"
	AvaII	0.9, 1.1	.64	.36	64	46.1	17.7	"

Note: Linkage disequilibrium not significant.

Table 179. Alpha-fetoprotein: AFP

Place/ Population	Enzyme	Alleles (kb)	Frequency +	Frequency -	n*	H	PE	Source
Caucasians	AvaII	1.5, 2.5	.42	.58	100	48.7	18.4	Murray et al. 1985
(USA)	PstI	12, 15	.88	.12	100	21.1	9.4	"

Note: Not polymorphic for BamHI, BclI, BglI, EcoRI, EcoRV, HaeIII, HgiAI, HincII, HindIII, HinfI, KpnI, MspI, MstII, PvuII, RsaI, SacI, SauIIIA, StuI, TaqI, XbaI, XmnI when 10 unrelated Caucasians were studied.

Table 180. Alpha-globin gene cluster: HBAC, AGC

Place/Population	XbaI					SacI					BglI				
	+	-	n*	H	PE	+	-	n*	H	PE	+	-	n*	H	PE
ENGLAND	.63	.37	30	46.6	17.9	.76	.24	38	36.5	14.9	.17	.83	42	28.2	12.1
MEDITERRANEAN	.53	.47	68	49.8	18.7	.80	.20	78	32.0	13.4	.13	.87	72	22.6	10.0
SAUDI ARABIA	.47	.53	32	49.8	18.7	.47	.53	38	49.8	18.7	.16	.84	38	26.9	11.6
INDIA	.61	.39	74	47.6	18.1	.78	.22	74	34.3	14.2	.11	.89	74	19.6	8.8
SOUTHEAST ASIA	.41	.59	32	50.0	18.7	.69	.31	32	42.8	16.8	.19	.81	32	30.8	13.0
NIGERIA	.50	.50	36	50.0	18.8	.42	.58	38	48.7	18.4	.03	.97	38	5.8	2.8
JAMAICA	.54	.46	50	49.7	18.7	.48	.52	96	49.9	18.7	.07	.93	72	13.0	6.1
MELANESIA IS	.63	.37	40	46.6	17.9	.08	.92	40	14.7	6.8	.38	.62	34	47.1	18.0
PAPUA NEW GUINEA	.40	.60	60	48.0	18.2	.02	.98	58	3.9	1.9	.58	.42	62	48.7	18.4

Table 180. Alpha-globin gene cluster: HBAC, AGC (cont'd)

Place/Population	IZHVR						PZ/Z					AccI				
	S	M	L	n*	H	PE	PZ	Z	n*	H	PE	+	-	n*	H	PE
ENGLAND	.02	.68	.30	40	44.7	19.1	.81	.19	42	30.8	13.0	.79	.21	42	33.2	13.8
MEDITERRANEAN	.08	.62	.30	78	51.9	25.5	.85	.15	78	25.5	11.1	.84	.16	64	26.9	11.6
SAUDI ARABIA	.11	.76	.13	38	39.3	20.6	.71	.29	38	41.2	16.4	.67	.33	36	44.2	17.2
INDIA	.08	.59	.33	74	53.7	26.2	.86	.14	74	24.1	10.6	.85	.15	74	25.5	11.1
SOUTHEAST ASIA	.16	.41	.43	32	62.1	32.9	.79	.21	29	33.2	13.8	.72	.28	25	40.3	16.1
NIGERIA	.21	.79		38	33.2	13.8	.55	.45	38	49.5	18.6	.90	.10	30	18.0	8.2
JAMAICA	.18	.81	.02	93	31.1	14.7	.48	.52	66	49.9	18.7	.74	.26	98	38.5	15.5
MELANESIA IS	.50	.50		40	50.0	18.8	.58	.42	40	48.7	18.4	.58	.42	40	48.7	18.4
PAPUA NEW GUINEA	.36	.64		58	46.1	17.7	.43	.57	58	49.0	18.5	.47	.53	62	49.8	18.7

Table 180. Alpha-globin gene cluster: HBAC, AGC (cont'd)

Place/Population	RsaI					PstI(1)					PstI(2)				
	+	-	n*	H	PE	+	-	n*	H	PE	+	-	n*	H	PE
ENGLAND	.48	.52	42	49.9	18.7	.08	.92	38	14.7	6.8	.11	.89	38	19.6	8.8
MEDITERRANEAN	.45	.55	74	49.5	18.6	.04	.96	78	7.7	3.7	.04	.96	72	7.7	3.7
SAUDI ARABIA	.32	.68	31	43.5	17.0	.03	.97	34	5.8	2.8	.06	.94	18	11.3	5.3
INDIA	.43	.57	75	49.0	18.5	.04	.96	72	7.7	3.7	.03	.97	60	5.8	2.8
SOUTHEAST ASIA	.31	.69	29	42.8	16.8	.03	.97	31	5.8	2.8	.03	.97	32	5.8	2.8
NIGERIA	.29	.71	79	41.2	16.4	.08	.92	38	14.7	6.8	.13	.87	38	22.6	10.0
JAMAICA	.19	.81	108	30.8	13.0	.04	.96	74	7.7	3.7	.08	.92	38	14.7	6.8
MELANESIA IS	.03	.97	40	5.8	2.8	.43	.57	40	49.0	18.5	.45	.55	20	49.5	18.6
PAPUA NEW GUINEA		1.00	60			.49	.51	77	50.0	18.7	.40	.60	58	48.0	18.2

Source: Higgs et al. 1986
Note: IZHVR refers to interzeta hypervariable region between the ζ_2 and $\Psi\zeta_1$ genes. L, M, and S refer to large, medium and small alleles, respectively. PZ and Z refer to the presence of pseudozeta and zeta-like sequence, respectively.

246

Table 181. Amylase, alpha, salivary: AMY1

Place/ Population	Enzyme	Alleles (kb)	Frequency +	Frequency −	n*	H	PE	Source
Japanese	PstI/BamHI[a]	5.7, 6.5	.55	.45	44	49.5	18.6	Ishizaki et al. 1985

Note: [a] Digested with two enzymes. Not polymorphic for BamHI, EcoRI, HindIII, and XhoI.

Table 182. Antithrombin III: AT3

Place/ Population	Enzyme	Alleles (kb)	Frequency +	Frequency −	n*	H	PE	Source
French	PstI	5/5.5, 10.5	.58	.42	50	48.7	18.4	Paslier et al. 1985
Greeks	"	"	.57	.43	44	49.0	18.5	Prochownik et al. 1983
Italians	"	"	.47	.53	58	49.8	18.7	"
Indians	"	"	.34	.66	38	44.9	17.4	"
Blacks	"	"	.59	.41	44	48.4	18.3	"

Table 183. Apolipoprotein A-I: APOA1

Place/ Population	Enzyme	Alleles (kb)	Frequency +	Frequency −	n*	H	PE	Source
Caucasians	ApaI	2.1, 3.5	.58	.42	600	48.7	18.4	Frossard et al. 1986

Note: Not polymorphic for AvaI, AvaII, BamHI, BanI, BanII, BglII, DraI, EcoRI, EcoRV, HaeIII, HgiAI, HinfI, HphI, NarI, NciI, NdeI, PstI, and RsaI.

Table 184. Apolipoprotein A-II: APOA2

Place/ Population	Enzyme	Alleles (kb)	Frequency +	Frequency −	n*	H	PE	Source
Caucasians	MspI	3.0, 3.7	.81	.19	174	30.8	13.0	Scott et al. 1985

Table 185. Apolipoprotein A-IV: APOA4

Place/Population	Enzyme	Alleles (kb)	Frequency +	Frequency −	n*	H	PE	Source
Caucasians	TaqI	1.6/2.0, 3.6	.98	.02	130	3.9	1.9	Coleman et al. 1986
	XbaI	9.8, 22	.77	.23	130	35.4	14.6	"
Mediterraneans	BglII	5.8, 6.4	.67	.33	132	44.2	17.2	Oettgen et al. 1986b
Blacks	"	5.8, 6.4	.26	.74	72	38.2	15.4	"

Note: A large portion of data for Mediterraneans and Blacks overlap with those in Table 189.1.

Table 186. Apolipoprotein B: APOB

Place/Population	Enzyme	Alleles (kb)	Frequency +	Frequency −	n*	H	PE	Source
Caucasians	HincII	7.1, 7.5	.12	.88	90	21.1	9.4	Darnfors et al. 1986
	PvuII	5.6, 8.1	.08	.92	88	14.7	6.8	"

Table 187.   Apolipoprotein C-II: APOC2

Haplotype (Fragment lengths in kb)	BglI (9, 12)	Taq I (3.5, 3.8)	Caucasians
1	−	−	.098
2	−	+	.348
3	+	−	.533
4	+	−	.022
n*			92
H			58.5
PE			30.9

Source: Wallis et al. 1984

Table. 188. Apolipoprotein C-III: APOC3

Place/Population	Enzyme	Alleles (kb)	Frequency +	Frequency −	n*	H	PE	Source
Mediterraneans	PvuII	3.6, 4.3	.07	.93	132	13.0	6.1	Oettgen et al. 1986a
Blacks	"	"		1.00	72			"

Note: Not polymorphic for BamHI, EcoRI, HindIII, BglII and KpnI. A large portion of data in this table overlap
      with those in Table 189.1.

Table 189.   Apolipoprotein AI-CIII-AIV: APOA1-APOC3-APOA4

Place/ Population	Enzyme	Alleles (kb)	Frequency +	Frequency -	n^*	H	PE	Source
Caucasians	TaqI	4.8, 8.6	.72	.28	88	40.3	16.1	Cohen et al. 1986
	AvaI	6, 15	.32	.68	34	43.5	17.0	"

Note: A large portion of data in this table overlap with those in Table 189.1.

Table 189.1 Apolipoprotein AI-CIII-AIV: APOA1-APOC3-APOA4

	Haplotype											Mediter- raneans	Amer. Blacks
	TaqI	XmnI	MspI	PstIA	SacI	PvuII	XbaIA	PstIB	PstIC	XbaIB	BglII		
1	+	-	+	+	-	-	+	-	+	-	+	.326	.119
2	+	-	+	+	-	+	-	-	-	-	+	.062	
3	+	-	-	+	+	-	+	+	-	+	-	.046	.015
4	+	+	+	+	-	-	+	-	-	-	-	.015	.239
5	+	-	+	+	-	+	-	-	-	-	-	.008	
6	+	-	+	+	-	-	+	-	+	+	+	.008	
7	+	+	+	+	-	-	+	-	+	-	+	.078	.015
8	+	-	+	+	-	-	+	-	-	+	+	.015	
9	+	-	+	+	+	-	-	+	-	-	-	.008	
10	+	-	+	-	-	-	+	-	+	-	+	.008	
11	+	-	+	+	-	-	+	-	-	-	-	.046	.075
12	+	-	+	+	-	-	-	-	-	-	+	.078	.075
13	-	-	+	+	-	-	+	-	-	-	+	.008	
14	+	-	+	+	-	-	+	+	-	+	-	.023	.030
15	-	-	+	+	-	-	+	-	-	-	-	.015	.194
16	+	-	+	-	-	-	+	-	-	-	-	.054	.015
17	-	-	+	+	-	-	-	-	+	-	+	.008	
18	+	-	+	+	-	-	-	-	+	-	+	.023	
19	+	-	+	+	-	-	+	-	-	-	+	.031	.015
20	+	-	+	+	+	-	+	+	-	+	-	.023	
21	+	-	+	+	-	-	-	-	-	-	-	.039	
22	+	+	+	+	-	-	+	-	+	+	+	.008	
23	-	-	+	-	-	-	+	-	-	-	-	.008	
24	+	-	+	+	+	-	+	-	+	+	-	.008	
25	+	-	-	+	-	-	+	-	+	-	+	.008	
26	+	+	+	+	-	-	-	-	-	-	+	.008	
27	+	-	+	+	-	-	+	-	+	-	-	.015	
28	+	+	+	+	-	-	+	-	+	-	-	.015	
29	-	+	+	+	-	-	+	-	+	-	-	.008	
30	+	+	+	-	-	-	+	-	-	-	+		.015
31	-	+	+	+	-	-	+	-	-	-	+		.015
32	+	-	+	+	-	-	+	-	-	+	-		.030
33	-	-	+	+	+	-	-	-	-	-	-		.015
34	+	+	+	+	-	-	+	-	-	+	-		.045
35	-	-	+	+	-	-	+	+	-	-	-		.015
36	-	-	+	+	-	-	+	-	-	+	-		.015
37	+	+	+	+	-	-	+	+	-	-	-		.015
38	-	-	-	+	+	-	+	-	-	-	-		.015
39	-	+	+	+	-	-	+	-	-	-	-		.015
40	+	-	-	+	-	-	+	+	-	-	-		.015
n^*												129	67
H												87.5	87.3
PE												75.5	75.5

Source: Antonarakis et al. 1986
Note: + and - denote the presence and absence of a restriction site, respectively.

Table 190. Beta-globin gene cluster (5'ϵ to 3'η): HBBC, BGC

	Haplotype					British	Cypriot	Itali-ans	Asian Indians	Thais	Chinese[a]	Afri-cans	Melan-esians	Polyn-esians
	HcII	HdIII	HdIII	HcII	HcII									
1	+	-	-	-	-	.43	.72	.64	.52	.91	.79	.05	.68	.78
2	-	+	-	+	+	.41	.17	.30	.25		.06	.10	.17	.11
3	-	+	+	-	+	.13	.10	.04	.14	.06	.12	.02	.02	
4	+	-	-	+	+								.06	
5	-	+	+	-	-	.03	.01	.02	.03					
6	-	+	+	+	+								.03	.02
7	+	+	-	+	+				.03				.01	.02
8	-	-	-	-	-				.01					
9	+	+	+	-	+				.01					
10	-	-	-	+	+				.01	.03	.02			
11	+	+	-	-	-				.01					
12	-	+	-	-	-							.03		
13	-	+	-	-	+							.20	.02	.07
14	-	-	-	-	+							.61	.01	
n*						37	82	50	111	32	80	61	173	55
H						62.9	44.3	49.8	64.5	16.7	35.8	57.4	50.3	37.4
PE						34.8	23.5	24.0	41.5	8.4	18.5	35.9	30.1	20.9

Source: Wainscoat et al. 1986; [a]Chan et al. 1986
Note: + and - denote the presence and absence of a restriction site, respectively.

Table 191. Coagulation factor VIIIC; hemophilia A: F8C, HEMA

Place/Population	Enzyme	Alleles (kb)	Frequency +	Frequency -	n*	H	Source
Caucasians	BclI	0.90, 1.17	.71	.29	133	41.2	Gitschier et al. 1985

Table 192. Coagulation factor IX; hemophilia B: F9, HEMB

Place/Population	Enzyme	Alleles (kb)	Frequency +	Frequency -	n*	H	Source
Caucasians	MspI	2.40, 5.8	.80	.20	74	32.0	Camerino et al. 1985
Japanese	"	"	1.00		81		Kojima et al. 1987
Caucasians	TaqI	1.30, 1.8	.29	.71	73	41.2	Camerino et al. 1984
Japanese	"	"	1.00		81		Kojima et al. 1987
Caucasians	XmnI	6.50, 11.5	.29	.71	72	41.2	Winship et al. 1984
Japanese	"	"	1.00		81		Kojima et al. 1987
Caucasians	Hinf1/DdeI	0.75, 0.8	.76	.24	74	14.9	Winship et al. 1984
Japanese	"	"	1.00		81		Kojima et al. 1987

250

Table 193. Collagen I alpha-1 polypeptide: COL1A1

Place/ Population	Enzyme	Alleles (kb)	Frequency +	Frequency -	n*	H	PE	Source
Caucasians	HindIII	7.0/6.9 13.9	.28	.72	34	40.3	16.1	Driesel et al. 1982b

Table 194. Collagen I alpha-2 polypeptide: COL1A2

Place/ Population	Enzyme	Alleles (kb)	Frequency +	Frequency -	n*	H	PE	Source
Caucasians	MspI	.5/1.6, 2.1	.86	.14	188	24.1	10.6	Tsipouras et al. 1984

Table 194.1 Collagen II alpha-2 polypeptide: COL2A2

	Haplotype BamHI	Haplotype EcoRI	Haplotype HindIII	Caucasians	Blacks
1	+	+	-	.617	.525
2	+	+	+	.367	.400
3	-	+	-		.025
4	+	-	-	.017	.050
n*				60	40
H				48.4	56.1
PE				20.0	27.4

Source: Eng & Storm 1985

Table 195. Growth hormone gene cluster: GHC

	Haplotype (Fragment lengths in kb) HincII (4.5, 6.7)	MspIA (3.6, 4.3)	MspIB (3.3, 3.9)	BglIIA (10.5, 13)	Northern Europeans	Mediter- raneans	U.S. Blacks
1	+	+	+	+	.03		.17
2	+	+	+	-	.20	.44	.12
3	+	-	-	+	.03		.04
4	-	+	+	+	.13	.12	.50
5	-	+	-	+	.10		
6	-	+	+	-	.03		.04
7	-	-	+	+	.03		.04
8	-	-	-	+	.43	.44	.08
n*					30	16	24
H					74.5	59.8	69.6
PE					52.6	30.5	48.2

Source: Chakravarti et al. 1984
Note: + and - denote the presence and absence of a restriction site, respectively.

Table 196. Hypoxanthine phosphoribosyltransferase: HPRT

Place/ Population	Enzyme	Allele (kb)			n*	H	Source
		12/25	22/18	22/25			
Caucasians	BamHI	.16	.07	.77	83	37.7	Nussbaum et al. 1983

Table 197. Immunoglobulin, heavy chain constant region of IgG (BamHI enzyme used): IGHG

	Haplotype (Fragment lengths in kb)	Caucasians (US)	Tunisians[a]
1	13.5, 12.5, 11.8, 9.4, 8.8	.44	.27
2	25.0, 12.5, 11.8, 10.0, 9.0	.46	.40
3	13.5, 12.5, 11.8, 9.0, 8.8	.02	.05
4	25.0, 12.5, 11.8, 10.0, 9.4	.07	.01
5	13.5, 12.5, 11.8, 10.0, 9.0		.19
6	13.5, 12.5, 11.8, 10.0, 9.4		.05
7	12.5, 11.8		.03
n*		41	78
H		59.0	72.5
PE		29.3	49.3

Source: Bech-Hansen et al. 1983; [a] Chaabani et al. 1986.

Table 198. Immunoglobulin, heavy chain constant region of IgM: IGHM

Enzyme	Allele (kb)	Caucasians	Enzyme	Allele (kb)	Caucasians
Sst A	7.5	.005	Sst D	3.7	.055
	7.4	.658		(*)[c]	.945
	6.9	.163			
				n*	184
	6.8	.130			
	D1-4[a]	.044		H	10.4
	n*	184		PE	4.9
	H	52.2	Sst E	2.75	.016
				2.7	.468
	PE	30.2		2.6	.005
				2.2	.511
Sst B	6.5	.011			
	6.3	.907		n*	184
	6.0	.082			
				H	52.0
	n*	184		PE	21.4
	H	17.1	Sst F	1.0	.995
				0.9	.005
	PE	8.2			
				n*	184

Continued

Table 198. Immunoglobulin, heavy chain constant region of IgM: IGHM (cont'd)

Enzyme	Allele (kb)	Caucasians	Enzyme	Allele (kb)	Caucasians
Sst C	5.5	.049			
	5.2	.005	H	1.0	
	4.9	.022			
	4.8	.908	PE	.5	
	4.1	.011			
	Null[b]	.005			
	n*	184			
	H	17.2			
	PE	8.0			

Source: Migone et al. 1983
Note: [a] Four different variants characterized by duplication. [b] "Null" allele. [c] An allele corresponding to a hypothetical band.

Table 199. Insulin: INS

	Haplotype					Whites	Blacks	Pima Indians
	RsaI	TaqI	5'FP	HincII	SacI	(US)	(US)	(US)
1	+	+	1	+	-	.114	.429	.333
2	-	+	1	+	-		.214	.433
3	+	-	1	+	-	.028	.071	.067
4	+	+	1	-	-	.286		
5	-	+	1	-	-	.114		
6	+	+	2	+	-	.057		
7	+	-	2	+	-	.028		
8	+	+	2	+	+	.057		
9	-	+	2	+	+	.028		
10	+	+	3	+	-		.071	.067
11	+	-	3	+	-		.036	
12	-	+	3	+	-	.114	.179	.100
13	+	+	3	-	-	.114		
14	-	+	3	-	-	.028		
15	+	-	3	-	-	.028		
n*						35	56	30
H						85.6	72.7	68.3
PE						71.8	50.6	43.5

Source: Chakravarti et al. 1986
Note: The length polymorphism (5'FP) was detected by using PvuII. + and - denote the presence and absence of a restriction site, respectively.

Table 200. Insulin receptor: INSR

Place/ Population	Enzyme	Alleles (kb)	Frequency +	-	Other	n*	H	PE	Source
Caucasians (US)	RsaI(1)	6.2, 7.0	.48	.52		92	49.9	18.7	Elbein et al. 1986
Blacks (US)	"	6.2, 7.0, 6.4	.67	.28	.06	58	46.9	22.7	"
Pima Indians (US)	"	6.2. 7.0, 3.4	.17	.48	.34	54	62.5	33.2	"

Continued

Table 200. Insulin receptor: INSR (cont'd)

Place/ Population	Enzyme	Alleles (kb)	Frequency			n*	H	PE	Source
			+	–	Other				
Caucasians (US)	RsaI(2)	2.2, 2.4	.77	.23		88	35.4	14.6	"
Blacks (US)	"	"	.63	.37		38	46.6	17.9	"
Pima Indians (US)	"	"	.63	.37		32	46.6	17.9	"
Caucasians (US)	BglII	3.4, 23.4	.17	.83		86	28.2	12.1	"
Blacks (US)	"	"	.26	.74		46	38.5	15.5	"
Pima Indians (US)	"	"	.02	.98		48	3.9	1.9	"
Caucasians (US)	SacI(1)	5.3, 5.8	.88	.12		98	21.1	9.4	"
Blacks (US)	"	"	.65	.35		48	45.5	17.6	"
Pima Indians (US)	"	"			1.00	58			"
Caucasians (US)	Sac(2)	7.4/2.4, 9.4	.12	.88		98	21.1	9.4	"
Blacks (US)	"	"	.02	.98		48	3.9	1.9	"
Pima Indians (US)	"	"	.02	.98		48	3.9	1.9	"

Table 201. Nerve growth factor, beta: NGFB

	Haplotype (Fragment lengths in kb)		Caucasians
	TaqI (4,3, 6.0)	BlgIII (1.4, 5.6)	
1	+	+	.65
2	–	+	.15
3	+	–	.20
4	–	–	
n*			138
H			51.5
PE			27.4

Source: Darby et al. 1985

Table 202. Parathyroid hormone: PTH

Population	Enzyme	Alleles (kb)	Frequency		n*	H	PE	Source
			+	–				
Caucasians	PstI	2.2, 2.8	.42	.58	90	48.7	18.4	Schmidtke et al. 1984
	TaqI	2.4, 2.5	.63	.37	94	46.6	17.9	"

Note: Not polymorphic for BamHI, EcoRI, HindIII, MspI, and XbaI when 10-16 individuals were examined.

Table 203. Phenylalanine hydroxylase: PAH

				Haplotype (Fragment lengths in kb)					Danes
	BglII (1.7, 3.6)	PvuII(a) (6, 19)	PvuII(b) (9.1, 11.5)	EcoRI (11, 17)	MspI (19, 23)	XmnI (6.5, 9.4)	HindIII (4.0,4.2/4.4)	EcoRV (25, 30)	(Denmark)
1	-	+	-	-	+	-	-	-	.348
2	-	+	-	-	+	-	+	+	.046
3	-	+	-	+	-	+	-	-	.030
4	-	+	-	+	-	+	+	+	.318
5	+	-	+	+	+	-	-	+	.106
6	+	-	+	+	+	-	-	-	
7	+	-	-	+	-	+	-	-	.106
8	-	+	-	+	+	-	-	+	.015
9	+	+	-	+	+	-	-	+	
10	-	+	-	+	+	-	-	-	.015
11	+	-	-	+	+	-	-	+	.015
12	-	+	-	-	+	-	=	+	
n*									66
H									75.2
PE									53.8

Source: DiLella et al. 1986; Chakraborty et al. 1987
Note: = indicates fragment length of 4.4 kb.

Table 204. Phosphoglycerate kinase 1: PGK1

Place/ Population	Enzyme	Allele freq +	Allele freq -	n*	H	Source
Mediterraneans	PstI	.55	.45	31	49.5	Hutz et al. 1984
Asiatic Indians	"	.69	.31	13	42.6	"
Blacks (US)	"	.59	.41	46	42.5	"

Table 205. Plasminogen: PLG

Place/ Population	Enzyme	Allele freq +	Allele freq -	n*	H	PE	Source
Caucasians (US)	RsaI	.25	.75	150	37.5	15.2	Murray et al. 1984a
Blacks (US)	"	.07	.93	30	13.0	6.1	"
Asians	"	.58	.42	30	48.7	18.4	"
Caucasians (US)	SacI	.31	.69	150	42.8	16.8	"
Blacks (US)	"	.15	.85	30	25.5	11.1	"
Asians	"		1.00	30			"
Caucasians (US)	MspI	.67	.33	150	44.2	17.2	"
Blacks (US)	"	.90	.10	30	18.0	8.2	"
Asians	"	1.00		30			"

Table 206. Proopiomelanocortin: POMC

| Haplotype (Fragment lengths in kb) | | | Caucasians |
SstI (10, 15)	RsaI (0.8, 1.9)		
1	+	+	.47
2	+	−	.28
3	−	+	.25
4	−	−	
n*			N.A.
H			63.8
PE			35.0

Source: Feder et al. 1985

Table 207. Urokinase gene: URK*

Place/ Population	Enzyme	Alleles (kb)	Frequency +	Frequency −	n*	H	PE	Source
Caucasians	BamHI	1.6, 7.0	.30	.70	82	42.0	16.6	Sebastio et al. 1985

Table 208. Alpha-globin deletion[a]: HBAD*

Place/Population	Haplotype ααα	Haplotype αα	Haplotype −α	n	H	PE	Source
EUROPE							
GREECE							
Cypriots		.930	.070	50	13.0	6.1	Pirastu et al. 1982
ITALY							'
Italians (Sardinia)		.819	.181	69	29.6	12.6	";
Italians (")		.873	.127	71	22.2	9.9	Galanello et al. 1984
AFRICA							
SENEGAL		.897	.103	29	18.5	8.4	Pagnier et al. 1984
UPPER VOLTA (Benin)		.859	.141	39	24.2	10.6	"
NORTH AMERICA							
UNITED STATES							
Blacks		.844	.156	211	26.3	11.4	Dozy et al. 1979
OCEANIA							
MELANESIA							
Papua New Guinea							
Highlands							
Madang (Bundi)		.951	.049	82	9.3	4.4	Flint et al. 1986
Chimbu		.978	.022	89	4.3	2.1	"

Continued

Table 208. Alpha-globin deletion[a]: HBAD* (cont'd)

Place/Population	Haplotype ααα	Haplotype αα	Haplotype -α	n	H	PE	Source
E. Highlands		.955	.045	157	8.6	4.1	"
S. Highlands		.971	.029	190	5.6	2.7	"
W. Highlands		.976	.024	21	4.7	2.3	"
Total		.965	.035	539	6.8	3.3	"
Coastal							
Northern		.317	.683	63	43.3	17.0	"
Southern		.781	.219	105	34.2	14.2	"
Eastern		.625	.375	20	46.9	17.9	"
Total		.610	.390	188	47.6	18.1	"
N. Solomon Is.		.550	.450	30	49.5	18.6	"
Vanuatu							
Espiritu Santo		.615	.385	178	47.4	18.1	"
Maewo		.672	.328	169	44.1	17.2	"
Penetecost		.725	.275	149	39.9	16.0	"
Malecula		.667	.333	21	44.4	17.3	"
Paama		.780	.220	25	34.3	14.2	"
Emae		.741	.259	85	38.4	15.5	"
Makura-Mataso		.911	.089	28	16.2	7.5	"
Efate		.750	.250	34	37.5	15.2	"
Tanna		.819	.181	224	29.6	12.6	"
Futuna		.946	.054	56	10.2	4.8	"
Aneityum		.924	.076	66	14.0	6.5	"
New Caledonia		.941	.059	84	11.1	5.2	"
Karkar Is.		.333	.667	27	44.4	17.3	Yenchitsomanus et al. 1985
POLYNESIA							
Polynesians	.042	.867	.092	300	23.8	12.1	Trent et al. 1985

Note: [a] A similar but different set of data are published by Yenchitsomanus et al. (1986).

Table 208.1 Alpha-globin deletion (subtypes of -α)[a]: HBAD*

Place/Population	$-\alpha^{3.7}$	$-\alpha^{4.2}$	n*	Source
ASIA				
MALAYASIA				
Malays		1.000	36	Lie-Injo et al. 1982
THAILAND				
Thais	.923	.077	39	Winchagoon et al. 1984
OCEANIA				
MELANESIA				
Papua New Guinea				
Highlands				
Madang (Bundi)		1.000	8	Flint et al. 1986
Chimbu		1.000	4	"
E. Highlands	.429	.571	14	"
S. Highlands	.364	.636	11	"
W. Highlands		1.000	1	"
Total	.263	.737	38	"
Coastal				
Northern	.116	.884	86	"
Southern	.587	.413	46	"
Eastern	.467	.533	15	"
Total	.306	.694	147	"
N. Solomon Is.	.519	.481	27	"
Vanuatu				
Espiritu Santo	.730	.270	137	"
Maewo	.856	.144	111	"

Continued

257

Table 208.1 Alpha-globin deletion (subtypes of $-\alpha$)[a]: HBAD* (cont'd)

Place/Population	$-\alpha^{3.7}$	$-\alpha^{4.2}$	n*	Source
Pantecost	.951	.049	82	"
Malecula	.571	.429	14	"
Paama	.727	.273	11	"
Emae	.378	.622	45	"
Makura-Mataso	.400	.600	5	"
Efate	.706	.294	17	"
Tanna	.605	.395	81	"
Futana	.500	.500	6	"
Aneityum	.900	.100	10	"
New Caledonia	.800	.200	10	"
Karkar Is.	.044	.956	23	Yenchitsomanus et al. 1985
Madang	.041	.959	122	"

Note: [a] A similar but different set of data are published by Yenchitsomanns et al. (1986).

Table 208.2 Alpha-globin deletion (subtypes of $-\alpha^{3.7}$): HBAD*

Population	$-\alpha^{3.7}$ I	$-\alpha^{3.7}$ II	$-\alpha^{3.7}$ III	$-\alpha^{3.7}$	$-\alpha^{4.2}$	n*	Source
Mediterraneans	.971	.029		1.000		35	Hill et al. 1985
Saudi Arabians	.964			.964	.036	28	"
Indians	.889			.889	.111	9	"
Southeast Asians	.796	.102		.898	.102	49	"
Jamaicans	.946	.036		.982	.018	56	"
Papua New Guineans	.106		.061	.167	.833	66	"
Vanuatu	.081		.588	.669	.331	148	"
Polynesians	.062		.938	1.000		32	"

Table 209. Gamma-globin deletion: HBGD*

Place/Population	Haplotype $\gamma\gamma\gamma$	Haplotype $\gamma\gamma$	Haplotype $-\gamma$	n	H	PE	Source
OCEANIA							
MELANESIA							
Vanuatu							
Espiritu Santo	.030	.860	.110	82	24.7	12.2	Hill et al. 1986
Maewo	.089	.813	.098	168	32.2	16.7	"
Pantecost	.040	.938	.022	136	11.8	6.0	"
Emae	.022	.860	.118	68	24.6	11.8	"
Central	.031	.943	.026	209	10.9	5.5	"
Tanna	.011	.968	.021	48	6.2	3.1	"
Futuna		.875	.125	54	21.9	9.7	"
Aneityum	.009	.893	.098	57	19.3	9.1	"
New Caledonia	.083	.900	.017	30	18.3	8.9	"
Total	.039	.896	.065	852	19.1	9.7	"

10   Miscellaneous

Table 210.   Acetylator system; isoniazid inactivation: AC

Place/Population	R	S	n	H	PE	Source
EUROPE						
FINLAND						
Finns	.202	.798	91	32.2	13.5	Mattila & Tiitinen 1967
Lapps/Saami	.293	.707	26	41.4	16.4	Tiitinen et al. 1967
SWEDEN	.132	.868	130	22.9	10.1	Hanngren et al. 1970
UNITED KINGDOM						
England	.211	.789	135	33.3	13.9	Evans 1969
ASIA						
ISRAEL						
Ashkenazi Jews	.206	.794	100	32.7	13.7	Szeinberg et al. 1961
AFGHANISTAN						
Pushtoon (Kabul)	.181	.819	85	29.6	12.6	Goedde et al. 1977c
INDIA						
South Indians	.221	.779	321	34.4	14.3	Gangadharam et al. 1961
Indians (Malaysia)	.146	.854	57	24.9	10.9	Ellard & Gammon 1977
BURMA	.390	.610	121	47.6	18.1	Smith & Tun 1968
MALAYSIA						
Malays (Singapore)	.352	.648	75	45.6	17.6	Ellard & Gammon 1977
THAILAND						
Thais (NE Thailand)	.333	.667	119	44.4	17.3	Kukongviriyapan et al. 1984
CHINA						
Chinese (Thailand)	.417	.583	47	48.6	18.4	"
Chinese (Malaysia)	.537	.463	386	49.7	18.7	Ellard & Gammon 1977
JAPAN						
Japanese	.663	.337	1,808	44.7	17.4	Sunahara et al. 1961
Ainu	.692	.308	86	42.6	16.8	"
KOREA						
Koreans (Seoul)	.678	.322	222	43.7	17.1	Kang & Lee 1973
Koreans (")	.463	.537	142	49.7	18.7	Goedde et al. 1986c
AFRICA						
LIBYA						
Arabs	.194	.806	40	31.3	13.2	Karim et al. 1981
EGYPT						
Copts & Moslems	.094	.906	50	17.0	7.8	Hashem et al. 1969
NIGERIA						
Nigerians (Nsukka, S. Nigeria)	.358	.642	80	46.0	17.7	Eze & Obidoa 1978
ETHIOPIA						
Amhara tr.	.089	.911	88	16.2	7.5	Russell & Russell 1971
SOUTHERN AFRICA						
San/Bushmen						
!Kung	.817	.183	30	29.9	12.7	Jenkins et al. 1974
Bantu/Nguni	.398	.602	270	47.9	18.2	Nurse et al. 1985(c)
NORTH AMERICA						
CANADA						
Eskimos	.785	.215	216	33.8	14.0	Armstrong & Peart 1960
UNITED STATES						
Whites	.232	.768	39	35.6	14.6	Hodgkin et al. 1979
Blacks	.370	.630	63	46.6	17.9	"
SOUTH AMERICA						
CHILE						
Atacameño (Toconao, N. Chile)	.551	.449	89	49.5	18.6	Goedde et al. 1984c
ECUADOR						
Shuara	.525	.475	68	49.9	18.7	Goedde et al. 1977a

Continued

Table 210. Acetylator system; isoniazid inactivation: AC (cont'd)

Place/Population	R	S	n	H	PE	Source
OCEANIA						
MELANESIA						
Papua New Guinea						
Mainland (Highland Provinces						
& Coastal Provinces)	1.000		114			Penketh et al. 1983
Solomon Is. (Northern)	.202	.798	11	32.2	13.5	"
POLYNESIA						
Polynesians	.670	.330	N.A.	44.2	17.2	Hayward 1975

Note: PE was computed under the assumption that the allele Ac^R is dominant to allele Ac^S.

Table 211. Beta-aminoisobutyric acid, urinary excretion: BAIB

Place/Population	High excretor (%)	n	Source
EUROPE			
ITALY	6.7	792	Calchi-Novati et al. 1954
UNITED KINGDOM			
England	9.6	345	Harris 1953-4
ASIA			
VIETNAM	26.2	187	Blumberg & Gartler 1961
CHINA	45.0	33	"
JAPAN	29.1	110	Ishimoto 1968
AFRICA			
EGYPT	59.0	117	Hashem & Khalifa 1969
TANZANIA	40.0	40	Blumberg & Gartler 1961
NORTH AMERICA			
ALASKA			
Eskimos	23.3	120	Allison et al. 1959
Athabascan Indians	56.0	25	"
UNITED STATES			
Whites (New York)	6.0	218	Gartler et al. 1956-7
Blacks (")	15.8	38	"
Apache Indians (Arizona)	59.0	110	Blumberg & Gartler 1961(c)
Papago " (")	50.0	42	"
Pima " (Arizona &	50.0	118	"
New Mexico)			
MIDDLE AMERICA			
MEXICO			
Zapotec Indians (Oaxaca)	35.4	277	Lasker et al. 1969
Tarascan " (Cherán)	26.0	119	"
HONDURAS			
Black Caribs	31.6	285	Gartler et al. 1956-7
OCEANIA			
MELANESIA			
New Guinea			
E. Highlands	8.6	70	Blumberg & Gartler 1961
MICRONESIA			
Marshall Is.	86.2	188	"
POLYNESIA			
Tahiti	12.0	46	"

Table 212. Cerumen (Ear wax type): CER*

Place/Population	W	w	n	H	PE	Source
EUROPE						
FINLAND	.46	.54	N.A.	49.7	18.7	Petrakis 1971
GERMANY	.824	.176	514	29.0	12.4	Matsunaga 1962
ASIA						
TURKEY (Asia Minor)	.475	.525	105	49.9	18.7	Petrakis et al. 1971
IRAN	.423	.577	54	48.8	18.5	"
AFGHANISTAN	.608	.392	78	47.7	18.2	"
INDIA						
Non-tribes	.522	.478	48	49.9	18.7	"
Tribes (Andhra Pradesh)	.614	.386	516	47.4	18.1	"
MALAYSIA						
Malays (Kuala Lumpur)	.481	.519	78	49.9	18.7	"
Aborigines	.739	.261	133	38.6	15.6	"
CHINA						
Mongolians (Khalkha)	.062	.938	1,099	11.6	5.5	Matsunaga 1962(c)
Chinese (Northern)	.021	.979	216	4.1	2.0	"
Chinese (Taiwan)	.140	.860	708	24.1	10.6	"
TAIWAN						
Aborigines	.465	.535	1,420	49.8	18.7	"
JAPAN						
Japanese	.085	.915	23,417	15.7	7.2	"
Ainu	.295	.705	185	41.6	16.5	Omoto 1970
KOREA	.076	.924	381	14.0	6.5	Matsunaga 1962(c)
NORTH AMERICA						
ALASKA						
Aleut/Eskimos	.304	.696	140	42.3	16.7	Petrakis 1969(c)
UNITED STATES						
Whites	.886	.114	368	20.2	9.1	"
Blacks (Ghana & Nigeria)	.931	.069	63[a]	12.8	6.0	Matsunaga 1962
Blacks (")	1.000		51			Petrakis 1971
Papago Indians (S. Arizona)	.227	.773	437	35.1	14.5	Niswander et al. 1970
MIDDLE AMERICA						
MEXICO						
Maya Indians	.243	.757	68	36.8	15.0	Petrakis 1969(c)
PANAMA						
Cuna Indians	.791	.209	90	33.1	13.8	"
SOUTH AMERICA						
PERU						
Quechua Indians	.124	.876	43	21.7	9.7	"
OCEANIA						
AUSTRALIA						
Aborigines (N. Territory)	.821	.179	498	29.4	12.5	Omoto 1973
MELANESIA						
Melanesian	.473	.527	732	49.9	18.7	Matsunaga 1962(c)
MICRONESIA						
Micronesian	.391	.609	458	47.6	18.1	"

Note: Allele W (wet) is dominant to allele w (dry). [a] The gene frequencies were estimated from 63 hybrid
individuals between American Blacks and Japanese.

Table 213.   Colorblindness: CB*

Place/Population	Colorblind (%)[a]	n	Source
EUROPE			
BELGIUM	7.5	9,540	Laet & van de Calseyde 1935
FRANCE	9.0	6,635	Kherumian & Pickford 1959
GERMANY	7.7	6,863	Schmidt 1936
GREECE	4.3	557	Kaplanoglou & Triantaphyllidis 1982
ITALY (Rome)	6.1	3,275	Malaspina et al. 1986
NETHERLANDS	8.0	3,168	Crone 1968
NORWAY	8.0	9,047	Waaler 1927
SWITZERLAND	9.0	1,000	Bally 1954
UNITED KINGDOM			
England (East coast)	6.7	43,278	Vernon & Straker 1943
England (Southwest)	9.3	6,648	"
England (Central)	8.6	14,455	"
Scotland (Eastern)	4.9	6,236	"
Isle of Man	5.6	303	Mitchell 1977
Orkney Is.	5.2	404	Boyce et al. 1973
YUGOSLAVIA			
Macedonians & Dalmatians	4.8	147	Fraser et al. 1966b
U.S.S.R.			
RUSSIA	9.2	1,343	Flekkel 1955
ASIA			
TURKEY (Asia Minor)	5.2	1,686	Ökte 1960
ISRAEL			
Ashkenazi Jews	9.1	778	Adam 1969(c)
Yemenite Jews	5.2	404	Kalmus et al. 1961
IRAN			
Kurds	8.1	504	Lightman et al. 1970
INDIA			
Punjabi (N. India)	10.8	56	Ray 1969
Gujarati (W. India)	4.0	25	"
S. Indians (Andhra Pradesh)	3.2	1,709	Murty & Vijayalaxmi 1974
Tribes (")	1.1	5,225	Naidu et al. 1978
Bengali (E. India)	3.9	226	Ray 1969
THAILAND			
Thais (Northern)	5.2	2,128	Flatz 1967
Thais (Bangkok)	5.6	1,658	Adam et al. 1969
CHINA			
Mongolians	7.2	291	Wang et al. 1983(c)
Han (Shanghai)	5.8	556	"
Hui	8.8	452	"
Tibetans	5.0	241	Tiwari 1969
Chinese (Bangkok)	5.7	669	Adam et al. 1969
JAPAN			
Japanese (Nagasaki)	4.4	1,206	Nemoto & Murao 1961
Ainu (Hokkaido)	.5	185[b]	Omoto 1970
KOREA			
Koreans (Seoul)	5.5	1,231	Kang et al. 1967b
Koreans (NE China)	4.6	519	Wang et al. 1983
AFRICA			
NIGERIA			
Hausa	2.1	380	Roberts 1967
ETHIOPIA			
Amhara tr.	6.3	142	Adam 1962
UGANDA			
Bantu	3.9	1,294	Adam et al. 1970
TANZANIA			
Hadza	1.4	209	Barnicot & Woodburn 1975
NORTH AMERICA			

Continued

Table 213. Colorblindness: CB* (cont'd)

Place/Population	Colorblind (%)[a]	n	Source
GREENLAND			
Eskimos	2.5	297	Skeller 1954
UNITED STATES			
Navajo Indians	1.1	535	Garth 1933
MIDDLE AMERICA			
MEXICO			
Maya Indians	3.6	778	Giles et al. 1968
Mestizo	5.7	795	Adam 1969(c)
SOUTH AMERICA			
BRAZIL			
Cayapo	.8	120	Salzano 1972
Caingang	2.3	301	"
CHILE			
Aymara	6.4	140	Cruz-Coke & Barrera 1969
COLOMBIA			
Tenza Village	2.5	790	Mueller & Weiss 1979
OCEANIA			
AUSTRALIA			
Aborigines (Full bloods)	1.9	4,455	Mann & Turner 1956
MELANESIA			
Papua New Guinea			
Papuans	4.6	3,685	Adam 1969
POLYNESIA			
Tonga	7.5	67	Beaglehole 1939

Note: [a] Percent of colorblind males. [b] One colorblind individual was suspected to be of Japanese origin.

Table 214. Debrisoquine oxidation; sparteine oxidation (S-defect): DOX*

Place/Population	Poor metabolizer (%)	n	Source
EUROPE			
SWEDEN	7.5	226	Steiner et al. 1985
SWITZERLAND	10.4	221	Küpfer & Preisig 1984
UNITED KINGDOM			
England	8.9	258	Evans et al. 1980
ASIA			
JAPAN (Tsukuba)	0.0	100	Nakamura et al. 1985
AFRICA			
GHANA	5.6	233	Woolhouse 1986
NIGERIA	8.6	128	Idle & Smith 1980
NORTH AMERICA			
CANADA			
Whites (Toronto)	7.2	83	Inaba et al. 1984
UNITED STATES			
Whites (Tennessee)	8.7	183	Nakamura et al. 1985

Table 215. Lactase activity: LAA*

Place/Population	Lactose absorber (%)	n	Source
EUROPE			
AUSTRIA	79.9	528	Rosenkranz et al. 1982
CZECHOSLOVAKIA	87.5	200	Flatz et al. 1982(c)
DENMARK	97.4	700	Gudmand-Höyer et al. 1969
FINLAND			
Finns	82.7	156	Sahi 1974
Lapps/Saami	58.4	519	Isokoski et al. 1981
FRANCE (Northern)	77.4	62	Flatz 1987(c)
GERMANY	85.2	1,805	Flatz et al. 1982
GREECE			
Greeks	62.5	16	Spanidou & Pertrakis 1972
Greeks	45.8	24	Doxiadis & Papageorgiadis 1973
HUNGARY	63.0	535	Czeizel et al. 1983
IRELAND	96.0	50	Fielding et al. 1981
ITALY (Northern)	49.9	383	Flatz 1987(c)
POLAND	62.6	275	Socha et al. 1984
SPAIN	85.3	265	Pena-Yanez et al. 1971
SWEDEN	99.0	400	Dahlqvist & Lindquist 1971
SWITZERLAND	83.3	18	Auricchio et al. 1963
UNITED KINGDOM			
England	94.7	75	Ho et al. 1982
YUGOSLAVIA (Solvenia)	64.7	153	Flatz 1987(c)
U.S.S.R.			
EUROPEAN PART			
Russians (Leningrad)	84.7	248	Flatz 1987(c)
ASIA			
TURKEY (Asia Minor)	28.8	470	Flatz et al. 1986a
ISRAEL			
Ashkenazi Jews	20.8	53	Gilat et al. 1970
JORDAN			
Bedouins	75.9	162	Flatz 1987(c)
SAUDI ARABIA			
Arabs	67.5	40	Cook & Al-Torki 1975
IRAN	31.4	105	Sadre & Karbasi 1979
AFGHANISTAN	17.4	270	Rahimi et al. 1976
PAKISTAN	40.1	414	Ahmad & Flatz 1984
INDIA			
Punjabi (N. India)	85.1	134	Tandon et al. 1977
N. Indians (New Delhi)	72.6	124	Tandon et al. 1981
S. Indians (Hyderabad)	59.7	72	Reddy & Pershad 1972
S. Indians (Trivandrum & Pondicherry)	33.3	60	Tandon et al. 1981
SRI LANKA	27.5	200	Senewiratne et al. 1977
BANGLADESH	19.2	234	Brown et al. 1979
INDONESIA			
Javanese	9.4	53	Flatz 1987(c)
THAILAND			
Thais (North)	0.0	149	"
Thais (Central)	2.9	279	"
VIETNAM			
Vietnamese (USA)	0.0	31	Anh et al. 1977
CHINA			
Mongolians (Inner Mongolia)	12.1	198	Wang et al. 1984
Han (N. China)	7.7	248	"
Kazakh (Xinjiang)	23.6	195	"
Chinese (Taiwan)	0.0	71	Sung & Shih 1972
JAPAN			
Japanese	15.2	66	Flatz 1987(c)
AFRICA			
EGYPT			

Continued

Table 215. Lactase activity: LAA* (cont'd)

Place/Population	Lactose absorber (%)	n	Source
Egyptians	27.4	570	Hussein et al. 1982
SUDAN			
Beja-Nomads	83.2	303	Bayoumi et al. 1982
Nilotics	25.5	282	"
NIGER			
Tuareg	87.3	118	Flatz et al. 1986b
SENEGAL	64.9	131	Flatz 1987(c)
NIGERIA			
Yoruba	2.4	41	Kretchmer et al. 1971
Fulani	42.4	33	"
CENTRAL AFRICAN REPUBLIC			
Bantu	5.4	112	Flatz 1987(c)
ETHIOPIA	10.3	58	Habte et al. 1973
KENYA			
Bantu	26.8	71	Pieters & van Rens 1973
SOUTH AFRICA			
San/Bushmen			
Tsumkwe !Kung	2.5	40	Nurse & Jenkins 1977
Khoi/Hottentot			
Keetmanshoop Nama	33.3	39	"
Negroes/Denasena	0.0	50	"
Bantu	21.7	115	Segal et al. 1983

NORTH AMERICA

GREENLAND			
Eskimos	12.0	25	Gudmand-Höyer & Jarnum 1969
ALASKA			
Eskimos & Indians	16.7	36	Duncan & Scott 1972
CANADA			
Amerindians (West Coast)	36.7	30	Leichter & Lee 1971
UNITED STATES			
Whites	84.0	50	Sheehy & Anderson 1965
Blacks	35.4	390	Flatz 1987(c)
Mexican Americans	45.5	11	Dill et al. 1972
"	51.5	33	Sowers & Winterfeldt 1975

MIDDLE AMERICA

MEXICO			
Mexicans (Tlaxcala)	26.2	401	Lisker et al. 1974

SOUTH AMERICA

BOLIVIA			
Aymara (La Paz)	37.7	122	Balanza & Taboada 1985
COLOMBIA			
Chami Indians	41.7	24	Alzate et al. 1969

OCEANIA

AUSTRALIA			
Aborigines	33.1	145	Flatz 1987(c)
MELANESIA			
Papau New Guinea			
Papuans	0.0	49	Cook 1979
Papuans (West Sepik)	22.9	35	Arnhold et al. 1981
Tribes (S. Highlands & Milne Bay Province)	13.6	59	Jenkins et al. 1981
Massim-Austronesian (Goodenough Is.)	17.2	29	Gibney et al. 1981

Note: The reliability of the frequency of lactose absorbers is not always high. For this reason two or more
different sets of data are included for some populations.

Table 216. Mephenytoin metabolism (M-defect): MEP*

Place/Population	Poor metabolizer (%)	n	Source
EUROPE			
SWITZERLAND	5.4	221	Küpfer & Preisig 1984
ASIA			
JAPAN (Tsukuba)	18.0	100	Nakamura et al. 1985
NORTH AMERICA			
CANADA			
Whites (Toronto)	4.2	118	Jurima et al. 1985
UNITED STATES			
Whites (Tennessee)	2.7	183	Nakamura et al. 1985

Table 217. Phenylthiocarbamide taste: PTC

Place/Population	T	t	n	H	PE	Source
EUROPE						
DENMARK	.416	.584	314	48.6	18.4	Mohr 1951-2
FINLAND						
Finns	.470	.530	760	49.8	18.7	Eriksson et al. 1970
Lapps/Saami	.468	.532	251	49.8	18.7	Eriksson et al. 1970
GERMANY (Freiberg)	.442	.558	93	49.3	18.6	Goedde & Ohligmacher 1965
GREECE (Almopia)	.457	.543	370	49.6	18.7	Kaplanoglou & Triantaphyll- idis 1982
ICELAND (Southwest)	.456	.544	1,221	49.6	18.7	Henke & Palsson 1977
IRELAND	.464	.536	783	49.7	18.7	Sunderland et al. 1971
ITALY	.480	.520	541	49.9	18.7	Maxia et al. 1975
NETHERLANDS	.468	.532	219	49.8	18.7	de Jong 1964
NORWAY	.448	.552	266	49.5	18.6	Merton 1958
POLAND	.461	.539	1,103	49.7	18.7	Modrzewska 1958
PORTUGAL	.590	.410	564	48.4	18.3	Cunha & Abreu 1956
SPAIN						
Spanish	.502	.498	306	50.0	18.7	Pons 1955
Basques	.338	.662	155	44.8	17.4	Valls Medina 1958
SWEDEN (S. Sweden)	.434	.566	200	49.1	18.5	Åkesson 1959
SWITZERLAND (Zurich)	.454	.546	544	49.6	18.6	Botsztejn 1942
UNITED KINGDOM						
England (Northern)	.476	.524	835	49.9	18.7	Cartwright & Sunderland 1967
Scotland (Northwest)	.457	.543	227	49.6	18.7	Mitchell et al. 1977(c)
Isle of Man	.473	.527	699	49.9	18.7	Mitchell et al. 1977
Orkney Is. (Westray)	.429	.571		49.0	18.5	Boyce et al. 1973
YUGOSLAVIA	.444	.556	459	49.4	18.6	Grünwald & Herman 1963
U.S.S.R.						
EUROPEAN PART						
Russians	.395	.605	486	47.8	18.1	Boyd & Boyd 1937
ASIAN PART						
Eskimos (Chukotskiy Pen.)	.505	.495	53	50.0	18.7	Rychkov & Sheremeteva 1972
Chukchi (")	.758	.242	68	36.7	15.0	"
ASIA						
TURKEY (Asia Minor)	.553	.447	2,000	49.4	18.6	Say et al. 1966
ISRAEL						
Yemenite Jews	.487	.513	498	50.0	18.7	Guttman et al. 1967
SYRIA						
Arabs	.396	.604	400	47.8	18.2	Mourant et al. 1976(c)

Continued

Table 217. Phenylthiocarbamide taste: PTC (cont'd)

Place/Population	T	t	n	H	PE	Source
IRAN	.476	.524	346	49.9	18.7	Lightman et al. 1970
PAKISTAN (Gilgit)	.460	.540	48	49.7	18.7	Chapman et al. 1972
INDIA						
Punjabi (N. India)	.434	.566	322	49.1	18.5	Sharma 1959
Gujarati (W. India)	.408	.592	200	48.3	18.3	Parikh et al. 1969
Bengali (E. India)	.447	.553	1,813	49.4	18.6	Das 1966
NEPAL						
Newar	.523	.477	589	49.9	18.7	Tiwari & Bhasin 1967
BURMA	.654	.346	300	45.3	17.5	Than-Than-Sint & Mya-Tu 1973
MALAYSIA						
Malays (Singapore)	.553	.447	50	49.4	18.6	Lugg & Whyte 1954
THAILAND	.689	.311	460	42.9	16.8	Boobphanirojana et al. 1970
PHILIPPINES						
Filipino	.800	.200	200	32.0	13.4	"
Negrito	.630	.370	73	46.6	17.9	Pascasio et al. 1974
CHINA						
Mongolians	.672	.328	521	44.1	17.2	Xu et al. 1982
Han	.689	.311	547	42.9	16.8	"
Hui (Linxia)	.641	.359	700	46.0	17.7	"
Dong	.679	.321	399	43.6	17.0	"
Uygur	.473	.527	2,271	49.9	18.7	Xu et al. 1982(c)
Kazak	.414	.586	1,089	48.5	18.4	"
Kirgiz	.351	.649	152	45.6	17.6	"
Uzbek	.449	.551	112	49.5	18.6	"
Tibetans (Lhasa)	.648	.352	613	45.6	17.6	Xu et al. 1982
Chinese (Singapore)	.859	.141	50	24.2	10.6	Lugg & Whyte 1954
JAPAN						
Japanese (Hokkaido)	.647	.353	1,625	45.7	17.6	Matsunaga et al. 1954
Ainu (")	.754	.246	232	37.1	15.1	Omoto 1970
KOREA						
Koreans (Seoul)	.612	.388	771	47.5	18.1	Kang et al. 1967a
Koreans (NE China)	.656	.344	573	45.1	17.5	Xu et al. 1982

AFRICA

Place/Population	T	t	n	H	PE	Source
LIBYA	.536	.464	65	49.7	18.7	Sunderland 1971
EGYPT	.540	.460	459	49.7	18.7	Boyd & Boyd 1937
SUDAN						
Sudanese (Northern)	.800	.200	100	32.0	13.4	Rife 1953
NIGERIA						
Yoruba	.662	.338	1,325	44.8	17.4	Scott-Emuakpor et al. 1975
CENTRAL AFRICAN REPUBLIC						
N'Zakara	.833	.167	360	27.8	12.0	Cresta 1964
ETHIOPIA						
Amhara tr.	.651	.349	123	45.4	17.6	Bat-Miriam et al. 1962
TANZANIA						
Hadza	.513	.487	118	50.0	18.7	Barnicot & Woodburn 1975
SOUTHERN AFRICA						
San/Bushmen						
!Kuboes (Hybrid people)	.676	.324	114	43.8	17.1	Jenkins 1972
Khoi/Hottentot						
Keetmanshoop Nama	.814	.186	115	30.3	12.8	"
Bantu/Zulu	.847	.153	86	25.9	11.3	Jenkins 1965

NORTH AMERICA

Place/Population	T	t	n	H	PE	Source
GREENLAND						
Eskimos	.269	.731	129	39.3	15.8	Alsbirk & Alsbirk 1972
ALASKA						
Eskimos	.493	.507	68	50.0	18.7	Allison & Blumberg 1959
CANADA						
Cree Indians (Alberta)	.855	.145	431	24.8	10.9	Matson 1940
Blackfeet Indians (")	.708	.292	129	41.3	16.4	Matson 1938
UNITED STATES						
Papago Indians (Arizona)	.880	.120	70	21.1	9.4	MacRoberts 1964

MIDDLE AMERICA

Continued

Table 217. Phenylthiocarbamide taste: PTC (cont'd)

Place/Population	T	t	n	H	PE	Source
MEXICO						
Mazateco (Oaxaca)	.639	.361	123	46.1	17.7	Novelo et al. 1964
Mayan Indians	.668	.332	379	44.4	17.3	Giles et al. 1968
PUERTO RICO	.669	.331	3,499	44.3	17.2	Thieme 1952
NICARAGUA						
Rama (Rama Kay)	.888	.112	79	19.9	9.0	de Stefano & Molieri 1976
JAMAICA						
Negroes	.722	.278	632	40.1	16.0	Terry 1950

SOUTH AMERICA

Place/Population	T	t	n	H	PE	Source
BRAZIL						
Caingang	.839	.161	77	27.0	11.7	Kalmus 1957
PERU						
Quechua	.813	.187	172	30.4	12.9	Garruto et al. 1975

OCEANIA

Place/Population	T	t	n	H	PE	Source
AUSTRALIA						
Aborigines (N. Territory)	.298	.702	152	41.8	16.5	Simmons et al. 1954
MELANESIA						
Karkar Is.						
Takia	.597	.403	172	48.1	18.3	Boyce et al. 1976
Solomon Is.	.675	.325	104	43.9	17.1	Willis & Booth 1968
MICRONESIA						
Caroline Is.	.668	.332	291	44.4	17.3	Simmons et al. 1953
POLYNESIA						
Cook Is.	.597	.403	215	48.1	18.3	Simmons et al. 1955

Note: Allele T (taster) is dominant to allele t (non-taster).

PART III

Maps of Gene Frequencies

11 Explanation of Maps

The following maps are presented to show the general patterns of the world distributions of gene frequencies for polymorphic loci. The number of populations considered for each map is about 100 or less, and only the populations that are native to each geographical area are considered. The loci of which the gene frequencies are close to 0 or 1 in most populations are not included. The loci for which gene frequency data are available only for a limited number of populations are also excluded.

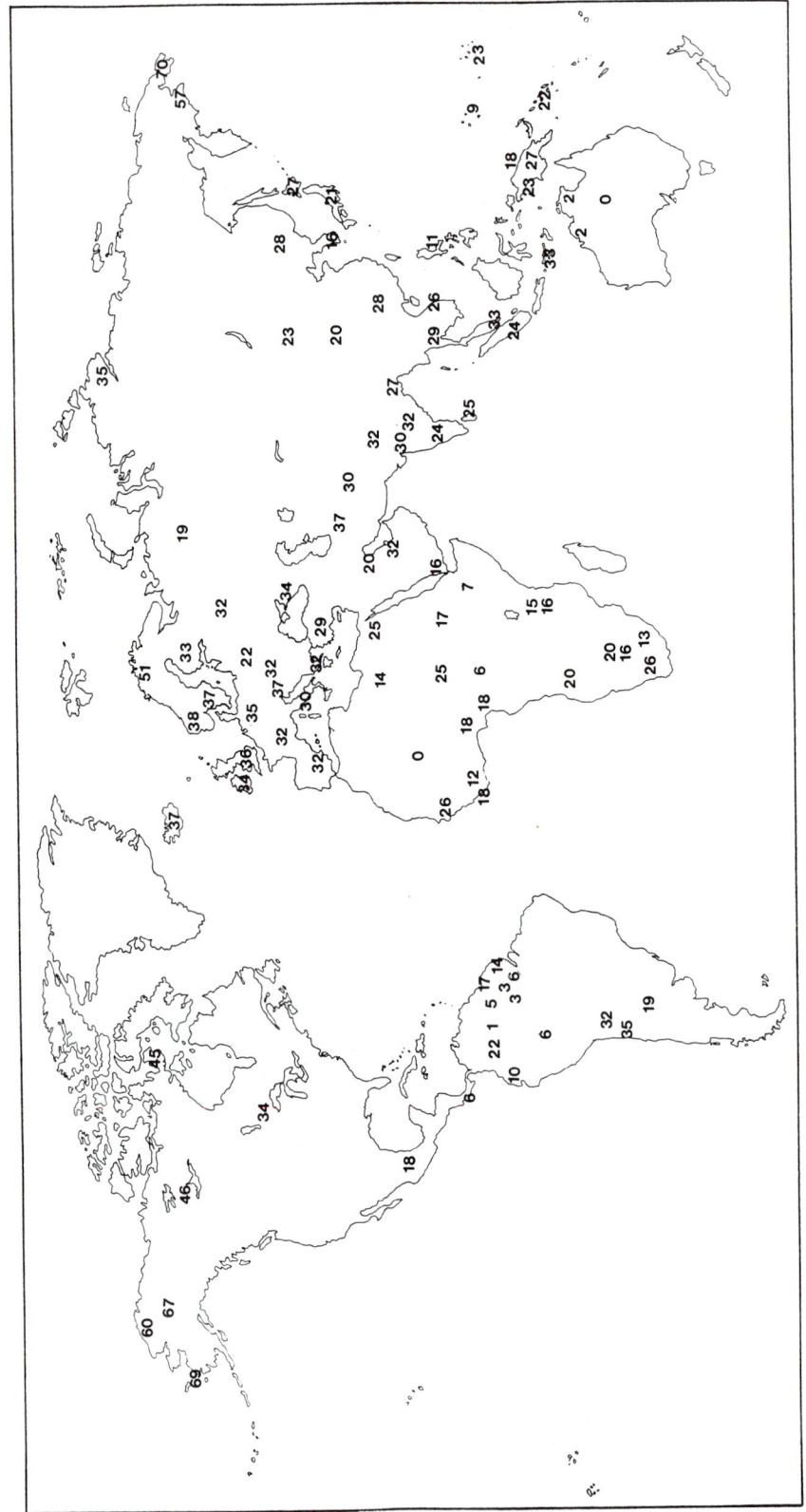

MAP 1. Distribution of allele *ACP1*[a] at the acid phosphatase 1 locus.

MAP 2. Distribution of allele *ACP1ᶜ* at the acid phosphatase 1 locus.

273

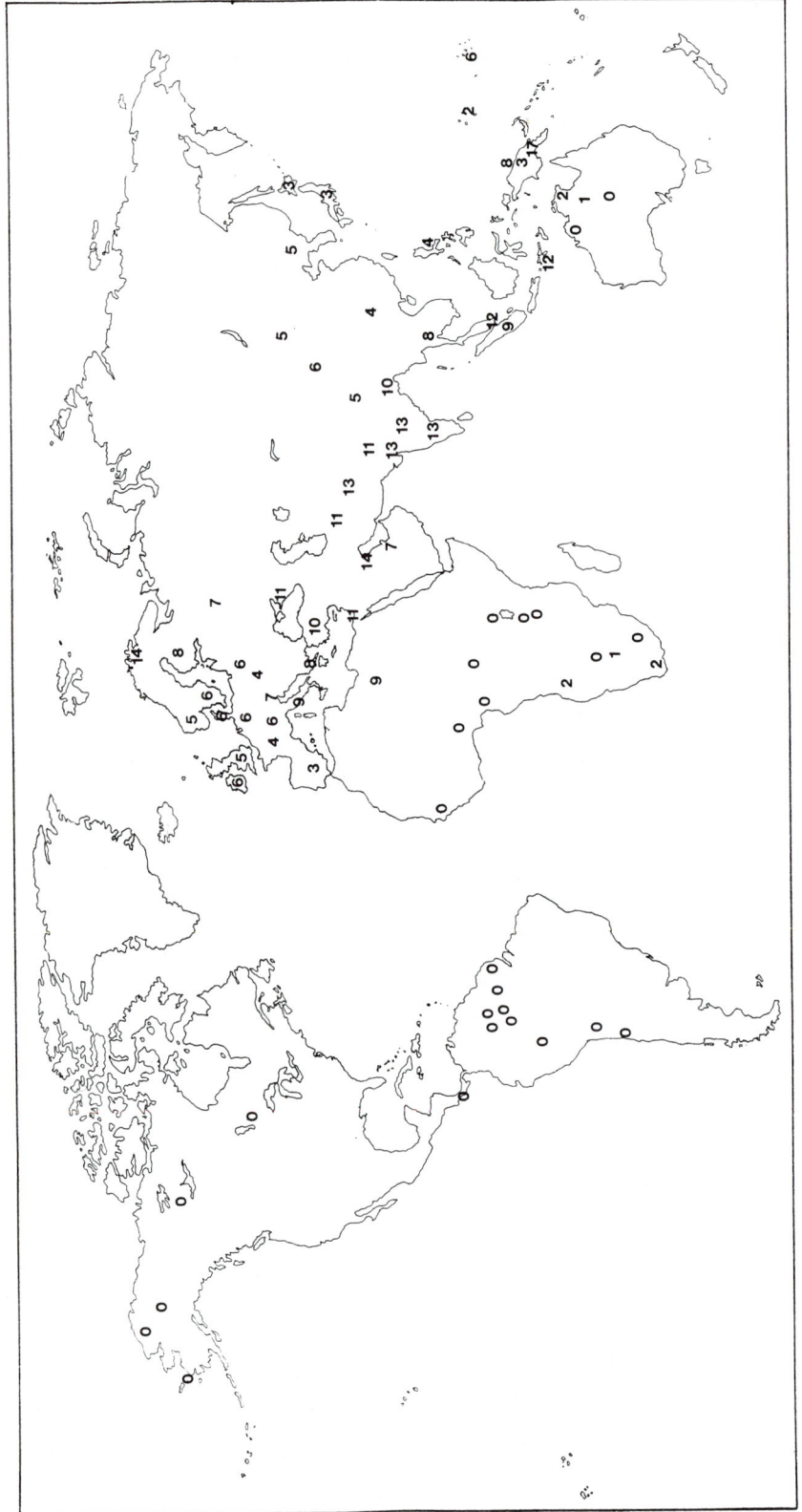

MAP 3. Distribution of allele *ADA²* at the adenosine deaminase locus.

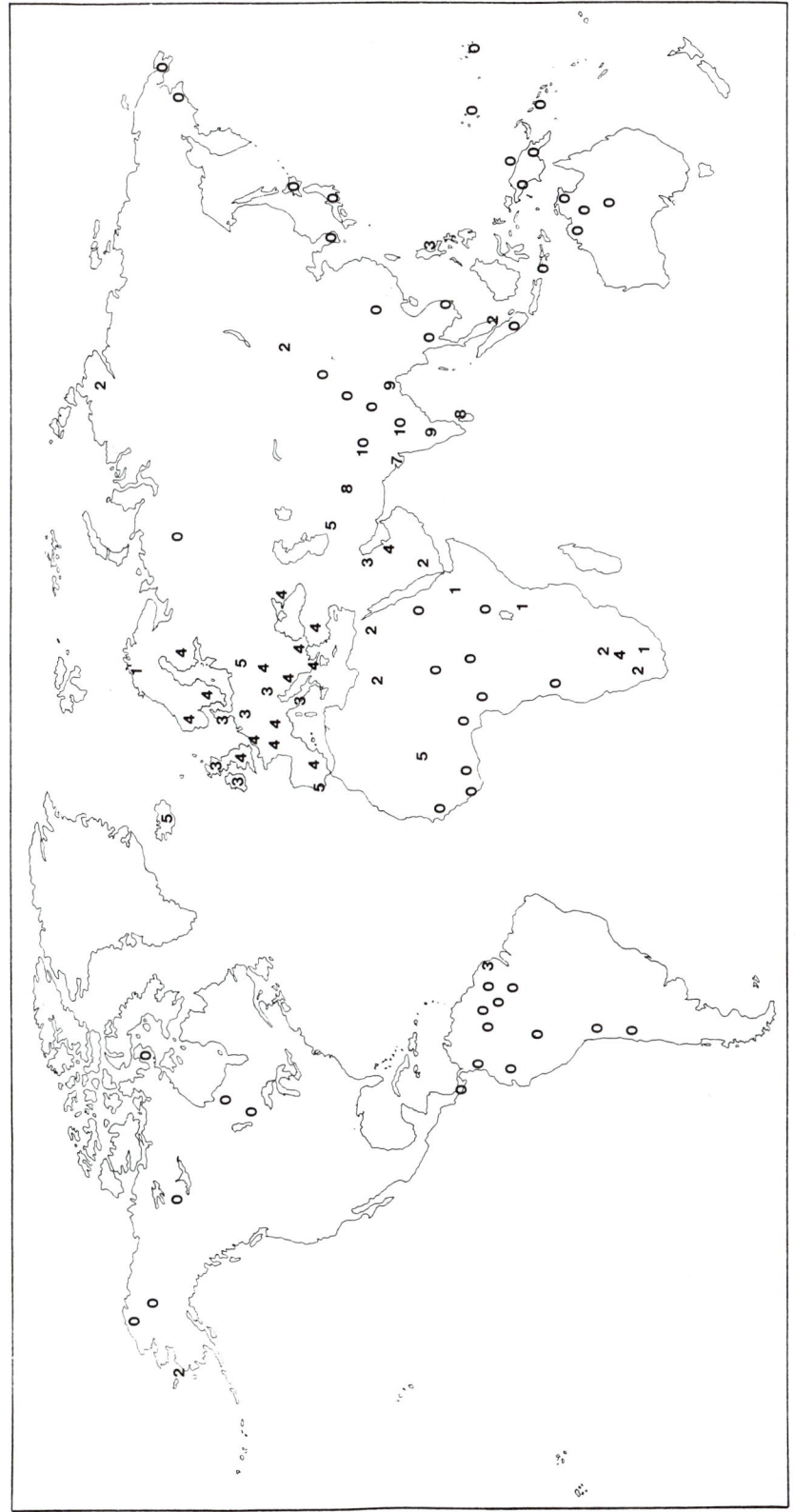

MAP 4. Distribution of allele AKI^2 at the adenylate kinase 1 locus.

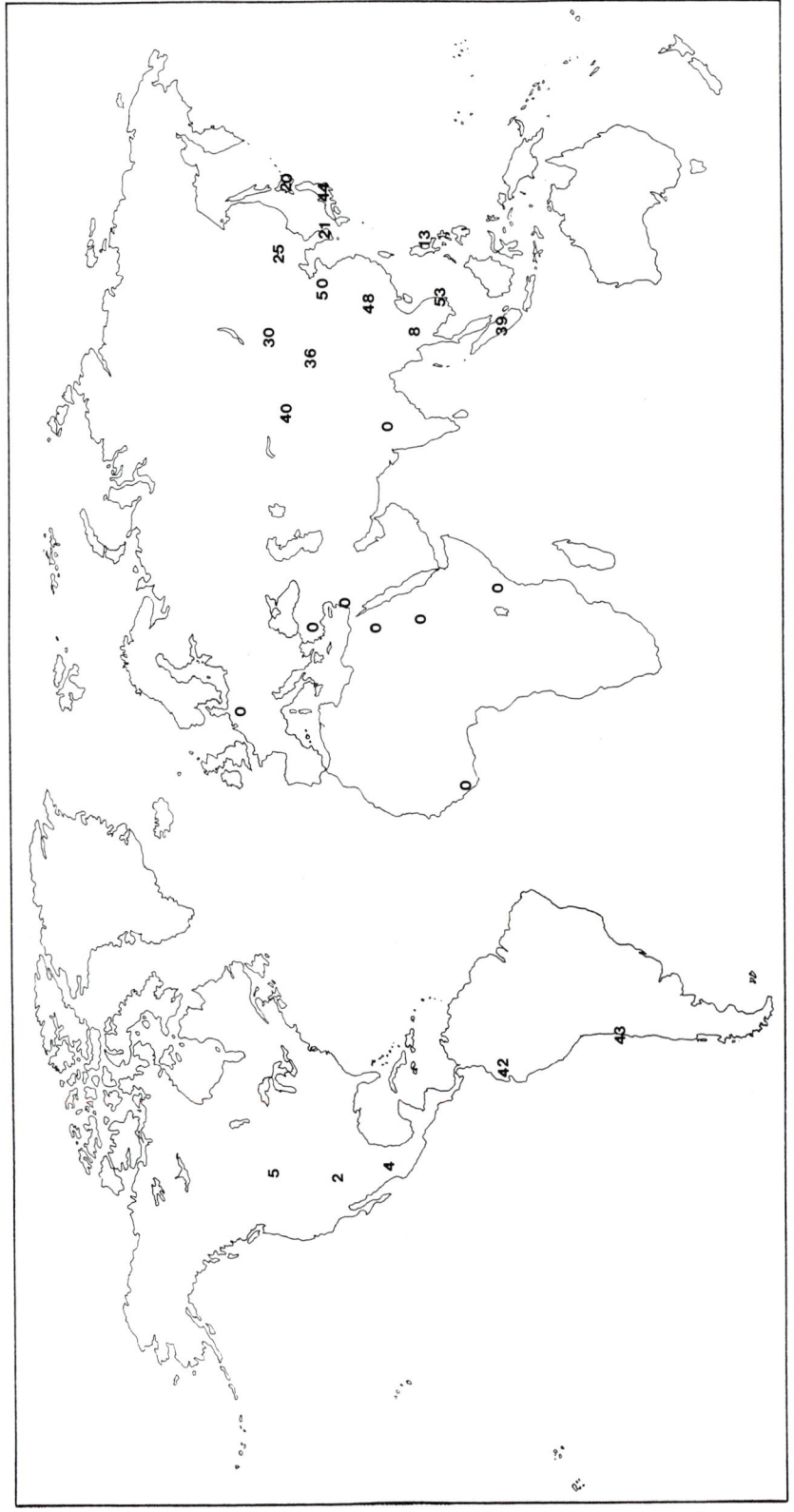

MAP 5. Distribution of the aldehyde dehydrogenase 2 deficiency.

MAP 6. Distribution of allele $CHE1^a$ at the cholinesterase 1 locus.

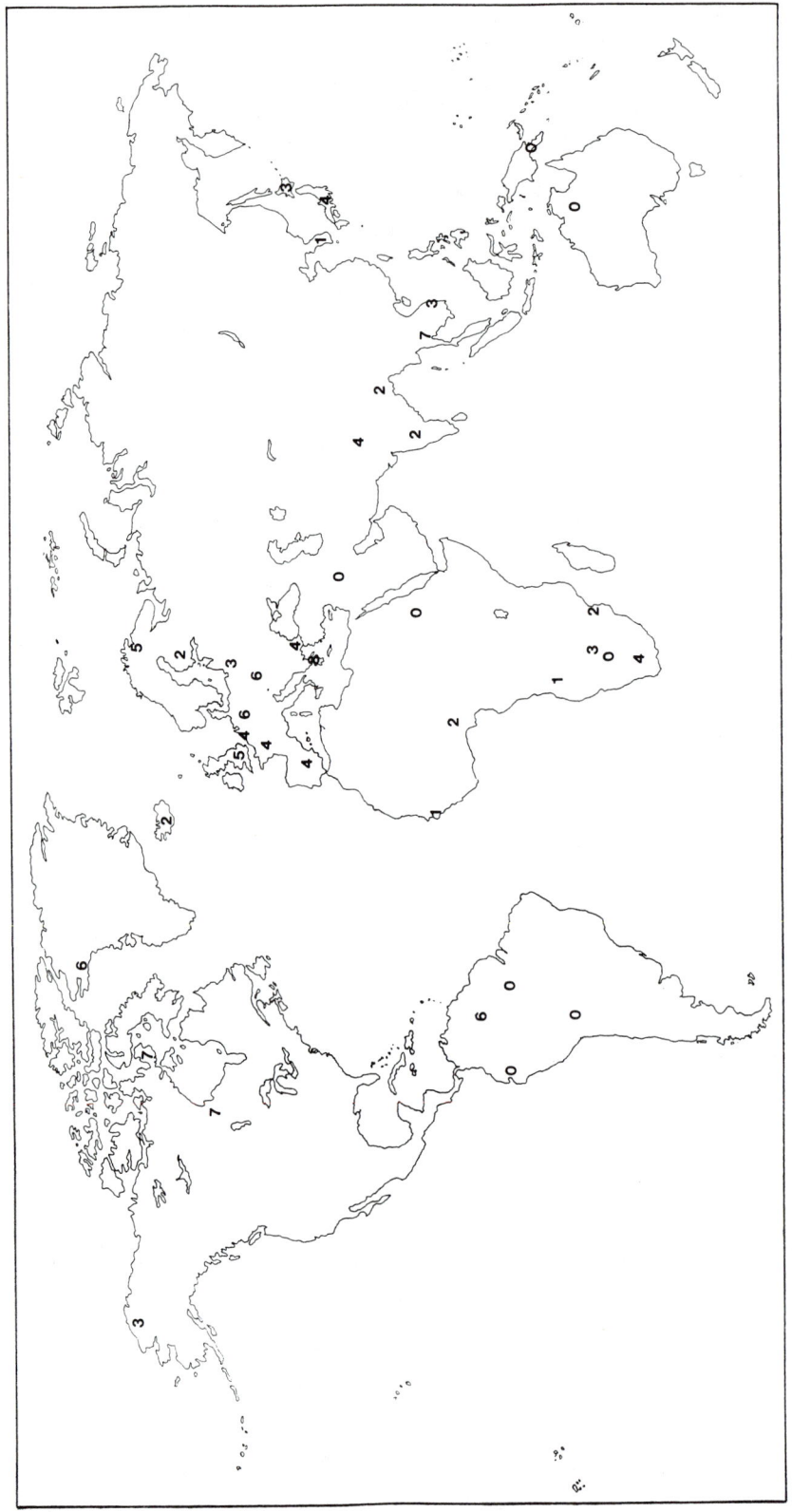

MAP 7. Distribution of allele *CHE2*⁺ at the cholinesterase 2 locus.

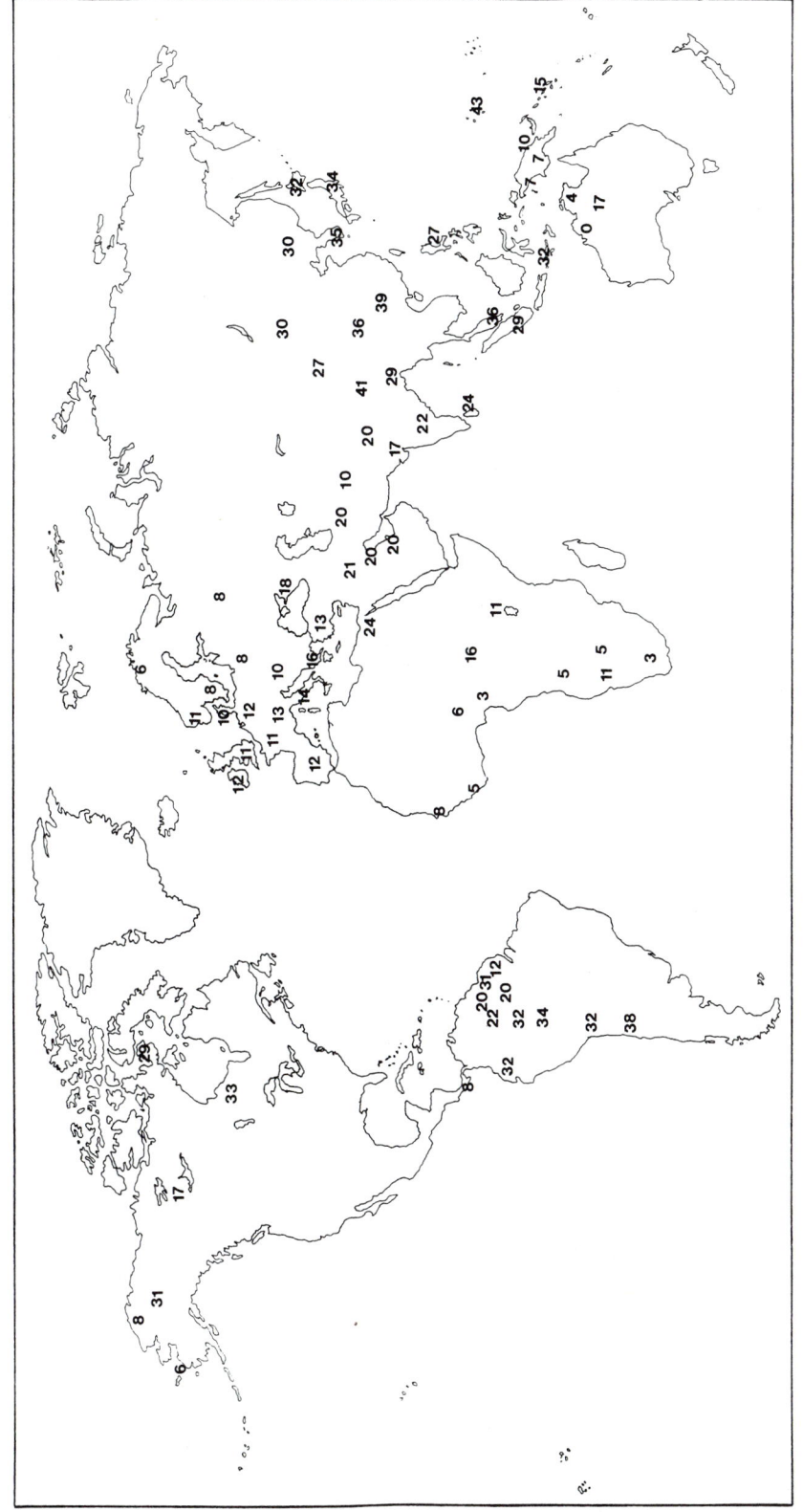

MAP 8. Distribution of allele *ESD*[2] at the esterase D locus.

279

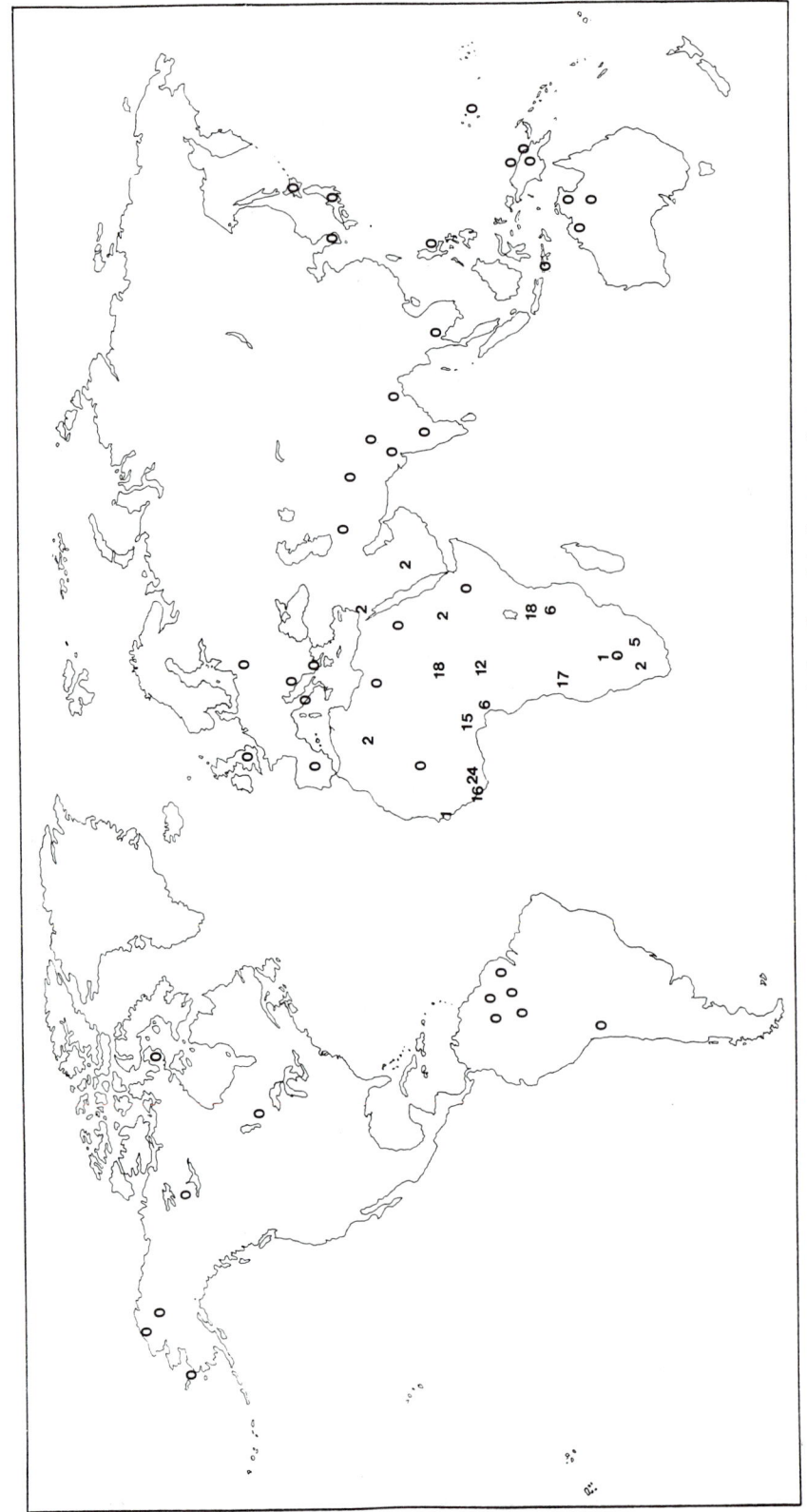

MAP 9. Distribution of allele A^- at the glucose-6-phosphate dehydrogenase locus.

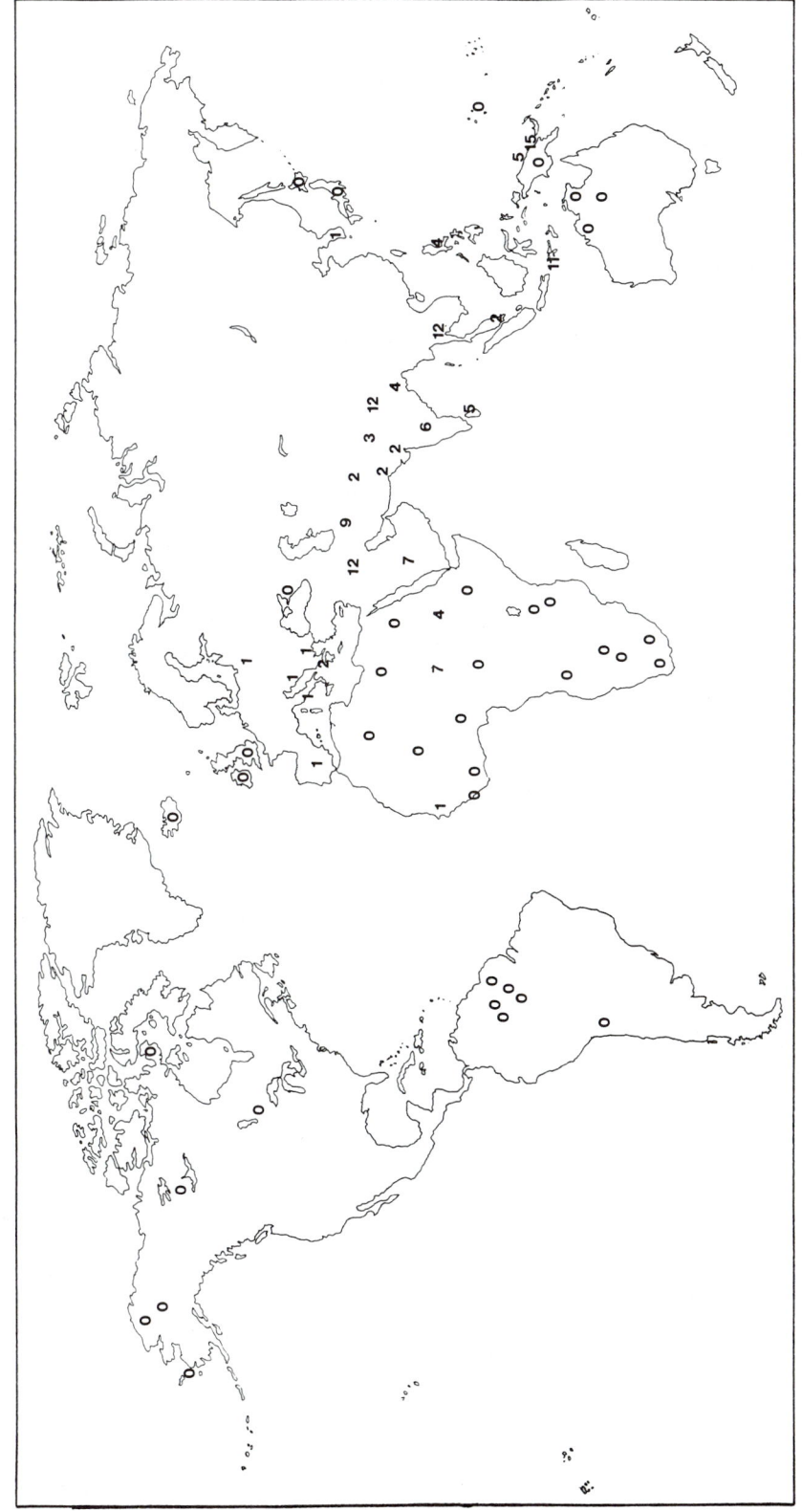

MAP 10. Distribution of allele B^- at the glucose-6-phosphate dehydrogenase locus.

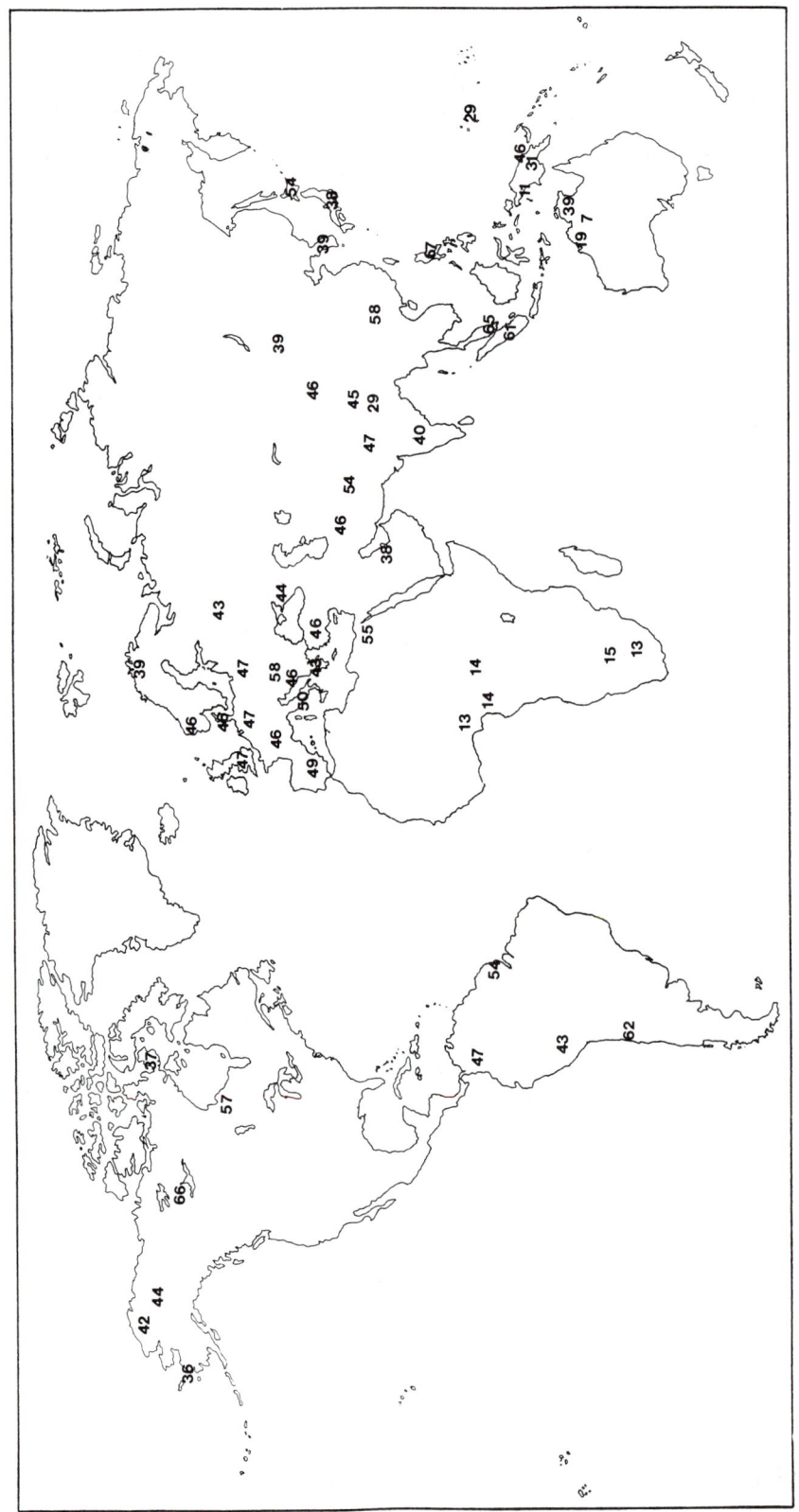

MAP 11. Distribution of allele *GPT1²* at the glutamic-pyruvate transaminase locus.

MAP 12. Distribution of allele *GPX1²* at the glutathione peroxidase 1 locus.

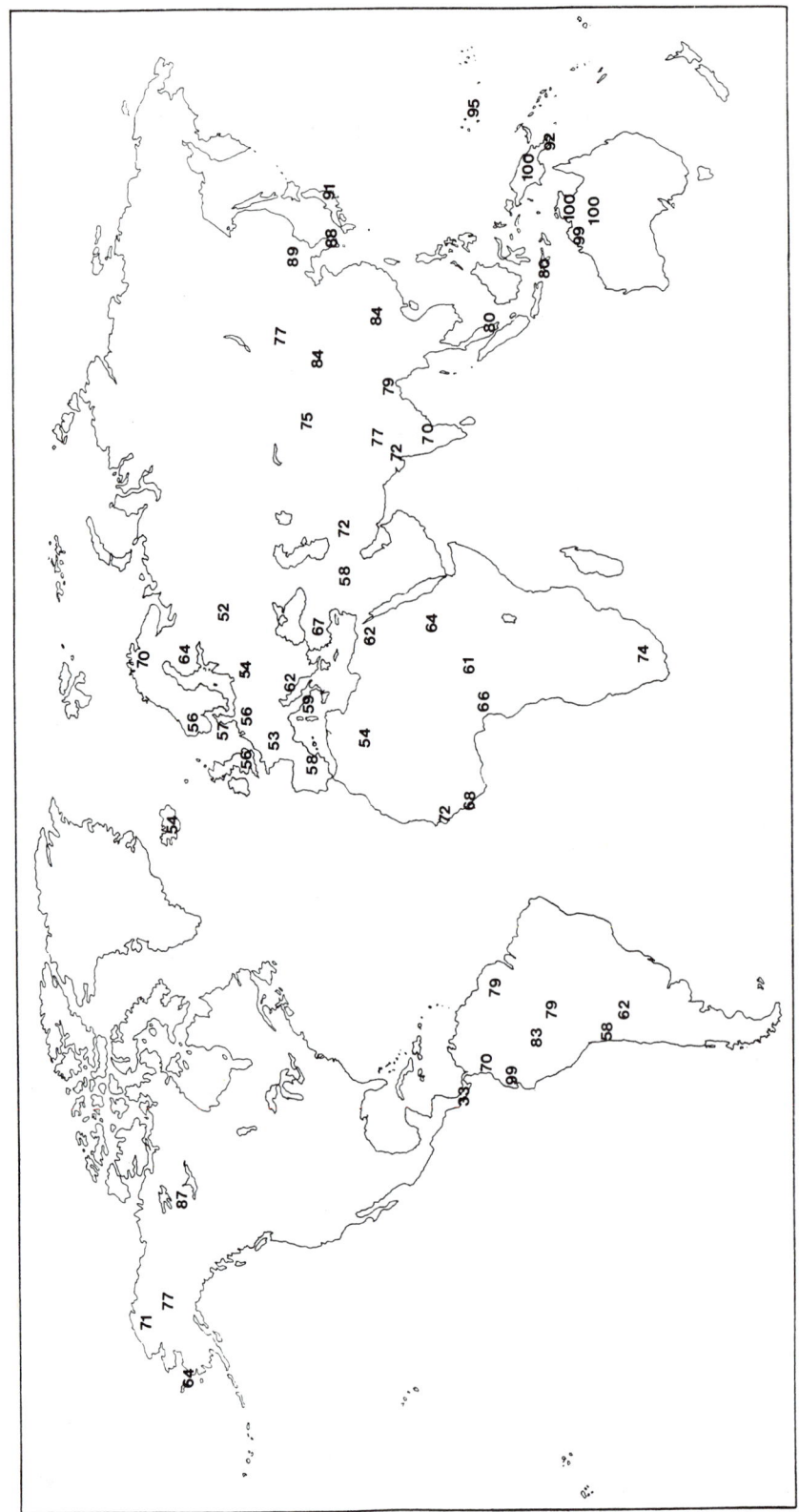

MAP 13. Distribution of allele *GLOI²* at the glyoxalase 1 locus.

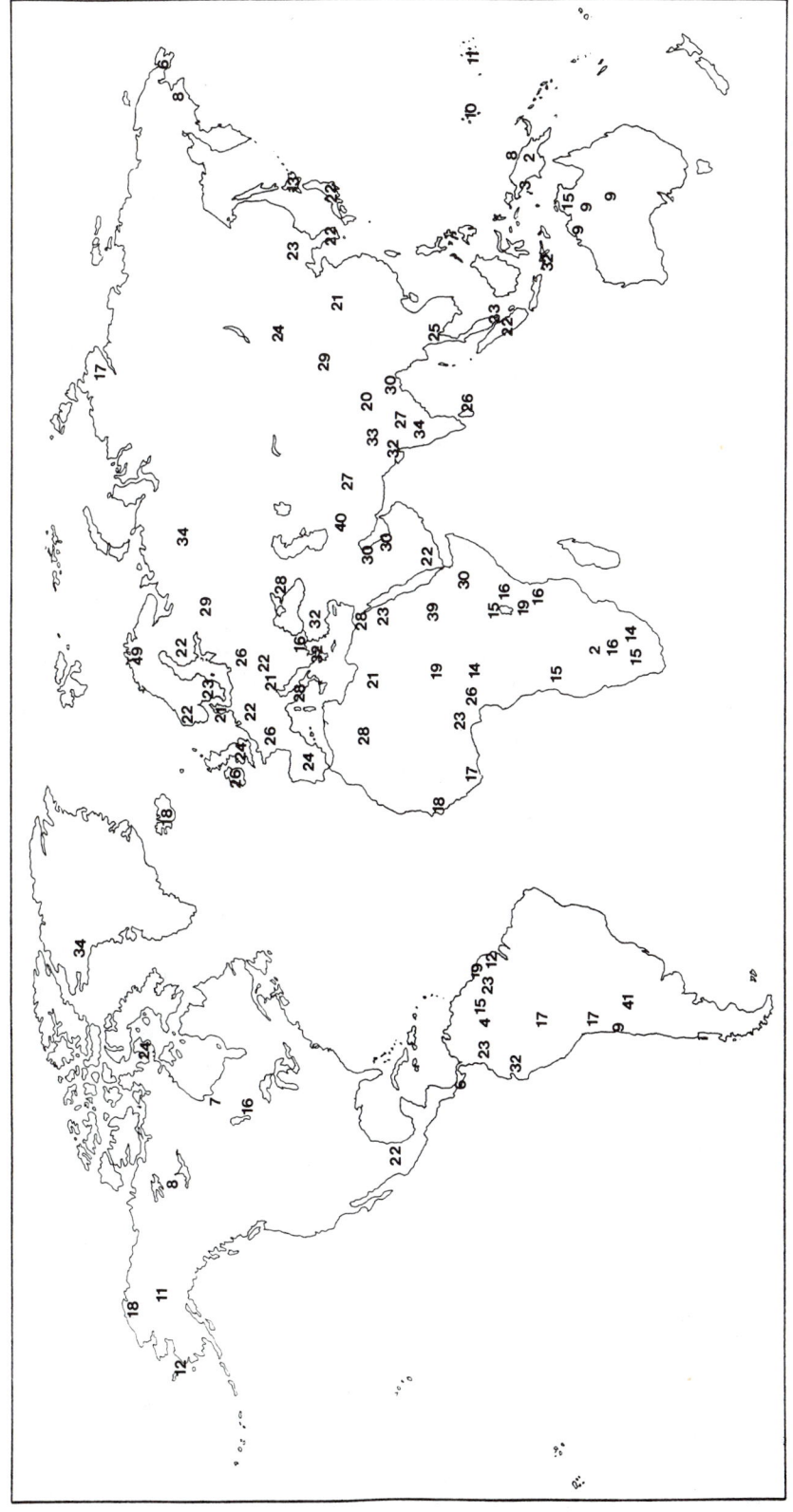

MAP 14. Distribution of allele $PGM1^2$ at the phosphoglucomutase 1 locus.

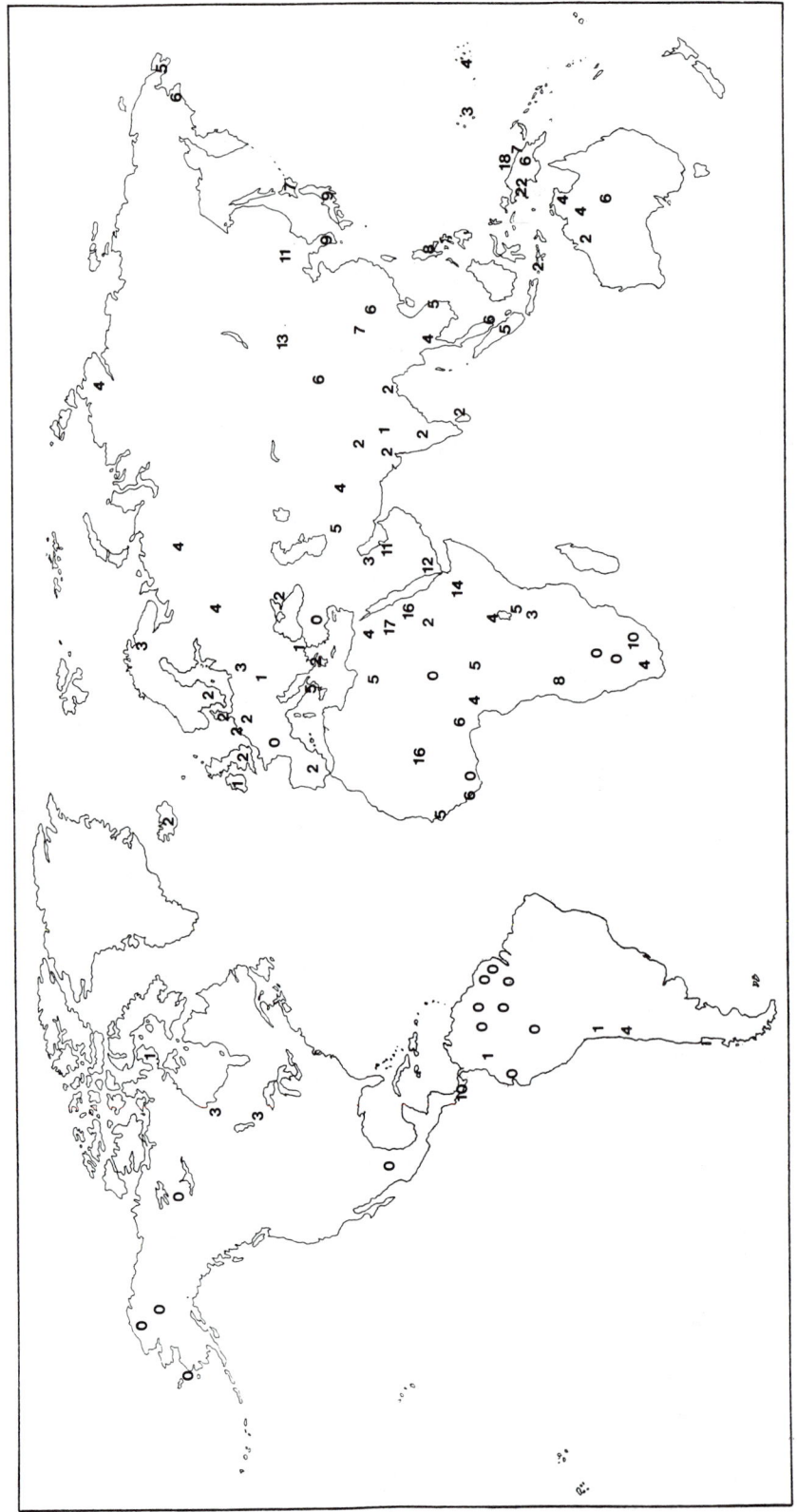

MAP 15. Distribution of allele *PGD*^{C} at the phosphogluconate dehydrogenase locus.

MAP 16. Distribution of allele PGP^2 at the phosphoglycolate phosphatase locus.

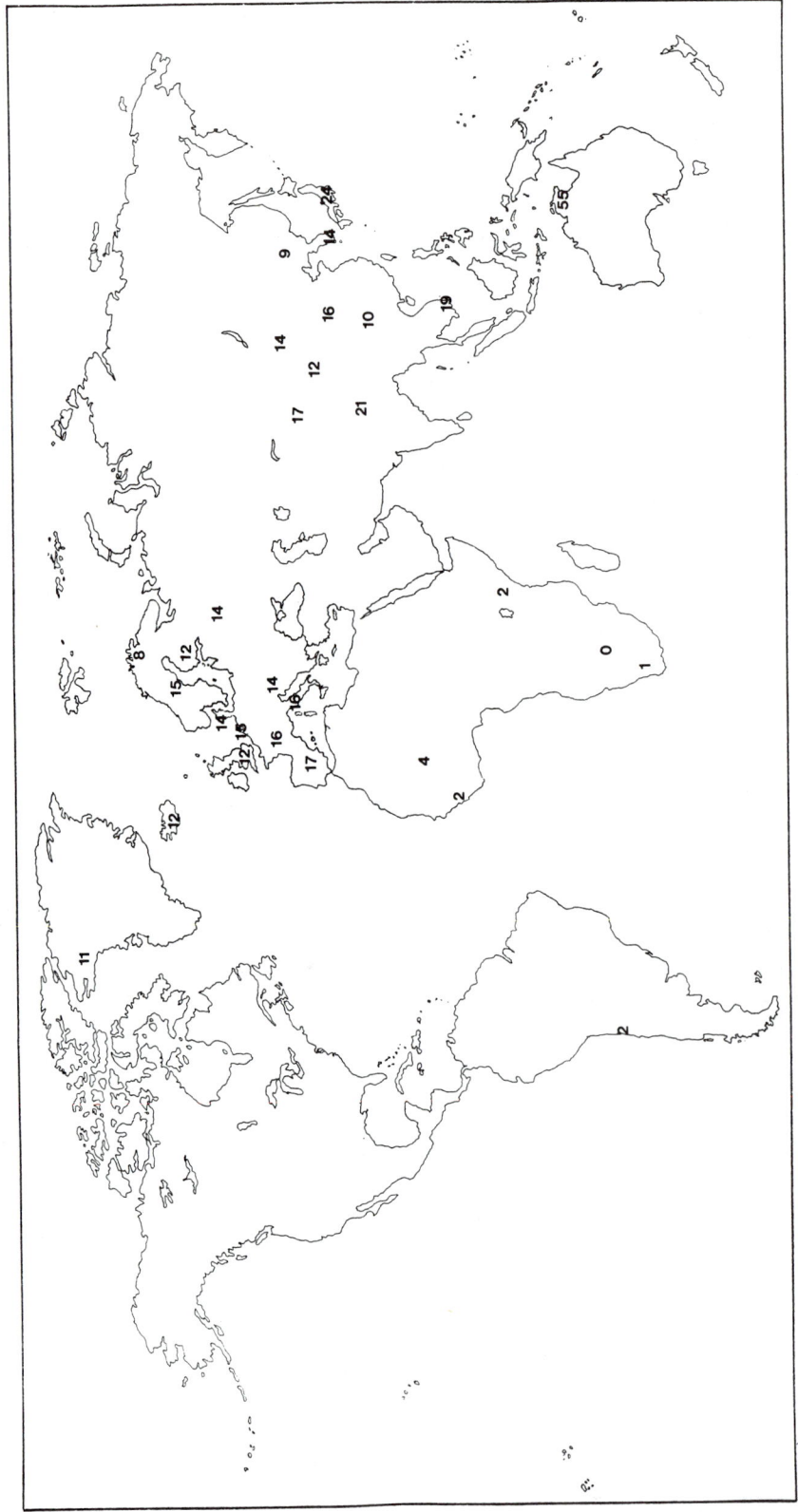

MAP 17. Distribution of allele M^2 at the alpha-1-antitrypsin (subtype) locus.

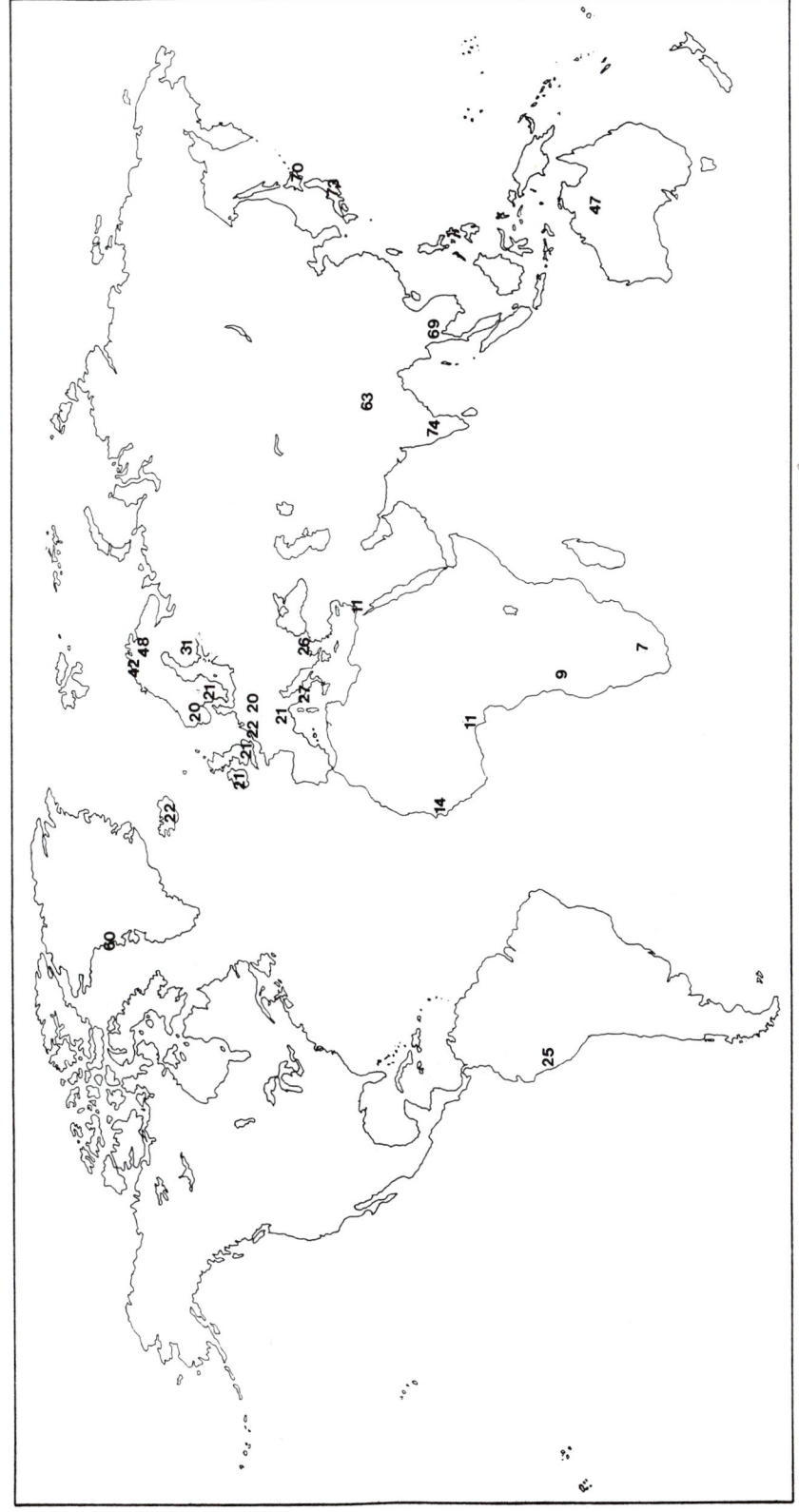

MAP 18. Distribution of allele Ag^x at the beta lipoprotein Ag system.

MAP 19. Distribution of allele CP^A at the ceruloplasmin locus.

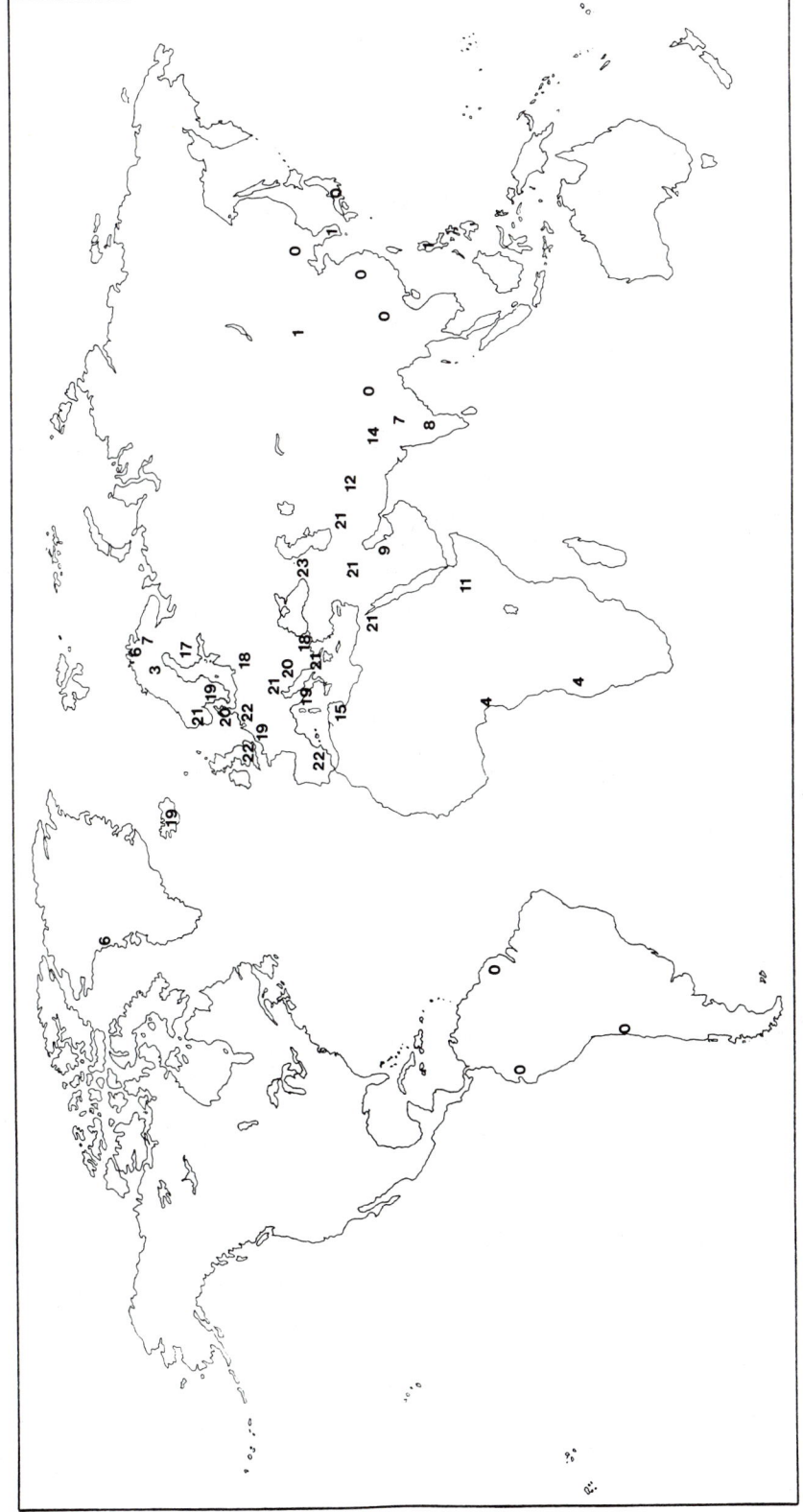

MAP 20. Distribution of allele $C3^F$ at the complement component 3 locus.

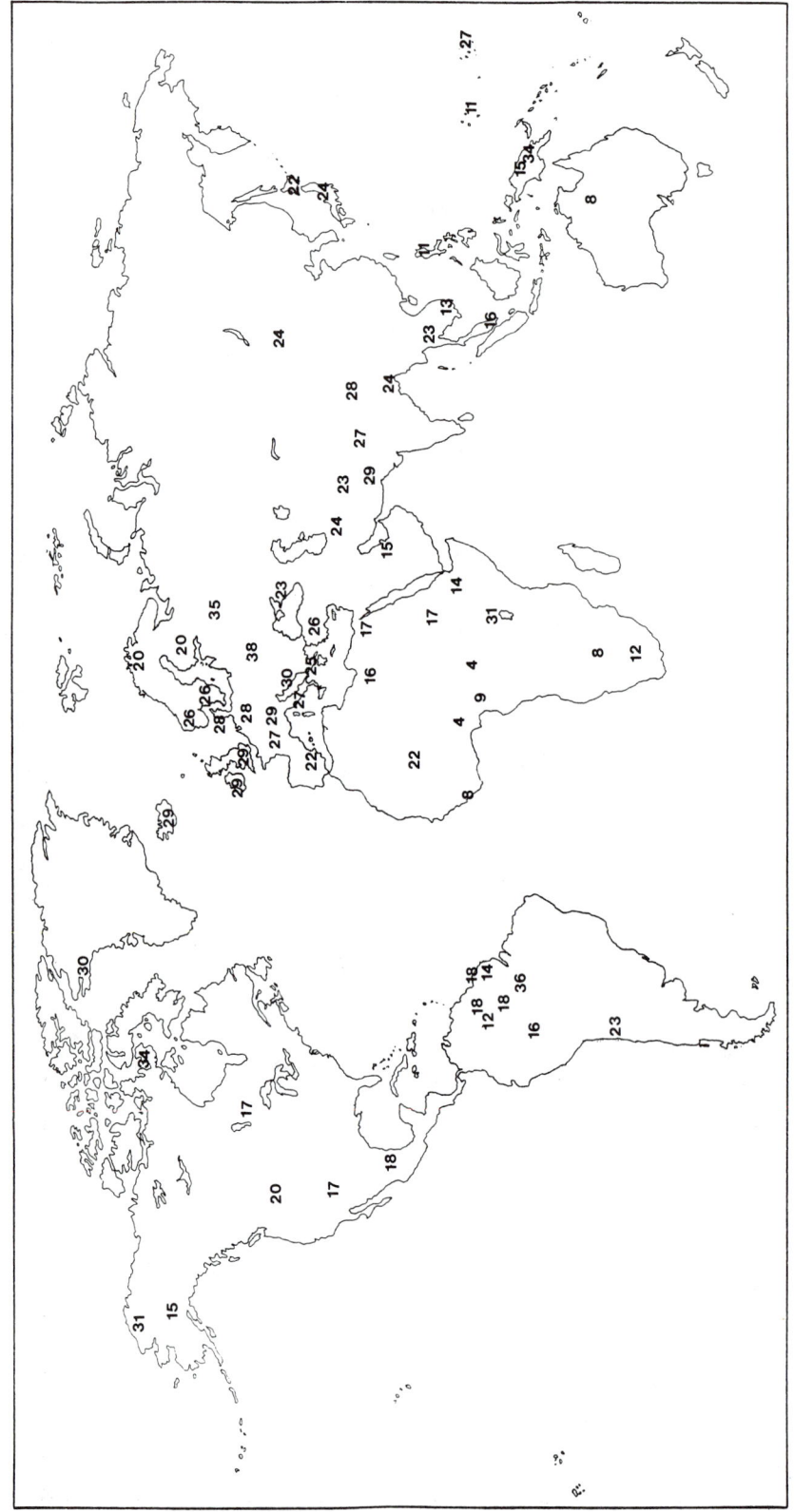

MAP 21. Distribution of allele GC^2 at the group specific component locus.

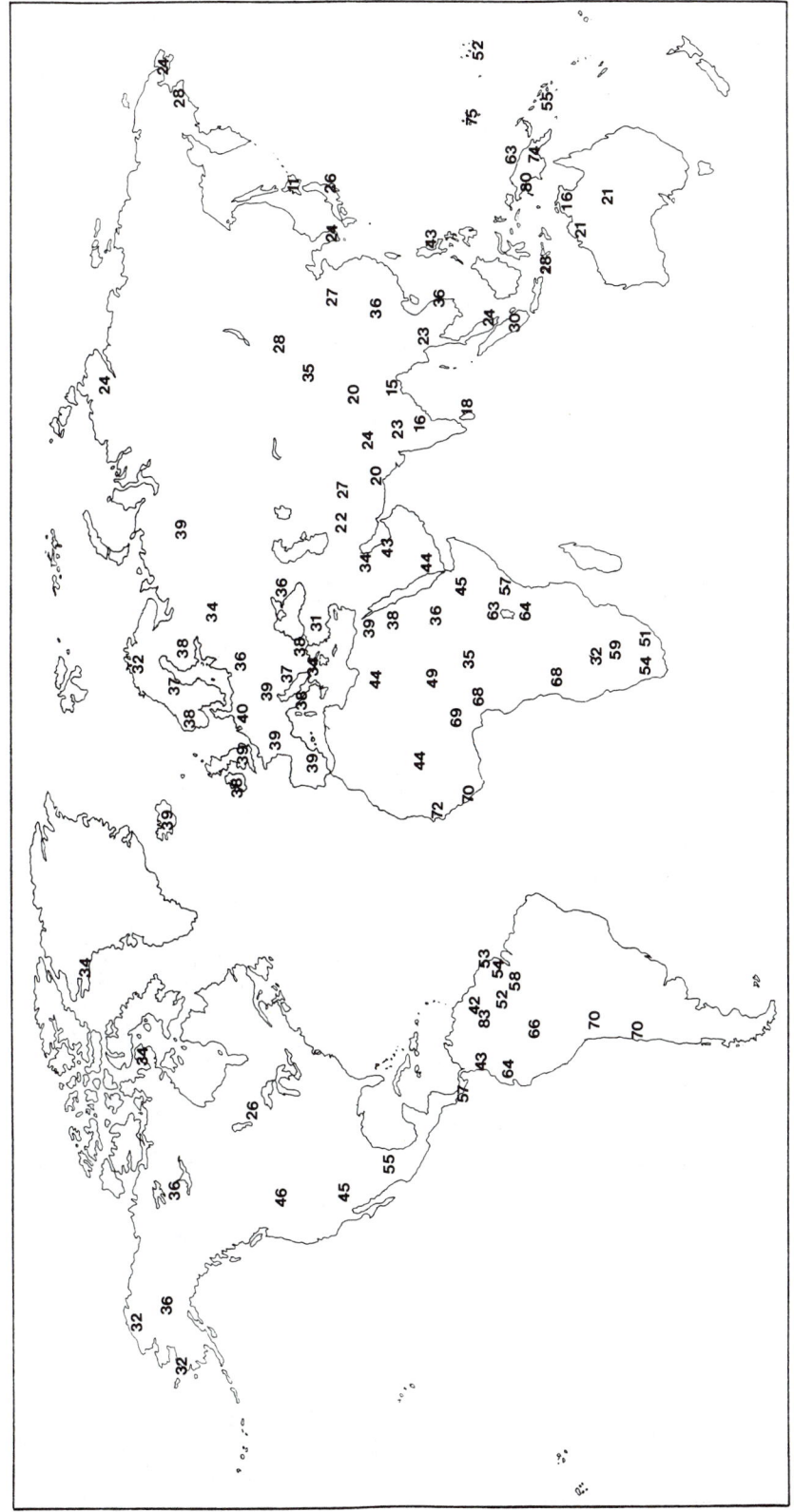

MAP 22. Distribution of allele Hp^1 at the haptoglobin-alpha locus.

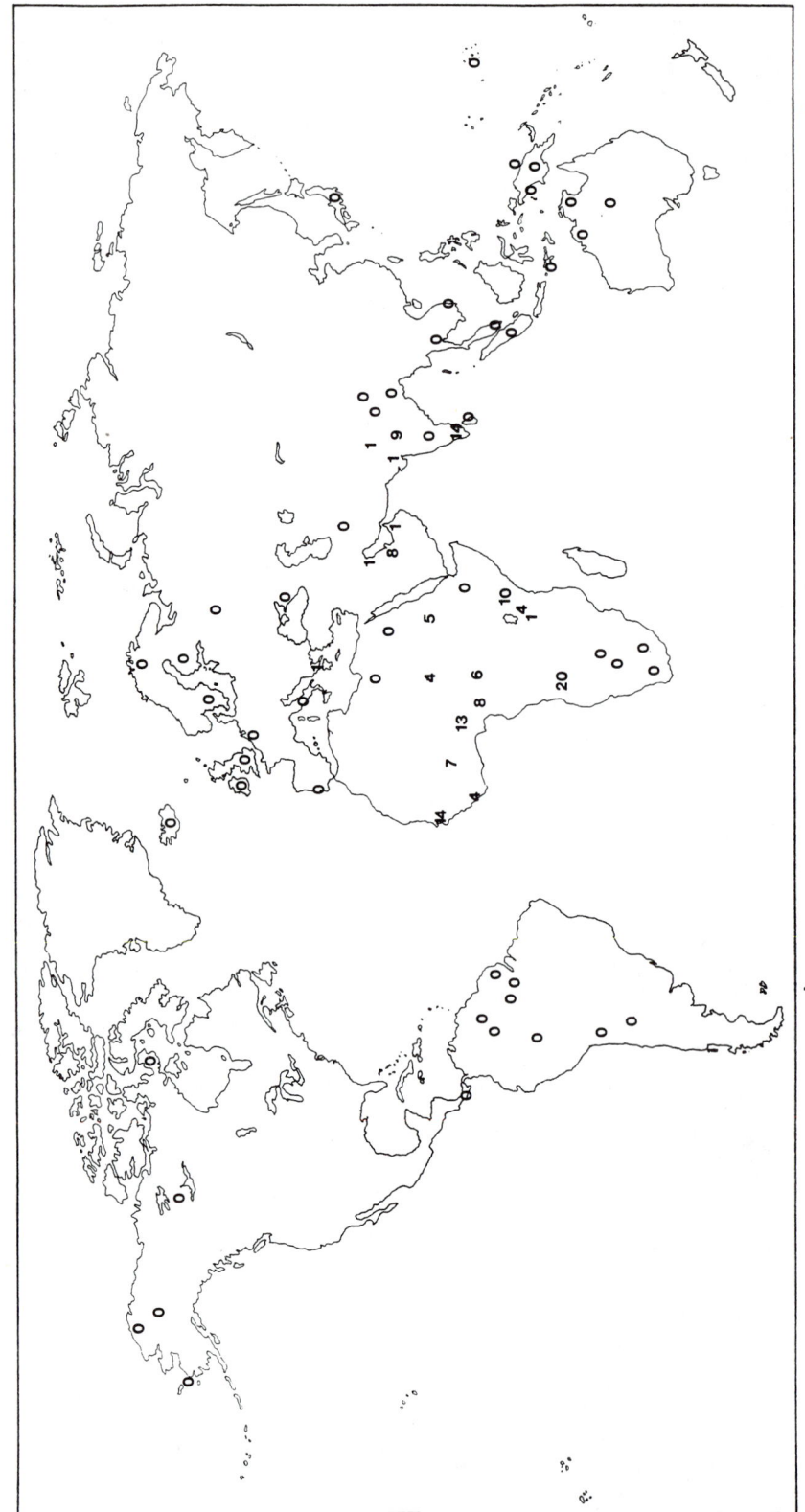

MAP 23. Distribution of allele *S* at the hemoglobin-beta locus.

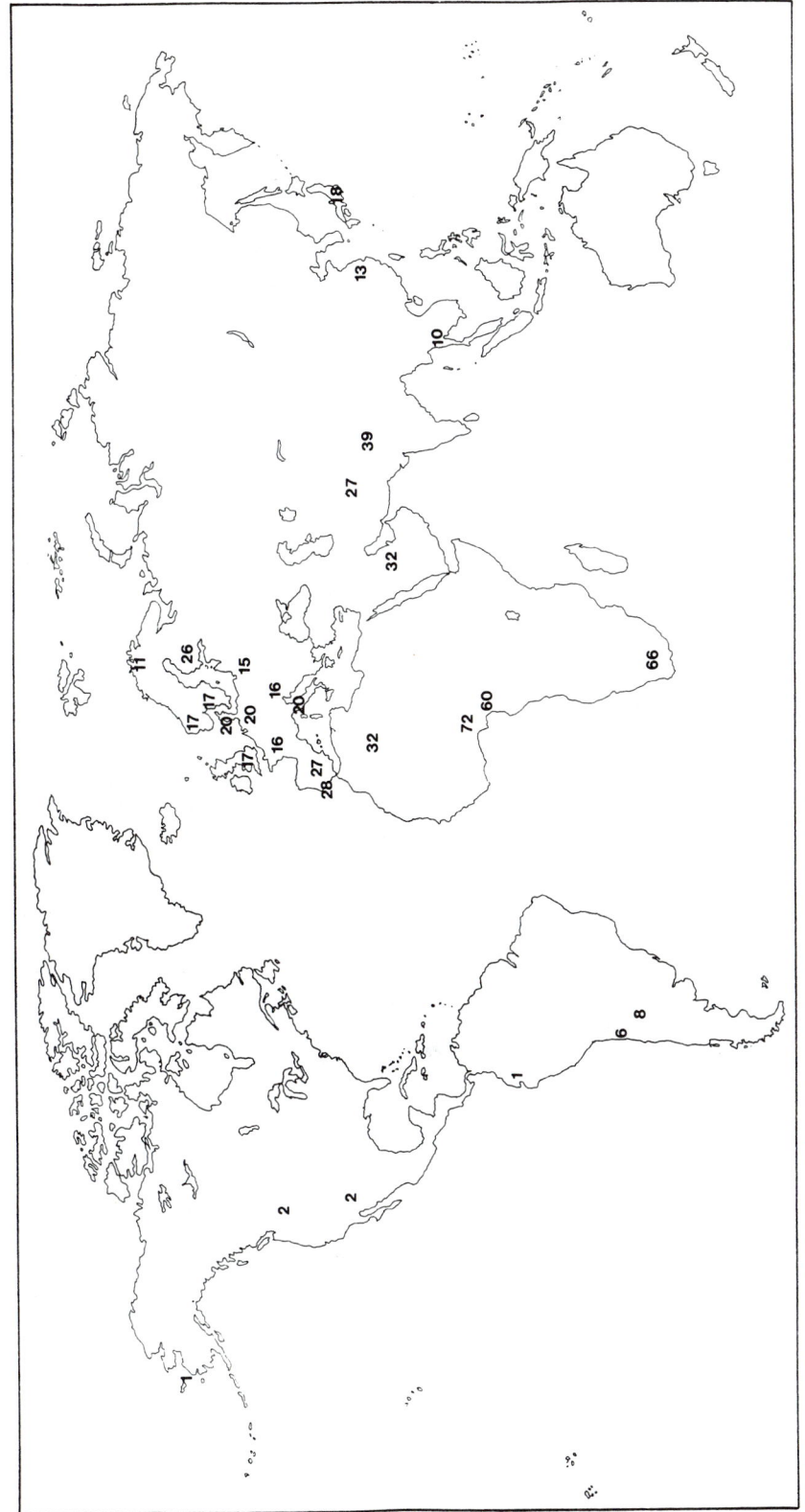

MAP 24. Distribution of allele BF^F at the proferdin factor B locus.

MAP 25. Distribution of allele TBG^S at the thyroxin-binding globulin locus.

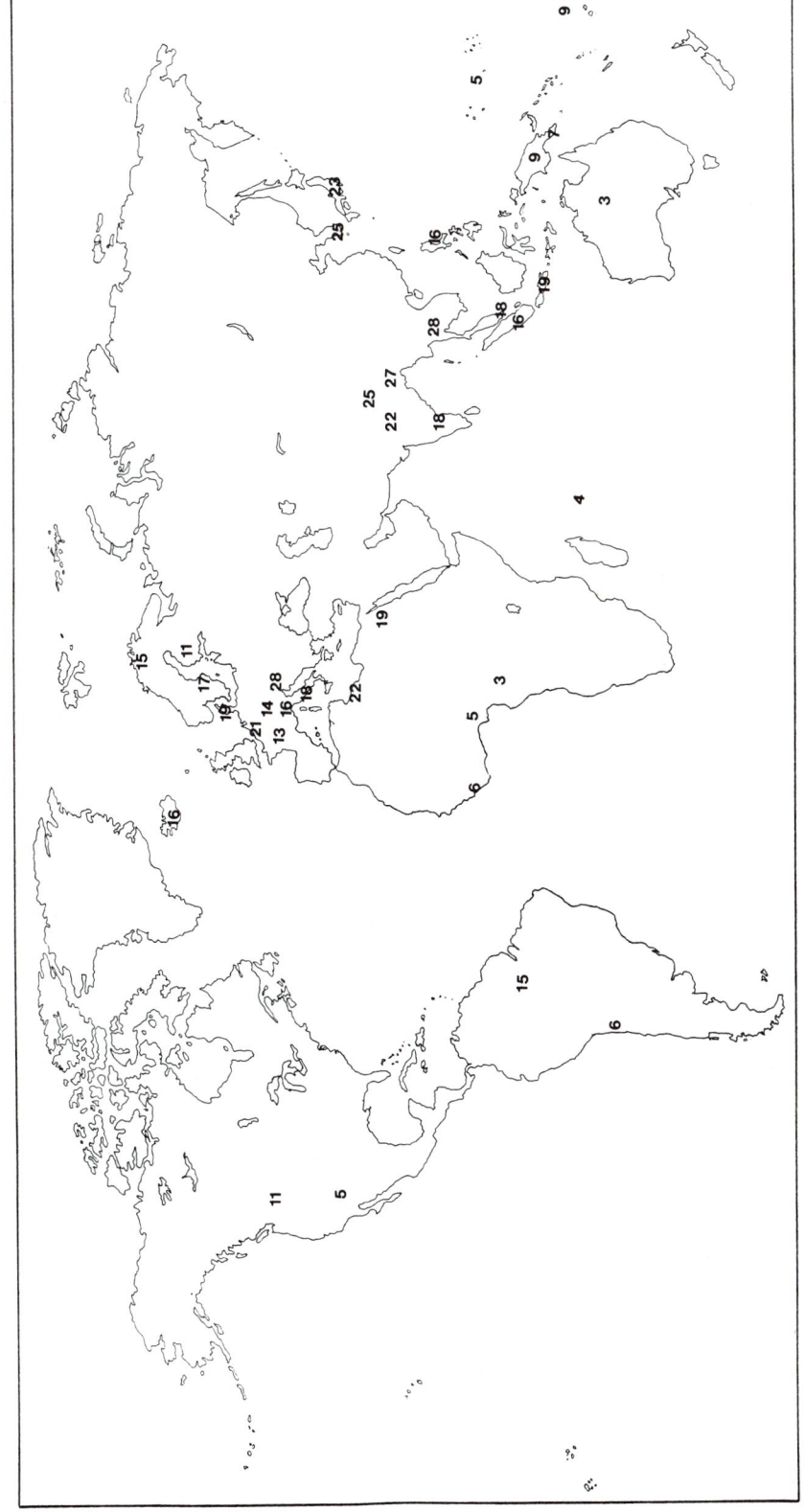

MAP 26. Distribution of allele C_2 at the transferrin (subtype) locus.

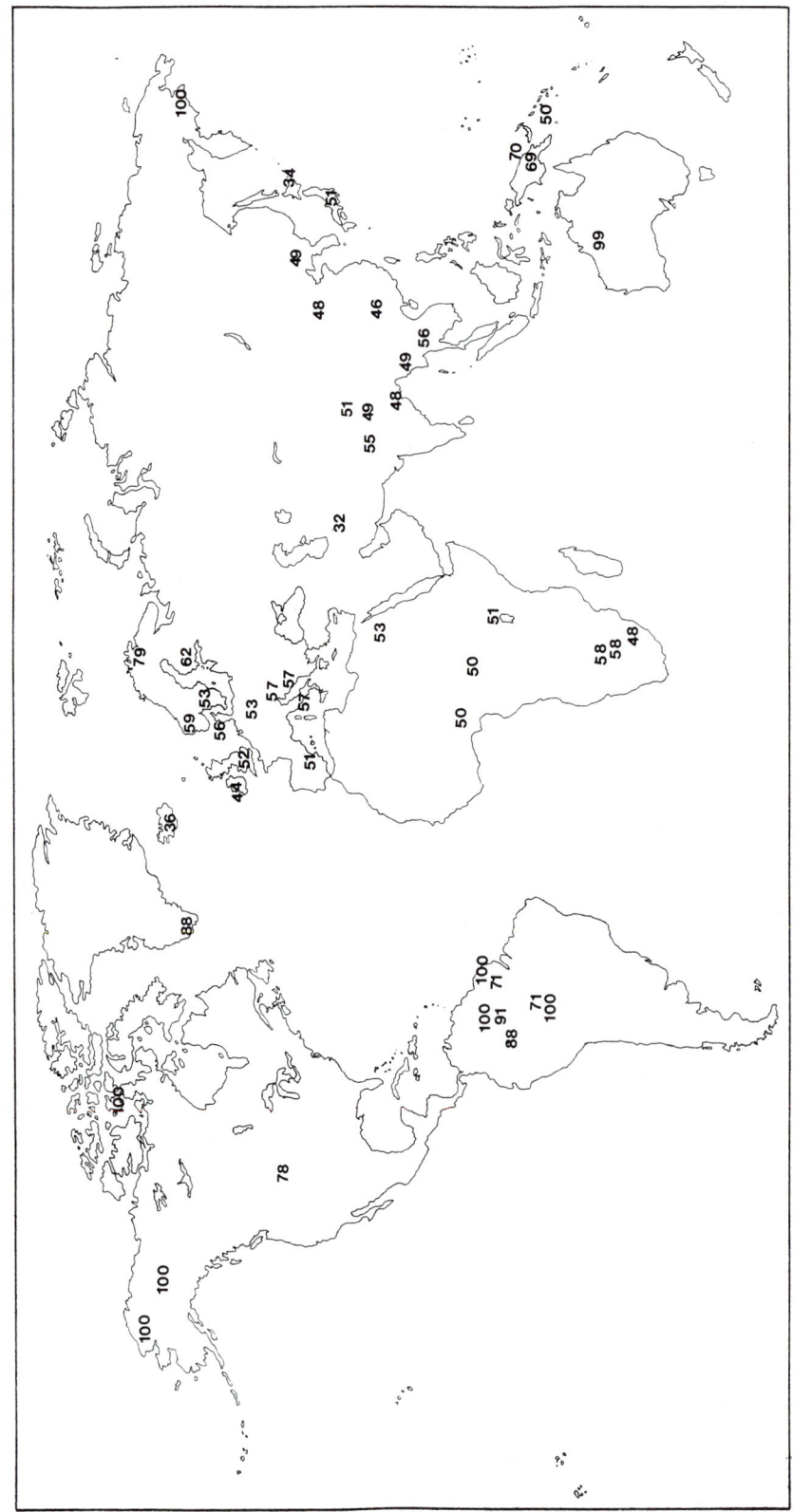

MAP 27. Distribution of allele *Se* at the ABH secretion locus.

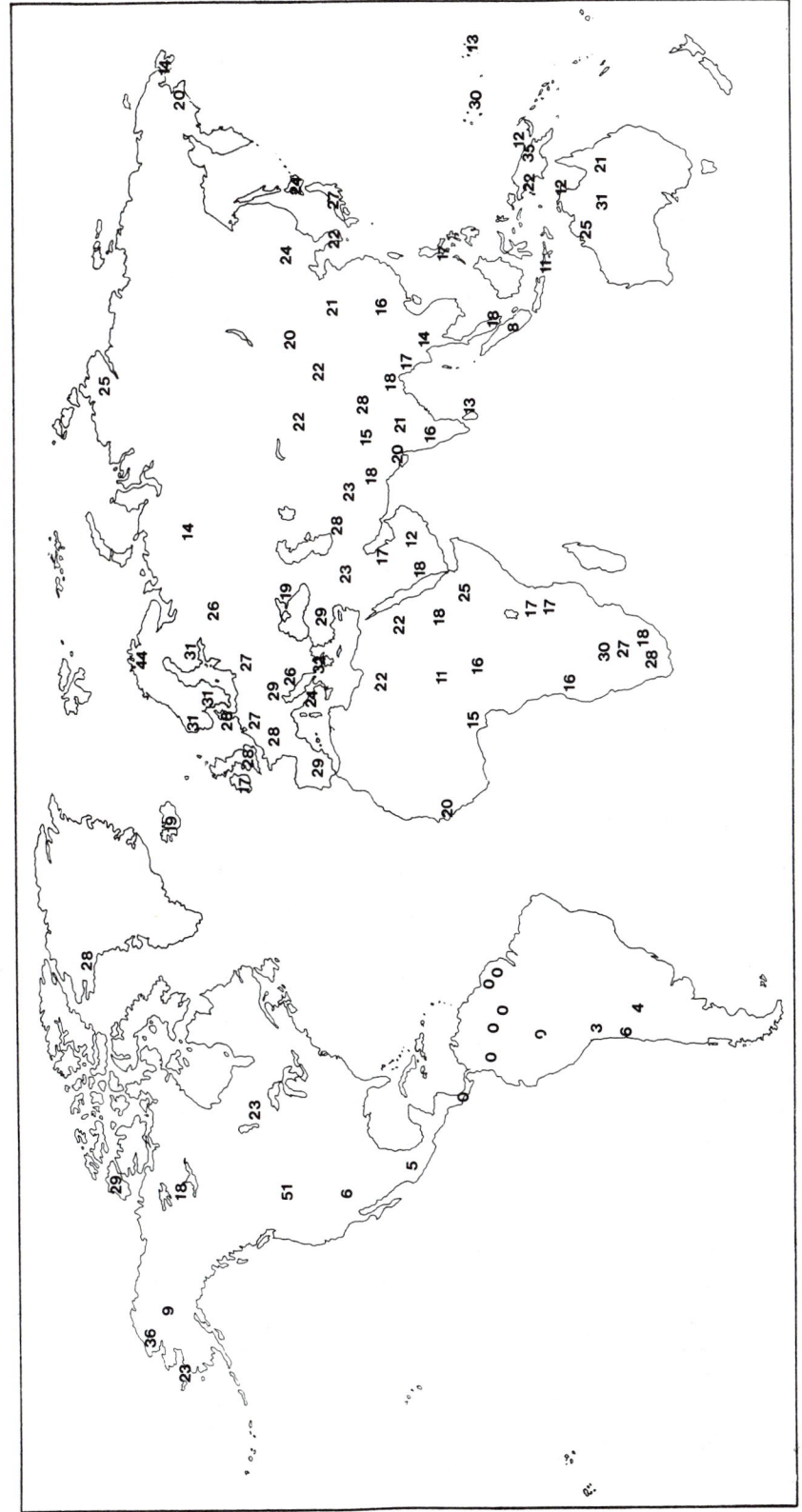

MAP 28. Distribution of allele *A* at the ABO system.

299

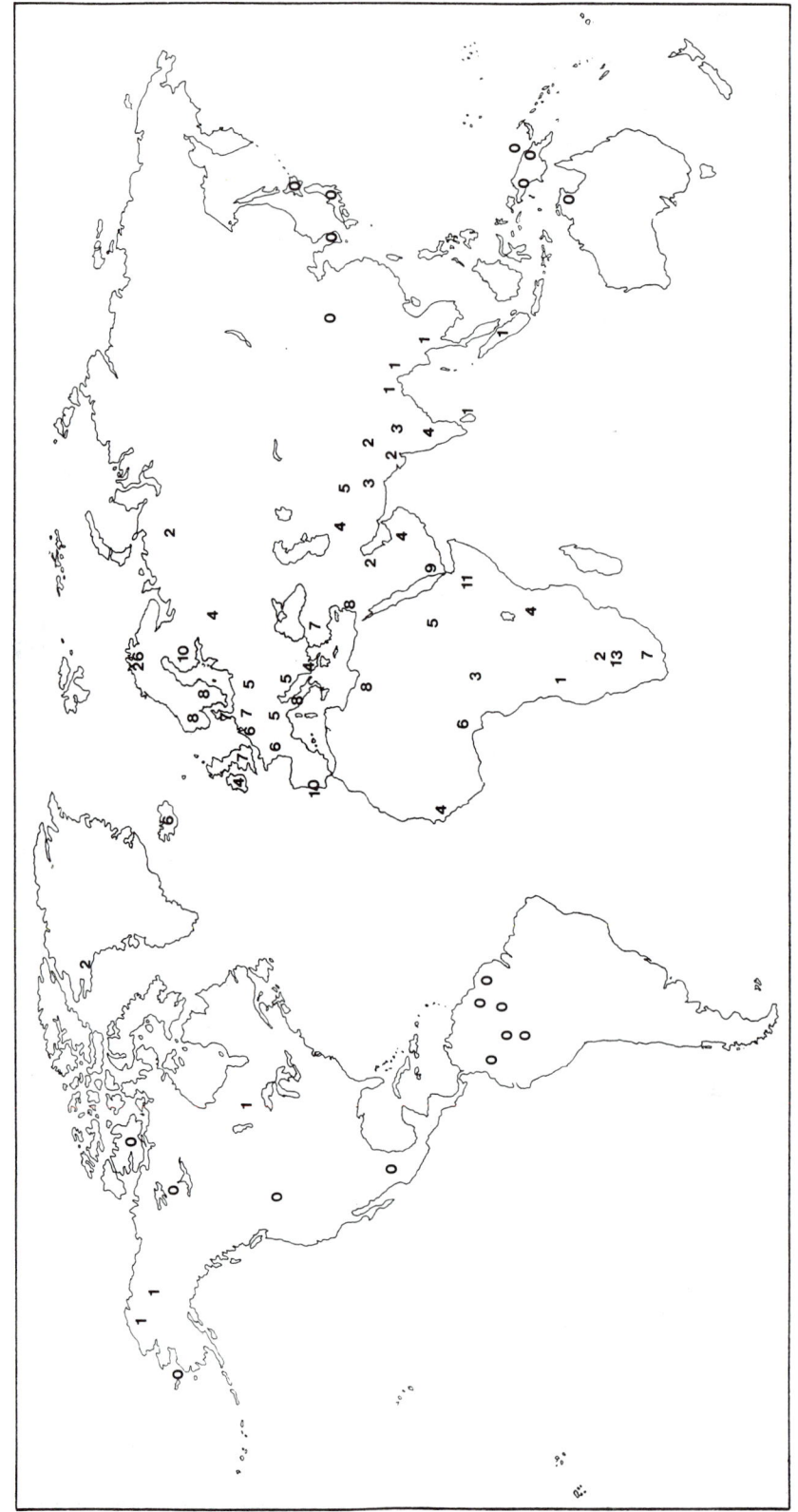

MAP 29. Distribution of allele A_2 at the ABO system.

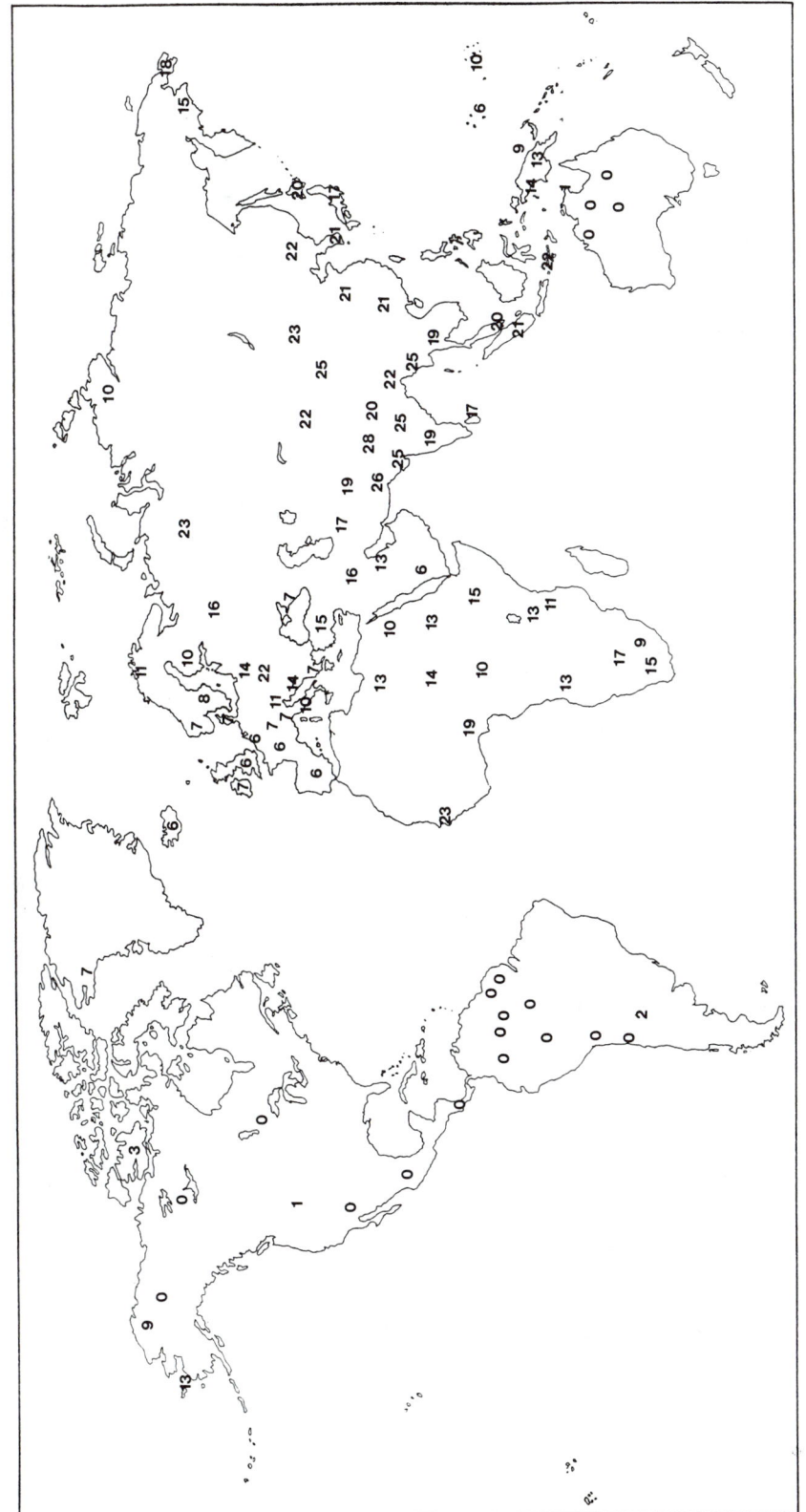

MAP 30. Distribution of allele *B* at the ABO system.

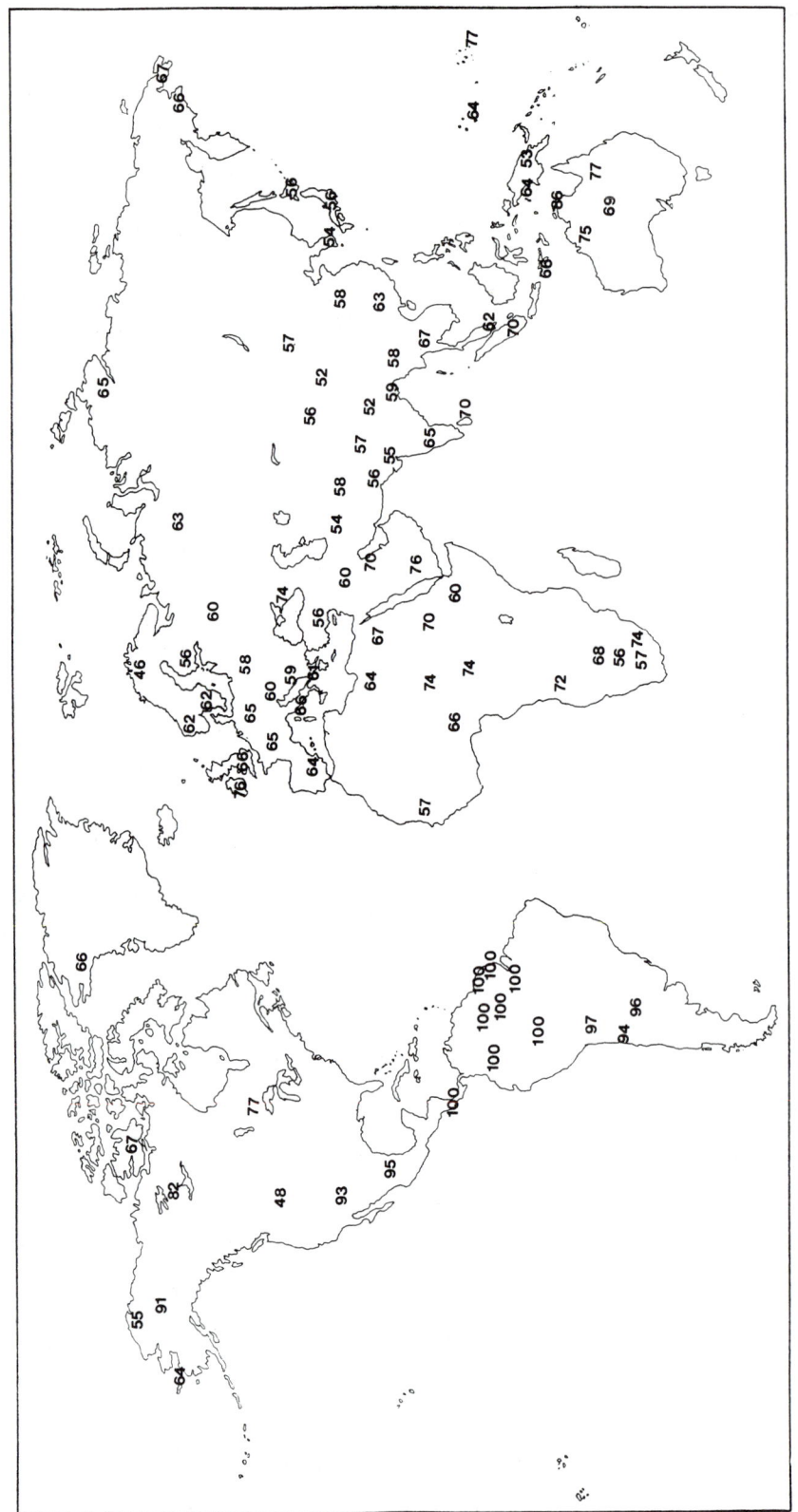

MAP 31. Distribution of allele *O* at the ABO system.

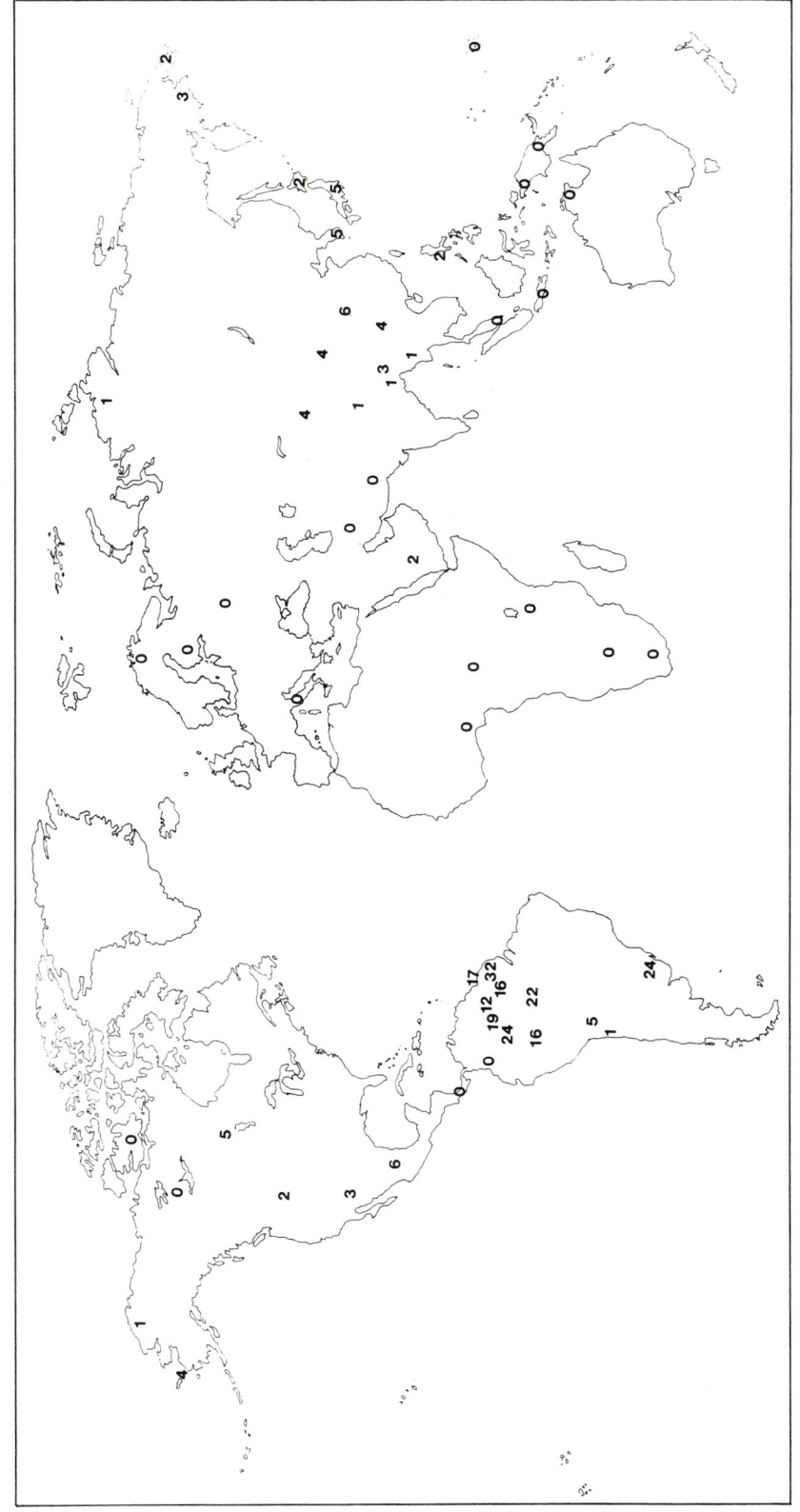

MAP 32. Distribution of allele D^a_i at the Diego system.

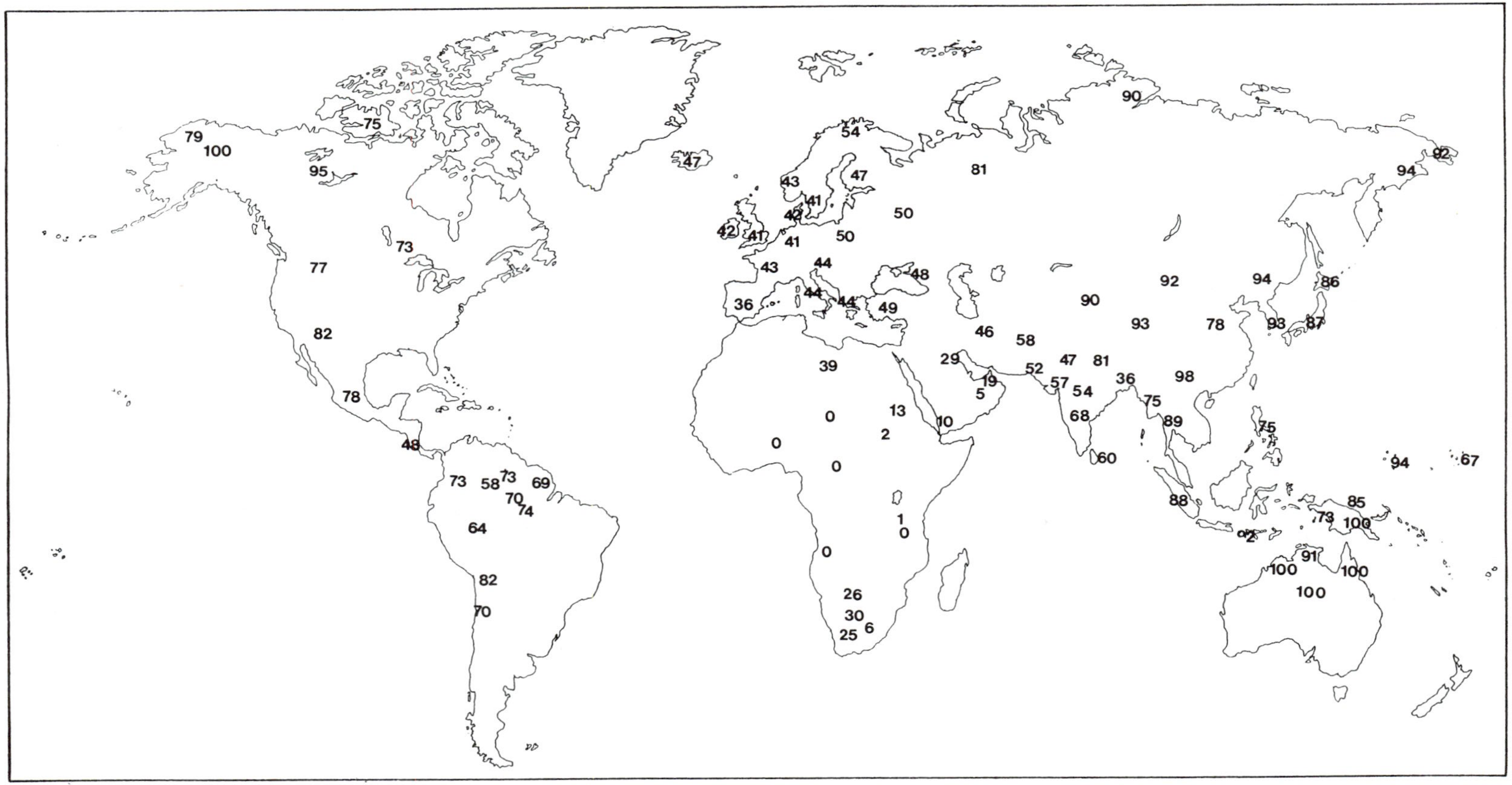

MAP 33. Distribution of allele Fy^a at the Duffy system.

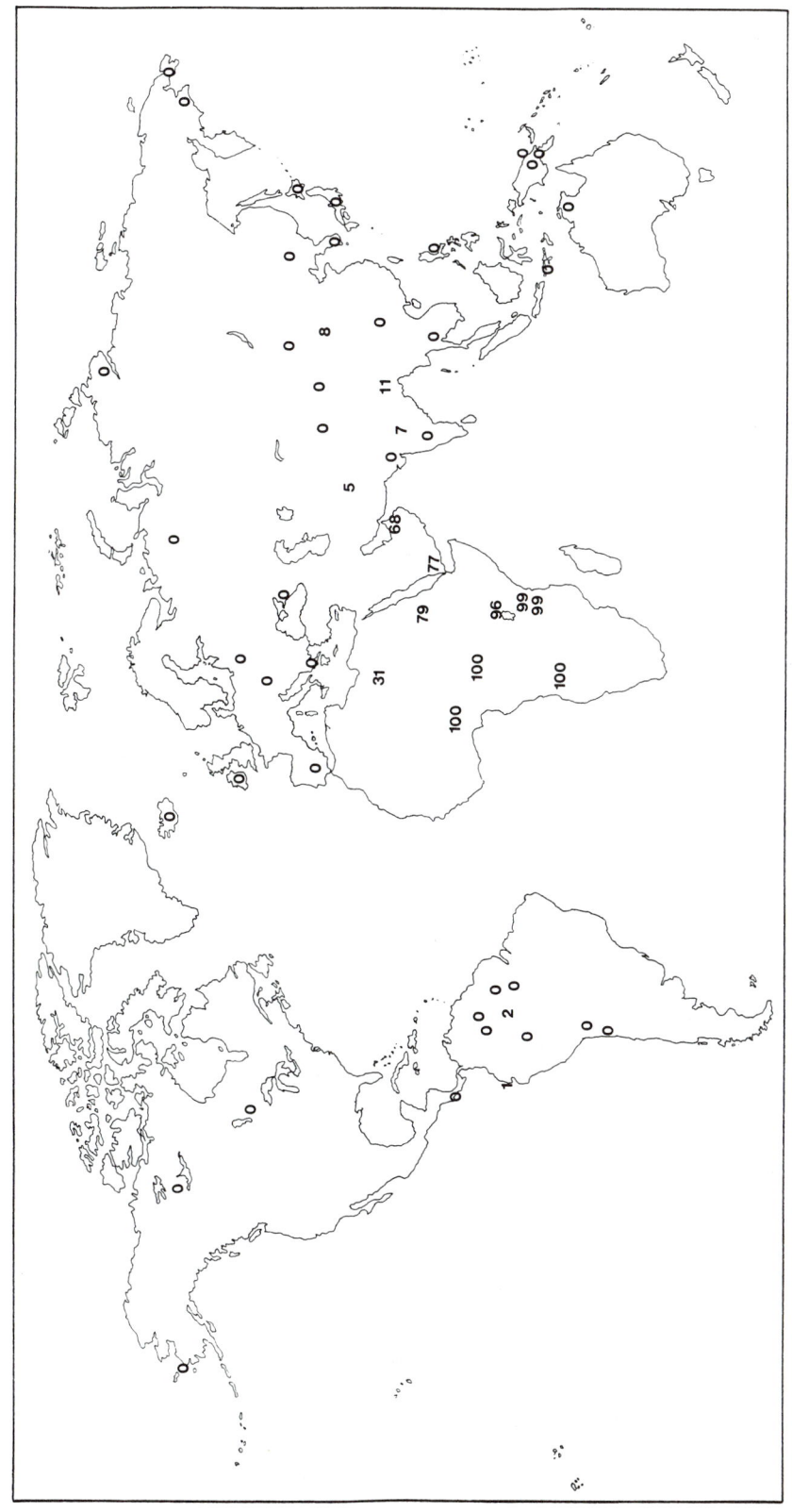

MAP 34. Distribution of allele *Fy* at the Duffy system.

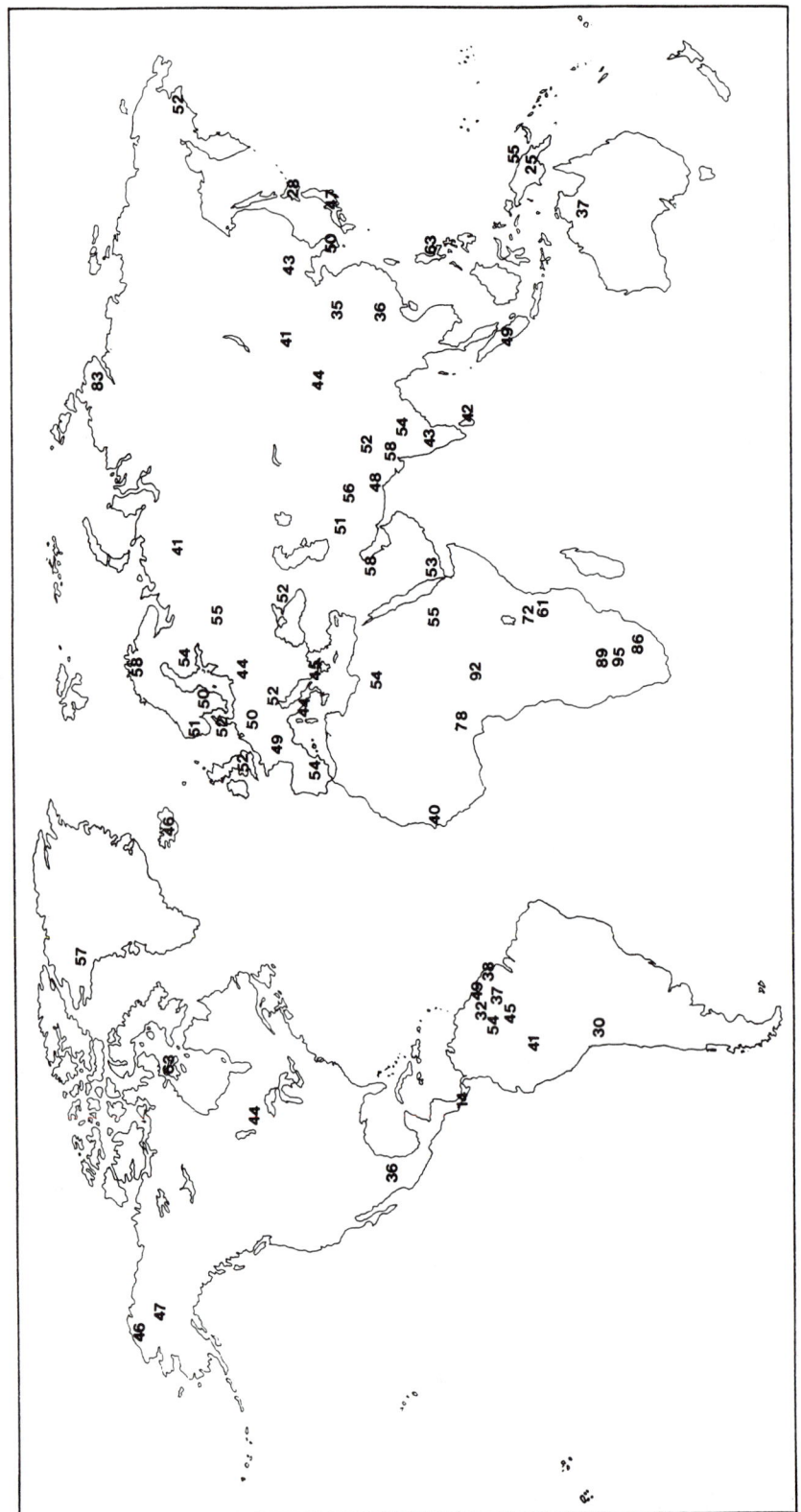

MAP 35. Distribution of allele Jk^a at the Kidd system.

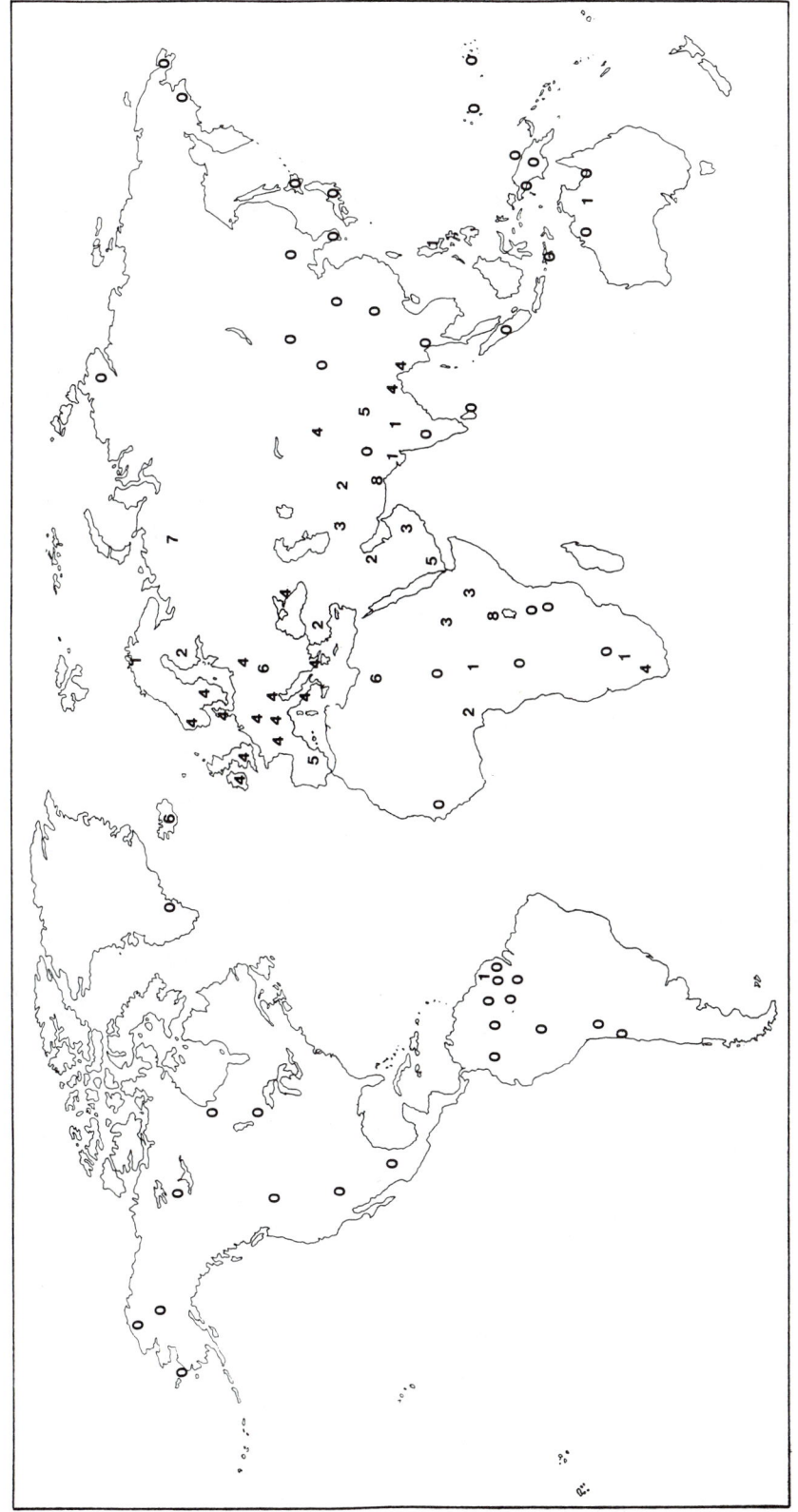

MAP 36. Distribution of allele *K* at the Kell system.

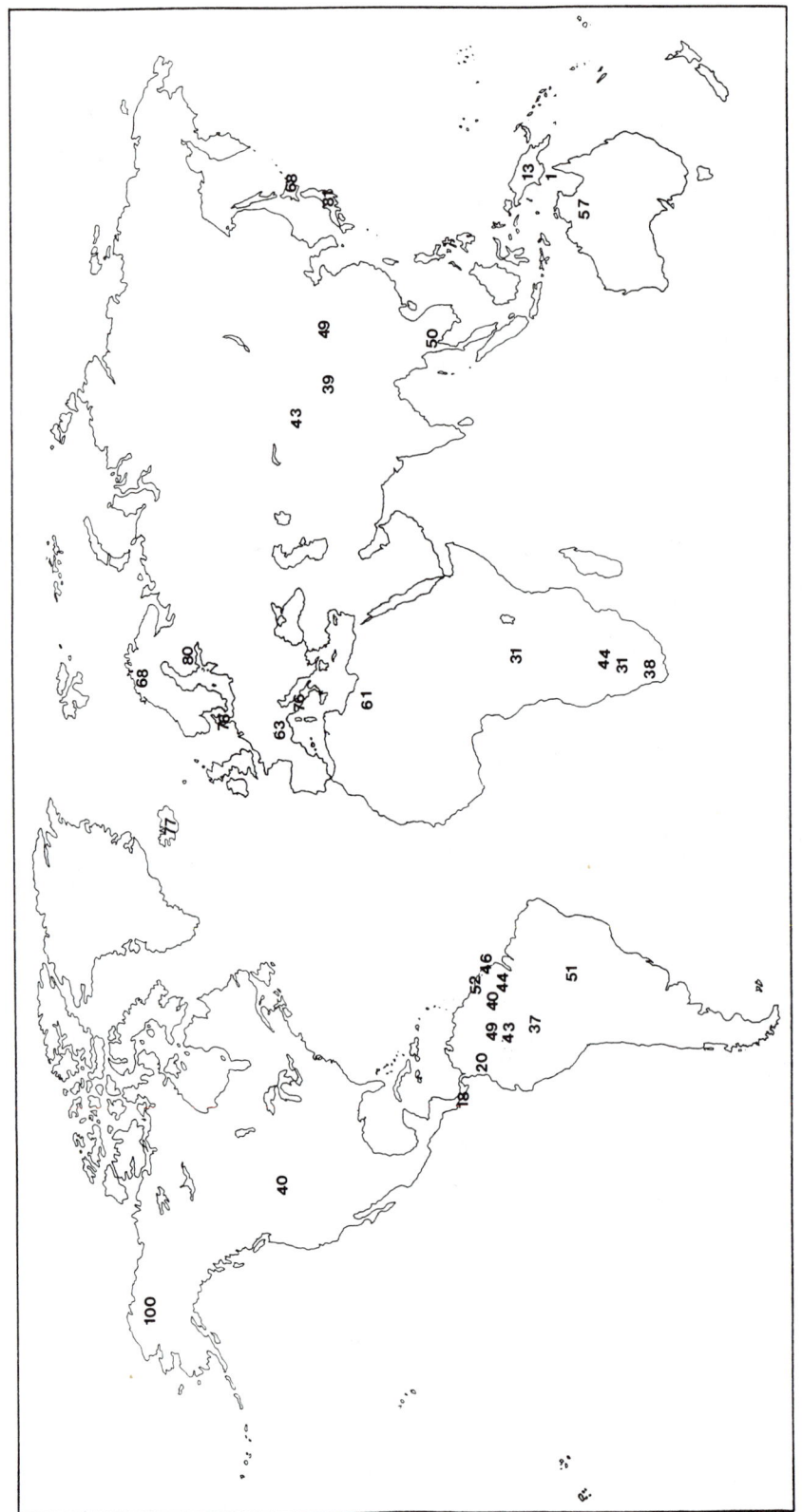

MAP 37. Distribution of allele *Le* at the Lewis system.

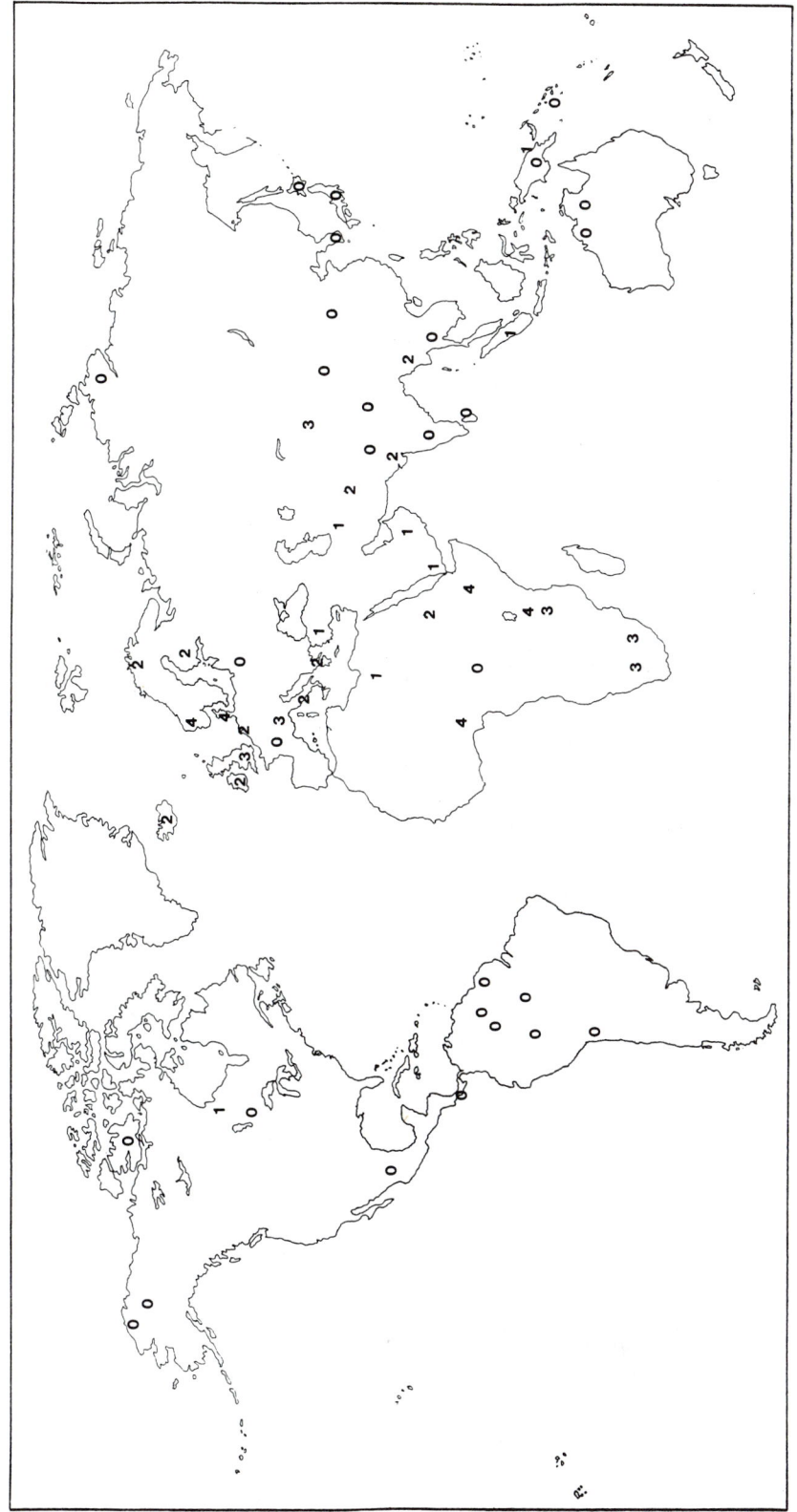

MAP 38. Distribution of allele Lu^a at the Lutheran system.

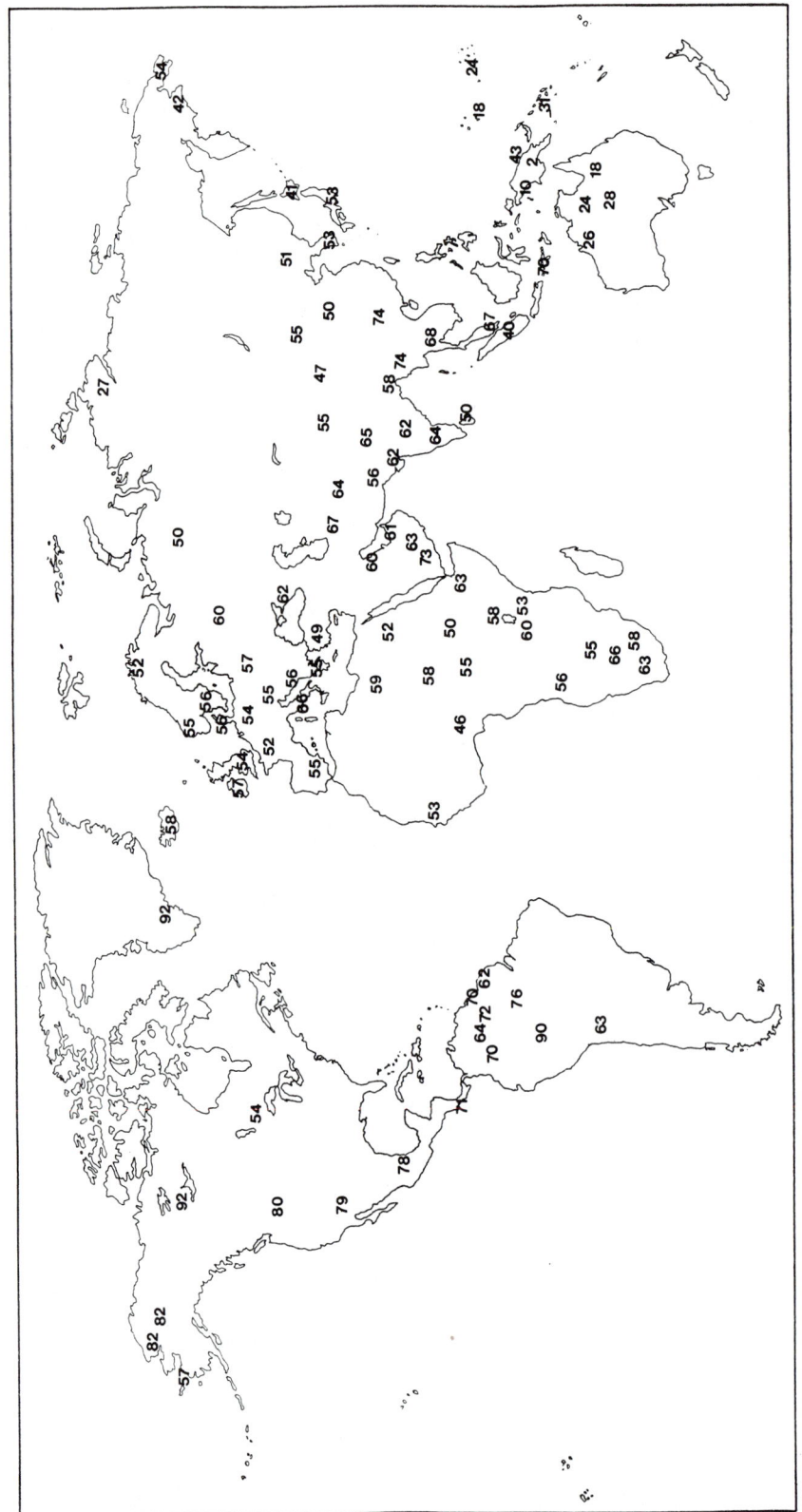

MAP 39. Distribution of allele *M* at the MNS system.

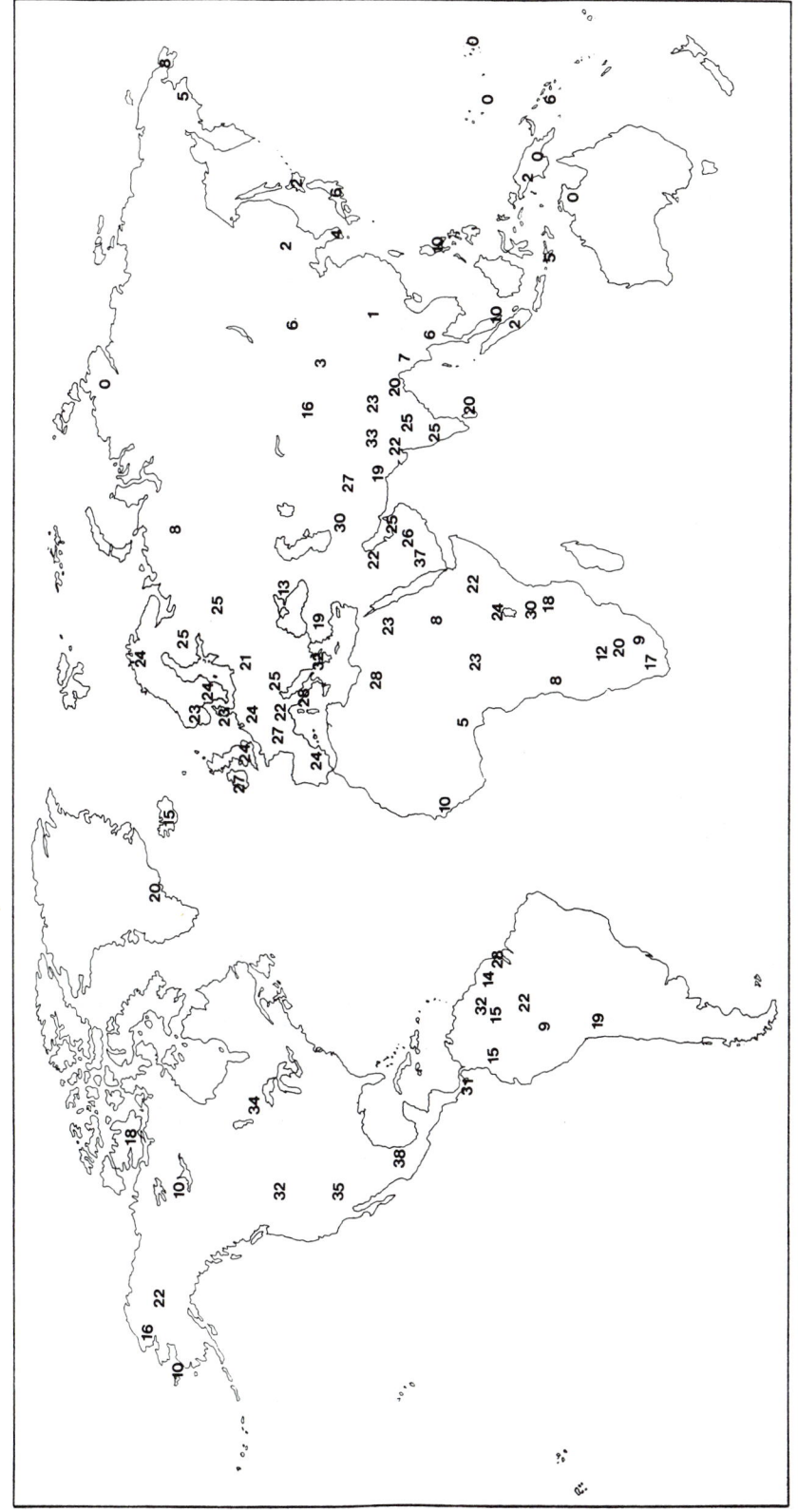

MAP 40. Distribution of allele *MS* at the MNS system.

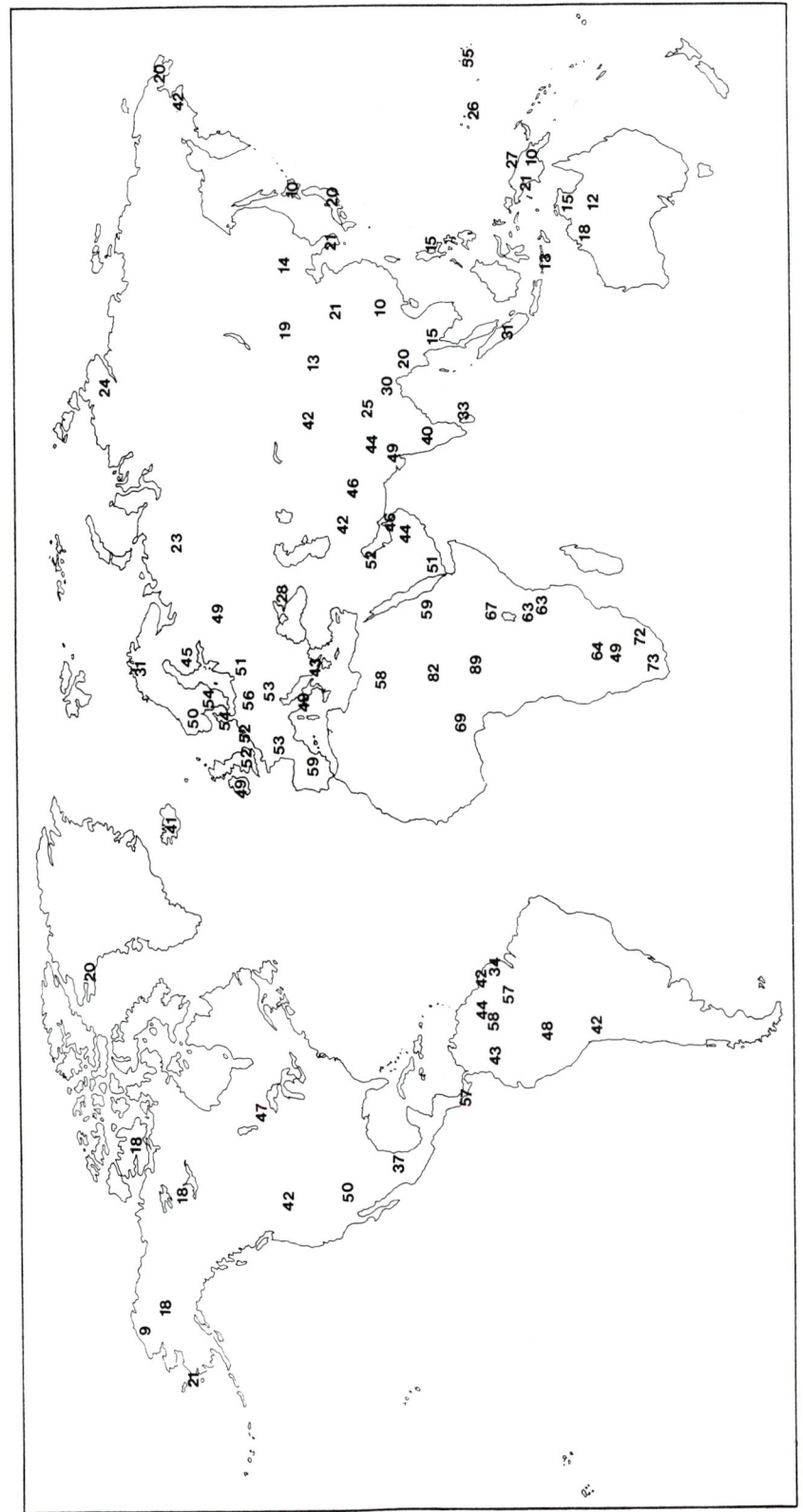

MAP 41. Distribution of allele P_1 at the P system.

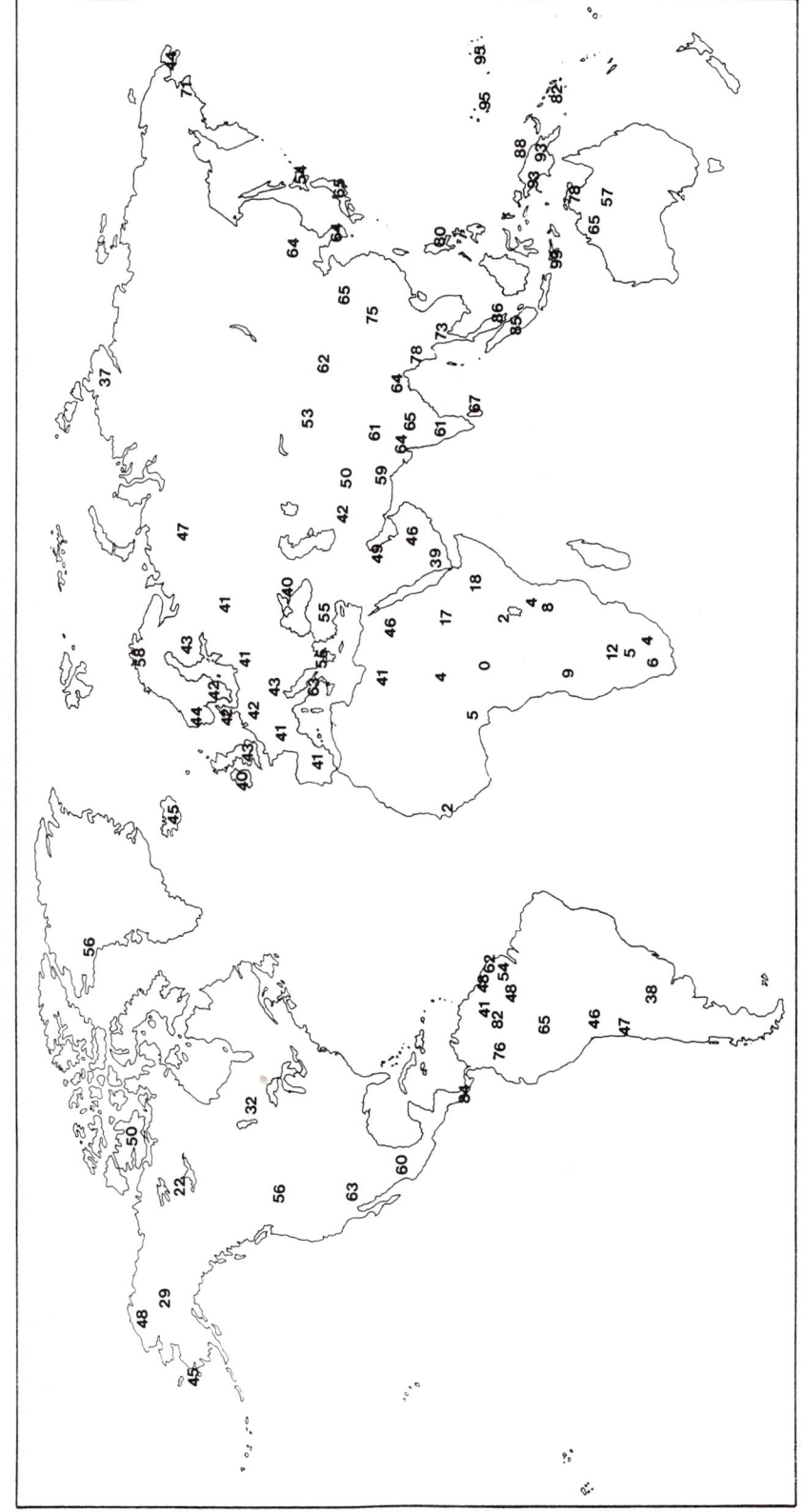

MAP 42. Distribution of allele *CDe* at the Rh system.

MAP 43. Distribution of allele *cDe* at the Rh system.

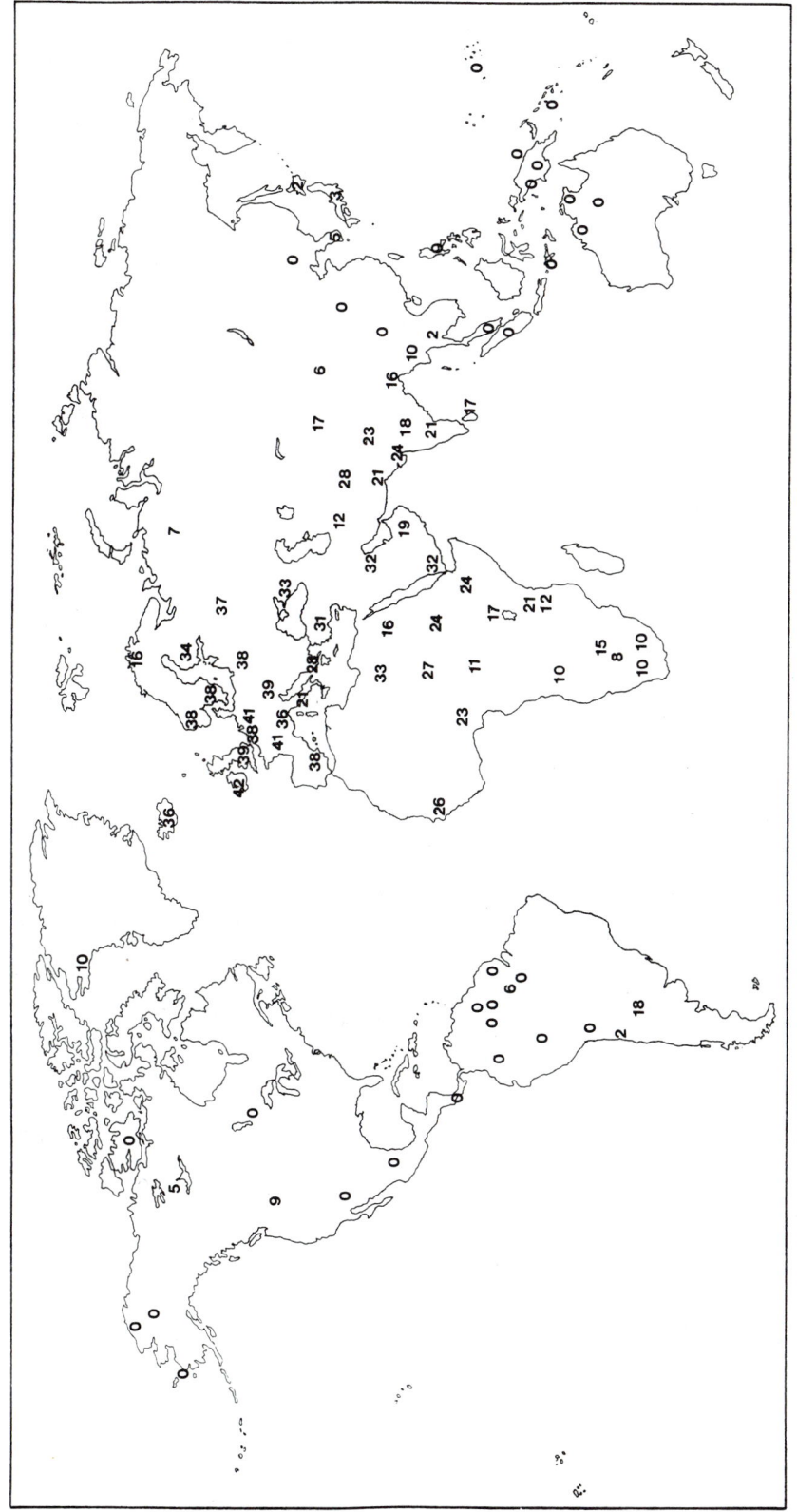

MAP 44. Distribution of allele *cde* at the Rh system.

315

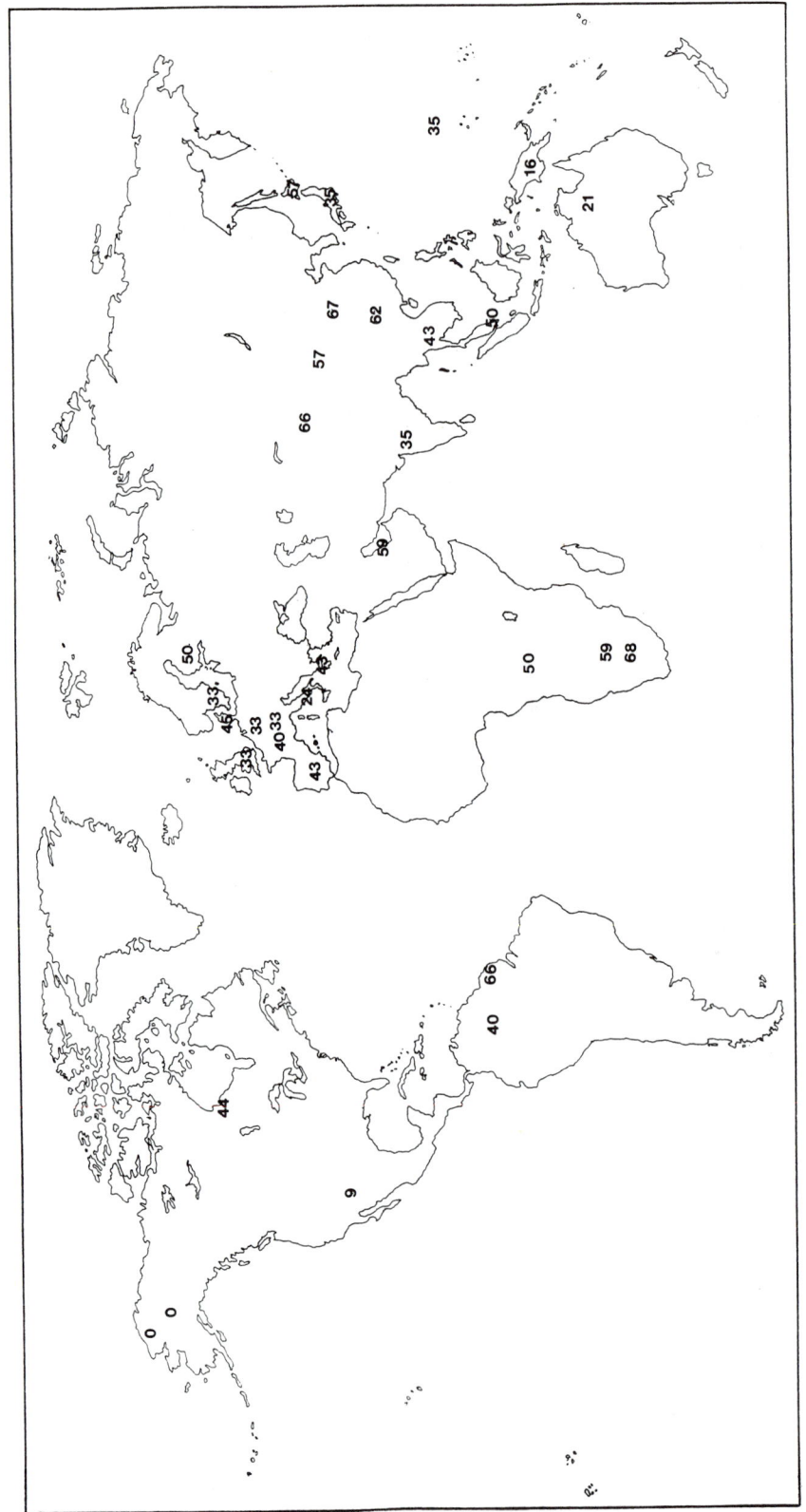

MAP 45. Distribution of allele *Xg* at the Xg system.

MAP 46. Distribution of allele *A1* at the HLAA locus.

MAP 47. Distribution of allele *B8* at the HLAB locus.

MAP 48. Distribution of allele *CW1* at the HLAC locus.

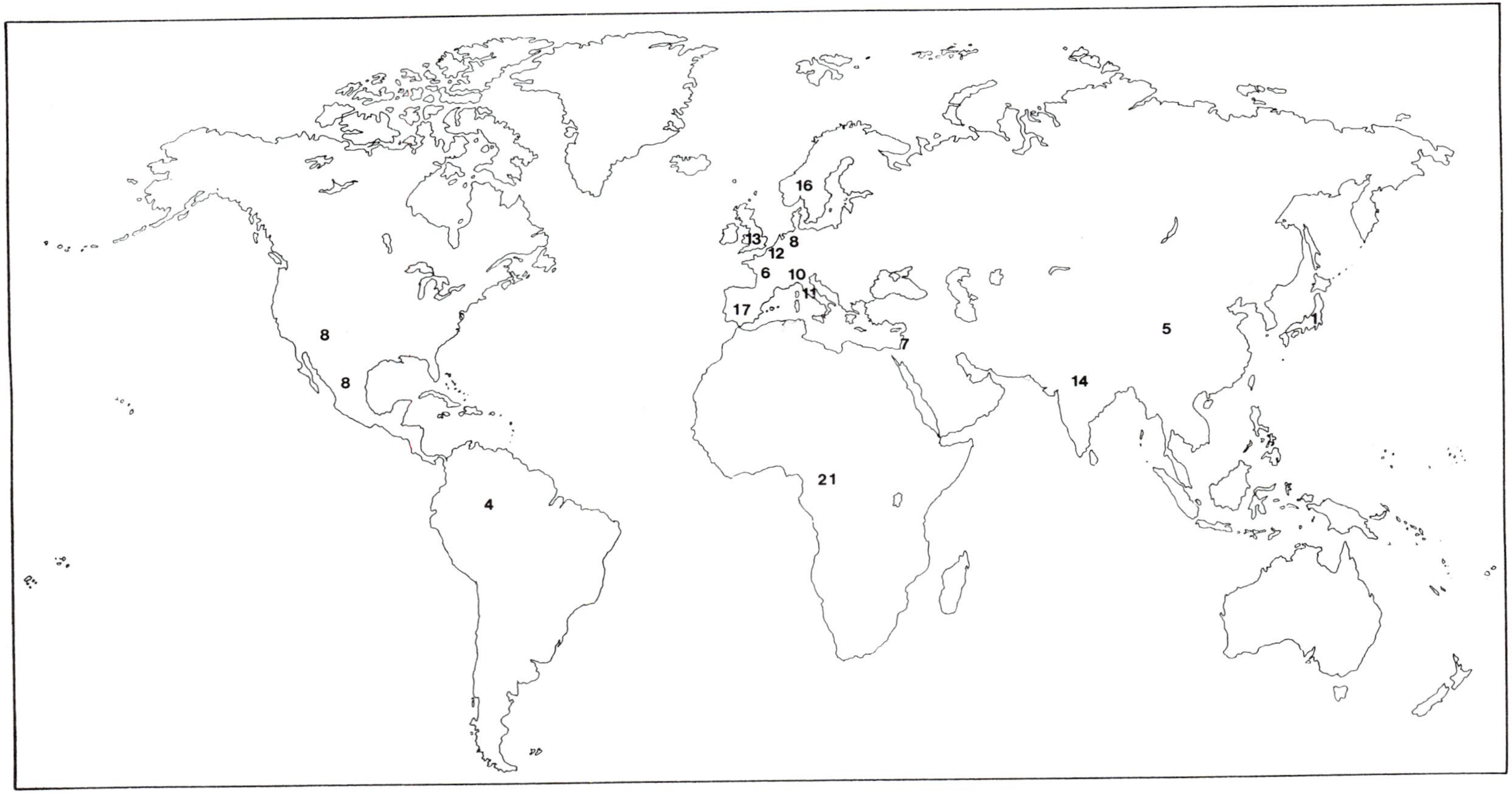

MAP 49. Distribution of allele *DR3* at the HLADR locus.

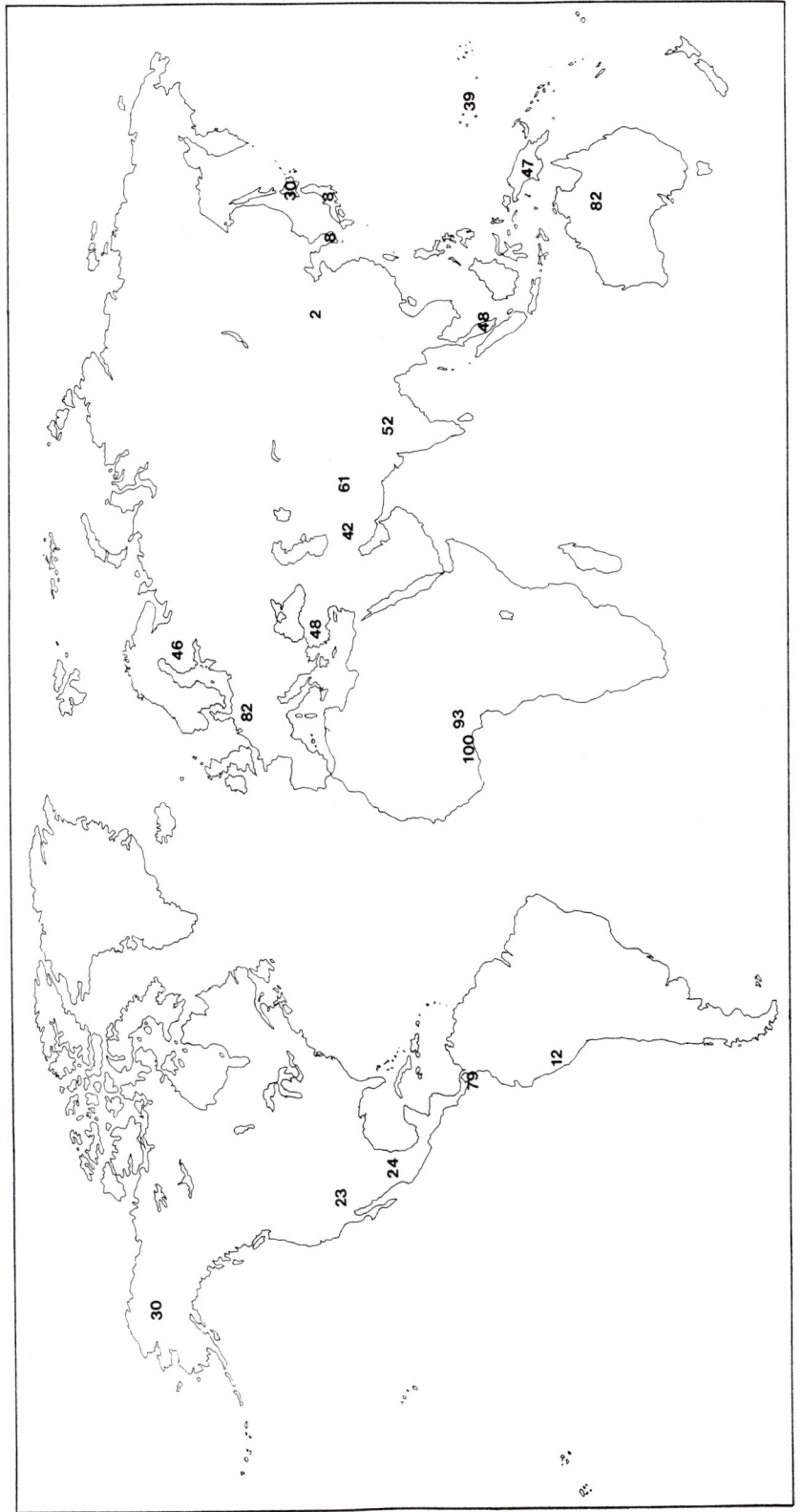

MAP 50. Distribution of allele *W* at the Cerumen locus.

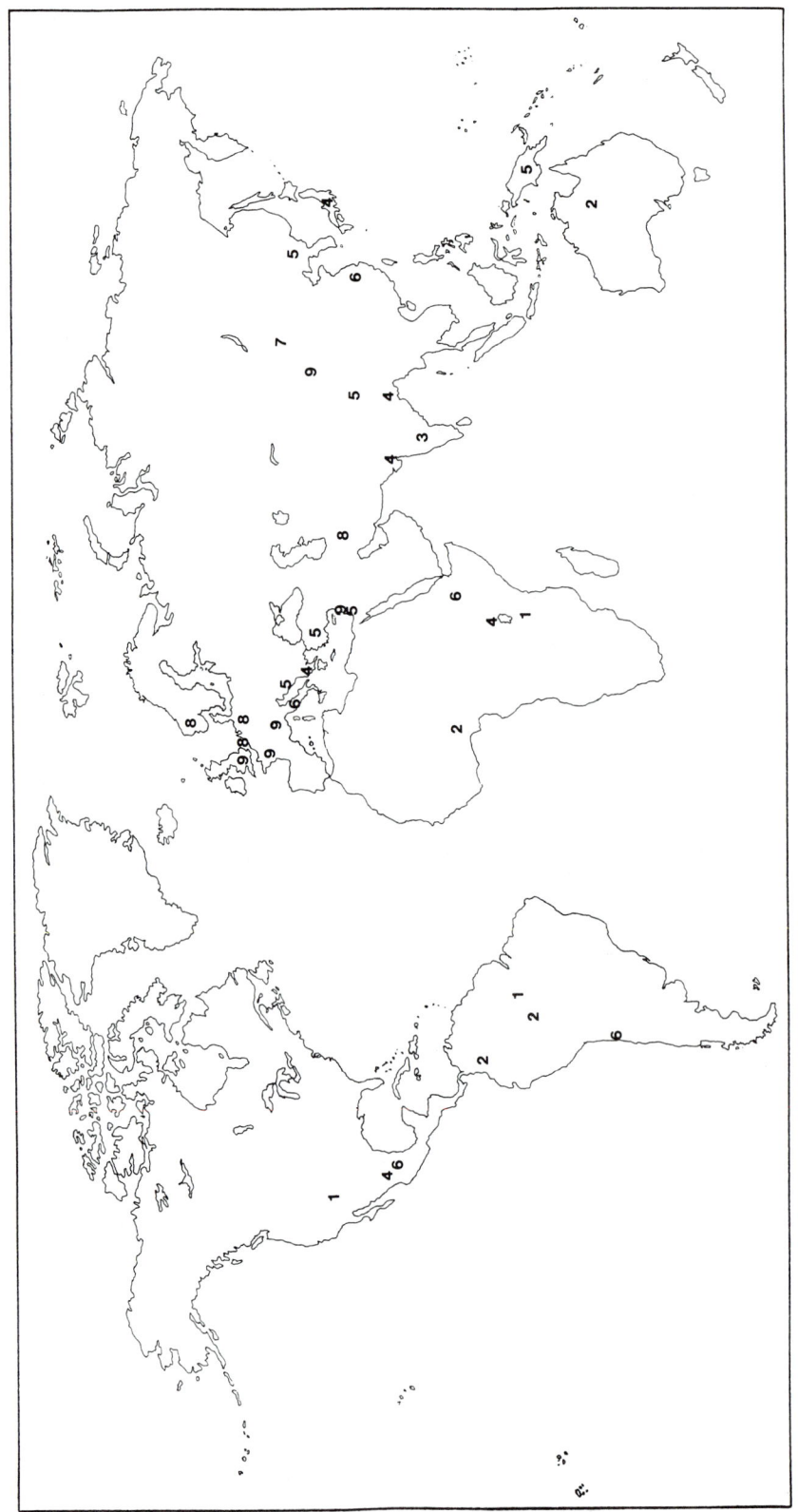

MAP 51. Distribution of colorblindness.

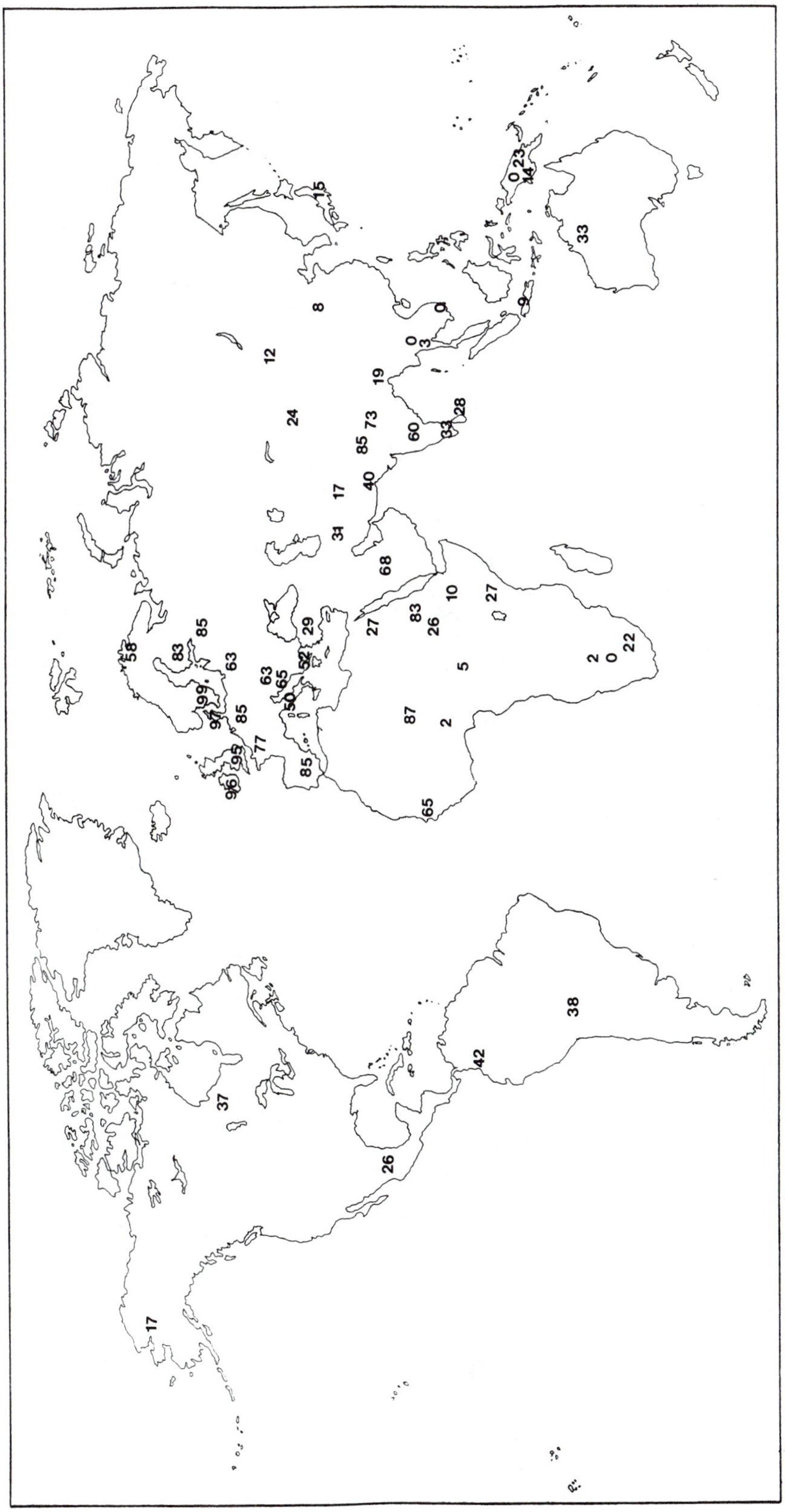

MAP 52. Distribution of lactose absorbers.

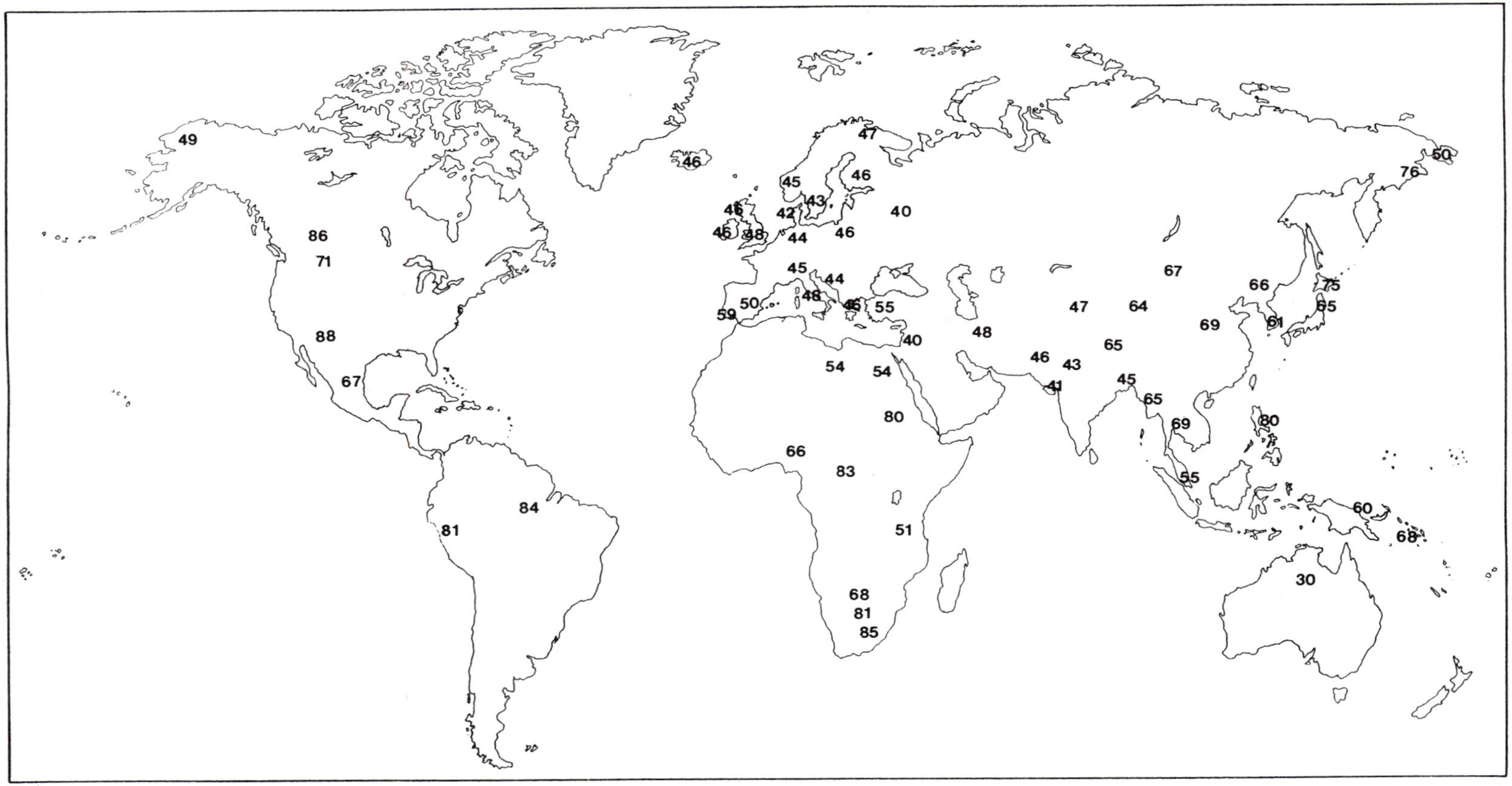

MAP 53. Distribution of allele T at the PTC locus.

Appendix A   Populations and Genetic Loci Examined

Place/Population	Genetic loci

EUROPE

EUROPEANS/ CAUCASIANS

(1) Enzymes: ACO1, ACO2, AK2, AK3, ALDH1, ALDH3, ALDH4, APRT, ARG1, ASL, CA1, CA2, DAMOX, DASOX, ESA, FH1, GABAT, GDA, GLUD, GOX, GPD1, GPD2, GPI, GST1, GUK1, ITPA, OTC, PEPB, PEPC, PEPD, PGP, PP, PSP, TPI. (2) Proteins: TBG, TBPA. (3) Blood groups: BEC*, CS, DO, JO, RM, VEN*.

AUSTRIA

(1) Enzymes: ACP1, ADA, AK1, GLO1, PGM1. (2) Proteins: C3, GC, HPA*, LP, PI(S), TF. (3) Blood groups: ABO, FY, JK, K, KA, MNS, P, RH, SE. (4) Miscellaneous: LAA*.

BELGIUM

(1) Enzymes: ACP1, ADA, AK1, ESD, GLO1, GPT1, PGM1, PGM2, PGD. (2) Proteins: AG, C3, GC, GC(S), HPA*, TF, TF(S). (3) Blood groups: MO. (4) Miscellaneous: CB.

BULGARIA

(1) Enzymes: ACP1, ADA, AK1, CHE1, CHE2, G6PD, LDHB, PEPA, PEPB, PGD, PGM1. (2) Proteins: AG, C3, GC, HPA*, LP, TF. (3) Blood groups: P, SE.

CZECHOSLOVAKIA

(1) Enzymes: ACP1, ADA, AK1, CHE1, CHE2, PGD, PGM1, PGM1(S), PGM3. (2) Proteins: HPA*, TF, (3) Blood groups: ABO, FY, K, MN, RH, WR, ZD. (4) Immunoglobulin systems: IGA2 (5) Miscellaneous: LAA*.

DENMARK

(1) Enzymes: ACP1, ADA, AK1, ESD, GALT, GLO1, GPT1, PGD, PGM1, PGM3, PON. (2) Proteins: BF, C3, GC, GC(S), HPA*, PI, PI(S), TF(S). (3) Blood groups: ABO, CHR, DH, FY, GE, JEN, JKN, K, LE, LU, MNS, P, RD, RH, SE, XG. (4) Immunoglobulin systems: IGHG1, KM. (5) DNA polymorphisms: PAH. (6) Miscellaneous: LAA*, PTC.

FAROE IS.

(1) Enzymes: ACP1, ADA, AK1, ESD, G6PD, GLO1, GPI, GPT1, LDH, MDH1, PGD, PGM1, PGM2, SOD1. (2) Proteins: HBB, HPA*, TF. (3) Blood groups: ABO, DI, FY, GE, GY, HE, JEN, JK, K, LU, MG, MI, MUR, NY, P, RD, RH, SW, VEL, VW, WR.

FINLAND
Finns

(1) Enzymes: ACP1, ADA, AK1, AMY1, CHE1, CHE2, ESD, FGH, GLO1, LDHA, PGM1, PGM1(S), PGM2, SOD1. (2) Proteins: AG, ALB, APOE, BF, C3, C4A, C4B, GC, GC(S), HBB, HPA*, HPA*(S), ORM1, PI, PI(S), TF, TF(S). (3) Blood groups: ABO, AN, DI, EN, FY, JK, K, LE, LU, MNS, NE, P, RH, SE, UL, VEL, XG,. (4) Immunoglobulin systems: IGHG1, KM. (5) Miscellaneous: AC, CER*, LAA*, PTC.

Lapps/Saami

(1) Enzymes: ADA, CHE1, CHE2, LDH, MDH1, PGM2, PGD, SOD1. (2) Proteins: AG, ALB, C3, HBB, HPA*(S), PI, PI(S). (3) Blood groups: SE, UL. (4) Immunoglobulin systems: IGHG1, KM. (5) Miscellaneous: AC, LAA*, PTC.

FRANCE
French

(1) Enzymes: ACP1, ADA, AK1, CHE1, CHE2, DIA1, ESD, FUCA, GLO1, GPT1, GST1, MDH1, PGD, PGM1, PGM1(S), PGM2, PON, SOD1. (2) Proteins: ALB, APOE, BF, BF(S), GC, GC(S), HPA*, ORM1, ORM2, PI, PI(S), TF, TF(S). (3) Blood groups: ABO, AU, DUZ*, FY, HEY, I, JK, K, LEK, LU, MNS, P, PLA*, RH, VEL, XG. (4) HLA systems: HLAA, HLAB, HLAC, HLADQ, HLADR. (5) Immunoglobulin systems: IGHG1. (6) DNA polymorphisms: AT3. (7) Miscellaneous: CB, LAA*.

Basques

(1) Enzymes: AK1, CHE1, CHE2, DIA1, ESD, GLO1, MDH1, PGD, PGM1, PGM1(S), PGM2, SOD1. (2) Proteins: BF, GC(S), HPA*. (3) Blood groups: FY, JK, MNS, P, RH. (4) Immunoglobulin systems: IGHG1.

GERMANY

(1) Enzymes: ACO1, ACO2, ACP1, ADA, ADH2, ADH3, AK1, ALAD, ALDH2, ALPP, AMY1, CHE1, CHE2, DIA1, DIA3, ESD, FUCA, GALT, GLO1, GOT2, GPT1, ME2, MPI, PGD, PGM1, PGM1(S), PGM2, PGM3, PGP, PKM2, PON, SAHH, SOD1, UMPK. (2) Proteins: AG, AHSG, APOA4, APOE, AT3, BF, BF(S), BG, C2, C3, C4A, C4B, C6, C8A, CP, F13A, F13B, GC, GC(S), HPA*, HPA*(S), HPB*, LP, PI, PI(S), TF, TF(S),. (3) Blood groups: ABO, FY, HEI*, IK, JK, K, LU, MNS, NY, P, PLA*, RH, SC, SE, XG. (4) HLA systems: HLAA, HLAB, HLAC, HLADQ, HLADR. (5) Immunoglobulin systems: IGHG1, KM. (6) DNA polymorphisms: GHC. (7) Miscellaneous: CB, CER*, LAA*, PTC.

GREECE

(1) Enzymes: ACP1, ADA, AK1, ALPP, CHE1, CHE2, DIA1, ESD, G6PD, GOT1, GOT2, GPI, GPT1, GSR, LDH, MDH1, NP, PGAMA, PGD, PGM1, PGM1(S), PGM2, SOD1. (2) Proteins: AG, ALB, BG, C3, CP, GC, GC(S), HBB, HPA*, HPA*(S), PI, TF, TF(S). (3) Blood groups: ABO, DI, FY, GE, GY, HE, JEN, JK, JS, K, LU, MG, MNS, MUR, NY, P, RD, RH, SW, VEL, VW, WR, XG, ZT. (4) Immunoglobulin

Place/ Population	Genetic loci

systems: KM. (5) DNA polymorphisms: APOC3, AT3, GHC, HBAC, HBAD*, PGK1. (6) Miscellaneous: CB, LAA*, PTC.

HUNGARY (1) Enzymes: ACP1, ADA, AK1, AMY2, CHE1, ESD, GLO1, GPT1, PGD, PGM1(S), PGP. (2) Proteins: BF, BG, C3, GC(S), HPA*, HPA*(S), LP, PI, PI(S), TF(S). (3) Blood groups: ABO, FY, MNS, RH. (4) Immunoglobulin systems: KM. (5) Miscellaneous: LAA*.

ICELAND (1) Enzymes: ACP1, AK1, ALPP, CHE1, CHE2, G6PD, GLO1, LDH, MDH1, PGD, PGM1, PGM2, SOD1. (2) Proteins: AG, BG, CP, C3, GC, GC(S), HBB, HPA*, LP, PI, PI(S), TF, TF(S). (3) Blood groups: ABO, DI, FY, JK, JS, K, LE, LU, MNS, P, RD, RH, SE, WR. (4) Immunoglobulin systems: IGHG1, KM. (5) Miscellaneous: PTC.

IRELAND (1) Enzymes: ACP1, ADA, AK1, ESD, G6PD, LDH, MDH1, PGD, PGM1, PGM2, SOD1. (2) Proteins: AG, BG, CP, GC, HBB, HPA*, HPA*(S), LP, PI, TF. (3) Blood groups: ABO, DI, FY, HE, K, LU, MNS, P, RH, SE, WR. (4) Immunoglobulin systems: IGHG1, KM. (5) Miscellaneous: LAA*, PTC.

ITALY (1) Enzymes: ACP1, ADA, AK1, ALAD, ALPP, AMY2, CA1, CA2, CHE1, DIA1, ESD, G6PD, GALK, GALT, GDH, GLO1, GPI, GPT1, GPX1, PGD, PGM1, PGM1(S), PGM2, PGM3, PGP, SAHH, SOD1, UMPK. (2) Proteins: AG, ALB, BF, C3, GC, GC(S), HBB, HBG, HPA*, HPA*(S), PI, PI(S), TF, TF(S). (3) Blood groups: ABO, DI, FY, JK, K, LE, LU, MNS, P, RH, SE, XG, WR. (4) HLA systems: HLAA, HLAB, HLAC, HLADR, HLADQ. (5) Immunoglobulin systems: IGA2, IGHG1, KM. (6) DNA polymorphisms: APOC3, APOA1-APOC3-APOA4, AT3, GHC, HBAC, HBAD*, HBBC, PGK1 (7) Miscellaneous: BAIB, CB, LAA*, PTC .

NETHERLANDS (1) Enzymes: ACP1, ADA, AK1, AMY1, CHE1, CHE2, DIA1, DIA4, ESD, GAPD, GLO1, GOT1, GPI, GPT1, GPX1, HK3, IDH1, LDH, MDH1, PEPA, PEPB, PEPC, PGD, PGK, PGM1, PGM2, SOD1. (2) Proteins: ALB, C3, DB, GC, HBB, HPA*, PA, PB, PI, PI(S), PR, TF. (3) Blood groups: ABO, BAK, CO, DI, FY, HOV, JK, K, KO, LAN, LE, LU, MNS, P, PLA*, RH, SE, XG. (4) HLA systems: HLAA, HLAB, HLAC, HLADR. (5) Immunoglobulin systems: IGA2, IGHG1, KM. (6) Miscellenous: CB, PTC.

NORWAY
  Norwegians (1) Enzymes: ACP1, ADA, AK1, CHE1, ESD, GLO1, GPT1, GPX1, PGM1, PGM3. (2) Proteins: AG, ALB, APOA4, BF, C2, C3, C6, C8A, C8B, F13B, GC, HPA*, HPA*(S), LD, LP, PI, TF, XM. (3) Blood groups: ABO, BP, CO, FY, JK, JN, K, LU, MNS, MO, NY, OL, P, PT, RH, RL, SE, SKJ*, TO, TR, WR, WU, XG. (4) Immunoglobulin systems: IGHG1, KM. (5) Miscellaneous: CB, PTC.

  Lapps/Saami (1) Enzymes: ACP1, ADA, AK1, ESD, GLO1, GPT1, PGM1, PGM2. (2) Proteins: AG, BF, C3, C6, GC, HPA*, HPA*(S), LP, PI, TF. (3) Blood groups: ABO, DI, FY, JK, K, LE, LU, MNS, NY, P, RH, SE, WR.

POLAND (1) Enzymes: ACP1, ADA, AK1, CHE2, ESD, FUCA, G6PD, GALT, GLO1, GPT1, LDH, MDH1, PGD, PGM1, PGM1(S), PGM2, PKM2. (2) Proteins: ALB, BF, C3, GC, HPA*, HPA*(S), TF. (3) Blood groups: ABO, FY, JK, K, LU, MNS, P, RH, SC, SE, XG. (4) Immunoglobulin systems: IGHG1, KM. (5) Miscellaneous: LAA*, PTC .

PORTUGAL
  Portuguese (1) Enzymes: ACP1, ADA, AK1, CHE1, CHE2, ESD, GLO1, GPT1, PGM1, SOD1. (2) Proteins: BF, C5, GC, HBB, HPA*, PI. (3) Blood groups: ABO, FY, K, MN, RH. (4) Immunoglobulin systems: IGHG1. (5) Miscellaneous: PTC.

SPAIN
  Spanish (1) Enzymes: ACP1, ADA, AK1, ALAD, CHE1, CHE2, ESD, G6PD, GLO1, GPT1, GSR, PGD, PGM1, PGM1(S), PGP, PKM2. (2) Proteins: BF, CP, C3, GC, HPA*, HPA*(S), PI, PI(S), TF. (3) Blood groups: ABO, FY, JK, K, MNS, P, RH, SE, XG. (4) HLA systems: HLAA, HLAB, HLAC, HLADQ, HLADR. (5) Miscellaneous: LAA*, PTC.

  Basques (1) Enzymes: ACP1, ADA, AK1, CHE1, CHE2, PGD, PGM1. (2) Proteins: BF, CP, C3, GC, HPA*, PI, TF. (3) Blood groups: ABO, FY, K, LE, LU, P, RH, SE, XG. (4) Miscellaneous: PTC.

SWEDEN
  Swedes (1) Enzymes: ACP1, ADA, AK1, ALPP, CHE1, ESD, LAPP, LDH, MDH1, MDH2, PGD, PGM1, PGM1(S), SOD1. (2) Proteins: AG, ALB, BF, C3, C6, GC, GC(S), HBB, HPA*, ORM1, PI, PI(S), TF, TF(S). (3) Blood groups: ABO, AN, FY, JK, K, MNS, NE, P, RH, SE, UL, XG. (4) HLA systems: HLAA, HLAB, HLAC, HLADQ, HLADR. (5) Immunoglobulin systems: IGHG1. (6) Miscellaneous: AC, DOX*, LAA*, PTC.

  Lapps/Saami (1) Enzymes: ACP1, ADA, AK1, ESD, PGD, PGM1, SOD1. (2) Proteins: C3, GC, HPA*, PI, PI(S), TF, TF(S). (3) Blood groups: ABO, DI, FY, K, MNS, P, RH,

326

Place/ Population	Genetic loci
SWITZERLAND	(1) Enzymes: ACP1, ADA, AK1, CAT, ESD, GALT, GLO1, GPT1, PGM1, PGM1(S), PGP. (2) Proteins: AG, BF, C3, GC, HPA*, PLG, TC2, TF(S). (3) Blood groups: ABO, CL, FY, K, LE, LU, MG, MNS, NE, NY, P, RH, SE, WR, XG, YT. (4) HLA systems: HLAA, HLAB, HLAC, HLADR. (5) Immunoglobulin systems: IGHG1, KM. (6) Miscellaneous: CB, DOX*, LAA*, MEP*, PTC.
UNITED KINGDOM England	(1) Enzymes: ACP1, ADA, ADH1, ADH3, AK1, ALPP, AMY1, CHE1, CHE2, DIA1, DIA3, ESD, FUCA, G6PD, GAA, GALT, GDH, GLO1, GOT1, GOT2, GPT1, HK3, LDHA, MDH1, ME2, NP, PEPA, PGD, PGM1, PGM1(S), PGM2, PGM3, PON, SOD1. (2) Proteins: AG, ALB, AT3, BF, BG, C2, C3, C5, C6, C7, GC, GC(S), HBB, HPA*, HPB*, PI, PI(S), PLG, TF. (3) Blood groups: ABO, AU, BP, BX, BY, CO, EN, EV*, FY, GD*, GE, GF, GO, HG, HT, JK, K, LAN, LEV*, LS, LU, MG, MNS, ORR*, P, PLA*, RDD*, RG, RH, RI, SC, SD, SE, SW, TR, V, VEL, WB, WR, XG, YT. (4) HLA systems: HLAA, HLAB, HLAC, HLADQ, HLADR. (5) Immunoglobulin systems: IGHG1, KM. (6) DNA polymorphisms: HBAC, HBBC. (7) Miscellaneous: AC, BAIB, CB, DOX*, LAA*, PTC.
Scotland	(1) Enzymes: ACP1, AK1, PGM1. (2) Proteins: APOE, HPA*, TF. (3) Blood groups: ABO, CL, SE, ST*, WR. (4) Miscellaneous: CB, PTC.
Isle of Lewis	(1) Enzymes: ACP1, ADA, AK1, G6PD, GPI, LDH, MDH1, PGD, PGM1, PGM2, SOD1. (2) Proteins: HPA*, HBB, TF. (3) Blood groups: ABO, FY, GE, GY, HE, JK, K, LU, MG, MNS, MUR, P, RD, RH, SW, VEL, VW, WR.
Isle of Man	(1) Enzymes: ACP1, ADA, AK1, ESD, PGD, PGM1. (2) Proteins: AG, HPA*, TF. (3) Blood groups: ABO, FY, K, LU, MNS, P, RH, SE. (4) Miscellaneous: CB, PTC.
Orkney Is.	(1) Enzymes: ADA, AK1, G6PD, GPI, LDH, MDH1, PGD, PGM1, SOD1. (2) Proteins: GC, HPA*, TF. (3) Blood groups: ABO, FY, K, P, RH. (4) Miscellaneous: CB, PTC.
N. Ireland	(1) Proteins: LP.
YUGOSLAVIA	(1) Enzymes: AK1, CHE1, G6PD, GLO1, GPT1. (2) Proteins: GC, HBG, HPA*. (3) Blood groups: ABO, MN, SE. (4) Miscellaneous: CB, LAA*, PTC.

U.S.S.R.

EUROPEAN PART

Russians	(1) Enzymes: ACP1, ADA, ESD, GLO1, GPT1, LDH, MDH1, PGD, PGM1, PGM2. (2) Proteins: ALB, C3, GC, HBA, HBB, HPA*, PI, PI(S), TF. (3) Blood groups: ABO, DI, FY, JK, K, MNS, P, RH. (4) Miscellaneous: CB, LAA*, PTC.
Abkhazians	(1) Enzymes: ACP1, ADA, AK1, CA1, CA2, ESD, G6PD, GLO1, GPI, GPT1, IDH1, LDHA, MDH1, PEPA, PEPB, PEPC, PGD, PGM1, PGM2, PGP. (2) Proteins: ALB, CP, GC, GC(S), HBA, HBB, HPA*, TF. (3) Blood groups: ABO, FY, JK, K, MNS, P, RH.
Maris	(1) Enzymes: ADA, CHE1, CHE2, LDH. (2) Proteins: HBB.

ASIAN PART
Eskimos	(1) Enzymes: ACP1, AK1, PGD, PGM1. (2) Proteins: HPA*, TF. (3) Blood groups: ABO, DI, FY, K, MNS, P, RH. (4) Miscellaneous: PTC.
Chukchi	(1) Enzymes: ACP1, AK1, PGD, PGM1. (2) Proteins: HPA*, TF. (3) Blood groups: ABO, DI, FY, JK, K, MNS, P, RH, SE. (4) Miscellaneous: PTC.
Nganasan	(1) Enzymes: ACP1, AK1, PGD, PGM1. (2) Proteins: HPA*, TF. (3) Blood groups: ABO, DI, FY, JK, K, LU, MNS P, RH.
Nentzi	(1) Enzymes: ACP1, AK1, PGD, PGM1. (2) Proteins: HPA*. (3) Blood groups: ABO, FY, JK, K, MNS, P, RH.

ASIA

TURKEY (Asia Minor)	(1) Enzymes: ACP1, ADA, AK1, ALDH2, CHE1, ESD, G6PD, GLO1, GPT1, PGD, PGM1, PGM2. (2) Proteins: GC, HPA*, TF. (3) Blood groups: ABO, FY, K, LU, MNS, RH. (4) Immunoglobulin systems: KM. (5) Miscellaneous: CB, CER*, LAA*, PTC.

Populations and genetic loci examined (cont'd)

Place/ Population	Genetic loci

ISRAEL

  Ashkenazi
    Jews
(1) Enzymes: ACP1, ADA, AK1, ALDH2, ARSA, CHE1, ESD, G6PD, GLO1, GPT1, PGM1, PGP. (2) Proteins: GC, HPA*. (3) Blood groups: ABO, DO, FY, JK, K, MN, P, SE, YT. (4) HLA systems: HLAA, HLAB, HLAC, HLADR. (5) Immunoglobulin systems: IGHG1, KM. (6) Miscellaneous: AC, CB*, LAA*.

  Yemenite
    Jews
(1) Enzymes: ACP1, ADA, AK1, CHE1, ESD, G6PD, GLO1, GPI, GPT1, LDH, MDH1, PGD, PGM1, PGP. (2) Proteins: AG, GC, HPA*, LP, TF. (3) Blood groups: ABO, DI, FY, HE, JK, JS, K, LU, MNS, P, RD, RH, SE, WR. (4) Immunoglobulin systems: IGHG1, KM. (5) Miscellaneous: CB, PTC.

  Habbanite Jews (1) Enzymes: ARSA. (2) Blood groups: DI, HE.

IRAQ (1) Enzymes: CHE2[a], ESD, G6PD, GLO1, SOD1. (2) Proteins: C3, GC(S). (3) Blood groups: ABO.

JORDAN (1) Enzymes: G6PD, LDH, PGD. (2) Proteins: ALB, BF, HPA*, TF(S). (3) Blood groups: ABO, RH. (4) Miscellaneous: LAA*.

KUWAIT

  Arabs
(1) Enzymes: ACP1, ADA, AK1, CAT, ESD, G6PD, LDH, MDH1, PGD, PGM1. (2) Proteins: HBB, HPA*, TF. (3) Blood groups: ABO, FY, JK, K, MNS, P, RH.

SAUDI ARABIA (1) Enzymes: ACP1, ADA, AK1, CHE1, ESD, G6PD, GPT1, HK3, LDH, MDH1, PGD, PGM1, PKM2, SOD1. (2) Proteins: ALB, BF, C3, CP, GC, HBB, HPA*, PI, TF. (3) Blood groups: ABO, DI, FY, HE, IN, JK, K, LU, MNS, P, RH, V. (4) DNA polymorphisms: HBAC, HBAD*. (5) Miscellaneous: LAA*.

UNITED ARAB EMIRATES

  Abu Dhabians (1) Enzymes: G6PD. (2) Proteins: HBB. (3) Blood groups: ABO, FY, MNS, P, VEL.

IRAN (1) Enzymes: ACP1, ADA, AK1, CA1, CA2, CAT, CHE1, DIA1, ESD, G6PD, GLO1, GOT1, GPI, GPT1, IDH1, LDH, MDH1, PEPA, PEPB, PEPC, PEPD, PGD, PGK, PGM1, PGM1(S), PGM2, PGP, SOD1. (2) Proteins: AG, ALB, BG, C3, CP, F13A, GC, GC(S), HBB, HPA*, HPA*(S), PI, TF. (3) Blood groups: ABO, DI, FY, IN, JK, K, LU, MNS, NY, P, RD, RH, SE, WR. (4) Immunoglobulin systems: IGHG1, KM. (5) Miscellaneous: CB, CER*, LAA*, PTC.

AFGHANISTAN (1) Enzymes: ACP1, ADA, AK1, CHE1, ESD, G6PD, GPI, GPT1, LDH, MDH1, PGD, PGM1, PGM2, SOD1. (2) Proteins: ALB, BF, BG, C3, CP, GC, HPA*, PI, TF. (3) Blood groups: ABO, FY, JK, K, LU, MNS, P, RH. (4) Miscellaneous: AC, CER*, LAA*.

PAKISTAN (1) Enzymes: CHE1, G6PD. (2) Proteins: BG, CP, GC, HPA*, PI, TF. (3) Blood groups: ABO, DI, FY, JK, JS, K, MNS, RH, V. (4) Miscellaneous: LAA*, PTC.

INDIA

  Indians
(1) Enzymes: ALDH2, CA1, CA2, TPI. (2) HLA systems: HLAA, HLAB, HLAC, HLADQ, HLADR. (3) DNA polymorphisms: AT3, HBAC, HBAD*, HBBC, PGK1.

  N. Indians
(1) Enzymes: ACP1, ADA, AK1, CHE1, CHE2, DIA1, ESD, G6PD, GLO1, GPI, GPT1, GPX1, LDH, MDH1, PEPA, PEPB, PGD, PGK, PGM1, PGM1(S), PGM2, SOD1. (2) Proteins: ALB, BF, BG, C3, CP, GC, GC(S), HBB, HPA*, PI, TBG, TF, TF(S). (3) Blood groups: ABO, FY, JK, K, LU, MNS, P, RH, SE. (4) Immunoglobulin systems: IGHG1, KM. (5) Miscellaneous: CB, LAA*, PTC.

  W. Indians
(1) Enzymes: ACP1, ADA, AK1, ALPP, ESD, G6PD, GLO1, GPI, GPX1, LDH, MDH1, PGD, PGM1, PGM2. PGP, SOD1. (2) Proteins: CP, HBB, HBG, HPA*, HPA*(S), PI, TF. (3) Blood groups: ABO, FY, IN, JK, K, LU, MNS, P, RH, XG. (4) Miscellaneous: CB, PTC.

  C. Indians
(1) Enzymes: ACP1, ADA, AK1, GPI, LDH, PGD, PGM1, PGM2. (2) Proteins: C3, HBB, HPA*. (3) Blood groups: ABO, FY, JK, K, MNS, RH.

  Bhil tr.
    (C. India)
(1) Enzymes: ACP1, ADA, AK1, ESD, G6PD, GPI, IDH1, MDH1, PGD, PGM1, PGM2, SOD1. (2) Proteins: HBB, HPA*, TF. (3) Blood groups: ABO, FY, K, MNS, P, RH, SE.

  S. Indians
    (Andhra Pr.)
(1) Enzymes: ACP1, ADA, AK1, ALPP, CHE2, DIA1, ESD, G6PD, GLO1, GOT1, GPI, GPT1, GPX1, IDH1, LDH, MDH1, NP, PEPA, PEPB, PEPC, PEPD, PGD, PGM1, PGM1(S), PGM2, SOD1. (2) Proteins: AG, ALB, C3, GC(S), HBB, HPA*, TBG, TF, TF(S). (3) Blood groups: ABO, FY, JK, K, LU, MNS, P, RH, V. (4) Immunoglobulin systems: IGHG1. (5) Miscellaneous: AC, CB, CER*, LAA*,

  S. Indians
    (Irula, etc)
(1) Enzymes: ACP1, AK1, G6PD, GPI, IDH1, LDHA, MDH1, NP, PEPA, PEPB, PEPC, PEPD, PGD, PGK, PGM1, PGM1(S), PGM2, SOD1. (2) Proteins: ALB, CP, GC, GC(S), HBB, HPA*, TBG, TF. (3) Blood

Place/ Population	Genetic loci
	groups: ABO, DI, FY, JS, LU, MNS, P, RH, SE, V. (4) Immunoglobulin systems: IGHG1. (5) Miscellaneous: CB, CER*.
E. Indians	(1) Enzymes: ACP1, ADA, AK1, ALPP, CHE1, CHE2, ESD, G6PD, GLO1, LDHA, MDH1, ME2, PEPA, PEPB, PGD, PGM1, PGM2, SOD1. (2) Proteins: BG, CP, GC, GC(S), HBB, HPA*, ORM1, PI, TF, TF(S). (3) Blood groups: ABO, DI, FY, K, MN, P, RH, SE. (4) Miscellaneous: CB, PTC.
Indians (Abroad)	(1) Enzymes: ACO1 ADH2, ADH3, ALDH3, AMY1, AMY2, FDH, FGH, GAA, GABAT, GDH, GLO1, GOT1, GOT2, GST1, GST2, GST3, GUK1, ITPA, ME1, NP, PGM3, PGP, PON, PP, SACPA, SACPB, SET1, SODA, SODB, UMPK. (2) Proteins: APOE, ME1. (3) Blood groups: GO, MIT, RI, V, XG. (4) Miscellaneous: AC.
SRI LANKA Sinhalese	(1) Enzymes: ACP1, ADA, AK1, ESD, G6PD, LDH, PGD, PGM1, PGM2, UMPK. (2) Proteins: HBB, HPA*, TF. (3) Blood groups: ABO, DI, FY, JK, JS, K, LU, MNS, P, RH, WR, V. (4) Immunoglobulin systems: IGHG1. (5) Miscellaneous: LAA*.
Veddah	(1) Proteins: HBB, HPA*, TF. (2) Blood groups: ABO, DI, FY, JS, K, LU, MNS, P, RH, V. (3) Immunoglobulin systems: IGHG1.
BANGLADESH Moslems	(1) Enzymes: ACP1, ADA, AK1, ESD, G6PD, GPI, LDHA, MDH1, PGD, PGM1, PGM2, SOD1. (2) Proteins: GC, HBB, HPA*, TF. (3) Blood groups: ABO, MNS, RH. (4) Miscellaneous: LAA*.
NEPAL	(1) Enzymes: ACP1, ADA, AK1, CA1, CA2, CAT, DPGM, ESD, G6PD, GOT1, GPI, GPT1, IDH1, LDH, MDH1, PEPA, PEPB, PEPC, PGAMA, PGD, PGK, PGM1, PGM2, SOD1. (2) Proteins: AHSG, GC(S), HBB, HPA*, ORM1, ORM2, PI, PI(S), TF, TF(S). (3) Blood groups: ABO, FY, GO, K, LU, MNS, P, RH, SE. (4) Miscellaneous: PTC.
BHUTAN	(1) Enzymes: ACP1, AK1, CHE2, G6PD, LDH, MDH1, PGD, PGM1. (2) Proteins: HBB, HPA*, TF. (3) Blood groups: ABO, DI, FY, JK, K, LU, MNS, P, RH, WR. (4) Immunoglobulin systems: IGHG1.
BURMA	(1) Enzymes: DAMOX, DASOX. (2) Proteins: C5, TF. (3) Blood groups: ABO, FY, K, LU, MNS, P, RH, SE. (4) Miscellaneous: AC, PTC.
MALAYSIA Malays	(1) Enzymes: ACO1, ACP1, ADA, AK1, ALPP, AMY1, AMY2, CA1, CA2, CHE1, DIA1, ESD, G6PD, GAA, GABAT, GDH, GLO1, GOT1, GOT2, GPI, GPT1, GPX1, HAGH, LDHA, MDH1, ME1, ME2, PEPA, PEPB, PGD, PGK, PGM1, PGM2, PGM3, PGP, SACPA, SACPB, SET1, SOD1, SODA, SODB, UMPK. (2) Proteins: ALB, APOE, GC, GC(S), HBB, HPA*, PI, TF, TF(S). (3) Blood groups: ABO, DI, JS, MNS, RH, V, XG. (4) Immunoglobulin systems: IGHG1. (5) DNA polymorphisms: HBAC, HBAD*. (6) Miscellaneous: AC, CER*, PTC.
Negrito/ Senoi Abor.	(1) Enzymes: G6PD, SOD1, SODA, SODB. (2) Proteins: ALB, HPA*, TF. (3) Blood groups: DI, FY, LU, MNS, P, SE. (4) Miscellaneous: CER.
INDONESIA Batak/ Indonesians	(1) Enzymes: ACP1, ADA, AK1, ALDH2, CA1, CA2, ESD, GLO1, GPI, GPT1, LDH, MDH1, PEPA, PEPB, PGD, PGK, PGM1, PGM1(S), PGM2, PGP. (2) Proteins: ALB, CP, F13A, GC(S), HBB, HPA*, PI, TBG, TF, TF(S) (3) Blood groups: ABO, DI, FY, IN, JK, K, LU, MNS, P, RH. (4) Miscellaneous: LAA*.
Balinese	(1) Enzymes: ACP1, ADA, AK1, CA1, CA2, DIA1, ESD, G6PD, GLO1, GOT1, GPI, IDH1, LDHA, MDH1, PEPA, PEPB, PEPC, PEPD, PGD, PGK, PGM1, PGM2, PGP, SOD1. (2) Proteins: GC(S), HBB, HPA*, HPA*(S), PI(S), TF, TF(S). (3) Blood groups: ABO, FY, K, LE, MNS, P, RH, SE.
THAILAND	(1) Enzymes: ACP1, ADA, AK1, ALAD, ALDH2, ALPP, CHE1, CHE2, ESD, G6PD, GPI, GPT1, PGD, PGM1, PGM1(S), PGP. (2) Proteins: AG, BF, C2, C3, GC, GC(S), HBB, HPA*, HPA*(S), PI, PI(S), PLG, TBG, TF, TF(S). (3) Blood groups: ABO, DI, DO, FY, GE, HE, I, IN, JS, K, LE, LU, MNS, P, RH ,SE, V, VEL, VW, WR, XG, YT. (4) Immunoglobulin systems: IGA2, IGHG1, KM. (5) DNA polymorphisms: HBAD*, HBBC. . (6) Miscellaneous: AC, CB, LAA*, PTC.
VIETNAM	(1) Enzymes: ACP1, AK1, ALDH2, CHE1, CHE2, G6PD, PGD. (2) Proteins: GC, HBB, HBG, HPA*, PI(S), TF. (3) Miscellaneous: BAIB, LAA*.
PHILIPPINES Filipino	(1) Enzymes: ACP1, ADA, AK1, ALDH2, ALPP, CA1, CHE1, ESD, G6PD, GOT1, GPI, GPT1, IDH1, LDHA,

Place/ Population	Genetic loci
	MDH1, MDH2, PGAMA, PGD, PGK, PGM1, PGM2, SOD1. (2) Proteins: BF, BG, GC, HPA*, HPA*(S), PGA, TF, TF(S). (3) Blood groups: ABO, DI, FY, JK, JS, K, MNS, P, RH. (4) Immunoglobulin systems: IGHG1, KM. (5) Miscellaneous: PTC.
Negrito	(1) Enzymes: ACP1, ADA, AK1, CA1, ESD, GLO1, GOT1, GPI, GPT1, LDH, MDH1, PGD, PGK, PGM1, PGM2, UMPK. (2) Proteins: C3, GC, HPA*, PI, TF. (3) Blood groups: ABO, DI, FY, K, JK, JS, LU, MI, MNS, P, RH, WR, XG. (4) Immunoglobulin systems: IGHG1, KM. (5) Miscellaneous: PTC.

CHINA

Place/ Population	Genetic loci
Mongolians	(1) Enzymes: ACP1, ADA, AK1, ALDH2, ESD, GLO1, GPT1, PGD, PGM1(S). (2) Proteins: C3, GC, HPA*, PI(S), TF. (3) Blood groups: ABO, FY, JK, K, MNS, P. (4) Miscellaneous: CB, CER*, LAA*, PTC.
Han	(1) Enzymes: ALDH2, GALT, PGM1(S). (2) Proteins: BF, C3, C6, C7, CP, GC(S), HBG, HPA*, PI(S). (3) Blood groups: ABO, DI, FY, IN, JK, K, LE, LU, MN, P, RH, SE, XG. (4) HLA systems: HLAA, HLAB, HLAC, HLADQ, HLADR. (5) Immunoglobulin systems: IGA2, IGHG1, KM. (6) DNA polymorphisms: ALB, HBBC. (7) Miscellaneous: CB, CER*, LAA*, PTC.
Hui	(1) Enzymes: ACP1, ADA, AK1, ALDH2, ESD, GLO1, GPT1, PGD, PGM1(S). (2) Proteins: HPA*, PI(S). (3) Blood groups: ABO, DI, FY, JK, K, LE, LU, MNS, P, RH, XG. (4) Miscellaneous: CB, PTC.
Dong	(1) Enzymes: ACP1, ADA, AK1, ALDH2, ESD, GLO1, GPT1, PGD, PGM1(S). (2) Proteins: HPA*, PI(S). (3) Blood groups: ABO, DI, FY, JK, LE, MNS, P, RH, SE, XG. (4) Miscellaneous: PTC.
Uygur	(1) Enzymes: ALDH2, GLO1. (2) Proteins: PI(S). (3) Blood groups: ABO, DI, FY, JK, K, LE, LU, MNS, P, RH, XG. (4) Miscellaneous: PTC.
Zhuang	(1) Enzymes: ACP1, ADA, AK1, ALDH2, ESD, GLO1, GPT1, PGD, PGM1(S). (2) Proteins: C3, HPA*, PI(S), TF. (3) Blood groups: ABO, DI, FY, JK, K, MNS, P, SE.
Tibetans	(1) Enzymes: ACP1, ADA, AK1, CA1, CA2, CAT, ESD, GOT1, GPI, GPT1, LDH, MDH1, PEPA, PEPB, PEPC, PGM1, PGM2, SOD1. (2) Proteins: AG, C3, GC, GC(S), HBB, HPA*, TF. (3) Blood groups: ABO, BP, CL, DI, FY, JN, K, LU, MG, MI, MNS, NY, P, PT, RH, SE, SKJ*, SW, VEL,VW, WB, WR, XG. (4) Miscellaneous: CB, PTC.
Chinese (Abroad)	(1) Enzymes: ACO1, ACO2, ACP1, ADA, ADH3, AK1, ALDH3, ALPP, AMY1, AMY2, CA1, CA2, CAT, CHE1, DIA1, ESD, FDH, FGH, G6PD, G6PDS*, GAA, GABAT, GDH, GLO1, GOT1, GOT2, GPI, GPT1, GST1, GST2, GST3, HAGH, IDH1, IDH2, LDHA, MDH1, MDH2, ME1, ME2, MPI, PEPA, PEPB, PGD, PGK, PGM1, PGM2, PGM3, PGP, PP, SACPA, SACPB, SET1, SOD1, SODA, SODB, UMPK. (2) Proteins: ALB, APOE, ATP*, BF, CON1, CON2 , DB , GC, GC(S), HBB, HPA*, HPA*(S), ORM1, PGA, PI, PIF, PR, TBG, TC2, TF, TF(S), VBRP*. (3) Blood groups: ABO, CL, DI, FY, GE, GO, K, MNS, P, PT, RH, SE, SKJ*, VEL, XG. (4) Immunoglobulin systems: IGHG1. (5) Miscellaneous: AC, BAIB, CB, CER*, LAA*, PTC.

TAIWAN

Place/ Population	Genetic loci
Toroko Aborigines	(1) Enzymes: ADA, AK1, ESD, GLO1, PGM1. (2) Proteins: C3, BF, GC, HPA*, TC2. (3) Blood groups: ABO, DI, FY, JK, K, LE, MNS, RH. (4) Immunoglobulin systems: IGA2. (5) Miscellaneous: CER*.

JAPAN

Place/ Population	Genetic loci
Japanese	(1) Enzymes: ACO1, ACP1, ADA, ADH2, ADH3, AK1, ALAD, ALDH2, ALPP, AMY1, CA1, CA2, CAT, CHE1 CHE2, DIA1, DIA3, ESD, FH1, G6PD, GABAT, GDH, GLO1, GOT1, GOT2, GPI, GPT1, GST1, HAGH, HK3, IDH1, LDHB, MDH1, MDH2, ME2, NP, PEPA, PEPB, PEPC, PEPD, PGD, PGK, PGM1, PGM1(S), PGM2, PGM3, PKM2, PXS*, SAHH, SDH, SOD1, TPI, UMPK. (2) Proteins: AG, AHSG, ALB, APOE, AT3, BF, C2, C3, C4A, C4B, C6, C7, C8A, CP, DB, F13A, F13B, FI, GC, GC(S), HPA*, HPA*(S), HBB, HBG, LC40P, LC49P, LC64P, LC100P, ORM1, PA, PB, PLG, PMF*, PI, PI(S), PR, TBG, TF, TF(S). (3) Blood groups: ABO, DI, DO, FY, GE, GY, I, JK, JR, K, LAN, LE, LU, MG, MNS, NY, OK, OS, P, PLA*, RH, SE, SW, TCA*, TS, UL, VEL, VW, WB, WR, XG, YUK. (4) HLA systems: HLAA, HLAB, HLAC, HLADQ, HLADR. . (5) Immunoglobulin systems: IGA2, IGHG1, KM. (6) DNA polymorphisms: ALB, AMY1, F9. (7) Miscellaneous: AC, BAIB, CB, CER*, DOX*, LAA*, MEP*, PTC.
Japanese (Abroad)	(1) Enzymes: FGH, G6PDS*, PGA, SACPA, SACPB, SAL1, SAL2, SET1, TC2. (2) Blood Groops: NFLD.
Ainu	(1) Enzymes: ACP1, ADA, AK1, ALDH2, CHE1, CHE2, DIA1, ESD, G6PD, GPI, GPT1, LDH, MDH1, PEPA, PEPB, PEPD, PGD, PGM1, PGM2. (2) Proteins: AG, ALB, GC, GC(S), HPA*, PI, TF. (3) Blood groups: ABO DI, FY, HE, JK, JS, K, LE, LU, MNS, P, RD, RH, SE, WR, XG. (4) Immunoglobulin

Place/ Population	Genetic loci
	systems: IGHG1, KM. (5) Miscellaneous: AC, CB, CER*, PTC.
KOREA Koreans	(1) Enzymes: ACP1, AK1, ALAD, ALDH2, AMY2, CHE2, ESD, G6PD, GLO1, GPT1, PGD, PGM1, PGM1(S), PGP. (2) Proteins: BF, BG, C2, C3, GC(S), HPA*, HPA*(S), PI, PI(S), PLG, TF(S), (3) Blood groups: ABO, DI, FY, JK, K, LE, LU, MNS, P, RH. (4) Immunoglobulin systems: IGHG1. (5) Miscellaneous: AC, CB, CER*, PTC.
Koreans (Abroad)	(1) Enzymes: ACP1, ADA, AK1, ALDH2, CAT, CHEI, ESD, GLO1, GPT1, PGD, PGM1(S). (2) Proteins: C3, CP, GC, GC(S), HPA*, LP, PI(S), TF. (3) Blood groups: ABO, DI, FY, K, JK, MNS, P, RH, SE. (4) Immunoglobulin systems: KM. (5) Miscellaneous: CB, PTC.

AFRICA

ALGERIA	(1) Enzymes: ACP1, CHE1, G6PD, GLO1, LDH, PGD, PGM1. (2) Proteins: BF, GC(S).
LIBYA	(1) Enzymes: ACP1, ADA, AK1, G6PD, PGD, PGM1. (2) Proteins: GC, HBB, HPA*. (3) Blood groups: ABO, FY, JK, K, LE, LU, MNS, P, RH. (4) Immunoglobulin systems: IGHG1. (5) Miscellaneous: AC, PTC.
EGYPT Egyptians	(1) Enzymes: ACP1, AK1, ALDH2, ESD, GLO1, GPT1, PGD, PGM1. (2) Proteins: AHSG, PI, C3, GC, HPA*, TF, TF(S). (3) Blood groups: MNS, P, RH, SE. (4) Immunoglobulin systems: IGHG1, KM. (5) Miscellaneous: AC, BAIB, LAA*, PTC.
Nubians	(1) Enzymes: ACP1, AK1, ESD, G6PD, GPI, IDH1, LDH, MDH1, PEPA, PEPB, PEPC, PGD, PGM1, PGM2. (2) Proteins: ALB, CP, HBB, HPA*, TF. (3) Blood groups: ABO.
SUDAN Sudanese	(1) Enzymes: ACP1, ALDH2, G6PD, GLO1, GPI, GSR, LDH, MDH1, PGD, PGM1, PGM2, PON, SOD1 (2) Proteins: ALB, CP, HBB, HPA*, TF. (3) Blood groups: ABO, FY, K, MNS, RH. (4) Miscellaneous: PTC.
Beja-Amarar	(1) Enzymes: ACP1, AK1, CHE2, G6PD, LDH, MDH1, PGD, PGM1. (2) Proteins: GC, HBB, HPA*, TF. (3) Blood groups: ABO, FY, K, JK, LU, MNS, P, RH. (4) Immunoglobulin systems: IGHG1 (5) Miscellaneous: LAA*.

CHAD
  Sara
    Majingay
	(1) Enzymes: ACP1, AK1, CHE1, G6PD, PGD, PGM1, PGM2. (2) Proteins: HPA*, HBB, TF. (3) Blood groups: ABO, FY, I, K, MN, P, RH

MALI	(1) Enzymes: ACP1, AK1, AMY1, CHE1, CHE2, DIA1, G6PD, LDH, PGD, PGM1, PGM2. (2) Proteins: GC, HPA*, PB, PI(S), TF.
SENEGAL Bedik, Dakar	(1) Enzymes: ACP1, ADA, AK1, CHE1, CHE2, G6PD, PGD, PGM1, PGM2. (2) Proteins: AG, GC(S), HBB, HPA*, TF. (3) Blood groups: ABO, AU, I, JK, K, MNS, P, RH, VW. (4) DNA polymorphisms: HBAD* (5) Miscellaneous: LAA*.
LIBERIA	(1) Enzymes: ACP1, AK1, ALAD, ALDH2, ESD, G6PD, GLO1, PGD, PGM1(S),. (2) Proteins: GC, HBB, HPA*, PI(S), TF(S) (3) Blood groups: MG.
IVORY COAST Tuka	(1) Enzymes: ACP1, AK1, CHE1, CHE2, G6PD, PGD, PGM1, PGM2. (2) Proteins: HPA*.
UPPER VOLTA	(1) Proteins: HBB. (2) DNA polymorphisms: HBAD.

NIGERIA
  Yoruba/
  Nigerians
	(1) Enzymes: ACO1, ACO2, ACP1, ADA, AK1, ALPP, APRT, CA1, CA2, CHE1, CHE2, ESD, FH1, G6PD, GOT1, GOT2, GPI, GPT1, GPX1, IDH1, LDHB, MDH1, PEPA, PEPB, PEPC, PEPD, PGD, PGM1, PGM2, PGM3, TPI. (2) Proteins: AG, BF, CP, GC, HBB, HPA*, HPA*(S), ORM1, ORM2, TC2, TF, TF(S). (3) Blood groups: ABO, DI, FY, HE, K, JK, JS, LU, MNS, P, RD, RH, SE, V, WR. (4) Immunoglobulin systems: IGHG1. (5) DNA polymorphisms: HBAC, HBBC. (6) Miscellaneous: AC, CB, CER*, DOX*, LAA*, PTC.

Nigerians (Abroad)	(1) Enzymes: GAA, ITPA, NP, PEPA, PP, PXS*. (2) Blood groups: BY, CS, GO, LS, RI, SW. (3) HLA systems: HLAA, HLAB, HLAC, HLADR. (4) DNA polymorphisms: AT3, HBBC.

Place/ Population	Genetic loci

CAMEROON — (1) Enzymes: ACP1, ADA, AK1, CHE1, ESD, G6PD, GLO1, GPT1, PGD, PGM1. (2) Proteins: BF, C3, GC, GC(S), HBB, HBD, HPA*, HPA*(S), PI, PI(S), TBG, TF, TF(S).

GAMBIA — (1) Enzymes: ESD, GLO1, (2) Proteins: C5, PLG.

CENTRAL AFRICAN REPUBLIC

Bantu — (1) Enzymes: GPX1, LDH. (2) Miscellaneous: LAA*

Pygmies — (1) Enzymes: ACP1, ADA, AK1, CA1, CA2, CHE1, DIA1, ESD, G6PD, GLO1, GOT1, GPI, GPT1, GPX1, LDH, MDH1, PEPA, PEPB, PEPC, PGM1, PGM2, PGP. (2) Proteins: GC, GC(S), HBB, HBD, HPA*, HPA*(S), TBG, TF. (3) Blood groups: ABO, DI, FY, I, JK, K, LAN, LU, MNS, P, RH, SE, WR, (4) Immunoglobulin systems: IGA2, IGHG1, KM

BURUNDI

Hutu — (1) Enzymes: ACP1, ADA, AK1, DIA1, DPGM, ESD, G6PD, GOT1, GPI, GPT1, IDH1, LDHB, MDH1, PGD, PGK, PGM1, PGM2. (2) Proteins: HBB, HPA*, TF. (3) Blood groups: ABO, JS, MNS, RH.

RWANDA

Hutu — (1) Enzymes: CA1, CA2, GLO1, GPX1, PEPB. (2) Proteins: HBD, TC2.

Pygmies — (1) Enzymes: GPX1. (2) Proteins: HBD, TC2.

CONGO — (1) Enzymes: CA2, DPGM, GPT1, GPX1, IDH1, PGAMA. (2) Proteins: GC(S), HPA*(S), PI(S), TF(S).

ZAIRE

Bantu, etc. — (1) Enzymes: ADA, GPX1. (2) Proteins: HBD, TBG. (3) Blood groups: ABO, DI, FY, JK, K, JS, LE, LU, MNS, P, RH, XG.

ETHIOPIA

Ethiopians — (1) Enzymes: CA2, CHE1, FH1, SOD1. (2) Proteins: C3, GC, TF. (3) Miscellaneous: LAA*.

Amhara tr. — (1) Enzymes: ACP1, AK1, G6PD, MDH1, PGD, PGM1. (2) Proteins: HBB, HPA*, TF. (3) Blood groups: ABO, FY, HE, JS, K, LU, MNS, P, RH, V. (4) Miscellaneous: AC, CB, PTC.

UGANDA — (1) Enzymes: ACP1, ADA, AK1, CA2, ESD, LDH, MDH1, PGD, PGM1, PGM2. (2) Proteins: GC, HPA*, TF. (3) Blood groups: FY, JS, K, MNS, P, RH, SE, V. (4) Miscellaneous: CB.

KENYA — (1) Enzymes: ALDH2, AMY1, DPGM, GPT1, IDH1, LDH, PGK, PGM1, UMPK. (2) Proteins: DB, HBB, HPA*, PA, PB, PI(S), PR, TF. (3) Miscellaneous: LAA*.

TANZANIA

Sandawe — (1) Enzymes: ACP1, ADA, AK1, G6PD, GPI, LDH, MDH1, PGD, PGM1, PGM2. (2) Proteins: HBB, HPA*, TF. (3) Blood groups: ABO, DI, FY, HE, JK, JS, K, LU, MNS, P, RD, RH, V, WR. (4) Immunoglobulin systems: IGHG1

Nyaturu — (1) Enzymes: ACP1, ADA, AK1, G6PD, GPI, LDH, MDH1, PGD, PGM1, PGM2. (2) Proteins: HBB, HPA*, TF. (3) Blood groups: ABO, DI, FY, HE, JK, JS, K, LU, MNS, P, RD, RH, V, WR. (4) Immunoglobulin systems: IGHG1

Hadza — (1) Enzymes: ACP1, AK1, LDH, MDH1, PGD, PGM1, PGM2. (2) Proteins: HPA*, LP, TF. (3) Blood groups: ABO, DI, FY, HE, JK, JS, K, LU, MNS, P, RH, V, WR. (4) Miscellaneous: CB, PTC.

ANGOLA

Njinga/
Negroes — (1) Enzymes: ACP1, ADA, AK1, CHE2 ESD, G6PD, PEPA, PEPB, PEPC, PEPD, PGD, PGM1, PGM2. (2) Proteins: AG, CP, C3, HBB, HPA*, TF. (3) Blood groups: ABO, FY, MNS, RH, HE, K.

MOZAMBIQUE

Bantu — (1) Enzymes: CHE1, CHE2, G6PD, IDH1, PGK. (2) Proteins: CP, BG, PI.

SOUTHERN AFRICA

Ambo
(SW Africa) — (1) Enzymes: CA1, CA2.

Place/ Population	Genetic loci
Basters/SW Africans (Namibia)	(1) Enzymes: ACP1, ADA, AK1, ESD, G6PD, PGD, PGM1, PGM2. (2) Proteins: HPA*, TF. (3) Blood groups: ABO, FY, HE, K, LE, MNS, P, RH, SE.
San/Bushmen	(1) Enzymes: ACP1, ADA, AK1, CA1, CA2, CHE2, ESD, G6PD, GPT1, IDH1, LDH, MDH1, PEPA, PEPB, PEPC, PEPD, PGD, PGK, PGM1, PGM1(S), PGM2. (2) Proteins: GC, GC(S), HBB, HBD, HPA*, PI(S), TBG, TF. (3) Blood groups: ABO, DI, FY, HE, JK, K, LE, MNS, P, RH, SE, V, XG. (4) Immunoglobulin systems: IGHG1, KM. (5) Miscellaneous: AC, LAA*, PTC.
Khoi/Hottentot	(1) Enzymes: ACP1, ADA, AK1, CHE2, G6PD, PEPA, PEPD, PGD, PGM1, PGM1(S), PGM2. (2) Proteins: HBB, HBD, HPA*, TF. (3) Blood groups: ABO, FY, JK, K, LE, MNS, P, RH, SE, V, XG. (4) Immunoglobulin systems: IGHG1, KM. (5) Miscellaneous: LAA*, PTC.
Negroes/Dama	(1) Enzymes: ACP1, ADA, AK1, CA1, CA2, CHE2, ESD, G6PD, GPT1, IDH1, PEPA, PEPB, PEPD, PGD, PGM1, PGM2. (2) Proteins: BF, BG, GC, HBB, HPA*, PI(S), TF. (3) Blood groups: ABO, FY, HE, K, LE, LU, MIT, MNS, P, RH, SE, V, XG. (4) Immunoglobulin systems: IGHG1. (5) Miscellaneous: LAA*.
Zulu/Bantu	(1) Enzymes: ACP1, ADA, AK1, CA1, CA2, ESD, FGH, G6PD, GLO1, GPT1, GPX1, LDH, MDH1, PEPA, PEPB, PGD, PGM1, PGM1(S), PGM2, PGP, SOD1. (2) Proteins: AG, ALB, CP, GC, GC(S), HBB, HBD, HPA*, ORM1, PI, PLG, TBG, TF. (3) Blood groups: ABO, FY, K, LU, MNS, P, RH, SE, V. (4)Immunoglobulin systems: IGA2, IGHG1, KM. (5) Miscellaneous: AC, LAA*, PTC.
Griqua	(1) Enzymes: ACP1, ADA, AK1, G6PD, PEPA, PEPB, PEPC, PEPD, PGD, PGM1, PGM2. (2) Proteins: HBB, HPA*, TF. (3) Blood groups: ABO, FY, HE, K, MNS, P, RH, SE.
TRISTAN DA CUNHA	(1) Enzymes: ACP1, ADA, AK1, CA1, CA2, ESD, G6PD, GPT1, GPX1, IDH1, PEPA, PEPB, PEPC, PEPD, PGD, PGM1, PGM2. (2) Proteins: HBB, HPA*, TF. (3) Blood groups: ABO, HE, MNS, RH, SE.

NORTH AMERICA

GREENLAND
Eskimos	(1) Enzymes: ADA, AK1, CHE1, CHE2, LDH, PGM1, PGM2, SOD1. (2) Proteins: AG, ALB, C3, GC, HBB, HPA*, LP, PI(S), TF. (3) Blood groups: ABO, FY, JK, K, MNS, P, RH, SE. (4) Immunoglobulin systems: IGHG1, KM. (5) Miscellaneous: CB, LAA*, PTC.

ALASKA
Eskimos (N. Alaska)	(1) Enzymes: ACP1, ADA, AK1, CA2, CHE1, CHE2, DPGM, ESD, G6PD, GLO1, GOT1, GPT1, PGD, PGM1, PGM1(S), RDS, UMPK. (2) Proteins: ALB, F13B, GC, GC(S), HBB, HPA*, HPX, TBG. (3) Blood groups: ABO, BE, DI, FY, JK, JR, JS, K, LE, LU, MG, MI, MNS, P, RH, SE, VEL, VW, WR, XG, YT. (4) Immunoglobulin systems: IGA2, IGHG1. (5) Miscellaneous: BAIB, CER, LAA*, PTC.
Eskimos (St.Lawrence Is.)	(1) Enzymes: ACP1, ADA, AK1, CA1, CA2, DPGM, ESD, G6PD, GALT, GLO1, GPD1, GPI, GPT1, IDH1, LDH, MDH1, NP, PEPA, PEPB, PEPC, PEPD, PGD, PGK, PGM1, PGM2, PGP, SOD1. (2) Proteins: ALB, BF, CP, F13B, HBA, HBB, HPA*, ORM1, ORM2, PLG, TF. (3) Blood groups: ABO, DI, FY, JK, K, MNS, P, RH. (4) Immunoglobulin systems: IGHG1, KM.
Athabaskan Indians	(1) Enzymes: ACP1, ADA, AK1, CHE1, DPGM, ESD, G6PD, GLO1, GOT1, GPT1, PGD, PGM1, RDS, UMPK. (2) Proteins: ALB, GC, HBB, HPA*. (3) Blood groups: ABO, BE, DI, FY, JK, JS, K, LU, MG, MI, MNS, P, RH, RD, S, SE, VEL, VW, WR, XG. (4) Immunoglobulin systems: IGA2, IGHG1. (5) Miscellaneous: BAIB, LAA*.

CANADA
Whites	(1) Enzymes: ACHE, GAA, PON. (2) Proteins: AHSG, APOE, C2. (3) Blood groups: ER, LW, MIT, NFLD, RE, WAL* (4) Miscellaneous: DOX*, MEP*.
Eskimos	(1) Enzymes: ACP1, ADA, AK1, ALDOA, ALPP, CHE1, CHE2, DIA1, DPGM, ENO1, ESD, G6PD, GAPD, GOT1, GPI, GPT1, GSR, HK3, IDH1, LDH, PEPA, PEPB, PEPC, PEPD, PFKL, PGAMA, PGD, PGK, PGM1, PGM2, PGM3, PKM2, SOD1. (2) Proteins: ALB, CP, GC, HBB, HPA*, HPA*(S), PI, TF. (3) Blood groups: ABO, DI, FY, JK, K, LU, MNS, P, RH, SE. (4) Immunoglobulin systems: IGHG1, KM. (5) Miscellaneous: AC.
Ojibwa	(1) Enzymes: ACP1, ADA, AK1, DIA1, G6PD, GOT1, GPI, GSR, IDH1, LDH, NP, PEPA, PEPB,

Place/ Population	Genetic loci
	PEPC, PFKL, PGD, PGK, PGM1, PGM2. (2) Proteins: ALB, CP, GC, HPA*, TF. (3) Blood groups: ABO, DI, FY, JK, K, LU, MI, MNS, NFLD, P, RH. (4) Immunoglobulin systems: IGHG1, KM.
Cree	(1) Enzymes: ACP1, ADA, AK1, ALPP, CHE2, DPGM, ENO1, ESD, G6PD, GOT1, GPI, GPT1, IDH1, NP, PEPA, PEPB, PEPC, PGAMA, PGD, PGK, PGM1, PGM2, PGM3, UMPK. (2) Proteins: ALB, GC, HPA*, TF. Blood groups: ABO, CO, DI, DO, FY, GE, GY, JK, K, LU, MG, MIT, MNS, NFLD, NY, RD, RH, SC, (3) XG, VEL. (4) Miscellaneous: PTC.
Dogrib	(1) Enzymes: ACP1, ADA, AK1, CA1, CA2, DPGM, ESD , G6PD, GALT, GAPD, GLO1, GPI, GPT1, IDH1, LDH, MDH1, NP, PEPA, PEPB, PEPC, PEPD, PGD, PGK, PGM1, PGM2, PGP, SOD1. (2) Proteins: ALB, CP, F13B, GC(S), HBB, HPA*, ORM1, ORM2, TF. (3) Blood groups: ABO, DI, FY, JK, K, MNS, P, RH. (4) Immunoglobulin systems: IGHG1, KM.

UNITED STATES
  Blackfeet
    Chippewa,
    etc.
	(1) Enzymes: ALDH2, GOT1. (2) Proteins: ALB, BF, GC, GC(S), HPA*, TF, TF(S). (3) Blood groups: ABO, DI, FY, JK, K, LE, LU, MNS, P, RH, SE, WR. (4) Miscellaneous: PTC.

Papago/Pima/ Navajo, etc.	(1) Enzymes: ADH2, ADH3, ALDH2, CHE1, GPX1, PGM1(S), PGP. (2) Proteins: ALB, BF, GC, GC(S), F13A, HPA*, HPX, PLG, TF, TF(S). (3) Blood groups: ABO, BI, CO, DI, DO, FY, GY, JK, K, MNS, P, RH, V, XG. (4) HLA systems: HLAA, HLAB, HLAC, HLADQ, HLADR. (5) Immunoglobulin systems: IGA2. (6) DNA polymorphisms: INS, INSR. (7) Miscellaneous: BAIB, CB, CER*, PTC.
Whites	(1) Enzymes: ACP1, ADA, ADH2, ADH3, AK1, ALDOA, ALPP, AMY1, AMY2, CA2, CA3, CDA, CHE1, CHE2, DIA1, DIA3, DPGM, ESB3, ESD, FUCA, G6PD, G6PDS*, GALK, GAPD, GLA, GLO1, GLUD, GOT1, GPT1, GPX1, HAGH, HEXA, HEXB, HPRT, IDH1, LDHA, MDH1, MDH2, ME1, ME2, PEPA, PEPD, PFKL, PGD, PGAMA, GM1, PGM1(S), PGM2, PNK, PON, PXL*, PXS*, SACPA, SACPB, SET1, SORD, UMPK. (2) Proteins: AG, APOA4, APOE, AT3, ATP, BF, C1R, C2, C3, C4A, C4B, C8A, C8B, CON1, CON2, CP, DB, F13A, F13B, GI, GC, GC(S), HBB, HBG, HF, HPA*, HPA*(S), HPX, LP,ORM1,ORM2, PA,PB, PCS,PE, PGA, PI, PI(S), PIF, PLG, PMF*, PMS*, PO, PR, PS, SAL1, SAL2, TC2, TBG, TF, TF(S), VBRP*, XM. (3) Blood groups: ABO, AUG*, BE, BI, CH*, CO, DP, EL, FY, GE, GN, GY, JK, JR, JS, K, KN, LU, MCC*, MG, MNS, OK, P, PEA*, PLA*, PLE1*, PLE2*, RD, RH, SF, TCA*, V, VW, WR. (4) HLA systems: HLAA, HLAB, HLAC, HLADQ, HLADR. (5) Immunoglobulin systems: IGA2, KM. (6) DNA polymorphisms: AFP, APOA1, APOA2, APOA4, APOB, APOC2, APOA1-APOC3-APOA4, COL1A1, COL1A2, COL2A2, F8C, F9 HPRT, IGHG, IGHM, INS, INSR, NGFB, PI, PLG, POMC, PTH, URK*. (7) Miscellaneous: AC, BAIB, CER*, DOX*, LAA*, MEP*.
Blacks	(1) Enzymes: ACP1, ADA, ADH2, ADH3, AK1, ALDOA, ALPP, AMY1, AMY2, CA2, CA3, CHE1, CHE2, DIA1, DPGM, ESA, ESB3, ESD, FH1, FUCA, G6PD, GLA, GLO1, GOT1, GPT1, GPX1, HAGH, HPRT, IDH1, LDHA, LDHB, MDH1, ME1, ME2, PEPA, PEPD, PGAMA, PGD, PGM1, PGM1(S), PGM2, PNK, SORD, UMPK. (2) Proteins: AG, APOA4, AT3, ATP*, BF, C1R, C2, C3, C6, C8A, CON1, CON2, CP, DB, F13A, F13B, G1, GC, GC(S), HBB, HBD, HBG, HPA*, HPA*(S), HPB*, HPX, LP, ORM1, ORM2, PA, PB, PCS, PE, PI, PI(S), PIF, PLG, PMS*, PO, PR, PS, TBG, TC2, TF, TF(S), XM, VBRP*. (3) Blood groups: ABO, AUG*, BE, BI, CO, DI, DO, FY, GE, GO, GY, JK, JS, K, LS, LU, MCC*, MG, MNS, OK, ORR*, P, PLA*, RH, SCᵇ, SF, TCA*, UL, V, VEL, WR, YT. (4) HLA systems: HLAA, HLAB, HLAC, HLADQ, HLADR. (5) Immunoglobulin systems: IGA2, KM. (6) DNA polymorphisms: ALB, APOC3, APOA1-APOC3-APOA4, COL2A2, GH, HBAD*, INS, INSR, PGK1, PLG. (7) Miscellaneous: AC, BAIB, LAA*.
Mexican- Americans	(1) Enzymes: ACP1, ADA, AK1, ESD, GLO1, GPT1, PGD, PGM1, PGM1(S), PGP. (2) Proteins: AHSG, F13A, GC(S), HPA*, PLG. (3) Blood groups: ABO, DI, FY, JK, K, MNS, OK, RH. (4) Miscellaneous: LAA*.
Orientals	(1) Enzymes: DPGM, IDH1, ATP*, PGAMA. (2) Proteins: DB, CON1, CON2, PIF, PLG, TC2. (3) Blood groups: JR, OK, SF, V.

MIDDLE AMERICA

MEXICO Am. Indians	(1) Enzymes: ACP1, CHE1, PGD, PGM1, PGM2. (2) Proteins: ALB, F13B, GC, HPA*, ORM1, ORM2, TF. (3) Blood groups: ABO, DI, FY, JK, K, LU, MNS, P, RH. (4) HLA systems: HLAA, HLAB, HLAC, HLADR. (5) Miscellaneous: BAIB, CB, CER*, PTC.
Mestizo	(1) Enzymes: ACP1, ALDH2, CHE1, PGM1. (2) Proteins: ALB, GC, HPA*, TF. (3) Blood groups: ABO, DI, FY, JK, K, LE, MNS, P, RH. (4) HLA systems: HLAA, HLAB, HLAC, HLADR. (5)

Place/ Population	Genetic loci
	Miscellaneous: CB, LAA*.

GUATEMALA/DOMINICA/
  HONDURAS
  Black Caribs

(1) Enzymes: ACP1, ADA, AK1, CA1, CA2, CHE2, ESD, IDH1, MDH1, PGD, PGM1, PGM1(S), PGM2. (2) Proteins: ALB, BF, CP, GC, GC(S), HBB, HPA*, PLG, TF. (3) Blood groups: ABO, DI, GY, K, MNS, MT, RH, YT. (4) Immunoglobulin systems: IGHG1, KM. (5) Miscellaneous: BAIB.

  Am. Indians

(1) Proteins: ALB, TC2.

COSTA RICA
  Guaymi

(1) Enzymes: ACP1, ADA, AK1, CA1, CA2, ESA1, ESD, GALT, GLO1, GPI, IDH1, LDHA, MDH1, NP, PEPA, PEPB, PEPC, PEPD, PGD, PGM1, PGM2, TPI. (2) Proteins: ALB, CP, HBB, HPA*, TF. (3) Blood groups: ABO, DI, FY, JK, LE, LU, MNS, P, RH.

PANAMA
  Blacks

(1) Enzymes: ACP1, ADA, AK1, ESD, GPI, IDH1, LDHA, MDH1, PEPB, PEPC, PGD, PGM1, PGM2. (2) Proteins: ALB, CP, HBB, HPA*, TBG, TF.

WEST INDIES

(1) Enzymes: ACP1, AK1, LDH, MDH1, PEPA, PEPB, PEPD, PGD, PGM1, PGM2, SOD1. (2) Proteins: ALB, CP, GC, HBB, HPA*, TF. (3) Blood groups: ABO, DI, FY, JK, K, LE, LS, MNS, RH, SE.

SOUTH AMERICA

ARGENTINA
  Mapuche

(1) Enzymes: ACP1, GLO1, PGM1, PGM2. (2) Proteins: ALB, BF, C3, CP, HBB, HPA*, TF. (3) Blood groups: ABO, RH.

BOLIVIA
  Aymara

(1) Enzymes: ACP1, ADA, AK1, ALDOA, CA1, CA2, CHE1, CHE2, DPGM, ESD, G6PD, GAPD, GPI, HK3, IDH1, LDH, MDH1, PEPA, PEPB, PEPC, PFKL, PGD, PGK, PGM1, PGM2, PKM2. (2) Proteins: ALB, CP, GC(S), HBA, HBB, HPA*, TF. (3) Blood groups: ABO, DI, FY, I, JK, JS, K, LU, MNS, P, RH. 4) Immunoglobulin systems: IGHG1, KM. (5) Miscellaneous: LAA*.

  Quechua/
    Siriono

(1) Enzymes: DIA1. (2) Proteins: GC(S).

BRAZIL
  Baniwa

(1) Enzymes: ACP1, ADA, AK1, CA1, CA2 ,ESA1, ESD, G6PD, GALT, GLO1, GPI, IDH1, LDH, MDH1, NP, PEPA, PEPB, PEPC, PGD, PGM1, PGM2, TPI. (2) Proteins: ALB, CP, GC, HBB, HPA*, TF. (3) Blood groups: ABO, DI, FY, JK, K, LE, MNS, P, RH, SE.

  Cayapo

(1) Enzymes: ACP1, ADA, AK1, ESA1, ESD, G6PD, GALT, GPI, IDH1, LDH, MDH1, PEPA, PEPB, PGD, PGM1, PGM2, SOD1. (2) Proteins: ALB, CP, GC, HBA, HBB, HPA*, TF. (3) Blood groups: ABO, DI, FY, JK, JS, K, LE, LU, MNS, P, RH, SE, WR. (4) Immunoglobulin systems: IGHG1, KM. (5) Miscellaneous: CB.

  Caingang

(1) Enzymes: ADA, AK1, CA1, CA2, ESA1, ESD, G6PD, PGD, PGM1, PGM2. (2) Proteins: ALB, CP, GC, GC(S), HBB, HPA*, TF, TF(S). (3) Blood groups: ABO, DI, FY, JK, K, MNS, P, RH, SE. (4) Miscellaneous: CB, PTC.

  Pacaás Novos

(1) Enzymes: ACP1, ADA, AK1, CA2, ESA1, ESD, G6PD, GLO1, PEPA, PEPB, PEPC, PGD, PGM1, PGM2. (2) Proteins: ALB, CP, GC(S), HBB, HPA*, TF. (3) Blood groups: ABO, FY, K, P, RH, SE.

  Parakana

(1) Enzymes: ADA, AK1, ESA1, ESD, G6PD, PGD, PGM1, PGM2. (2) Proteins: ALB, CP, GC, HBB, HPA*, TF. (3) Blood groups: ABO, DI, FY, JK, K, MNS, P.

  Macushi

(1) Enzymes: ACP1, ADA, AK1, CA1, CA2, ESA1, ESD, GALT, GLO1, GPI, IDH1, LDH, MDH1, NP, PEPA, PEPB, PGD, PGM1, PGM2, TPI. (2) Proteins: ALB, CP, GC, GC(S), HBA, HBB, HPA*, TF. (3) Blood groups: ABO, DI, FY, JK, K, LE, MNS, P, RH, SE.

  Wapishana

(1) Enzymes: ACP1, ADA, AK1, CA1, CA2, ESA1, ESD, GALT, GPI, IDH1, LDH, MDH1, NP, PEPA, PEPB, PGD, PGM1, PGM2, TPI. (2) Proteins: ALB, CP, GC, HBA, HBB, HPA*, TF. (3) Blood groups: ABO, DI, FY, JK, K, LE, MNS, P, RH.

  Ticuna

(1) Enzymes: ACP1, ADA, AK1, ESD, GALT, GLO1, GPI, LDH, MDH1, NP, OTC, PEPA, PEPB, PEPC,

Place/ Population	Genetic loci

PEPD, PGD, PGM1, PGM2, PP, TPI. (2) Proteins: ALB, CP, GC, HBB, HPA*, TF. (3) Blood groups: ABO, DI, FY, JK, K, LE, LU, MNS, P, RH.

Xavante, Bahia, etc. (1) Enzymes: ADH1, ADH3, CA1, CHE1, CHE2. (2) Proteins: APOE.

CHILE

Atacameno (1) Enzymes: ACP1, ADA, AK1, ALAD, ALDH2, AMY2, CHE1, ESD, GLO1, GPT1, PGD, PGM1, PGM1(S), PON. (2) Proteins: BF, C3, GC, GC(S), HPA*, PI(S), PLG, TF, TF(S). (3) Blood groups: ABO, DI, FY, K, MNS, RH. (4) Miscellaneous: AC.

Aymara (1) Enzymes: GPT1, LDHA. (2) Miscellaneous: CB.

COLOMBIA

Noanama/ Colombian (1) Enzymes: ACP1, AK1, GLO1, GPT1, GPX1, LDH, MDH1, PEPA, PEPB, PGD, PGM1,PGP, SOD1. (2) Proteins: ALB, BF, C3, C4A, C4B, CP, HPA*, ORM1, ORM2, TF. (3) Blood groups: ABO, DI, FY, K, LE, MNS, P, RH. (4) Miscellaneous: CB, LAA*.

ECUADOR

Shuara, etc. (1) Enzymes: ACP1, AK1, ALDH2, CHE1, CHE2, ESD, GPT1, PGD, PGM1. (2) Proteins: BF, BG, C3, GC, HPA*, PI, TF. (3) Miscellaneous: AC.

Waorani (1) Enzymes: ACP1, AK1, CA1, CA2, DIA1, ESD, GLO1, GPI, IDH1, LDHA, MDH1, PEPA, PEPB, PGD, PGK, PGM1, PGM2, SOD1.

FRENCH GUIANA

Wayampi (1) Enzymes: ACP1, ADA, AK1, ESD, GPI, GPT1, IDH1, LDH, MDH1, PEPA, PEPB, PGD, PGM1, PGM2, SOD1. (2) Proteins: ALB, C3, GC, HBB, PI, TF. (3) Blood groups: ABO, DI, FY, I, JK, K, LE, MNS, P, RH, SE, XG.

PERU

Quecha etc. (1) Enzymes: GOT1, GPT1, GPX1, GSR, PEPE, PGAMA, PGK. (2) Proteins: AG. (3) Miscellaneous: CER*, PTC.

SURINAM

Trio (1) Enzymes: ACP1, G6PD, GPX1, PGD, PGM1, PGM2. (2) Proteins: ALB, GC, HPA*, TF. (3) Blood groups: ABO, DI, FY, JK, K, LE, LU, MNS, P, RH.

Wajana (1) Enzymes: GPX1.

VENEZUELA

Makiritare (1) Enzymes: ACP1, ADA, AK1, CA1, CA2, CHE1, CHE2, ESA1, ESD, G6PD, GALT, GPI, IDH1, LDH, MDH1, NP, PEPA, PEPB, PGD, PGM1, PGM2, SOD1, TPI. (2) Proteins: ALB, CP, GC, HBA, HBB, HPA*, LP, TF. (3) Blood groups: ABO, DI, FY, JK, JS, K, LE, LU, MNS, P, RH, SE, WR. (4) Immunoglobulin systems: IGHG1, KM.

Yanomama (1) Enzymes: ACP1, ADA, AK1, CA1, CA2, ESA1, ESD, GALT, G6PD, GPI, IDH1, LDH, MDH1, NP, PEPA, PEPB, PGM1, PGM2, PGD, SOD1, TPI. (2) Proteins: AG, ALB, CP, GC, HPA*, HBA, HBB, LP, TF. (3) Blood groups: ABO, DI, FY, JK, JS, K, LE, LU, MG, MNS, P, RH, SE, VW, WR, XG. (4) Immunoglobulin systems: IGHG1, KM.

OCEANIA

AUSTRALIA

Caucasians (1) Enzymes: FDH, GCTG, GST1, GST2, IDH2, MPI, PP. (2) Proteins: APOE, BF(S), F2, F13A, F13B. (3) Blood groups: MIL.

Aborigines (1) Enzymes: CHE1, CHE2, FGH, GPX1, MPI, PP. (2) Proteins: AG, F2, F13A.

Aborigines (N. Terr.) (1) Enzymes: ACO1, ACO2, ACP1, ADA, AK1, CA1, CA2, DIA1, ESD, FGH, FH1, FH2, G6PD, GAA, GLO1, GOT1, GPI, GPT1, HAGH, IDH1, IDH2, LDH, MDH1, MDH2, PEPA, PEPB, PGD, PGK, PGM1, PGM2, PGM3, SOD1. (2) Proteins: ALB, C5, CP, GC, GC(S), HBB, HPA*, HPA*(S), PI(S), TBG, TF. (3) Blood groups: ABO, BY, DI, FY, JK, K, LE, LU, MNS, P, RH, XG, WB. (4) Immunoglobulin systems: IGHG1, KM. (5) Miscellaneous: CB, CER*, LAA*, PTC.

Aborigines (C. Aust.) (1) Enzymes: ACO1, ACP1, ADA, AK1, CA1, CA2, GAA, GPI, LDH, MDH1, PEPA, PEPB, PGD, PGK, PGM1, PGM1(S), PGM2, PGP. (2) Proteins: HBB, HPA*, TF, TF(S). (3) Blood groups: ABO, MN.

Place/ Population	Genetic loci

Aborgines (W. Aust.)

(1) Enzymes: ACP1, ADA, AK1, CA1, CA2, DIA1, ESD, G6PD, GLO1, GOT1, GPI, GPT1, IDH1, LDH, MDH1, PEPA, PEPB, PEPC, PEPD, PGD, PGK, PGM1, PGM2, PGP, SOD1. (2) Proteins: HBB, HPA*, HPA*(S), TBG, TF. (3) Blood groups: ABO, FY, K, LU, MN, P, RH, SE. (4) Immunoglobulin systems: IGA2.

MELANESIA

Melanesians

(1) Enzymes: DPGM. (2) Proteins: F13A, F13B. (3) DNA polymorphisms: HBAC, HBBC, HBAD*, HBGD*. (4) Miscellaneous: BAIB, CER*.

W. New Guinea

Asmat/New Guineans

(1) Enzymes: ACP1, AK1, CA1, CA2, ESD, GOT1, GPI, GPT1, IDH1, LDH, MDH1, PEPB, PEPC, PGAMA, PGD, PGK, PGM1, PGM2, SOD1. (2) Proteins: ALB, CP, GC, HBB, HPA*, TF. (3) Blood groups: ABO, DI, FY, K, MNS, P, RH, WB, XG.

Papua New Guinea

Papuans

(1) Enzymes: ACO1, DIA1, FGH, GPX1, HAGH, IDH2, MDH2, MPI, PGP. (2) Proteins: C5, GC(S), ORM1, ORM2. (3) Blood groups: JK, LU, RD, SE. (4) Immunoglobulin systems: IGA2, IGHG1, KM. (5) DNA polymorphisms: HBAC, HBAD*. (6) Miscellaneous: AC, CB, LAA*.

Gainj & Kalam (N. Central Highlands)

(1) Enzymes: ACP1, ADA, AK1, CA1, CA2, ESA1, ESB, ESD, G6PD, GALT, GLO1, GOT1, GPI, GPT1, HK1, HK2, IDH1, LDH, MDH1, NP, PEPA, PEPB, PEPC, PEPD, PGD, PGK, PGM1, PGM2, SOD1, TPI. (2) Proteins: ALB, CP, GC, HBB, HPA*, TF. (3) Blood groups: ABO, DI, FY, GE, JK, K, LE, MN, P, RH.

Jimi Valley (W.Highlands)

(1) Enzymes: ACP1, ADA, AK1, G6PD, GPI, LDH, MDH1, PGM1, PGM2, PGD. (2) Proteins: F13A, GC, HPA*, TF, TF(S). (3) Blood groups: ABO, FY, GE, K, MNS, P, RD, RH, WR.

Yagaria, etc. (E. High-lamds)

(1) Enzymes: ACO1, ACP1, ADA, AK1, ESD, FH1, FH2, G6PD, GAA, GPI, GPT1, LDH, MDH1, MDH2, PGD, PGK, PGM1, PGP . (2) Proteins: GC, HPA*, TBG, TF, TF(S). (3) Blood groups: ABO, DI, FY, GE, HE, JS, K, LU, MNS, P, RD, RH, VW, WR.

Port Moresby

(1) Enzymes: ACO1, ACO2, ALPP, FH1, FH2, IDH2, GAA, MDH2, MPI, PGM1(S), PGM3, PGP, PP. (2) Proteins: TBG, TFC(S).

Fuyuge (C. High-lands)

(1) Enzymes: ACP1, ADA, AK1, DIA1, G6PD, GOT1, GPI, GPT1, IDH1, LDH, MDH1, PEPA, PEPB, PEPC, PEPD, PGD, PGK, PGM1, PGM2, SOD1. (2) Proteins: ALB, CP, HBB, HPA*, TF. (3) Blood groups: ABO, FY, GE, K, MNS, P, RH.

Daga (Bay (Province)

(1) Enzymes: ACP1, ADA, AK1, CA1, CA2, CHE2, ESD, G6PD, GLO1, GPT1, MDH1, PEPA, PEPB, PGD, PGM1(S), PGM2. (2) Proteins: HBB, HPA*, TF. (3) Blood groups: ABO, FY, HE, K, MNS, P, RH. (4) Miscellaneous: LAA*.

Karkar Is

Takia/ Waskia

(1) Enzymes: ACP1, ADA, AK1, ESD, G6PD, GPI, GPT1, GLO1, LDH, MDH1, PGD, PGK, PGM1, PGM2, PGP. (2) Proteins: F2, HPA*, TF. (3) Blood groups: ABO, DI, FY, GE, HE, I, JK, JS, K, LU, MNS, P, RD, RH, SE, VW, WR, ZT. (4) DNA polymorphisms: HBAD*. (5) Miscellaneous: PTC.

Manus Is.

(1) Enzymes: ACP1, ADA, AK1, G6PD, GPI, LDH, MDH1, PEPA, PEPB, PGD, PGK, PGM1, PGM2, SOD1. (2) Proteins: ALB, CP, HBB, HPA*, PI, TF. (3) Blood groups: ABO, FY, GE, K, MN, P, RH.

Buka Is.

(1) Enzymes: ACP1, ADA, ESD, GPT1, LDH, MDH1, PGD, PGK, PGM1, PGM2, SOD1. (2) Proteins: F13A, HPA*, TF. (3) Blood groups: ABO, FY, GE, K, MNS, RH.

Solomon Is.

(1) Enzymes: ACP1, AK1, DIA1, ESD, GOT1, GPI, GPT1, IDH1, LDH, MDH1, PEPB, PEPC, PEPD, PGD, PGK, PGM1, PGM2, SOD1. (2) Proteins: ALB, C5, GC(S), HPA*, TF. (3) Blood groups: ABO, GE, LU, MNS, RH, SE, VEL. (4) Immunoglobulin systems: IGHG1, KM. (5) Miscellaneous: AC, PTC.

Fiji Is.

(1) Enzymes: PGM1(S). (2) Proteins: BF, C2, C6, F2, F13A.

MICRONESIA

Micronesians

(1) Enzymes: ESA, GPX1, PGM1(S). (2) Proteins: BF, C6, TBG, TF(S). (3) Blood groups: JK, VEL, XG. (4) Miscellaneous: CER*.

W. Caroline

(1) Enzymes: ACP1, ADA, AK1, CA1, CA2, ESD, GPI, GPT1, HAGH, LDHA, MDH1, PEPA, PEPB, PGD, PGK, PGM1, PGM2, PGP, SOD1. (2) Proteins: GC, GC(S), TF. (3) Blood groups: FY, JK, K, MNS, P, RH. (4) Immunoglobulin systems: IGHG1, KM. (5) Miscellaneous: PTC.

337

Populations and genetic loci examined (cont'd)

Place/ Population	Genetic loci
E. Caroline	(1) Enzymes: ACP1, AK1 ESD, GPT1, G6PD, GLO1, GPT1, PGD, PGM1, PGM2. (2) Proteins: HPA*. (3) Blood groups: ABO.
Marshall Is.	(1) Enzymes: ACP1, ADA, AK1, ALDOA, CA1, CA2, DPGM, GALT, GPI, IDH1, LDHA, MDH1, NP, PEPA, PEPB, PGD, PGM1, PGM2, TPI. (2) Proteins: ALB, CP, GC, HBB, HPA*, TF. (3) Blood groups: ABO, DI, FY, K, MNS, P, RH, SE. (4) Miscellaneous: BAIB.
Naru Is.	(1) Enzymes: HAGH.
NEW ZEALAND Blood donors Christchurch	(1) Proteins: APOE.
POLYNESIA Polynesians	(1) Enzymes: ALPP, GPX1, MDH2. (2) Blood groups: MIL. (3) DNA polymorphisms: HBAD*, HBBC. (4) Miscellaneous: AC, BAIB, CB,
Samoans	(1) Enzymes: ACP1, ADA, AK1, CA1, CA2, DIA1, ESD, FGH, G6PD, GLO1, GOT1, GPI, GPT1, HAGH, IDH1, LDH, MDH1, PEPA, PEPB, PEPC, PEPD, PGD, PGK, PGM1, PGM1(S), PGM2, PGP, SOD1. (2) Proteins: ALB, BF, C2, C6, CP, F2, F13A, GC(S), HBB, HPA*, TBG, TF, TF(S). (3) Blood groups: ABO, FY, GE, I, JK, K, MNS, P, RH.
Cook Is.	(1) Enzymes: PGM1(S), PGP. (2) Proteins: F13A, GC, GC(S), TBG, TF(S). (3) Blood groups: DI, VEL, SE. (4) Miscellaneous: PTC
Easter Is.	(1) Enzymes: CHE2. (2) Proteins: C3, LP, XM. (3) Blood groups: LU. (4) Immunoglobulin systems: IGA2

Note: [a]Jews. [b]Blacks in Canada.

Appendix B   Gene Symbols and Genetic Loci

Gene symbol	Genetic locus	Table	Gene symbol	Genetic locus	Table
AAT1	Alanine aminotransferase; glutamic pyruvate transaminase	42		Apolipoprotein AI-CIII-AIV (DNA)	189.1
ABO	ABO	141	APOB(D)	Apolipoprotein B (DNA)	186
AC	Acetylator system; isoniazid inactivation	210	APOC2(D)	Apolipoprotein C-II (DNA)	187
			APOC3(D)	Apolipoprotein C-III (DNA)	188
ACHE	Acetylcholinesterase	1	APOE	Apoliopoprotein E	82
ACO1	Aconitase 1, soluble	5	APRT	Adenine phosphoribosyltrans-ferase	75
ACO2	Aconitase 2, mitochondrial	75			
ACP1	Acid phosphatase 1	2	ARG1	Arginase, liver	75
ADA	Adenosine deaminase	6	ARSA	Arylsulfatase A	75
ADH1	Alcohol dehydrogenase I, alpha polypeptide	75	ASL	Argininosuccinate lyase	75
			AT3	Antithrombin III	80
ADH2	Alcohol dehydrogenase I, beta polypeptide	8	AT3(D)	Antithrombin III (DNA)	182
			ATP*	Anodal tear protein	79
ADH3	Alcohol dehydrogenase I, gamma polypeptide	9	AU	Auberger	142
			AUG*	August	167
AFP(D)	Alpha-fetoprotein (DNA)	179	BAIB	Beta-aminoisobutyric acid, urinary excretion	211
AG	Beta lipoprotein, Ag system	83			
AHSG	Alpha-2-HS-glycoprotein	78	BAK	Bak	168
AK1	Adenylate kinase 1, soluble	7	BE	Berrens	167
AK2	Adenylate kinase 2, mito-chondrial	75	BEC*	Becker	167
			BF	Properdin factor B; glycine-rich beta-glycoprotein	127
AK3	Adenylate kinase 3, mito-chondrial	75			
			BF	Properdin factor B; glycine-rich beta-glycoprotein (subtypes)	127.1
ALAD	Aminolevulinate dehydratase, delta	13			
ALB	Albumin	76	BG	Beta-2-glycoprotein I, Bg system	86
ALB(D)	Albumin (DNA)	177	BI	Biles	167
ALDH1	Aldehyde dehydrogenase 1, cytosolic	75	BP	Bishop	167
			BX	Box	167
ALDH2	Aldehyde dehydrogenase 2, mitochondrial	10	BY	Batty	167
			C1R	Complement component C1r, subcomponent	91
ALDH3	Aldehyde dehydrogenase 3	11			
ALDH4	Aldehyde dehydrogenase 4	75	C2	Complement component 2	92
ALDOA	Aldolase A	75	C3	Complement component 3	93
ALPP	Alkaline phosphatase, placental	12	C4A	Complement component 4A	94
			C4B	Complement component 4B	95
AMY1	Amylase, alpha; salivary	14	C5	Complement component 5	96
AMY1(D)	Amylase, alpha, salivary (DNA)	181	C6	Complement component 6	97
AMY2	Amylase, alpha; pancreatic	15	C7	Complement component 7	98
AN	Ahonen	167	C8A	Complement component 8; alpha-gamma polypeptide	99
APOA4	Apolipoprotein A-IV	81			
APOA1(D)	Apolipoprotein A-I (DNA)	183	C8B	Complement component 8; beta polypeptide	100
APOA2(D)	Apolipoprotein A-II (DNA)	184			
APOA4(D)	Apolipoprotein A-IV (DNA)	185	CA1	Carbonic anhydrase I	16
APOA1-APOC3-APOA4			CA2	Carbonic anhydrase II	17
	Apolipoprotein AI-CIII-AIV (DNA)	189	CA3	Carbonic anhydrase III	18
APOA1-APOC3-APOA4			CAT	Catalase	19

Gene symbol	Genetic locus	Table	Gene symbol	Genetic locus	Table
CB*	Colorblindness	213	ESB	Esterase B, erythrocyte	75
CDA	Cytidine deaminase	22	ESB3	Esterase B3, leukocyte	28
CER*	Cerumen	212	ESD	Esterase D; S-formyl-	29
CFH	Factor H	102		glutathione hydrolase	
CH*	Chido	167	EV*	Evans	167
CHE1	Cholinesterase (serum) 1;	20	F2	Coagulation factor II;	88
	pseudocholinesterase 1			prothrombin	
CHE2	Cholinesterase 2;	21	F8C(D)	Coagulation factor VIIIC;	191
	pseudocholinesterase 2			hemophilia A	
CHR	Chr	167	F9(D)	Coagulation factor IX;	192
CL	Caldwell	167		hemophilia B	
CO	Colton	143	F13A	Coagulation factor XIII-A	89
COL1A1(D)	Collagen I alpha-1 polypeptide	193	F13B	Coagulation factor XIII-B	90
COL1A2(D)	Collagen I alpha-2	194	FDH	Formaldehyde dehydrogenase	75
	polypeptide		FGH	S-formylglutathione hydro-	31
COL2A2(D)	Collagen II alpha-2	194.1		lase; esterase D	
	polypeptide		FH1	Fumarate hydratase, soluble	75
CON1	Salivary protein CON1	128	FH2	Fumarate hydratase, mito-	75
CON2	Salivary protein CON2	129		chondrial	
CP	Ceruloplasmin	87	FI	Factor I	103
CS	Cs	144	FR	Fr	167
DAMOX	D-amino acid oxidase	75	FUCA	Fucosidase, alpha-L	32
DASOX	D-aspartate oxidase	75	FY	Duffy (Fy^a)	147
DB	Double-band parotid salivary	101	FY	Duffy (Fy^a, Fy^b)	147.1
	protein		G1	Parotid salivary glycoprotein	121
DBP	Group specific component;	104	G6PD	Glucose-6-phosphate dehydro-	36
	vitamin D binding protein			genase	
DBP	Group specific component;	104.1	G6PDS*	Glucose-6-phosphate dehydro-	37
	vitamin D binding protein			genase, salivary; hexose-6-	
	(subtypes)			phosphate dehydrogenase	
DH	Duch	167	GAA	Glucosidase, alpha, acid	39
DI	Diego	145	GABAT	Gamma-aminobutyric acid	34
DIA1	Diaphorase NADH	23		transaminase;	
DIA3	Diaphorase 3 (sperm)	24		Gaba-transaminase	
DIA4	Diaphorase 4	25	GALK	Galactokinase	75
DO	Dombrock	146	GALT	Galactose-1-phosphate	33
DOX*	Debrisoquine oxydation	214		uridyl transferase	
	(S-defect);sparteine		GAPD	Glyceraldehyde-3-phosphate	75
	oxidation (S-defect)			dehydrogenase	
DP	Dupuy	167	GBG	Glycine-rich beta-glycopro-	127
DPGM	2,3 Diphosphoglycerate mutase	26		tein; properdin factor B	
DUZ*	Duzo	168	GBG	Glycine-rich beta-glycopro-	127.1
EL	Elridge	167		tein; properdin factor B	
EN	Envelope	167		(subtypes)	
ENO1	Enolase 1; phosphopyruvate	75	GC	Group specific component;	104
	hydratase			vitamin D binding protein	
ER*	Er	167	GC	Group specific component;	104.1
ESA^a	Arylesterase; paraoxonase;	55		vitamin D binding protein	
	esterase A			(subtypes)	
ESA1	Esterase A1; A$_{1-3}$	27	GCTG	Gamma-glutamyl cyclotransferase	75

Gene symbol	Genetic locus	Table	Gene symbol	Genetic locus	Table
GD*	Good	167	HE	Henshaw	149
GDA	Guanine deaminase	75	HEI*	Heibel	167
GDH	Glucose dehydrogenase	35	HEMA(D)	Hemophilia A; coagulation factor VIII (DNA)	191
GE	Gerbich	148			
GF	Griffiths	167	HEMB(D)	Hemophilia B; coagulation factor IX (DNA)	192
GHC(D)	Growth hormone gene cluster (DNA)	195			
GLA	Galactosidase, alpha	75	HEXA	Hexosaminidase-A; beta-N-acetyl-glucosaminidase A	75
GLO1	Glyoxalase I	47			
GLO2	Glyoxalase II; hydroxyacyl glutathione hydrolase	49	HEXB	Hexosaminidase-B; beta-N-acetyl-glucosaminidase B	75
GLUD	Glutamate dehydrogenase	75	HEY	Hey	167
GN	Gonsowski	167	HF	Factor H	102
GO	Gonzales	167	HG	Hughes	167
GOT1	Glutamic-oxaloacetic transaminase 1, soluble	40	HK1	Hexokinase I	75
			HK2	Hexokinase II	75
GOT2	Glutamic-oxaloacetic transaminase 2, mitochondrial	41 / 41	HK3	Hexokinase III	48
GOX	Glycolate oxidase	75	HLAA	HLA-A histocompatibility type	169
GPD1	Glycerol-3-phosphate dehydrogenase A	75	HLAB	HLA-B histocompatibility type	170
GPD2	Glycerol-3-phosphate dehydrogenase B	75	HLAC	HLA-C histocompatibility type	171
GPI	Glucose phosphate isomerase; phosphohexose isomerase; phosphoglucose isomerase	38	HLADQ	HLA-DQ histocompatibility type	172
			HLADR	HLA-DR histocompatibility type	167
GPT1	Glutamic pyruvate transaminase	42	HOV	Hov	167
GPX1	Glutathione peroxidase	43	HPA*	Haptoglobin, alpha	105
GSR	Glutathione reductase	44	HPA*	Haptoglobin, alpha (subtypes)	105.1
GST1	Glutathione-S-transferase-1	45	HPB*	Haptoglobin, beta	139
GST2	Glutathione-S-transferase-2	46	HPRT	Hypoxanthine phosphoribosyl-transferase	75
GST3	Glutathione-S-transferase-3	75			
GUK1	Guanylate kinase 1	75	HPRT(D)	Hypoxanthine phosphoribosyl-transferase (DNA)	196
GY	Gregory	167			
H6PD	Hexose-6-phosphate dehydrogenase; glucose-6-phosphate dehydrogenase, salivary	37	HPX	Hemopexin	109
			HT	Hunt	167
			I	I	167
HAGH	Hydroxyacyl glutathione hydrolase; glyoxalase II	49	IDH1	Isocitrate dehydrogenase, soluble	50
HBA	Hemoglobin, alpha	139	IDH2	Isocitrate dehydrogenase, mitochondrial	75
HBAC(D)	Alpha-globin gene cluster (DNA)	180			
HBAD*	Alpha-globin deletion	208	IGA2	Immunoglobulin Am2 system	174
HBAD*	Alpha-globin deletion (subtypes of -α)	208.1	IGHG(D)	Immunoglobulin, heavy chain constant region of IgG (DNA)	197
HBAD*	Appha-globin deletion (subtypes of -$\alpha^{3.7}$)	208.2	IGHG1	Immunoglobulin Gm1 system	175
			IGHM(D)	Immunoglobulin, heavy chain constant region of IgM (DNA)	198
HBB	Hemoglobin, beta	106			
HBBC(D)	Beta-globin gene cluster (DNA)	190	IGKC	Immunoglobulin Km (Inv) system	176
HBD	Hemoglobin, delta	107			
HBG	Hemoglobin, gamma	108	IN	Indian antigen	167
HBGD*	Gamma-globin deletion	209	INS(D)	Insulin (DNA)	199

Gene symbol	Genetic locus	Table	Gene symbol	Genetic locus	Table
INSR(D)	Insulin receptor (DNA)	200	MNS	MNS, MN	156
ITPA	Inosine triphosphatase A	75	MO	Moen	167
JEN	Jensen	167	MPI	Mannose phosphate isomerase	75
JK	Kidd (Jk^a)	152	MT	Martin	167
JK	Kidd (Jk^a, Jk^b)	152.1	MUR	Mur	167
JN	Jn	167	NE*	Ne	157
JO	Jobbins	167	NFLD	Newfoundland	167
JR	Jr	167	NGFB(D)	Nerve growth factor, beta (DNA)	201
JS	Sutter	163	NP	Nucleoside phosphorylase	54
K	Kell	151	NY	Nyberg	167
KA	Kamhuber	167	OK	Ok	167
KEL	Kell	151	OL	Ol (Oldeide)	167
KM	Immunoglobulin Km (Inv) system	176	ORR*	Orriss	167
KN	Knops-Helgeson	167	ORM1	Orosomucoid 1; alpha-1-acid glycoprotein (locus 1)	114
KO	Ko	168			
LAA*	Lactase activity	215	ORM2	Orosomucoid 2; alpha-1-acid (locus 2)	115
LAN	Lan	167			
LAPP	Leucine aminopeptidase, placentral	75	OS	Os	167
LC40P	Lymphocyte cytosol 40k polypeptide	110	OTC	Ornithine transcarbamylase	75
			P	P	158
LC49P	Lymphocyte cytosol 49k polypeptide	111	PA	Parotid acidic protein Pa	116
			PAH(D)	Phenylalanine hydroxylase (DNA)	203
LC64P	Lymphocyte cytosol 64k polypeptide	112	PALB	Thyroxine-binding prealbumin	139
			PB	Parotid basic protein Pb	117
LC100P	Lymphocyte cytosol 100k polypeptide	113	PCS	Proline-rich salivary protein Pc	126
			PE	Salivary protein Pe	130
LD	Beta lipoprotein, Ld system	84	PEA*	PE	167
LDHA	Lactate dehydrogenase A	51	PEPA	Peptidase A	56
LDHB	Lactate dehydrogenase B	51	PEPB	Peptidase B	57
LE	Lewis	153	PEPC	Peptidase C	58
LEK	Lek	168	PEPD	Peptidase D	59
LES	Lewis	153	PEPE	Peptidase E	75
LEV*	Levy	167	PFKL	Phosphofructokinase, liver type	75
LP	Beta lipoprotein, Lp system	85	PGA	Pepsinogen	123
LS	Lewis II	167	PGAMA	Phosphoglycerate mutase A; phosphoglyceric acid mutase	75
LU	Lutheran	154			
LW	LW	167	PGD	Phosphogluconate dehydrogenase	64
MCC*	McCoy	155	PGI	Phosphohexose isomerase; glucose phosphate isomerase; phosphoglucose isomerase;	38
MDH1	Malate dehydrogenase, soluble	52			
MDH2	Malate dehydrogenase, mitochondrial	75	PHI	Phosphohexose isomerase; glucose phosphate isomerase; phosphoglucose isomerase;	38
ME1	Malic enzyme 1, cytoplasmic	75			
ME2	Malic enzyme 2, mitochondrial	53			
MEP*	Mephenytoin metabolism (M-defect)	216	PGK	Phosphoglycerate kinase	65
			PGK1(D)	Phosphoglycerate kinase 1 (DNA)	204
MG	Mg	167	PGM1	Phosphoglucomutase 1	61
MI	Miltenberger	167	PGM1	Phosphoglucomutase 1 (subtypes)	61.1
MIL	Milne	167	PGM2	Phosphoglucomutase 2	62
MIT	Mit	167	PGM3	Phosphoglucomutase 3	63
MN	MN, MNS	156	PGP	Phosphoglycollate phosphatase	66

Gene symbol	Genetic locus	Table	Gene symbol	Genetic locus	Table
PI	Protease inhibitor; alpha-1-antitrypsin	77	SAHH	S-adenosylhomocysteine hydrolase	70
PI	Protease inhibitor; alpha-1-antitrypsin (subtypes)	77.1	SAL1	Salivary protein SAL 1	132
PI(D)	Alpha-1-antitrypsin; protease inhibitor	178	SAL2	Salivary protein SAL II	133
PIF	Parotid isoelectric focusing protein	118	SC	Scianna	167
			SD	Sd	161
PK3	Pyruvate kinase 3	68	SDH	Succinate dehydrogenase	75
PKM2	Pyruvate kinase 3	68	SE	ABH secretion	140
PLA*	PL^{A1}, Zw	168	SET1	Esterase, salivary	30
PLAP	Alkaline phosphatase, placental	12	SF	Stoltzfus	162
PLE1*	PL^{E1}	168	SKJ*	Skjelbred	167
PLE2*	PL^{E2}	168	SOD1	Superoxide dismutase 1, soluble; indophenol oxidase; tetrazolium oxidase	71
PLG	Plasminogen	124			
PLG(D)	Plasminogen (DNA)	205	SODA	Superoxide dismutase-A, salivary	75
PMF*	Parotid middle-band protein (fast)	119	SODB	Superoxide dismutase-B, salivary	72
			SORD	Sorbitol dehydrogenase	75
PMS*	Parotid middle-band protein (slow)	120	ST*	Stobo	167
			SW	Swann	167
PNK	Pyridoxine kinase (activity)	67	TBG	Thyroxine-binding globulin	134
PO	Salivary protein Po	131	TBPA	Thyroxine-binding prealbumin	139
POMC(D)	Proopiomelanocortin (DNA)	206	TCA*	Tc	167
PON	Paraoxonase; arylesterase	55	TC2	Transcobalamin II	135
PP	Pyrophosphatase, inorganic	75	TF	Transferrin	136
PPH	Phosphopyruvate hydratase; enolase 1	75	TF	Transferrin (subtypes)	136.1
			TO	Torkildsen	167
PR	Proline-rich salivary protein Pr	125	TPI	Triosephosphate isomerase	73
PRB3	Parotid salivary glycoprotein	121	TR	Traversu	167
PS	Parotid size variant	122	TS, TSU*	Tsunoi	167
PSP	Phosphoserine phosphatase	75	TTR	Thyroxine-bonding prealbumin	139
PT	Peters	167	UL	Karhula, antigen	150
PTC	Phenylthiocarbamide taste	217	UMPK	Uridine monophosphate kinase	74
PTH(D)	Parathyroid hormone (DNA)	202	URK(D)*	Urokinase (DNA)	207
PXL*	Peroxidase, leukocyte	75	V	V	164
PXS*	Peroxidase, salivary	60	VBRP*	Vitamin B$_{12}$ binding (R) protein	137
RD	Radin	167	VEL	Vel	167
RDD*	Raddon	167	VEN*	Ven	167
RDS	Rhodanese	69	VW	Vw (Verweyst)	167
RE	Reid	167	WAL*	Waldner	167
RG	Rodgers	160	WB	Webb	167
RH	Rhesus	159	WD	Waldner	167
RI	Ridley	167	WR	Wright	167
RL	Rosenlund	167	WU	Wulfsberg	167
RM	Rm (Romunde)	167	XG	Xg	165
SACPA	Acid phosphatase-A, salivary	3	XM	Xm system	138
			YT	Yt (Cartwright)	166
SACPA	Acid phosphatase-B, salivary	4	YUK	Yuk	168
			ZD	Zd	167
			ZT	Marriott	167

Note: The gene symbols assigned by us are marked with *. The symbol D in parentheses refers to DNA polymorphism. [a] Some authors (e. g. Barrentes et al. 1982) have used gene symbol ESA for the ESA1 locus.

References to Data Tables

Abe, K. and Akiyama, K. (1983) Mitochondrial malic enzyme polymorphism in Japanese population. Jpn. J. Hum. Genet. 28:11-15.

Abe, S, Hiraiwa, K, and Sebetan, I. M. (1986) Genetic polymorphism of α_2HS glycoprotein and group-specific component in Egyptian. Jpn. J. Hum. Genet. 31:187.

Adam, A. (1962) A survey of some genetical characters in Ethiopian tribes. VII. Color-vision. Amer. J. Phys. Anthrop. 20:194-195.

Adam, A. (1969) A further query on color blindness and natural selection. Soc. Biol. 16:197-202.

Adam, A., Mweisigye, E., and Tabani, E. (1970) Ugandan colorblindness revisited. Amer. J. Phys. Anthrop. 32:59-64.

Adam, A., Puenpatom, M., Davivongs, V., and Wangspa, S. (1969) Anomaloscopic diagnoses of red-green blindness among Thais and Chinese. Hum. Hered. 19:509-513.

Aebi, H. (1966) The investigation of inherited enzyme deficiencies with special reference to acatalasia. In: Proc. Third Intl. Congr. Genet., Chicago (Crow, J. F. and Neel, J. V., eds.), Johns Hopkins Univ. Press, Baltimore, pp.189-205.

Agarwal, D. P., Benkmann, H. G., and Goedde, H. W. (1974) Levels of serum β_1C/β_1A-globulin (C3) and its polymorphism in leprosy patients and healthy controls from Ethiopia and Mali. Humangenetik 21:355-359.

Agarwal, D. P., Goedde, H. W., Schloot, W., Flatz, G., and Rhode, R. (1973) A note on atypical serum cholinesterase and genetic factors in leprosy. Hum. Hered. 23:374-380.

Ageheim, H. and Bergström, J. (1972) Adenosine deaminase polymorphism in a Swedish population. Acta Genet. Med. Gemellol. 21:135-138.

Agosti Romero, L., Ikin, E. W., and Mourant, A. E. (1950) Les groupes sanguins ABO, MNS, et Rh des Galiciens (Espagne N. D.). Rev. Hémat. 5:325-328. Cited from Mourant et al. (1976).

Ahmad, M. and Flatz, G. (1984) Prevalence of primary adult lactose malabsorption in Pakistan. Hum. Hered. 34:69-75.

Akaishi, S. and Kudo, T. (1975) Blood groups. In: Anthropological and Genetic Studies on the Japanese (Watanabe, S., Kondo, S., and Matsunaga, E., eds.) Univ. Tokyo Press, Tokyo, pp.77-107.

Akbari, M. T., Papiha, S. S., Roberts, D. F., and Farhud, D. D. (1984) Serogenetic investigations of two populations of Iran. Hum. Hered. 34:371-377.

Åkesson, H. (1959) Taste deficiency for phenylthiourea in southern Sweden. Acta Genet. Med. Gemellol. 8:431-433.

Akiyama, K., Nakamura, S., and Abe, K. (1984) Gene frequencies of S-adenosylhomocysteine hydrolase (SAHH) in a Japanese population. Hum. Genet. 68:191-192.

Aksoy, M., Ikin, E. W., Mourant, A. E., and Lehmann, H. (1958) Blood groups, haemoglobins, and thalassaemia in Turks in southern Turkey and Eti-Turks. Brit. Med. J. ii, 937-939.

Al-Agidi, S. K., Papiha, S. S., Shukri, S. M. (1980) Glyoxalase I (GLO; EC 4.4.1.5): Gene frequency variation in Iraq. Hum. Hered. 30:259-261.

Alberdi, F., Allison, A. C., Blumbert, G. S., Ikin, E. W., and Mourant, A. E. (1957) The blood groups of the Spanish Basques. J. R. Anthrop. Inst. 87:217-219.

Allbrook, D., Barnicot, N. A., Dance, N., Lawler, S. D., Marshall, R., and Mungai, J. (1965) Blood groups, haemoglobin and serum factors of the Karamojo. Hum. Biol. 37:217-237.

Allison, A. C. and Blumberg, B. S. (1959) Ability to taste phenylthiocarbamide among Alaskan Eskimos and other populations. Hum. Biol. 31:352-359.

Allison, A. C., Blumberg, B. S., and Gartler, S. M. (1959) Urinary excretion of β-amino-isobutyric acid in Eskimo and Indian populations of Alaska. Nature 183:118-119.

Allison, A. C., Broman, B., Mourant, A. E., and Ryttinger, L. (1956) The blood groups of the Swedish Lapps. J. R. Anthrop. Inst. 86:87-94.

Allison, A. C., Hartmann, O., Brendemoen, O. J., and Mourant, A. E. (1952) The blood groups of the Norwegian Lapps. Acta Path. Microbiol. Scand. 31:334-338. Cited from Mourant et al. (1976).

Al-Khafaji, S. D. and Al-Rubeai, M. A. F. (1976) The frequencies of the ABO and Rh(D) blood groups in the Kurdish population of Iraq. Ann. Hum. Biol. 3:189-191.

Al-Nassar, K. E., Conneally, P. M., Palmer, C. G., and Yu, P.-O. (1981) The genetic structure of the Kuwaiti population. I. Distribution of 17 markers with genetic distance analysis. Hum. Genet. 57:192-198.

Alper, C. A. (1976) Inherited structural polymorphism in human C2: Evidence for genetic linkage between C2 and Bf. J. Exp. Med. 144:1111-1115.

Alper, C. A., Boenisch, T., and Watson, L. (1972) Genetic polymorphism in human glycine-rich beta glycoprotein. J. Exp. Med. 135:68-80.

Alper, C. A., Marcus, D., Raum, D., Petersen, B. H., and Spira, T. J. (1983) Genetic polymorphism in C8 β-chains: Evidence for two unlinked genetic loci for the eighth component of human complement (C8). J. Clin. Invest. 72:1526-1531.

Alper, C. A. and Rosen, F. S. (1976) Genetics of the complement system. In: Advances in Human Genetics (Harris, H. and Hirschhorn, K., eds.) 7:141-188.

Alsbirk, K. E. and Alsbirk, P. H. (1972) PTC taste sensitivity in Greenland Eskimos from Umanaq. Hum. Hered. 22:445-452.

Altay, C., Say, B., and Tuncbilek, E. (1974) Frequency of red cell adenosine deaminase and 6-phosphogluconate dehydrogenase in a sample of the Turkish population. Hum. Hered. 24:306-308.

Alter, A. A., Gelb, A. G., Chown, B., Rosenfield, R. E., and Cleghorn, T. E. (1967) Gonzales (Go[a]), a new blood group character. Transfusion 7:88-91.

Altland, K., Bucher, R., Kim, T. W., Busch, H., Brockelmann, C., and Goedde, H. W. (1969) Population genetic studies on pseudocholinesterase polymorphism in Germany, Czechoslovakia, Finland and among Lapps. Humangenetik 8:158-161.

Altland, K., Epple, F., and Goedde, H. W. (1967) Pseudocholinesterase variants in Thailand and Japan. Humangenetik 4:127-129.

Altukhov, Y. P., Khil'chevskaya, R. I., and Shurkhal, A. V. (1981). Levels of polymorphism and heterozygosity of the Russian population of Moscow: Data on 22 gene loci specifying blood proteins. Soviet Genetics 17:391-399.

Alzate, H., González, H., and Guzmán, J. (1969) Lactose intolerance in South American Indians. Amer. J. Clin. Nutr. 22:122-123.

Amorim, A. and Siebert, G. (1982) Glutamate pyruvate transaminase, esterase D, glyoxalase 1, and phosphoglucomutase 1 polymorphisms in Porto District (Portugal). Hum. Hered. 32:298-300.

Ananthakrishnan, R., Beck, W., and Walter, H. (1974) ABO incompatibility and its interaction with haptoglobins and placental alkaline phosphatase. Humangenetik 23:31-43.

Ananthakrishnan, R., Walter, H., and Tsacheva, L. (1972) Red cell enzyme polymorphisms in Bulgaria. Humangenetik 15:186-190.

Andersen, P. H. (1948) The blood group system L. A new blood group L$_2$. A case of epistasy within the blood groups. Acta Path. Microbiol. Scand. 25:728-731. Cited from Mourant et al. (1976).

André, R., Salmon, C., and Malassenet, R. (1956) A propos de trois exemples d'anticorps anti Jk[a] acquis par transfusions observés en association avec des anticorps anti Fy[a], S, c, E. Notes cliniques, sérologiques et anthropologiques. Rev. Hémat. 11:495-502. Cited from Mourant et al. (1976).

André, R., Salmon, C., Malassenet, R., and Philippon, S. (1954) Accidents de transfusion et iso-immunisations foeto-maternelles dus au système de groupe sanguin Kell-Cellano (Données étiologiques et sérologiques, conclusions pratiques). Sem. Hop. Paris 30:3703-3707. Cited from Mourant et al. (1976).

Angelini Rota, M., Atella, P., and Baglioni, G. (1961) Contributo alla conoscenza della distribuzione degli antigeni dei sistemi ABO, MN, Rh nella popolazione del Lazio. Acta Genet. Med. Gemellol. 10:212-237. Cited from Mourant et al. (1976).

Angelopoulos, B., Kalos, A., and Danopoulos, E. (1967) Transferrin variants in Greeks. J. Med. Genet. 4:31-32.

Angelopoulos, B., Tsoukantas, A., and Danopoulos, E. (1966) Distribution of haptoglobin subtypes in Greeks. J. Med. Genet. 3:276-278.

Anh, N. T., Thuc, T. K., and Welsh, J. D. (1977) Lactose malabsorption in adult Vietnamese. Amer. J. Clin. Nutr. 30:468-469.

Antonarakis, S. E., Oettgen, P., Chakravarti, A., Halloran, S. L., Hudson, R. R., Feisee, L., and Karathanasis, S. K. (1986) Organisation and DNA polymorphisms of the human apolipoprotein AI-CIII-AIV gene cluster. Personal communication.

Apeshiotis, F. and Bender, K. (1986) Evidence that S-formylglutathione hydrolase and esterase D polymorphisms are identical. Hum. Genet. 74:176-177.

Applewhaite, F., Ginsberg, V., Gerena, J., Cunningham, C. A., and Gavin, J. (1967) A very frequent red cell antigen, At[a]. Vox Sang. 13:444-445.

Archimandris, A., Fertakis, A., Stathopoulou, R., Kalos, A., and Angelopoulos, B. (1975) Distribution of Gm and Inv factors in two samples of the Greek population. Acta Genet. Med. Gemell. 24:329-331.

Arends, T., Davies, D. A., and Lehmann, H. (1967) Absence of variants of usual serum pseudocholinesterase (Acylcholine acylhydrolase) in South American Indians. Acta Genet. Stat. Med. 17:13-16.

Arends, T., Weitkamp, L. R., Gallango, M. L., Neel, J. V., and Schultz, J. (1970) Gene frequencies and microdifferentiation among the Makiritare Indians. II. Seven serum protein systems. Amer. J. Hum. Genet. 22:526-532.

Armstrong, A. R. and Peart, H. E. (1960) A comparison between the behavior of Eskimos and non-Eskimos to the administration of isoniazid. Amer. Rev. Resp. Dis. 81:588-594.

Arnaúd, P., Chapuis-Cellier, C. Vittoz, P., and Creyssel, R. (1977) Alpha-1-antitrypsin phenotypes in Lyon, France. Hum. Genet. 39:63-68.

Arnaúd, P., Galbraith, R. M., Faulk, W. P., and Black, C. (1979) Pi phenotypes of alpha$_1$-antitrypsin in southern England: Identification of M subtypes and implications for genetic studies. Clin. Genet. 15:406-410.

Arnhold, R. G., Perman, J. A., and Nurse, G. T. (1981) Persistent high intestinal lactase activity in Papua-New Guinea. The breath hydrogen test in a Sepik population. Ann. Hum. Biol. 8:481-484.

Arvilommi, H. (1972) Studies on serum Pi- and C3- polymorphism in Finland. Acta Path. Microbiol. Scand. Sec. B. (Suppl) 234:1-50.

Arvilommi, H., Berg, K., and Eriksson, A. W. (1973) C'3 types and their inheritance in Finnish Lapps, Maris (Cheremisses) and Greenland Eskimos. Humangenetik 18:253-259.

Arzhelas, L. K. and Resnikova, M. N. (1968) Distribution of Kidd antigene (IK[a]) among Moscow population (In Russian). Vopr. Antrop. 29:162-165. Cited from Mourant et al. 1976.

Asakawa, J., Satoh, C., Takahashi, N., Fujita, M., Kaneko, J., Goriki, K., Hazama, R., and Kageoka, T. (1984) Electrophoretic variants of blood proteins in Japanese. III. Triosephosphate isomerase. Hum. Genet. 68:185-188.

Asakawa, J., Takahashi, N., Rosenblum, B. B., and Neel, J. V. (1985) Two-dimensional gel studies of genetic variation in the plasma proteins of Ameridians and Japanese. Hum. Genet. 70:222-230.

Ashton, G. C. and Simpson, N. E. (1966) C5 types of serum cholinesterase in a Brazilian population.

Amer. J. Hum. Genet 18:438-447.

Atkin, J. and Rundle, A. T. (1974) Serum β_2-glycoprotein 1 phenotype frequencies in an English population. Humangenetik 21:81-84.

Auricchio, S., Rubino, A., Landolt, M., Semenza, G., and Prader, A. (1963) Isolated intestinal lactase deficiency in the adult. Lancet ii:324-326.

Awdeh, Z. L. and Alper, C. A. (1980) Inherited structural polymorphism of the fourth component of human complement. Proc. Natl. Acad. Sci. USA 77:3576-3580.

Azen, E. A. (1972) Genetic polymorphism of basic proteins from parotid saliva. Science 176:673-674.

Azen, E. A. (1976) Genetic polymorphism of human anodal tear protein. Biochem. Genet. 14:225-235.

Azen, E. A. (1977) Salivary peroxidase (SAPX): Genetic modification and relationship to the proline-rich (Pr) and acidic (Pa) proteins. Biochem. Genet. 15:9-29.

Azen, E. A. and Denniston, C. (1974) Genetic polymorphism of human salivary proline-rich proteins: Further genetic analysis. Biochem. Genet. 12:109-120.

Azen, E. A. and Denniston, C. (1979) Genetic polymorphism of vitamin B_{12} binding (R) proteins of human saliva detected by isoelectric focusing. Biochem. Genet. 17:909-920.

Azen, E. A. and Denniston, C. (1980) Polymorphism of Ps (parotid size variant) and detection of a protein (PmS) related to the Pm (parotid middle band) system with genetic linkage of Ps and Pm to G1, Db, and Pr genetic determinants. Biochem. Genet. 18:483-501.

Azen, E. A. and Denniston, C. (1981) Genetic polymorphism of PIF (parotid isoelectric focusing variant) proteins with linkage to the PPP (parotid proline-rich protein) gene complex. Biochem. Genet. 19:475-485.

Azen, E. A., Hurley, C. K., and Denniston, C. (1979) Genetic polymorphism of the major parotid salivary glycoprotein (G1) with linkage to the genes for Pr, Db, and Pa. Biochem. Genet. 17:257-279.

Azen, E. A. and Yu, P.-L. (1984a) Genetic polymorphism of CON 1 and CON 2 salivary proteins detected by immunologic and concanavalin A reactions on nitrocellulose with linkage of CON 1 and CON 2 genes to the SPC (salivary protein gene complex). Biochem. Genet. 22:1-19.

Azen, E. A. and Yu, P.-L. (1984b) Genetic polymorphisms of Pe and Po salivary proteins with probable linkage of their genes to the salivary protein gene complex (SPC). Biochem. Genet. 22:1065-1080.

Azevêdo, E. S., DaSilva, M. C. B. O., and Travares-Neto, J. (1975) Human alcohol dehydrogenase ADH_1, ADH_2, and ADH_3 loci in a mixed population of Bahia, Brazil. Ann. Hum. Genet. 39:321-327.

Azim, A. A., Kamel, K., Gaballah, M. F., Sabry, F. H., Ibrahim, W., Selim, O., and Moafy, N. (1974) Genetic blood markers and anthropometry of the populations in Aswan governorate, Egypt. Hum. Hered. 24:12-23.

Badakere, S. S., Parab, B. B., and Bhatia, H. M. (1974) Further observations on the In^a (Indian) antigen in Indian populations. Vox Sang. 26:400-401.

Badakere, S. S., Vasantha, K., Bhatia, H. M., Ala, F., Clarke, V. A., Moesri, R., Sommai, S., and Amin, A. B. (1980) High frequency of In^a antigen among Iranians and Arabs. Hum. Hered. 30:262-263.

Bagster, I. A. and Parr, C. W. (1976) Human erythrocyte glyoxalase I polymorphism. J. Physiol. 256:56P-57P.

Bajatzadeh, M., Neumann, S., and Walter, H. (1969a) Pseudocholinesterases and human red cell acid phosphatases in Koreans. Humangenetik 7:91-92.

Bajatzadeh, M. and Walter, H. (1969a) Blood and serum group typings in Koreans. Hum. Hered. 19:514-523.

Bajatzadeh, M. and Walter, H. (1969b) Studies on the population genetics of the ceruloplasmin polymorphism. Humangenetik 8:134-136.

Bajatzadeh, M. and Walter, H. (1969c) Investigations on distribution of blood and serum groups in Iran. Human Biol. 41:401-415.

Bajatzadeh, M., Walter, H., and Pálsson, J. (1969b) Phosphoglucomutase (EC.2.7.5.1) and adenylate kinase (E.C. 2.7.4.3) typings in Koreans and Irish. Humangenetik 7:353-355.

Balakrishnan, C. R. and Ashton, G. C. (1974) Polymorphisms of human salivary proteins. Amer. J. Hum. Genet. 26:145-153.

Balanza, E. and Taboada, G. (1985) The frequency of lactase phenotypes in Aymara children. J. Med. Genet. 22:128-130.

Ball, P. A. J. (1962) Influence of the secretor and Lewis genes on susceptibility to duodenal ulcer. Brit. Med. J. ii, 948-950.

Ballowitz, L., Fielder, H., Hoffman, C., and Pettenkofer, H. (1968) "Heibel", a new rare human blood group antibody revealed by a haemolytic disease of the newborn. Vox Sang. 14:307-309.

Bally, C. (1954) Untersuchungen über die Verkehrstüchtigkeit farbensinngestörter Knaben. Z. Unfallmed. Berufskr 47:100. Cited from Post (1971).

Banerjee, B., Saha, N., Daoud, Z. F., Khalaf, F. H., and Qudah, H. (1981) A genetic study of the Jordanians. Hum. Hered. 31:65-69.

Bargagna, M. and Abbagnale, L. (1982) Isoelectric focusing of human red cell phosphoglucomutase (PGM_1). Phenotype distribution in the population of Tuscany and two hereditary variants. Hum. Genet. 61:242-245.

Barker, R. F. and Hopkinson, D. A. (1977) The genetic and biochemical properties of the D-amino acid oxidases in human tissues. Ann. Hum. Genet. 41:27-42.

Barker, R. F. and Hopkinson, D. A. (1978) Genetic polymorphism of human phosphoglycolate phosphatase (PGP). Ann. Hum. Genet. 42:143-151.

Barnicot, N. A., Garlick, J. P., Adam, A., and Bat-Miriam, M. (1962) A survey of some genetical characters in Ethiopian tribes. III. Haptoglobins and Transferrins. Amer. J. Phys. Anthrop. 20:175-178.

346

Barnicot, N. A., Krimbas, C., McConnell, R. B., and Beaven, G. H. (1965) A genetical survey of Sphakiá, Crete. Hum. Biol. 37:274-298.

Barnicot, N. A. and Woodburn, J. C. (1975) Colour-blindness and sensitivity to PTC in Hadza. Ann. Hum. Biol. 2:61-68.

Barrantes, R., Smouse, P. E., Neel, J. V., Mohrenweiser, H. W., and Gershowitz, H. (1982) Migration and genetic infrastructure of the central American Guaymi and their affinities with other tribal groups. Amer. J. Phys. Anthrop. 58:201-214.

Basu, A., Namboodiri, K. K., Weitkamp, L. R., Brown, W. H., Pollitzer, W. S., and Spivey, M. A. (1976) Morphology, serology, dermatoglyphics, and microevolution of some village populations in Haiti, West Indies. Hum. Biol. 48:245-269.

Bat-Miriam, M., Adam, A., and Hananel, Z. (1962) A survey of some genetical characters in Ethiopian tribes. VI. Taste thresholds for phenylthiourea. Amer. J. Phys. Anthrop. 20:190-193.

Battaglini, P. F., Ranque, J., Bridonneau, C., Salmon, C., and Nicoli, R. M. (1964) Étude de facteur VEL dans la population marseillaise à propos d'un cas d'immunisation anti-VEL. 10th Congr. Int. Soc. Blood Transfus., Stockholm. ii:309-311. Cited from Mourant et al. (1976).

Battista, N., Stout, T. D., Lewis, M., and Kaita, H. (1980) A new rare blood group antigen - 'Mit'. Voz Sang. 39:331-334.

Baur, M. P. and Danilovs, J. A. (1980) Population analysis of HLA-A, B, C, DR, and other genetic markers. In: Histocompatibility Testing 1980 (Terasaki, P. I., ed.), UCLA Tissue Typing Laboratory, Los Angeles, pp. 955-993.

Baur, M. P., Neugebauer, M. Deppe, H., Sigmund, M., Luton, T., Mayr, W. R. and Albert, E. D. (1984) Population analysis on the basis of deduced haplotypes from random families. In: Histo-compatibility Testing 1984 (Albert, E. D., Baur, M. P., and Mayr, W. R., eds.), Springer-Verlag, New York, pp.333-341.

Bayoumi, R. A., Flatz, S. D., Kühnau, W., and Flatz, G. (1982) Beja and Nilotes: Nomadic pastoralist groups in the Sudan with opposite distributions of the adult lactase phenotypes. Amer. J. Phys. Anthrop. 58:173-178.

Bayoumi, R. A. L., Omer, A., Samuel, A. P. W., Saha, N., Sebai, Z. A., and Sabaa, H. M. A. (1979) Haemoglobin and erythrocytic glucose-6-phosphate dehydrogenase variants among selected tribes in Western Saudi Arabia. Trop. Geogr. Med. 31:245-252.

Beaglehole, E. (1939) Tongan colour-vision. Man 39:170-172. Cited from Post (1962).

Bech-Hansen, N. T., Linsley, P. S., and Cox, D. W. (1983) Restriction fragment length polymorphisms associated with immunoglobulin C_7 genes reveal linkage disequilibrium and genomic organization. Proc. Natl. Acad. Sci. USA 80:6952-6956.

Beckman, G. (1973) Population studies in northern Sweden. VI. Polymorphism of superoxide dismutase. Hereditas 73:305-310.

Beckman, G. and Beckman, L. (1969) The placental alkaline phosphatase polymorphism. Variations in Hawaiian subpopulations. Hum. Hered. 19:524-529.

Beckman, G. and Beckman, L. (1977) Population studies in northern Sweden. VIII. Ethnic heterogeneity and prenatal selection in the esterase D polymorphism. Hum. Hered. 27:403-407.

Beckman, G., Beckman, L., and Cedergren, B. (1971) Population studies in northern Sweden. II. Red cell enzyme polymorphism in the Swedish Lapps. Hereditas 69:243-248.

Beckman, G. and Beckman, L., and Magnùsson, S. S. (1972) Placental alkaline phosphatase phenotypes and pre-natal selection. Hum. Hered. 22:473-480.

Beckman, G., Beckman, L., and Nordenson, I. (1980a) Alpha$_1$-antitrypsin phenotypes in northern Sweden. Hum. Hered. 30:129-135.

Beckman, G., Beckman, L., and Sikström, C. (1980b) Transferrin C subtypes in different ethnic groups. Hereditas 92:189-192.

Beckman, G. and Christodoulou, C. (1974) Variants of soluble and mitochondrial malate dehydrogenase in the Hawaiian, Nigerian and Swedish populations. Hum. Hered. 24:294-299.

Beckman, G. and Jóhannsson, E. O. (1967) Distribution of placental alkaline phosphatase types in the Icelandic population. Acta Genet. Stat. Med. 17:413-417.

Beckman, L. (1959) A contribution to the physical anthropology and population genetics of Sweden. Hereditas 45:189.

Beckman, L., Beckman, G., Christodoulou, C., and Ifekwunigwe, A. (1967) Variations in human placental alkaline phosphatase. Acta Genet. Stat. Med. 17:406-412.

Beckman, L., Beckman, G., and Frölander, N. (1983) Population studies in Northern Sweden. XII. The haptoglobin polymorphism. Hum. Hered. 371-376.

Beckman, L., Björling, G., and Christodoulou, C. (1966) Multiple molecular forms of leucine aminopeptidase in man. Acta Genet. Stat. Med. 16:223-230.

Beckman, L., Broman, B., Jonsson, B., and Mellbin, T. (1959) Further data on the blood groups of the Swedish Lapps. Acta Genet. Stat. Med. 9:1-8.

Beckman, L., Brönnestam, R., Cedergren, B., and Lidén, S. (1974) HLA antigens, blood groups, serum groups and red cell enzyme types in psoriasis. Hum. Hered. 24:496-506.

Beckman, L. and Holmgren, G. (1961) Transferrin variants in Lapps and Swedes. Acta Genet. Stat. Med. 11:106-110.

Beckman, L., Holmgren, G., and Martensson, E. H. (1961) Transferrin variants in the Swedish population. Proc. 2nd Int. Cong. Hum. Genet., Rome. p.737.

Beckman, L. and Mellbin, T. (1959) Haptoglobin types in the Swedish Lapps. Acta Genet. Stat. Med. 9:306-309.

Bender, K., Frank, R., and Hitzeroth, H. W. (1977) Glyoxalase I polymorphism in South African Bantu-speaking Negroids. Hum. Genet. 38:223-226.

347

Benkmann, H-G., Bogdanski, P., and Goedde, H. W. (1983) Polymorphism of delta-aminolevulinic acid dehydratase in various populations. Hum. Hered. 33:62-64.

Benkmann, H-G., Goedde, H. W., Agarwal, D. P. Flatz, G., Rahimi, A., Kaifie, S., and Delbrück, H. (1980) Properdin factor B polymorphism in Afghanistan. Hum. Hered. 30:39-43.

Benkmann, H-G., Paik, Y. K., Chen, L. Z., and Goedde, H. W. (1986) Polymorphism of 6-PGD in South Korea: a new genetic variant 6-PGD Korea. Hum. Genet. 74. 204-205.

Beretta, M., Barberio, C., Ranzani, G., and Bertolotti, E. G. (1977) An analysis of red cell enzymatic markers in the province of Bologna (Italy). Hum. Hered. 27:352-355.

Berg, K. (1965) A new serum type in man-the Ld system. Vox Sang. 10:513-527.

Berg, K. (1968) The Lp system. Ser. Haemat. 1:111-136. Cited from Mourant et al. (1976).

Berg, K. (1969) Genetic studies of the adenylate kinase (AK) polymorphism. Hum. Hered. 19:239-248.

Berg, K. (1974) Studies of polymorphic traits for the characterization of populations: The populations of Scandinavia. In: Genetic Polymorphisms and Diseases in Man (Ramot, B., ed.), Academic Press, New York, pp. 21-29.

Berg, K. and Bearn, A. G. (1966) An inherited X-linked system in man. The Xm system. J. Exp. Med. 123:379-397. Cited from Mourant et al. (1976).

Berg, K. and Eriksson, A. W. (1971) Genetic marker systems in Arctic populations. I. Lp and Ag data on the Greenland Eskimos. Hum. Hered. 21:129-133.

Berg, K. and Eriksson, A. W. (1973a) Genetic marker systems in Arctic populations. V. The inherited Ag(x) serum lipoprotein antigen in Finnish Lapps. Hum. Hered. 23:241-246.

Berg, K. and Eriksson, A. W. (1973b) Genetic marker systems in Arctic populations. VI. Polymorphism of C'3 in Icelanders. Hum. Hered. 23:247-250.

Berg, K., and Eriksson, A. W. (1973c) Genetic marker systems in Artic populations. VII. Genetic variation in serum lipoproteins in Icelanders. Hum. Hered. 23:251-256.

Beringer, M. (1967) Statistische Auswertung von Blut- und Serumgruppenbestimmungen bei Fällen von strittiger Abstammung. Zürich, Dr. thesis. Cited from Tills et al. (1983a).

Bernal, J. E., Papiha, S. S., Keyeux, G., Lanchbury, J. S., and Mauff, G. (1985) Complement polymorphism in Colombia. Ann. Hum. Biol. 12:261-265.

Bernini, L. F. (1986) Hemoglobin, haptoglobin, and transferrin. In: African Pygmies (Cavalli-Sforza, L. L., ed.), Academic Press, New York, pp. 231-246.

Bernstein, S. C., Bowman, J. E., and Noche, L. K. (1980) Population studies in Cameroon. Hum. Hered. 30:251-258.

Bertin, T., Harris, J. E., Ferrell, R. E., and Schull, W. J. (1978) The Nubians of Kom Ombo: Serum and red cell protein types. Hum. Hered. 28:66-71.

Bertrams, J., Hintzen, U., Schlicht, V., Schoeps, S., Gries, F. A., Louton, T. K., and Bauer, M. P. (1984) Gene and haplotype frequencies of the fourth component of complement (C4) in type 1 diabetics and normal controls. Immunobiol. 166:335-344.

Beutler, E. and Kuhl, W. (1972) Biochemical and electrophoretic studies of α-galactosidase in normal man, in patients with Fabry's disease, and in Equidae. Amer. J. Hum. Genet. 24:237-249.

Beutler, E., West, C., and Beutler, B. (1974) Electrophoretic polymorphism of glutathione peroxidase. Ann. Hum. Genet. 38:163-169.

Bhasin, M. K. (1970) The blood groups of the Newars of Nepal. Hum. Biol. 42:369-376.

Bhasin, M. K. and Fuhrmann, W. (1972) Geographic and ethnic distribution of some red cell enzymes. Humangenetik 14:204-223.

Bhatia, H. M. (1963) Frequency of sex-linked blood group Xg^a in Indians in Bombay: Preliminary study. Indian J. Med. Sci. 17:491-492.

Bhattacharjee, P. N. (1956) A genetic survey in the Rarhi Brahmin and the Muslim of West Bengal: A_1A_2BO, MN, Rh blood groups, ABH secretion, sickle-cell, P.T.C. taste, middle phalangeal hair and colour blindness. Bull. Dep. Anthrop. India. 5:18-28. Cited from Mourant et al. (1976).

Bhattacharyya, S. P. and Saha, N. (1984a) Mitochondrial malic enzyme polymorphism among different ethnic groups in Singapore. Hum. Hered. 34:393-395.

Bhattacharyya, S. P. and Saha, N. (1984b) Glutathione-S-transferase polymorphism (loci 1 and 2) in Singapore Chinese - evidence of additional allele in each locus. Amer. J. Hum. Genet. 36:162S.

Bhattacharyya, S. P., Saha, N., and Wee, K. P. (1985) γ-aminobutyric acid transaminase (GABAT) polymorphism among ethnic groups in Singapore - with report of a new allele. Amer. J. Hum. Genet. 37:358-361.

Bias, W. B., Light-Orr, J. K., Krevans, J. R., Humphrey, R. L., Hamill, P. V. V., Cohen, B. H., and McKusick, V. A. (1969) The Stoltzfus blood group, a new polymorphism in man. Amer. J. Hum. Genet. 21:552-558.

Bird, G. W. G., Jayaram, T. K., Ikin, E. W., Mourant, A. E., and Lehmann, H. (1957) The blood groups and haemoglobin of the Gorkhas of Nepal. Amer. J. Phys. Anthrop. 15:163-169.

Bissbort, S., Bender, K., Wienker, T. F., and Grzeschik, K. H. (1983) Genetics of human S-adenosylhomocysteine hydrolase. A new polymorphism in man. Hum. Genet. 65:68-71.

Bjarnason, O., Bjarnason, V., Edwards, J. H., Fridriksson, S., Magnusson, M., Mourant, A. E., and Tills, D. (1973) The blood groups of Icelanders. Ann. Hum. Genet. 36:425-455. Cited from Steinberg and Cook (1981).

Black, F. L., Salzano, F. M., Layrisse, Z., Franco, M. H. L. P., Harris, N. S., and Weimer, T. A. (1980) Restriction and persistence of polymorphisms of HLA and other blood genetic traits in the Parakana Indians of Brazil. Amer. J. Phys. Anthrop. 52:119-132.

Blake, N. M. (1976) Glutamic pyruvic transaminase and esterase D types in the Asian-Pacific area. Hum. Genet. 35:91-102.

Blake, N. M. (1978) Genetic variants of carbonic anhydrase in the Asian-Pacific area. Ann. Hum. Biol.

5:557-568.

Blake, N. M. (1979) Genetic variation of red cell enzyme systems in Australian Aboriginal populations. Occasional Papers in Human Biology (Australian Institute of Aboriginal Studies, Canberra) 2:39-82.

Blake, N. M. (1984) Placental enzymes: A population genetic study. Acta Anthropogenet. 8:199-207.

Blake, N. M., Hawkins, B. R., Kirk, R. L., Bhatia, K., Brown, P., Garruto, R. M., and Gajdusek, D. C. (1983) A population genetic study of the Banks and Torres Islands (Vanuatu) and of the Santa Cruz Islands and Polynesian Outliers (Solomon Islands). Amer. J. Phys. Anthrop. 62:343-361.

Blake, N. M. and Hayes, C. (1980) A population genetic study of phosphoglycolate phosphatase. Ann. Hum. Biol. 7:481-484.

Blake, N. M. and Kirk, R. L. (1972) Personal communication. Cited from Williams and Hopkinson (1975).

Blake, N. M., Kirk, R. L., and Baxi, A. J. (1970) The distribution of some enzyme group systems among Marathis and Gujaratis in Bombay. Hum. Hered. 20:409-416.

Blake, N. M., Kirk, R. L., and Bonham, D. C. (1971b) Placental alkaline phosphatase types in Maoris and Polynesian Islanders. N. Z. Med. J. 74:170-172.

Blake, N. M., Kirk, R. L., and Fliegner, J. R. H. (1969a) Placental alkaline phosphatase types in Papuans and Fijians. Med. J. Aust. 2:342-344.

Blake, N. M., Kirk, R. L., and Matsumoto, H. (1969b) Placental alkaline phosphatase types in Japanese. Jpn. J. Hum. Genet. 13:243-248.

Blake, N. M., Kirk, R. L., McDermid, E. M., Omoto, K., and Ahuja, Y. R. (1971a) The distribution of serum protein and enzyme group systems among north Indians. Hum. Hered. 21:440-457.

Blake, N. M., Kirk, R. L., and Mehra, B. (1969c) Placental alkaline phosphatase types in Malaysia. Hum. Hered. 19:20-24.

Blake, N. M., Kirk, R. L., and Osathanondh, V. (1968) Placental alkaline phosphatase types in Thailand. Med. J. Aust. 2:1042-1045.

Blake, N. M., McDermid, E. M., Kirk, R. L., Ong, Y. W., and Simons, M. J. (1973a) The distribution of red cell enzyme groups among Chinese and Malays in Singapore. Singapore Med. J. 14:2-8.

Blake, N. M. and Omoto, K. (1975) Phosphoglucomutase types in the Asian-Pacific area: a critical review including new phenotypes. Ann. Hum. Genet. 38:251-273.

Blake, N. M., Omoto, K., Kirk, R. L., and Gajdusek, D. C. (1973b) Variation in red cell enzyme groups among populations of the Western Caroline Islands, Micronesia. Amer. J. Hum. Genet. 25:413-421.

Blake, N. M. and Spargo, R. M. (1986) Population genetic studies in the Kimberley of Western Australia. Hum. Hered. 36:286-298.

Bloomfield, L., Rowe, G. P., and Green, C. (1986) The webb (Wb) antigen in south Wales donors. Hum. Hered. 36:352-356.

Blumberg, B. S. and Gartler, S. M. (1961) The urinary excretion of β-aminoisobutyric acid in Pacific populations. Hum. Biol. 33:355-362.

Blumberg, B. S., Ikin, E. W., and Mourant, A. E. (1961) The blood groups of the pastoral Fulani of Northern Nigeria and the Yoruba of Western Nigeria. Amer. J. Phys. Anthrop. 19:195-201. Cited from Mourant et al. (1976).

Blumberg, B. S., Workman, P. L., and Hirschfeld, J. (1964) Gamma-globulin, group specific, and lipoprotein groups in a U.S. white and Negro population. Nature 202:561-563.

Blume, K. G., Löhr, G. W., Praetsch, O., and Rüdiger, H. W. (1968) Beitrag zur Populationsgenetik der Pyruvatkinase menschlicher Erythrocyten. Humangenetik 6:261-265.

Blundell, G., Frazer, A., Cole, R. B., and Nevin, N. C. (1975) Alpha$_1$-antitrypsin phenotypes in Northern Ireland. Ann. Hum. Genet. 38:289-294.

Board, P. G. (1979) Genetic polymorphism of the A subunit of human coagulation factor XIII. Amer. J. Hum. Genet. 31:116-124.

Board, P. G. (1980a) Electrophoretic investigation of γ-glutamyl-cyclotransferase from human erythrocytes. Hum. Hered. 30:248-250.

Board, P. G. (1980b) Genetic polymorphism of the B subunit of human coagulation factor XIII. Amer. J. Hum. Genet. 32:348-353.

Board, P. G. (1980c) Genetic polymorphism of human erythrocyte glyoxalase II. Amer. J. Hum. Genet. 32:690-694.

Board, P. G. (1981) Biochemical genetics of glutathione-S-transferase in man. Amer. J. Hum. Genet. 33:36-43.

Board, P. G. (1983) Further electrophoretic studies of erythrocyte glutathione peroxidase. Amer. J. Hum. Genet. 35:914-918.

Board. P. G. and Castle, S. (1982) Electrophoretic studies of coagulation factor XIII and fibronectin. In: Factor XIII and Fibronectin (Egbring, R. and Klingemann, H. G. eds.), Medizinische Verlagsgesellschaft, Marburg, pp.69-78. Cited from Kamboh, M. I. and Ferrell, R. E. (1986a).

Board, P. G. and Coggan, M. (1981) Polymorphism of the A subunit of coagulation factor XIII in the Pacific region. Description of new phenotypes. Hum. Genet. 59:135-136.

Board, P. G. and Coggan, M. (1986) Genetic heterogeneity of S-formylglutathione hydrolase. Ann. Hum. Genet. 50:35-39.

Board, P. G., Coggan, M., and Pidcock, M. E. (1982) Genetic heterogeneity of human prothrombin (FII). Ann. Hum. Genet. 46:1-9.

Boerwinkle, E., Saha, N., and Utermann, G. (1986) Personal communication.

Boerwinkle, E., Visvikis, S., Welsh, D., Steinmetz, J., Hanash, S. M., and Sing, C.F. (1987) The use of measured genotype information in the analysis of quantitative phenotypes in man. II.The role of the apolipoprotein E polymorphism in determining levels, variability and covariability of cholesterol, betalipoprotein and triglycerides in a sample of unrelated individuals. Amer. J. Med. Genet. 27:567-582.

349

Boev, P. and Popwassilev, I. (1969) Zur Häufigkeit der Blutgruppen ABO, MN und P sowie der Serumgruppen Hp und Gm in Bulgarien. Antropol. Anz. 31:184-188. Cited from Mourant et al. 1976.

Böhme, A., Cleve, H., Schönitzer, D., Reissigl, H., Kazda, S., and Müller, W. (1983) α_1-antitrypsin (Pi) types and subtypes in the Tyrolean population. Hum. Genet. 63:193-194.

Boizard, B. and Wautier, J-L. (1984) Lek[a], a new platelet antigen absent in Glanzmann's thrombasthenia. Vox Sang. 46:47-54.

Boman, H. (1981) Distribution of the E_1[a] gene among Norwegian blood donors studied by an automated screening method. Hum. Hered. 31:308-311.

Bonazzi, L. (1968) On a rare genetic variation of plasma albumin: Bisalbuminaemia. Clin. Chem. Acta 20:362-363.

Bonné, B., Ashbel, S., Modai, M., Godber, M. J., Mourant, A. E., Tills, D., and Woodhead, B. G. (1970) The Habbanite isolate. I. Genetic markers in the blood. Hum. Hered. 20:609-622.

Bonné-Tamir, B. (1975) Cited from Tills et al. (1983a).

Boobphanirojana, P., Chetanasilpin, M., Saengudom, C., and Flatz, G. (1970) Phenylthiocarbamide taste thresholds in the population of Thailand. Humangenetik 10:329-334.

Booth, P. B., Faogali, J. L., Kirk, R. L., and Blake, N. M. (1977) HLA types, blood groups, serum protein, and red cell enzyme types among Samoans in New Zealand. Hum. Hered. 27:412-423.

Booth, P. B. and McLoughlin, K. (1972) The Gerbich blood group system, especially in Melanesians. Vox Sang. 22:73-84.

Booth, P. B., Mourant, A. E., Tills, D., Kopec, A. C., Warlow, A., Teesdale, P., Hornabrook, R. W., Crane, G. G., and Saave, J. J. (1981) Genetic surveys from the Central, Morobe and Northern Districts, Papua New Guinea. Ann. Hum. Biol. 8:435-445.

Booth, P. B., Tills, D., Warlow, A., Kopec, A. C., Mourant, A. E., Teesdale, P., and Hornabrook, R. W. (1982) Red cell antigen, serum protein and red cell enzyme polymorphisms in Karkar Islanders and inhabitants of the adjacent north coast of New Guinea. Hum. Hered. 32:385-403.

Bosron, W. F. and Li, T.-K. (1981) Genetic determinants of alcohol and aldehyde dehydrogenases and alcohol metabolism. Seminars in Liver Disease 1:179-188.

Botsztejn, C. (1942) Zur Kenntnis des Geschmacksblindheit gegenüber Phenylthiocarbamid (PTC) in der Zürcher Bevölkerung und deren Erbgang. Arch. Klaus-Stift. VererbForsch. 17:109-123. Cited from Mourant et al. (1976).

Bottini, E., Lucarelli, P., Palmarino, R., Spennati, G. F., and Orzalesi, M. (1971) Alkaline phosphatase polymorphism of the human placenta in people of Negro and European origins living in Connecticut. Hum. Biol. 43:1-6.

Bottini, E., Lucarelli, P., Palmarino, R., Spennati, G. F., and Reynaud, G. (1970) Placental alkaline phosphatase polymorphism in some Italian populations. Humangenetik 11:62-65.

Bouali, M., Dehay, C., Benajam, A., Poirier, J. C., Degos, L., and Marcelli-Barge, A. (1981) HLA-A, B, C, Bf, and glyoxalase I polymorphisms in a sample of the kabyle population (Algeria). Tissue Antigens 17:501-506.

Bouloux, C., Gomila, J., and Langaney, A. (1972) Hemotypology of the Bedik. Hum. Biol. 44:289-302.

Bowen, P., O'Callaghan, F. and Lee, C. S. N. (1971) Serum protein polymorphisms in Indians of Western Canada. Hum. Hered. 21:242-253.

Bowman, J. E., Carson, P. E., Frischer, H., and Garay, A. L. (1966) Genetics of starch-gel electrophoretic variants of human 6-phosphogluconic dehydrogenase: Population and family studies in the United States and in Mexico. Nature 210:811-813.

Bowman, J. E., Carson, P. E., Frischer, H., Powell, D., Colwell, E. J., Legters, L. J., Cottingham, A. J., Boone, S. C., and Hiser, W. W. (1971) Hemoglobin and red cell enzyme variation in some populations of the Republic of Vietnam with comments on the malaria hypothesis. Amer. J. Phys. Anthrop. 34:313-324.

Boyd, W. C. and Boyd, L. G. (1937) Sexual and racial variations in ability to taste phenyl-thio-carbamide, with some data on the inheritance. Ann. Eugen. 8:46-51.

Boyce, A. J., Harrison, G. A., Platt, C. M., and Hornabrook, R. W. (1976) Association between PTC taster status and goitre in a Papua New Guinea population. Hum. Biol. 48:769-773.

Boyce, A. J., Holdsworth, V. M. L., and Brothwell, D. R. (1973) Demographic and genetic studies in the Orkney Islands. In: Genetic Variation in Britain (Roberts, D. F. and Sunderland, E., eds.), Taylor & Francis Ltd., London, pp. 109-128.

Boyd, W. C. and Boyd, L. G. (1937) New data on blood groups and other inherited factors in Europe and Egypt. Amer. J. Phys. Anthrop. 23:49-70.

Boyd, W. C. and Boyd, L. G. (1954) The blood groups in Pakistan. Amer. J. Phys. Anthrop. 12:393-405.

Boyer, S. H., Fainer, D. C., and Watson-Williams, E. J. (1963a) Lactate dehydrogenase variant from human blood: Evidence for molecular subunits. Science 141:642-643.

Boyer, S. H., Rucknagel, D., Weatherall, D. J., and Watson-Williams, W. J. (1963b) Further evidence for linkage between the β and δ loci governing human hemoglobin and the population dynamics of linked genes. Amer. J. Hum. Genet. 15:438-448.

Bradbrook, I. D., Grant, A., and Adinolfi, M. (1971) Ag(x) and Ag(y) antigens in studies of paternity cases in the United Kingdom. Hum. Hered. 21:493-499.

Braend, M., Efremov, G., Fagerhol, M. K., and Hartmann, O. (1965) Albumin and transferrin variants in Norwegians. Hereditas 53:137-142.

Brazier, D. M. and Goldsmith, K. L. G. (1968) Frequency of certain Gm and Inv factors in the United Kingdom. Nature 219:193. Cited from Steinberg and Cook (1981).

Breguet, G., Ney, R., Grimm, S. L., Hope, S. L., Kirk, R. L., Blake, N. M., Narenda, I. B., and Toha, A. (1982a) Genetic survey of an isolated community in Bali, Indonesia. I. Blood groups, serum proteins and hepatitis B serology. Hum. Hered. 32:52-61.

Breguet, G., Ney, R., Kirk, R. L., and Blake, N. M. (1982b) Genetic survey of an isolated community in Bali, Indonesia. II. Haemoglobin types and red cell isozymes. Hum. Hered. 32:308-317.

Brendemoen, O. J., (1950) Further studies of agglutination and inhibition in the Le^a-Le^b systems. J. Lab. Clin. Med. 36:335-341. Cited from Mourant et al. (1976).

Brinkmann, B., Hoppe, H. H., Hennig, W., and Koops, E. (1971) Red cell enzyme polymorphisms in a Northern German population. Hum. Hered. 21:278-288.

Brinkmann, B., Krukenberg, P., and Brinkmann, M. (1972) Gene frequencies of soluble glutamic-pyruvic-transaminase in a Northern German population (Hamburg). Humangenetik 16:355-356.

Brinkmann, B., Reiter, J., and Krüger, O. (1973) Genhäufigkeiten einiger Enzympolymorphismen in Mittelmeerländern. Humangenetik 20:141-146.

Brocteur, J., Hoste, B., and André, A. (1980) Plasma protein and enzyme polymorphisms in Belgium. Hum. Hered. 30:221-224.

Broman, P., Grundin, R., and Lins, P. E. (1971) The red cell acid phosphatase polymorphism in Sweden: Gene frequencies and application to disputed paternity. Acta Genet. Med. Gemellol. 20:77-81.

Brönnestam, R. (1973) Studies of the C3 polymorphism. Distribution of C3 phenotypes in different areas of Sweden. Hum. Hered. 23:361-369.

Brönnestam, R., Beckman, L., and Cedergren, B. (1971) Genetic polymorphism of the complement component C3 in Swedish Lapps. Hum. Hered. 21:267-271.

Brown, H. B., Parry, L., Khatun, M., and Ahmed, G. (1979) Lactose malabsorption in Bangladeshi village children: relation with age, history of recent diarrhea, nutritional status, and breast feeding. Amer. J. Clinc. Nutr. 32:1962-1969.

Brown, K. S. and Johnson, R. S. (1970) Populations studies on southwestern Indian tribes. III. Serum protein variations of Zuni and Papago Indians. Hum. Hered. 20:281-286.

Buchanan, D. I., Patterson, M., and Turc, J. M. (1983) Diego antibodies. Transfusion 23:80.

Büchi, E. C. (1959) Blut, Geschmack und Farbensinn bei den Kurumba (Nilgiri, Südindien). Arch. Klaus-Stift. Vererb-Forsch. 34:310-316. Cited from Mourant et al. (1976).

Budtz-Olsen, O. E. (1958) Haptoglobins and haemoglobins in Australian Aborigines, with a simple method for the estimation of haptoglobins. Med. J. Australia. ii:689-693. Cited from Mourant et al. (1976).

Budyakov, O. S. 1966. Determination of A$_1$ and A$_2$ subgroups in thin blood by means of phytagglutinins (In Russian). Vop. Anthrop. 22:120-122. Cited from Mourant et al. 1976.

Bütler, R., Metaxas--Bühler, M., Rosin, S., and Wandrey, R. (1959) Untersuchungen über Haptoglobingruppen von Smithies. Schweitz. Med. Wschr. 89:1041-1043. Cited from Mourant et al. (1976).

Caeiro, B. and Rey, D. (1985) Genetic heterogeneity of delta-aminolevulinate dehydrase and phosphoglycolate phosphatase in north-west Spain. Hum. Hered. 35:21-24.

Calchi-Novati, C., Ceppellini, R., Bianco, I., Silvestroni, E., and Harris, H. (1954) β-aminoisobutyric acid excretion in urine. A family study in an Italian population. Ann. Eugen. 18:335-336. Cited from Mourant et al. (1976).

Caldwell, K., Blake, E. T., and Sensabaugh, G. F. (1976) Sperm diaphorase: Genetic polymorphism of a sperm-specific enzyme in man. Science 191:1185-1187.

Camerino, G., Grzeschik, K. H., Jaye, M., De La Salle, H., Tolstochev, P., Lecocq, J. P., Heilig, R., and Mandel, J. L. (1984) Regional localization on the human X chromosome and polymorphism of the coagulation factor IX gene (hemophilia B locus). Proc. Natl. Acad. Sci. USA 81:498-502.

Camerino, G., Oberlé, I., Drayna, D., and Mandel, J. L. (1985) A new MspI restriction fragment length polymorphism in the hempphilia B locus. Hum. Genet. 71:79-81.

Camoens, H., Monn, E., and Berg, K. (1972) Genetic marker systems in Arctic populations. III. Polymorphism of red cell adenosine deaminase (ADA) in Norwegian Lapps. Hum. Hered. 22:561-565.

Campillo, F. L., Gallardo, L. E., Senra, A. (1973) Distribution of the Kell blood groups in the Spanish population. Hum. Hered. 23:499-500.

Cantor, R. M. and Kaback, M. M. (1985) Sandhoff disease (SHD) heterozygote frequencies (HF) in North American (NA) Jewish (J) and Non-Jewish (NJ) populations: Implications for carrier (C) screening. Amer. J. Hum. Genet. 37:A48.

Carfagna, M., Gaudio, L., Patricolo, M. R., and Spadacenta, F. (1976) Pancreatic amylase polymorphism: Another example of a distinctive gene frequency among Sardinians. Hum. Hered. 26:59-65.

Carracedo, A. and Concheiro, L. (1982) PGM subtypes in Galicia (NW Spain). Hum. Hered. 32:133-135.

Carracedo, A. and Concheiro, L. (1983) Enzyme polymorphisms in Galicia (NW Spain). Hum. Hered. 33:160-162.

Carro-Ciampi, G., Kadar, D., and Kalow, W. (1981) Distribution of serum paraoxon hydrolyzing activities in a Canadian population. Can. J. Physiol. Pharmacol. 59:904-907.

Carter, N. D. (1972) Carbonic anhydrase II polymorphism in Africa. Hum. Hered. 22:539-541.

Cartwright, R. A. (1976) Unifactorially inherited attributes of the population of Holy Island, Northumberland. Ann. Hum. Biol. 3:351-362.

Cartwright, R. A., Bethel, I. L., Hargreaves, H., Izatt, M., Jolly, J., Mitchell, R. J., Sawhney, K. S., Smith, M., Sunderland, E., and Teasdale, D. (1976) The red blood cell esterase D polymorphism in Europe and Asia. Hum. Genet. 33:161-166.

Cartwright, R. A., Hargreaves, H. J., and Sunderland, E. (1977) Serum protein and isoenzyme polymorphisms from Nottingham, England. Hum. Biol. 49:629-640.

Cartwright, R. A. and Sunderland, E. (1967) Phenylthiocarbamide (PTC) tasting ability in populations in the north of England: with a note on endemic goitre. Acta Genet. Stat. Med. 17:211-221.

Casado, H. F. (1975) Contributión al conocimiento del sistema de antigenos eritrocitarios P en una muestra de la población española. IV Cong. Nac. Genét. Hum., Zaragosa. Cited from Tills et al. (1983a).

Castle, S. L. and Board, P. G. (1982) Electrophoretic investigation of formaldehyde dehydrogenase from human tissues. Hum. Hered. 32:222-224.

Castle, S. L. and Board, P. G. (1985) An extended survey of the genetic polymorphism at the human coagulation factor XIII: A subunit structural locus. Hum. Hered. 35:101-106.

Cavalli-Sforza, L. L., Zonta, L. A., Nuzzo, F., Bernini, L., De Jong, W. W., Meera Khan, P., Ray, A. K., Went, L. N., Siniscalco, M., Nijenhuis, L. E., van Loghem, E., and Modiano, G. (1969) Studies on African pygmies. I. A pilot investigation of Babinga Pygmies in the Central African Republic (with an analysis of genetic distances). Amer. J. Hum. Genet. 21:252-274.

Cazal, P., Graafland, R., and Mathieu, M. (1951) Les groupes sanguins chez les Gitans de France. 4th Int. Congr. Blood Transfus., Lisbon, pp. 356-364. Cited from Mourant et al. (1976).

Cedergren, B., Nordenson, I., and Beckman, L. (1983) Population studies in northern Sweden. XI. The Duffy blood group polymorphism. Hum. Hered. 33:365-370.

Ceppellini, R. (1954) On the genetics of secretor and Lewis characters: a family study. 5th Int. Cong. Blood Transfus., Paris, pp. 207-211. Cited from Mourant et al. (1976).

Chaabani, H., Bech-Hansen, N. T., and Cox, D. W. (1986) Restriction fragment length polymorphisms associated with immunoglobulin heavy chain gamma genes in Tunisians. Hum. Genet. 73:110-113.

Chahal, S. M. S. and Papiha, S. S. (1981) A population genetic study of the Jat Sikhs, Punjab, India. J. Indian Anthrop. Soc. 16:251-260.

Chakraborty, R., Gershowitz, H., Ferrell, R. E., Barton, S. A., and Schull, W. J. (1985) Immunoglobulin Gm and Km) allotypes in the Aymara of Chile and Bolivia. Ann. Hum. Biol. 12:533-543.

Chakraborty, R., Lidsky, A. S., Daiger, S. P., Güttler, F., Sullivan, S., DiLella, A. G., and Woo, S. L. C. (1987) Polymorphic DNA haplotypes at the human phenylalanine hydroxylase locus and their relationship with phenylketonuria. Hum. Genet. (in press)

Chakravarti, A., Elbein, S. C., and Permutt, M. A. (1986) Evidence for increased recombination near the human insulin gene: Implication for disease association studies. Proc. Natl. Acad. Sci. USA 83:1045-1049.

Chakravarti, A., Phillips. J. A. III., Mellits, K. H., Buetow, K. H., and Seeburg, P. H. (1984) Patterns of polymorphism and linkage disequilibrium suggest independent origins of the human growth hormone gene cluster. Proc. Natl. Acad. Sci. USA 81:6085-6089.

Chan, K. T. (1962) The ABO blood group frequency distribution of Singapore based on a blood donor sample. Singapore Med. J. 3:3-15.

Chan, V., Chan, T. K., Cheng, M. Y., Leung, N. K., Kan, Y. W. and Todd, D. (1986) Characteristics and distribution of β thalassemia haplotypes in South China. Hum. Genet. 73:23-26.

Chandanayingyong, D., Bejrachandra, S., Metaseta, P., and Pongsataporn, S. (1979) Further study of Rh, Kell, Duffy, P, MN, Lewis and Gerbich blood groups of the Thais. SE Asian J. Trop. Med. & Pub. Hlth. 10:209-211.

Chandanayingyong, D., Sasaki, T. T., and Greenwalt, T. J. (1967) Blood groups of the Thais. Transfusion 7:269-276.

Chapman, J. A., Grant, I. S., Taylor, G., Mahmud, K., Sardar-ul-Mulk, and Shahid, M. A. (1972) Endemic goitre in the Gilgit Agency, West Pakistan. With an appendix of dermatoglyphics and taste-testing. Phil. Trans. 263:459-491.

Char, K. S. N. and Rao, P. R. (1986) Glyoxylase I phenotypes in some endogamous populations of Andhra Pradesh, India. Hum. Hered. 36:123-125.

Charlesworth, D. (1972) Starch-gel electrophoresis of four enzymes from human red blood cells: glyceraldehyde-3-phosphate dehydrogenase, fructoaldolase, glyoxalase II and sorbitol dehydrogenase. Ann. Hum. Genet. 35:477-484.

Charlionet, R., Sesboüé, R., Morcamp, C., Lefebvre, F., and Martin, J. P. (1981) Genetic variants of serum alpha-1-antitrypsin (Pi types) in Normans: Common Pi M subtypes and new phenotypes. Hum. Hered. 31:104-109.

Chaudhuri, S., Mukherjee, B., Ghosh, J., and Roychoudhury, A. K. (1967) Blood groups of the Chinese in Calcutta. Nature 213:1245.

Chen, K. H., Cann, H., Chen, T. C., van West, B., and Cavalli-Sforza, L. (1985) Genetic markers of an aboriginal Taiwanese population. Amer. J. Phys. Anthrop. 66:327-337.

Chen, L. and Du, R. (1984) The incidence of aldehyde dehydrogenase deficiency in different Chinese minorities. Ann. Report Inst. Genet., Academ. Sinica, p. 60.

Chen, L. and Yan, Y. (1984) The distribution of the phosphoglucomutase-1 (PGM_1) subtypes in several Chinese nationalities. Acta Anthrop. Sinica 3:285-289.

Chen, S.-H., Anderson, J. E., and Giblett, E. R. (1971) 2, 3-Diphosphoglycerate mutase: Its demonstration by electrophoresis and the detection of a genetic variant. Biochem. Genet. 5:481-486.

Chen, S.-H., Anderson, J., Giblett, E. R., and Lewis, M. (1974) Phosphoglyceric acid mutase: Rare genetic variants and tissue distribution. Amer. J. Hum. Genet. 26:73-77.

Chen, S.-H., Fossum, B. L. G., and Giblett, E. R. (1972a) Genetic variation of the soluble form of NADP-dependent isocitric dehydrogenase in man. Amer. J. Hum. Genet. 24:325-329.

Chen, S.-H. and Giblett, E. R. (1971) Genetic variation of soluble glutamic-oxaloacetic transaminase in man. Amer. J. Hum. Genet. 23:419-424.

Chen, S.-H. and Giblett, E. R. (1972) Phosphoglycerate kinase: Additional variants and their geographic distribution. Amer. J. Hum. Genet. 24:229-230.

Chen, S.-H., Giblett, E. R., Anderson, J. E., and Fossum, B. L. G. (1972b) Genetics of glutamic-pyruvic transaminase: Its inheritance, common and rare variants, population distribution, and differences in catalytic activity. Ann. Hum. Genet. 35:401-409.

352

Chen, S.-H., Giblett, E. R., and Motulsky, A. G. (1973) Some red cell enzyme phenotype frequencies in Chinese. Humangenetik 17:341-343.

Chern, C. J. and Beutler, E. (1976) Biochemical and electrophoretic studies of erythrocyte pyridoxine kinase in White and Black Americans. Amer. J. Hum. Genet. 28:9-17.

Chih-chuan, L., Qi, Z., Ying, Q., and Wang, L. (1983) Types and subtypes of haptoglobin in the Chinese population. Hum. Genet. 63:175-177.

Chin, J. (1964) Absence of Di^{a+} in Malayan aborigines. Nature 201:1039.

Chopra, V. P. (1970) Studies on serum groups in the Kumaon region, India. Humangenetik 10:35-43.

Chown, B. and Lewis, M. (1952) Personal communication. Cited from Mourant et al. (1976).

Chown, B. and Lewis, M. (1959) The blood group genes of the Copper Eskimo. Amer. J. Phys. Anthrop. 17: 13-18.

Chown, B. and Lewis, M. (1962) The blood groups and secretor status of three small communities in Alaska. Oceania 32:211-218. Cited from Mourant et al. (1976).

Chown, B. and Lewis, M. (1960) The blood group and secretor genes of the Eskimo on Southampton Island. Bull. Nat. Mus. Can. No. 180, pt.i, pp.181-190. Cited from Mourant et al. (1976).

Chown, B. and Lewis, M. (1962) The blood groups and secretor status of three small communities in Alaska. Oceania 32:211-218.

Christiansen, R. and Sachs, W. (1981) Verteilung und Erbgang der Merkmale des erythrozyten-Isoenzymes der Phosphoglykolatphosphatase (PGP) in Schleswig-Holstein. Forensic Sci. int. 18:267. Cited from Caeiro and Rey (1985).

Clark, P. (1982) Alpha-1-protease inhibitor phenotypes in Australia. Hum. Hered. 32:225-227.

Clegg, E. J., Tills, D., Warlow, A., Wilkinson, J., and Marin, A. (1985) Blood group variation in the Isle of Lewis. Ann. Hum. Biol. 12:345-361.

Cleghorn, T. E. (1960) The frequency of the Wr^a, By and Mg blood group antigens in blood donors in the south of England. Vox Sang. 5:556-560.

Cleghorn, T. E. (1961) The occurrence of certain rare blood group factors in Britain. Thesis – Sheffield. Cited from Mourant et al. (1976).

Cleve, H. (1966) Die Verteilung der Haptoglobin-Untergruppen in einer Stichprobe gesunder Blutspender aus Hessen. Humangenetik 2:115-118.

Cleve, H. (1974) The variants of the group-specific component – A review of their distribution in human populations. In: Genetic Polymorphisms and Diseases in Man (Ramot et al., eds.), Academic Press, New York, pp. 7-29.

Cleve, H. and Deicher, H. (1965) Haptoglobin "Marburg": Untersuchungen über eine seltene erbliche Haptoglobin-Variante mit zwei verschiedenen Phänotypen innerhalb einer Familie. Humangenetik 1:537-550.

Cleve, H., Patutschnick, W., Nevo, S., and Wendt, G. G. (1978) Genetic studies on the Gc subtypes. Hum. Genet. 44:117-122.

Cleve, H., Ramot, B., and Bearn, A. G. (1962) Distribution of the serum group-specific components in Israel. Nature 195:86-87.

Cleve, H. and Vierucci, A. (1965) Distribution of Gc-types in northern Italy. Acta Genet. Stat. Med. 15: 243-247.

Coates, P. M. and Cortner, J. A. (1986) Genetic polymorphism of esterase B3 in human leukocytes. Ann. Hum. Genet. 50:207-216.

Coates, P. M. and Simpson, N. E. (1972) Genetic variation in human erythrocyte acetylcholinesterase. Science 175:1466-1467.

Cohen, P. T. W. and Omenn, G. S. (1972) Human malic enzyme: High frequency polymorphism of the mitochondrial form. Biochem. Genet. 7:303-311.

Cohen, T., Karathanasis, S. K., Kazazian, H. H., Jr., and Antonarakis, S. E. (1986) DNA polymorphic sites in the human ApoAI-CIII-AIV cluster: TaqI and Ava I. Nucleic Acids Res. 14:1924.

Coleman, R. T., Dillan, N. A., Lim, D. W., Malloy, M. J., Kane, J. P., and Frossard, P. M. (1986) TaqI and XbaI RFLPs detected with a human apo IV (Apo4) cDNA probe. Nucleic Acids Res. 14:7818.

Constans, J., Gouaillard, C., and Breguet, G. (1986) Serum protein polymorphism in Bali (Indonesia). Ann. Hum. Biol. 13:537-545.

Constans, J., Hazout, S., Garruto, R. M., Gajdusek, D. C., and Spees, E. K. (1985) Population distribution of the human vitamin D binding protein: Anthropological considerations. Amer. J. Phys. Anthrop. 68:107-122.

Constans, J., Kühnl, P., Viau, M., and Spielmann, W. (1980a) A new procedure for the determination of transferrin^C (Tf^C) subtypes by isoelectric focusing. Hum. Genet. 55:111-114.

Constans, J. and Salzano, F. M. (1980) Gc and transferrin isoelectric focusing subtypes among Brazilian Indians. J. Hum. Evol. 9:489-494.

Constans, J., Viau, M., and Gouaillard, C. (1980b) Pi^{M4}: An additional Pi^M subtype. Hum. Genet. 55: 119-121.

Constans, J., Viau, M., Gouaillard, C., and Clerc, A. (1981a) Haptoglobin polymorphism among Saharian and West African groups: Haptoglobin phenotype determination by radioimmunoelectrophoresis on Hp O samples. Amer. J. Hum. Genet. 33:606-616.

Constans, J., Viau, M., Jaeger, G., and Palisson, M. J. (1981b) Gc, Tf, Hp, subtype and α^1-antitrypsin polymorphisms in a pygmy Bi-Aka sample. Hum. Hered. 31:129-137.

Contreras, M., Armitage, S. E., and Stebbing, B. (1984a) The MNSs antigen Ridley (Ri^a). Vox Sang. 46:360-365.

Contreras, M., Armitage, S. E., and Lubenko, A. (1984b) The need for a panel of PL(A1 negative) donors. Br. J. Haematol. 58:192.

Contreras, M., Lubenko, A., Armitage, S., Cleghorn, T., and Jenkins, J. (1980) Frequency and inheritance of the Bx^a (Box) antigen. Vox Sang. 39:225-228.

353

Cook, G. C. (1979) Intestinal lactase status of adults in Papua New Guinea. Ann. Hum Biol. 1:55-58.

Cook, G. C. and Al-Torki, M. T. (1975) High intestinal lactase concentrations in adult Arabs in Saudi Arabia. Br. Med. J. 3:135-136.

Cook, P. J. L. (1975) The genetics of α_1-antitrypsin: a family study in England and Scotland. Ann. Hum. Genet. 38:275-287.

Cooke, K. B., Cleghorn, T. E., and Lockey, E. (1961) Two new families with bisalbuminaemia: An examination of possible links with other genetically controlled variants. Biochem. J. 81:39.

Cooper, A. J., Blumberg, B. S., Workman, P. L., and McDonough, J. R. (1963) Biochemical polymorphic traits in a U.S. white and Negro population. Amer. J. Hum. Genet. 15:420-428.

Corbo, R. M. (1986) Personal communication.

Corbo, R. M., Palmarino, R., Spennati, G. F., Pascone, R., and Lucarelli, P. (1980) Human placental phosphoglucomutase locus 3 studies in the Italian population. Jpn. J. Hum. Genet. 25:325-328.

Corbo, R. M., Spennati, G. F., Scacchi, R., Palmarino, R., Della Penna, M. R., and Berrelli, P. (1981) A survey of serum protein and enzyme polymorphisms in the district of L'Aquila (Italy). Hum. Hered. 31:167-171.

Corcoran, P. A., Allen, F. H. Jr., Allison, A. C., and Blumberg, B. S. (1959) Blood groups of Alaskan Eskimos and Indians. Amer. J. Phys. Anthrop. 17:187-193.

Corney, G., Fisher, R. A., Cook, P. J. L., Noades, J., and Robson, E. B. (1977) Linkage between α fucosidase and the rhesus blood group. Ann. Hum. Genet. 40:403-405.

Cox, D. W., Andrews, B. J., and Wills, D. E. (1986) Genetic polymorphism of α_2HS-glycoprotein. Amer. J. Hum. Genet. 38:699-706.

Cox, D. W., Simpson, N. E., and Jantti, R. (1978) Group-specific component, alpha$_1$-antitrypsin and esterase D in Canadian Eskimos. Hum. Hered. 28:341-350.

Cox, D. W., Woo, S. L. C., and Mansfield, T. (1985) DNA restriction fragments associated with α_1-antitrypsin indicate a single origin for deficiency allele PIZ. Nature. 316:79-81.

Crawford, M. H. (1967) Personal communication. Cited from Mourant et al. (1976).

Crawford, M. H., Gonzalez, N. L., Schanfield, M. S., Dykes, D. D., Skradski, K., and Polesky, H. F. (1981) The Black Caribs (Garifuna) of Livingston, Guatemala: Genetic markers and admixture estimates. Hum Biol. 53:87-103.

Crawford, M. H., Leyshon, W. C., Brown, K., Lees, F., and Taylor, L. (1974) Human biology in Mexico. II. A comparison of blood group, serum and red cell enzyme frequencies and genetic distances of the Indian populations of Mexico. Amer. J. Phys. Anthrop. 41:251-268.

Cresta, M. (1964) Antropologia morfologica e sierologica dei N'Zakara della Repubblica Centrafricana. Ric. Sic. (biol.) 34:131-142. Cited from Mourant et al. (1976).

Crone, R. A. (1968) Incidence of known and unknown color vision defects: A study of 6526 secondary school pupils in Amsterdam. Ophthalmologica (Basel) 155:37-55. Cited from Post (1971).

Crosti, N., Serra, A., Cagiano-Malvezzi, D., and Tagliaferri, G. (1976) The rare allele SOD-A^2 in the Italian population. Ann. Hum. Biol. 3:343-350.

Cruz, J. M., Bender, K., Burckhardt, K., Küppers, F., Benkmann, H.-G., and Goedde, H. W. (1973) Genetic studies of some red cell and serum protein polymorphisms in the population of Vilarinho da Furna (Portugal). Trabalhos do Instituto de Antropologia. Porto, No. 15. 56:3-15. Cited from Tills et al. (1983a).

Cruz-Coke, R. and Barrera, R. (1969) Colour-blindness among Aymara in Chile. Amer. J. Phys. Anthrop. 31:229-230.

Cumming, A. M. and Robertson, F. W. (1984) Polymorphism at the apolipoprotein-E locus in relation to risk of coronary disease. Clin. Genet. 25:310-313.

Cunha, A., Xavier da, and Abreu, M. D. A. (1956) A sensibilidade gustativa da feniltiocarbamida em Portugeses. Contr. Antrop. Portug. 6:85-96. Cited from Mourant et al. (1976).

Curtain, C. C., van Loghem, E., Fudenberg, H. H., Tindale, N. B., Simmons, R. T., and Doherty, R. L., and Vos, G. (1972) Distribution of the immunoglobulin markers at the IgG1, IgG2, IgG3, IgA2, and K-chain loci in Australian Aborigines: Comparison with New Guinea populations. Amer. J. Hum. Genet. 24:145-155.

Czeizel, A., Flatz, G., and Flatz, S. D. (1983) Prevalence of primary adult lactose malabsorption in Hungary. Hum. Genet. 64:398-401.

Dahlqvist, A. and Lindquist, B. (1971) Lactose intolerance and protein malnutrition. Acta Paediat. Scand. 60:488-494. Cited from Flatz (1986).

Daiger, S. P., Labowe, M. L., Parsons, M., Wang, L., and Cavalli-Sforza, L. L. (1978) Detection of genetic variation with radioactive ligands. III. Genetic polymorphisms of transcobalamin II in human plasma. Amer. J. Hum. Genet. 30:202-214.

Daiger, S. P., Rummel, D. P., Wang, L., and Cavalli-Sforza, L. L. (1981) Detection of genetic variation with radioactive ligands. IV. X-linked, polymorphic genetic variation of thyroxine-binding globulin (TBG). Amer. J. Hum. Genet. 33:640-648.

Daniels, G. L., Judd, W. J., Moore, B. P. L., Neitzer, G., Ouellet, P., Plantos, M., and Verrette, S. (1982) A 'new' high frequency antigen Er^a. Transfusion 22:189-193.

Darby, J. K., Feder, J., Selby, M., Riccardi, V., Ferrell, R., Siao, D., Goslin, K., Rutter, W., Shooter, E. M., and Cavalli-Sforza, L. L. (1985) A discordant sibship analysis between β-NGF and neurofibromatosis. Amer. J. Hum. Genet. 37:52-59.

Darnborough, J., Dunsford, I., and Wallace, J. A. (1969) The En^a antigen and antibody: A genetical modification of human red cells affecting their blood grouping reactions. Vox Sang. 17:241-255.

Darnfors, C., Nilsson, J., Protter, A. A., Carlsson, P., Talmud, P. J., Humphries, S. E., Whalstörm, J., Wiklund, O., and Bjursell, G. (1986) RFLPs for the human apolipoprotein B gene: HincII and

PvuII. Nucl. Acids Res. 14:7135

Das, S. R. (1966) Application of phenylthiocarbamide taste character in the study of racial variation (data on world taste gene distribution). J. Indian Anthrop. Soc. 1:63-80.

Das, S. R., Mukherjee, D. P., and Bhattacharjee, P. N. (1967) Survey of the blood groups and PTC taste among the Rajbanshi caste of West Bengal (ABO, MNS, Rh, Duffy and Diego). Acta Genet. Stat. Med. 17:433-445.

Das, S. R., Mukherjee, B. N. and Das, S. K. (1974) Caste variation in the distribution of placental alkaline phosphatase genes among the Hindus of West Bengal. Ann. Hum. Biol. 1:65-71.

Das, S. R., Mukherjee, B. N., Das, S. K., Blake, N. M., and Kirk, R. L. (1970) The distribution of some enzyme group systems among Bengalis. Indian J. Med. Res. 58:866-875.

Daveau, M., Rivat, L., Lalouel, J. M., Langaney, A., Roberts, D. F., and Simons, M. J. (1980) Frequencies of Gm and Km allotypes in the population of Singapore, Sri Lanka and Punjabis in North India. Hum. Hered. 30:237-244.

David, V., Fauchet, R., Phengsavath, H., and Le Gall, J. Y. (1983) Properdin factor B (Bf) polymorphism: subtyping of SS phenotypes. Hum. Genet. 64:189-190.

Davidson, R. G. and Cortner, J. A. (1967) Mitochondrial malate dehydrogenase: A new genetic polymorphism in man. Science 157:1569-1571.

Davidson, R. G., Fildes, R. A., Glen-Bott, A. M., Harris, H., Robson, E. B., and Cleghorn, T. E. (1965) Genetical studies on a variant of human lactate dehydrogenase (subunit A). Ann. Hum. Genet. 29:5-17.

Davignon, J., Sing, C. F., Lussier-Cacan, S. Bouthillier, D. (1984) Xanthelasma, latent dyslipoproteinemia and atherosclerosis: contribution of apo E polymorphism. In: Latent Dyslipoproteinemia and Atherosclerosis (De Gennes, J. L., Polonowsky, J. and Paoletti, R., eds.), Raven Press, New York, pp. 213-223.

Davrinche, C., Rivat, C., and Rivat-Peran, L. (1981a) Human properdin factor B: Gene frequency study in an African Negroid population (Niger). Hum. Hered. 31:304-307.

Davrinche, C., Rivat, C., Rivat-Peran, L., Helel, A. N., Boukef, K., Lefranc, M. P., and Lefranc, G. (1981b) Genetic variants of human C3 and properdin factor B in a population from Tunisia. Hum. Hered. 31:299-303.

de Córdoba, S. R. and Rubinstein, P. (1984) Genetic polymorphism of human factor H (β1H). J. Immunol. 132:1906-1908.

de Jong, W. W. W. (1964) Smaakproven met phenylthiocarbamide (P.T.C.) bij Nederlandse schoolkinderen. Het verband met endemische krop. Geneesk. Bl. 50:349-384.

de Natale, A., Cahan, A., Jack, J. A., Race, R. R., and Sanger, R. (1955) V: a 'new' Rh antigen, common in Negroes, rare in white people. J. Amer. Med. Ass. 159:247-250. Cited from Race & Sanger (1975).

de Soyza, K. (1978) Polymorphism of human salivary amylase: A preliminary communication. Hum. Genet. 45:189-192.

de Stefano, G. F. and Molieri, J. J. (1976) P.T.C. tasting among three indian groups of Nicaragua. Amer. J. Phys. Anthrop. 44:371-374.

de Veber, L. L., Clark, G. W., Hunking, M., and Stroup, M. (1971) Maternal anti-LW. Transfusion 11:33-35.

Destro-Bisol, G., Briziobello, A., Adriani, A., and Spedini, G. (1986a) Frequencies of the GPX_1^T (or GPX_1^{*2}) and CA_{II}^2 alleles in some Congo populations. Hum. Hered. 36:58-61.

Destro-Bisol, G., Menchicchi, F., Ranalletta, D., and Spedini. G. (1986b) EsD in Negro and Caucasian populations: Is the EsD^5 a 'Caucasian allele'? Hum. Hered. 36:154-157.

Detter, J. C., Anderson, J. E., and Giblett, E. R. (1970a) NADH diaphorase: An inherited variant associated with normal methemoglobin reduction. Amer. J. Hum. Genet. 22:100-104.

Detter, J. C., Stamatoyannopoulos, G., Giblett, E. R., and Motulsky, A. G. (1970b) Adenosine deaminase: Racial distribution and report of a new phenotype. J. Med. Genet. 7:356-357.

Detter, J. C., Ways, P. O., Giblett, E. R., Baughan, M. A., Hopkinson, D. A., Povey, S., and Harris, H. (1968) Inherited variations in human phosphohexose isomerase. Ann. Hum. Genet. 31:329-338.

Dewald, G. and Rittner, C. (1979) Polymorphism of the second component of human complement (C2). Vox Sang. 37:47-54.

Dewey, W. J. and Mann, J. D. (1967) Xg blood group frequencies in some further populations. J. Med. Genet. 4:12-15.

DiLella, A. G., Marvit, J., Lidsky, A. S., Güttler, F., and Woo, S. L. C. (1986) Tight linkage between a splicing mutation and a specific DNA haplotype in phenylketonuria. Nature 322:799-803

Dill, J. E., Levy, M., Wells, R. F., and Weser, E. (1972) Laçtase deficiency in Mexican-American males. Amer. J. Clin. Nutr. 25:869-870.

Dimo-Simonin, N., Brandt-Casadevall, C., and Gujer, H.R. (1985) Gene frequencies of plasminogen in Switzerland. Hum. Hered. 35:343-345.

Dinçol, G., Erdem, S., and Aksoy, M. (1976) Transferrin types in Turkish people. Hum. Hered. 26:349-350.

Dissing, J. (1973) Opløselig glutamat-pyruvat-transaminase (GPT): en dansk populations undersøgelse og anvendelsen i faderskabssager. Nordisk Rättsmedicinsk förenings förhandlingar. 5. mötet, pp. 177-185 (Lund 1973). Cited from Olaisen, O. (1975).

Dissing, J. and Eriksen, B. (1984) Human red cell esterase D polymorphism in Denmark, its use in paternity cases and the description of a new phenotype. Hum. Hered. 34:148-155.

Dobosz, T. (1983) Distribution of red cell enzyme polymorphisms in south-west Poland. Hum. Hered. 33:55-57.

Does, J. A. V., D'Amaro, J., Leeuwen, A. V., Meera Khan, P., Bernini, L. F., van Loghem, E., Nijenhuis, L., Steen, G. V., Rood, J. J. V., Rubinstein, P. (1973) HLA typing in Chilean Aymara Indians. In: Histocompatibility Testings 1972 (Dausset, J. and Colombani, J., eds.), Munksgaard,

Copenhagen, pp. 391-395.

Domenici, R., Giari, A., Bargagna, M., and Weidinger, S. (1986) Distribution of C3 and Bf allotypes in Tuscany (Italy). Hum. Hered. 36:330-332.

Donald, L. J. (1976) Genetical variation of placental alkaline phosphatase in Canada. In: Protides of the Biological Fluids, 24th Colloquium (Peeters, H., ed.), Pergamon Press, Oxford, pp. 95-98.

Donald, L. J. (1977) Placental enzyme polymorphisms in Canadian populations. II. Phosphoglucomutase. Hum. Hered. 27:280-284.

Donegani, J. A., Ibrahim, K. A., Ikin, E. W., and Mourant, A. E. (1950) The blood groups of the people of Egypt. Heredity 4:377-382.

Douglas, R., Jacobs, J., Hoult, G. E., and Staveley, J. M. (1962) Blood groups, serum genetic factors and haemoglobins in Western Solomon Islanders. Transfusion 2:413-418.

Douglas, R., Jacobs, J., McCarthy, D. D., and Staveley, J. M. (1966) Blood group, serum genetic factors, and hemoglobins in Cook Islanders. II. Rarotonga. Transfusion 6:324-326.

Douglas, R., Jacobs, J., Sherliker, J., and Staveley, J. M. (1961) Blood groups, serum genetic factors, and haemoglobins in Gilbert Islanders. N. Z. Med. J. 60:146-152. Cited from Mourant et al. (1976).

Douglas, R. and Staveley, J. M. (1959) The blood groups of Cook Islanders. J. Polynes. Soc. 68:14-20. Cited from Mourant et al. (1976).

Doxiadis, S. and Papageorgiadis, G. (1973) Lactose intolerance in Greeks. Lancet 1. 271.

Dozy, A. M., Kan, Y. W., Embury, S. H., Mentzer, W. C., Wang, W. C., Lubin, B., Davis, J. R., Jr., and Koenig, H. M. (1979) α-Globin gene organisation in blacks precludes the severe form of α-thalassaemia. Nature 280:605-607.

Driesel, A. J., Bierotte, E. and Röhrborn, G. (1982a) Human GOT_M phenotypes in Western Germany (Düsseldorf region). Hum. Hered. 32:145-146.

Driesel, A. J., Schumacher, A. M., and Flavell, R. A. (1982b) A Hind III restriction site polymorphism in the human collagen α1 (I)-like gene on chromosome No. 7. Hum. Genet. 62:175-176.

Ducos, J., Ruffié, J., Colombies, P, Marty, Y., and Ohayon, E. (1965) I antigen in leukaemic patients. Nature 208:1329-30.

Duncan, I. W. and Scott, E. M. (1972) Lactose intolerance in Alaskan Indians and Eskimos. Amer. J. Clin. Nutr. 25:867-868.

Duncan, I. W., Scott, E. M. and Wright, R. C. (1974) Gene frequencies of erythrocytic enzymes of Alaskan Eskimos and Athabaskan Indians. Amer. J. Hum. Genet. 26:244-246.

Dunn, D. S., Madhoo, B., and Turnbull, R., and Jenkins, T. (1986) Alpha-1-antitrypsin variation in southern Africa. Hum. Hered. 36:238-242.

Dykes, D. D., Crawford, M. H., and Polesky, H. F. (1983a) Population distribution in North and Central America of PGM_1 and Gc subtypes as determined by isoelectric focusing (IEF). Amer. J. Phys. Anthrop. 62:137-145.

Dykes, D. D., DeFurio, C. M. and Polesky, H. F. (1982) Transferrin (Tf) subtypes in US Amerindians, Whites and Blacks using thin-layer agarose gels: Report on a new variant Tf^{C8}. Electrophoresis 3:162-164.

Dykes, D. D., Miller, S. A., and Polesky, H. F. (1984) Distribution of $α_1$-antitrypsin variants in a US white population. Hum. Hered. 34:308-310.

Dykes, D., Nelson, M. and Polesky, H. (1983b) Distribution of plasminogen allotypes in eight populations of the Western Hemisphere. Electrophoresis 4:417-420.

Dykes, D. D. and Polesky, H. F. (1985) FXIIIA phenotyping by isoelectric focusing and immunoblotting: Gene frequencies in a population of U.S. Whites and Blacks. Electrophoresis 6:521-523.

Dykes, D. D., Polesky, H. F., and Crawford, M. H. (1981) Properdin factor B (Bf) distribution in North and Central American populations. Electrophoresis 2:320-323.

Ebeli-Struijk, A. C., Wurzer-Figurelli, E. M., Ajmar, F., and Meera Khan, P. (1976) The distribution of esterase D variants in different ethnic groups. Hum. Genet. 34:299-306.

Eckerson, H. W., Wyte, C. M., and La Du, B. N. (1983) The human serum paraoxonase/arylesterase polymorphism. Amer. J. Hum. Genet. 35:1126-1138.

Edinger, H., Schloot, W., and Goedde, H. W. (1975) Zum Polymorphismus der sauren Erythrozytenphosphatasen; Untersuchungen zur formalen Genetik und zur Populationsgenetik in Thailand. Z. Morph. Anthrop. 66:217-231. Cited from Tills et al. (1983a).

Edwards, Y. H. and Hopkinson, D. A. (1979) The genetic determination of fumarase isozymes in human tissues. Ann. Hum. Genet. 42:303-313.

Edwards, Y. H., Hopkinson, D. A., and Harris, H. (1971) Inherited variants of human nucleoside phosphorylase. Ann. Hum. Genet. 34:395-408.

Edwards, Y. H., Potter, J. E., and Hopkinson, D. A. (1979) A comparison of biochemical properties of the human diaphorase (DIA_3) isozymes determined by the common alleles DIA^1_3, DIA^2_3 and DIA^3_3. Ann. Hum. Genet. 42:293-302.

Edwards-Moulds, J. M. and Alperin, J. B. (1986) Studies of the Diego blood group among Mexican-Americans. Transfusion 26:234-236.

Efremov, G. and Braend, M. (1964) Serum albumin: polymorphism in man. Science 146:1679-1680.

Ehnholm, C. (1969) The distribution of haptoglobin subtypes in the Finnish population. Hum. Hered. 19:222-226.

Ehnholm, C. and Eriksson, A. W. (1969) Haptoglobin subtypes among Finnish Skolt Lapps. Ann. Med. Exp. Fenn. 47:52-54.

Ehnholm, C., Lukka, M., Kuusi, T., Nikkilä, E., and Utermann, G. (1986) Apolipoprotein E polymorphism in the Finnish population: gene frequencies and relation to lipoprotein concentrations. J. Lipid Res. 27:227-235.

Eiberg, H. and Mohr, J. (1981) Genetics of paraoxonase. Ann. Hum. Genet. 45:323-330.

Eiberg, H. and Mohr, J. (1986) Identity of the polymorphisms for esterase D and S-formylglutathione hydrolase in red blood cells. Hum. Genet. 74:174-175.

Elbein, S. C., Corsetti, L., Ullrich, A., and Permutt, M. A. (1986) Multiple restriction fragment length polymorphisms at the insulin receptor locus: A highly informative marker for linkage analysis. Proc. Natl. Acad. Sci. USA 83:5223-5227.

el Dewi, S. (1951) The Rh types and their clinical effects in Egypt. J. Egypt. Med. Ass. 34:283-291. Cited from Mourant et al. (1976).

el Hassan, A. M., Godber, M. G., Kopec, A. C., Mourant, A. E., Tills, D., and Lehmann, H. (1968) The hereditary blood factors of the Beja of the Sudan. Man 3:272-283.

el-Hazmi, M. A. F., al-Swailem, A. R., al-Faleh, F. Z., and Warsy, A. S. (1986) Frequency of glucose-6-phosphate dehydrogenase, pyruvate kinase and hexokinase deficiency in the Saudi population. Hum. Hered. 36:45-49.

Ellard, G. A. and Gammon, P. T. (1977) Acetylator phenotyping of tuberculosis patients using matrix isoniazid or sulphadimidine and its prognostic significance for treatment with several intermittent isoniazid-containing regimens. Br. J. Clin. Pharmac. 4:5-14.

Eng, C. E. L. and Strom, C. M. (1985) Analysis of three restriction fragment length polymorphisms in the human type II procollagen gene. Amer. J. Hum. Genet. 37:719-732.

Erdem, S., Aksoy, M., and Çetingil, A. I. (1966) Distribution of haptoglobin types in Turkish people. Nature 210:315-316.

Eriksen, B. (1979) Human red cell glyoxalase I polymorphism in Denmark and its application to paternity cases. Hum. Hered. 29:265-271.

Eriksen, B. and Dissing, J. (1980) Human red cell galactose-1-phosphate uridyltransferase (EC 2.7.7.12): Electrophoretically determined polymorphism in Denmark and its use in paternity cases. Hum. Hered. 30:27-32.

Eriksson, A. W. (1974) Genetic polymorphisms in Finno-Ugrian populations: Finns, Lapps, and Maris. In: Genetic Polymorphisms and Diseases in Man (Ramot, B., ed.), Academic Press, New York, pp. 30-44.

Eriksson, A. W., Fellman, J., Forsius, H., and Lehmann, W. (1970) Phenylthiocarbamide tasting ability among Lapps and Finns. Hum. Hered. 20:623-630.

Eriksson, A. W., Fellman, J., Kirjarinta, M., Eskola, M-R., Singh, S., Benkman, H. G., Goedde, H. W., Mourant, A. E., Tills, D., and Lehmann, W. (1971a) Adenylate kinase polymorphism in populations in Finland (Swedes, Finns and Lapps) in Maris and in Greenland Eskimos. Humangenetik 12:123-130.

Eriksson, A. W., Kirjarinta, M., Fellman, J., Eskola, M-R., and Lehmann, W. (1971b) Adenosine deaminase polymorphism in Finland (Swedes, Finns and Lapps), the Mari Republic (Cheremisses) and Greenland (Eskimos). Amer. J. Hum. Genet. 23:568-577.

Eriksson, A. W., Kirjarinta, M., Lehtosalo, T., Kajanoja, P., Lehmann, W., Mourant, A. E., Tills, D., Singh, S., Benkmann, H. G., Hirth, L., and Goedde, H. W. (1971c) Red cell phosphoglucomutase polymorphism in Finland - Swedes, Finns, Finnish Lapps, Maris (Cheremisses) and Greenland Eskimos and segregation studies of PGM_1 types in Lapp families. Hum. Hered. 21:140-153.

Eriksson, A. W., Partanen, K., Frants, R. R., Pronk, J. C., and Kostense, P. J. (1986) ABH secretion polymorphism in Icelanders, Åland Islanders, Finns, Finnish Lapps, Komi and Greenland Eskimos: a review and new data. Ann. Hum. Biol. 13:273-285.

Escallon, M. H. (1987) Genetic Studies of the Human α_1-Acid Glycoprotein (Orosomucoid). Ph. D. Thesis, Univ. of Texas, Houston, USA.

Escallon, M. H., Ferrell, R. E., and Kamboh, M. I. (1987) Genetic studies of low abundance human plasma proteins. VI. Evidence for a second orosomucoid structural locus (ORM2) expressed in plasma. Amer. J. Hum. Genet. (in press).

Evans, D. A. P. (1969) An improved and simplified method of detecting the acetylator phenotype. J. Med. Genet. 6:405-407.

Evans, D. A. P. (1986) Acetylation. In: Ethnic Differences in Reactions to Drugs and Xenobiotics (Kalow, W., Goedde, H. W., and Agarwal, D. P., eds.), Alan Liss, New York, pp. 209-241.

Evans, D. A. P., Mahgoub, A., Solan, T. P., Idle, J. R., and Smith, R. L. (1980) A family and population study of the genetic polymorphism of debrisoquine oxidation in a white population. J. Med. Genet. 17:102-105.

Eze, L. C. and Obidoa, O. (1978) Acetylation of sulfamethazine in a Nigerian population. Biochem. Genet. 16:1073-1077.

Facchini, F., Gruppioni, G., and Rivalta, G. (1973) Richerche sui sistemi Lewis e Secretore nella popolazione bolognese. Trasfus. Sangue 18:405-420. Cited from Tills et al. (1983a).

Fagerhol, M. K. (1968) Pi-system: Genetic variants of serum alpha-1-antitrypsin. Series Haematologica 11:153-161.

Fagerhol, M. K., Eriksson, A. W., and Monn, E. (1969) Serum Pi types in some Lappish and Finnish populations. Hum. Hered. 19:360-364.

Fagerhol, M. K. and Tenfjord, O. W. (1968) Serum Pi types in some European, American, Asian and African populations. Acta Path. Microbiol. Scand. 72:601-608. Cited from Piantelli et al. (1978).

Farhud, D. D., Ananthakrishnan, R., Walter, H., and Loser, J. (1973) Electrophoretic investigation of some red cell enzymes in Iran. Hum. Hered. 23:263-266.

Farhud, D. D. and Walter, H. (1972) Hp subtypes in Iranians. Hum. Hered. 22:184-189.

Farhud, D. D. and Walter, H. (1973) Polymorphism of C'3 in German, Bulgarian, Iranian and Angola population. Humangenetik 17:161-164.

Feder, J., Gurling, H. M. D., Darby, J., and Cavalli-Sforza, L. L. (1985) DNA restriction fragment analysis of the proopimelanocortin gene in schizophrenia and bipolar disorders. Amer. J. Hum.

Genet. 37:286-294.

Ferrell, R. E., Bertin, T., Barton, S. A., Rothhammer, F., and Schull, W. J. (1980) The multinational Andean genetic and health program. IX. Gene frequencies and rare variants of 20 serum proteins and erythrocyte enzymes in the Aymara of Chile. Amer. J. Hum. Genet. 32:92-102.

Ferrell, R. E., Bertin, T., and Schull, W. J. (1981a) An electrophoretic study of glycolytic enzymes in a human population living at high altitude: The Aymara of northern Chile and western Bolivia. Hum. Genet. 56:397-399.

Ferrell, R. E., Bertin, T., Young, R., Barton, S. A., Murillo, F., and Schull, W. J. (1978a) The Aymara of western Bolivia. IV. Gene frequencies for eight blood groups and 19 protein and erythrocyte enzyme systems. Amer. J. Hum. Genet. 30:539-549.

Ferrell, R. E., Chakraborty, R., Gershowitz, H., Laughlin, W. S., and Schull, W. J. (1981b) The St. Lawrence Island Eskimos: Genetic variation and genetic distance. Amer. J. Phys. Anthrop. 55:351-358.

Ferrell, R. E., Nunez, A., Bertin, T., Labarthe, D. R., and Schull, W. J. (1978b) The Blacks of Panama: Their genetic diversity as assessed by 15 inherited biochemical systems. Amer. J. Phys. Anthrop. 48:269-276.

Ferrell, R. E., Salamatina, N. V., Dalakishvili, S. M., Bakuradze, N. A., and Chakraborty, R. (1985) A population genetic study in the Ochamchir region, Abkhazia, SSR. Amer. J. Phys. Anthrop. 66:63-71.

Ferrell, R. E., Ueda, N., Satoh, C., Tanis, R. J., Neel, J. V., Hamilton, H. B., Inamizu, T., and Baba, K. (1977) The frequency in Japanese of genetic variants of 22 proteins. I. Albumin, ceruloplasmin, haptoglobin, and transferrin. Ann. Hum. Genet. 40:407-418.

Fertakis, A., Tsourapas, A., Douratsos, D., and Angelopoulos, B. (1974) Pi phenotypes in Greeks. Hum. Hered. 24:313-316.

Fielder, H. and Pettenkofer, H. (1968) Ein "neuer" Phänotyp im Isoenzymsystem der Phosphoglukomutasen des Menschen (PGM$_1$0). Blut 18:33-34.

Fielding, J. F., Harrington, M. G., and Fottrell, P. F. (1981) The incidence of primary hypolactasia amongst the native Irish. Ir. J. Med. Sci. 150:276-277.

Fisher, R. A., Turner, B. M., Dorkin, H. L., and Harris, H. (1974) Studies on human erythrocyte inorganic pyrophosphatase. Ann. Hum. Genet. 37:341-353.

Flatz, G. (1967) Farbsehstörungen in der Bevölkerung Nordthailands. Humangenetik 3:328-330.

Flatz, G. (1987) Genetics of lactose digestion in humans. Adv. Hum. Genet. 16:1-77.

Flatz, G., Henze, H. J., Palabiyikoglu, E., Dagalp, K., and Türkkan, T. (1986a) Distribution of the adult lactase phenotypes, lactase repression and persistence in Turkey. Trop. Geogr. Med. (in press.)

Flatz, G., Howell, J. N., Doench, J., and Flatz, S. D. (1982) Distribution of physiological adult lactase phenotypes, latose absorber and malabsorber, in Germany. Hum. Genet. 62:152-157.

Flatz, G., Pik, C., and Sringam, S. (1965) Haemoglobin E and beta-thalassemia: their distribution in Thailand. Ann. Hum. Genet. 29:151-170.

Flatz, G., Saengudom, C., and Sanguanbhokhai, T. (1969) Lactose intolerance in Thailand. Nature 221:758-759.

Flatz, G., Schildge, C., and Sekou, H. (1986b) Distribution of adult lactase phenotypes in the Tuareg of Niger. Amer. J. Hum. Genet. 38:515-520.

Flatz, G. and Tantachamroon, T. (1970) Glucose-6-phosphate dehydrogenase in the population of northern Thailand: Evidence for two common electrophoretic variants with deficient enzyme activity. Humangenetik 10:335-339.

Fleischer, E. A. and Monn, E. (1970) Haptoglobin types and subtypes in Lappish and non-Lappish Norwegians. Amer. J. Hum. Genet. 22:105-108.

Flekkel, A. B. (1955) Concerning the problem of differential diagnosis of colorblindness. Dokl. Akad. Nauk SSSR, Otd. 1:57-60. Cited from Post (1971).

Flint, J., Hill, A. V. S., Bowden, D. K., Oppenheimer, S. J., Sill, P. R., Serjeantson, S. W., Bana-Koiri, J., Bhatia, K., Alpers, M. P., Boyce, A. J., Weatherall, D. J., and Clegg, J. B. (1986) High frequencies of α-thalassaemia are the result of natural selection by malaria. Nature 321:744-750.

Fox, J. A. and Taswell, H. F. (1969) Anti-Gn[a], a new antibody reacting with a high-incidence erythrocyte antigen, Transfusion 9:265-269.

Fox, M. H., Weyer, S. M., Thurmon, T. F., and Berenson, G. S. (1981) Genetically controlled enzymatic variation in a southern, biracial, semi-rural community. Hum. Hered. 31:138-151.

Frank, S., Schmidt, R. P., and Baugh, M. (1970) Three new antibodies to high-incidence antigenic determinants (anti-El, anti-Dp and anti-So). Transfusion 12:254-257.

Frants, R. R. and Eriksson, A. W. (1976) α_1-antitrypsin: Common subtypes of Pi M. Hum. Hered. 26:435-440.

Frants, R. R. and Eriksson, A. W. (1978) Reliable classification of six Pi M subtypes by separator isoelectric focusing. Hum. Hered. 28:201-209.

Fraser, G. R., Giblett, E. R., Lee, T. C., and Motulsky, A. G. (1965) Blood and serum groups in Taiwan. J. Med. Genet. 2:21-23.

Fraser, G. R., Giblett, E. R., and Motulsky, A. G. (1966a) Population genetic studies in the Congo. III. Blood groups (ABO, MNSs, Rh, Js[a]). Amer. J. Hum. Genet. 18:546-552. Cited from Mourant et al. (1976).

Fraser, G. R., Giblett, E. R., Stransky, E., and Motulsky, A. G. (1964) Blood groups in the Philippines. J. Med. Genet. 1:107-109. Cited from Mourant et al. (1976).

Fraser, G. R., Grünwald, P., Kitchin, F. D., and Steinberg, A. G. (1969a) Serum polymorphisms in Yugoslavia. Hum. Hered. 19:57-64.

Fraser, G. R., Grünwald, P., and Stamatoyannopoulos, G. (1966b) Glucose-6-phosphate dehydrogenase

(G6PD) deficiency, abnormal haemoglobins, and thalassemia in Yugoslavia. J. Med. Genet. 3:35-41.

Fraser, G. R., Steinberg, A. G., DeFaranas, B., Mayo, O., Stamatoyannopoulos, G., and Motulsky, A. G. (1969b) Gene frequencies at loci determining blood-group and serum-protein polymorphisms in two villages of northwestern Greece. Amer. J. Hum. Genet. 21:46-60.

Fraser, G. R., Volkers, W. S., Bernini, L. F., van Loghem, E., Meera Khan, P., and Nijenhuis, L. E. (1974) Gene frequencies in a Dutch population. Hum. Hered. 24:435-448.

Fráter-Schröder, M., Hitzig, W. H., and Bütler, R. (1979) Studies on transcobalamin. 1. Detection of transcobalamin II isoproteins in human serum. Blood 53:193-203.

Friedman, R. D., Merritt, A. D. and Rivas, M. L. (1975) Genetic studies of human acidic salivary protein (Pa). Amer. J. Hum. Genet. 27:292-303.

Fröhlander, N. and Ljungberg, B. (1986) Serum protein groups in renal cell carcinoma. Hum. Hered. 36:119-122.

Frossard, P. M., Coleman, R., Funke, H., and Assman, G. (1986) ApaI RFLP 5.4 kb 5' to the human apolipoprotein AI (APO A1) gene. Nucleic Acids Res. 14:1922.

Fujita, M., Satoh, C., Asakawa, J., Nagahata, Y., Tanaka, Y., Hazama, R., and Goriki, K. (1985a) Electrophoretic variants of blood proteins in Japanese. V. Ceruloplasmin. Jpn. J. Hum. Genet. 30:43-50.

Fujita, M., Satoh, C., Asakawa, J., Nagahata, Y., Tanaka, Y., Hazama, R., and Krasteff, T. (1985b) Electrophoretic variants of blood proteins in Japanese. VI. Transferrin. Jpn. J. Hum. Genet. 30:191-200.

Fünfhausen, G. and Gremplewski, K. (1967) Die Verteilung des Blutgruppenantigens BU[a] in Berlin. Z. Ärztl. Fortbild. 61:769. Cited from Mourant et al. (1976).

Furuhata, T., Nakajima, H., Ikemoto, S., and Nagata, H. (1961) Frequencies of Rh and Kidd blood types, and private and public antigens among the Japanese. Proc. Japan Acad. 37:319-323.

Furuhjelm, U., Myllylä, G., Nevanlinna, H. R., Nordling, S., Pirkola, A., Gavin, J., Gooch, A., Sanger, R., and Tippett, P. (1969) The red cell phenotype En(a−) and anti-En[a]: Serological and physicochemical aspects. Vox Sang. 17:256-278.

Furuhjelm, U., Nevanlinna, H. R., Gavin, J., and Sanger, R. (1972) A rare blood group antigen An[a] (Ahonen). J. Med. Genet. 9:385-391.

Furuhjelm, U., Nevanlinna, H. R., Nurkka, R., Gavin, J., Tippett, P., Gooch, A., and Sanger, R. (1968) The blood group antigen Ul[a](Karhula). Vox Sang. 15:118-124.

Gajdusek, D. C., Leyshon, W. C., Kirk, R. L., Blake, N. M., Keats, B., and McDermid, E. M. (1978) Genetic differentiation among populations in Western New Guinea. Amer. J. Phys. Anthrop. 48:47-64.

Gall, J. Y. L., Gall, M. L., Godin, Y., and Serre, J. L. (1982) A study of genetic markers of the blood in four Central African population groups. Hum. Hered. 32:418-427.

Galanello, R., Maccioni, L., Ruggeri, R., Perseu, L., and Cao, A. (1984) Alpha thalassaemia in Sardinian newborns. Brit. J. Haematol. 58:361-368.

Gallango, M. L. and Suinaga, R. (1978) Uridine monophosphate kinase polymorphism in two Venezuelan populations. Amer. J. Hum. Genet. 30:215-218.

Gangadharam, P. R. J., Bhatia, A. L., Radhakrishna, S., and Selkon, J. B. (1961) Rate of inactivation of isoniazid in South Indian patients with pulmonary tuberculosis. Bull. World Health. Orgn. 25:765-777. Cited from Mourant et al. (1975).

Garcia, S. C., Moragón, A. C., and López-Fernández, M. E. (1979) Frequency of glutathione reductase, pyruvate kinase and glucose-6-phosphate dehydrogenase deficiency in a Spanish population. Hum. Hered. 29:310-313.

Garlick, J. P. (1958) Personal communication. Cited from Mourant et al. (1976).

Garlick, J. P. and Barnicot, N. A. (1957) Blood groups and haemoglobin variants in Nigerian (Yoruba) schoolchildren. Amer. J. Hum. Genet. 21:420-425.

Garruto, R. M., Hoff, C., Baker, P. T., and Jacobi, H. J. (1975) Phenotypic variation in ABO and Rh blood groups, PTC tasting ability, and lingual rotation among Southern Peruvian Quechua Indians. Hum. Biol. 47:193-199.

Garry, P. J. (1977) Atypical (E_1^a) and fluoride-resistant (E_1^f) cholinesterase genes: Absent in a native American Indian population. Hum. Hered. 27:433-436.

Garth, T. R. (1933) The incidence of color-blindness among races. Science 77:333-334.

Gartler, S. M., Firschein, I. L., and Gidaspow, T. (1956-7) Some genetical and anthropological considerations of urinary β-aminoisobutyric acid excretion. Acta Genet. Stat. Med. 6:435-46. Cited from Mourant et al. (1976).

Gedde-Dahl, T. and Berg, K. (1965) Linkage in man: The Inv and the Lp serum type systems. Nature 208:1126.

Gedde-Dahl, T., Jr., Natvig, J. B. and Gundersen. S. K. (1971) Inheritance of Gm(g) and a gene complex Gm[a]Gm[g] weak. Clin. Genet. 2:356-366.

Geerdink, R. A., Bartstra, H. A., and Hopkinson, D. A. (1974a) Phosphoglucomutase (PGM_2) variants in Trio Indians from Surinam. Hum. Hered. 24:40-44.

Geerdink, R. A. Bartstra, H. A., and Schillhorn van Veen, J. M. (1974b) Serum proteins and red cell enzymes in Trio and Wajana Indians from Surinam. Amer. J. Hum. Genet. 26:581-587.

Geerdink, R. A., Nijenhuis, L. E., van Loghem, E., and Sjoe, E. L. F. (1974c) Blood groups and immunoglobulin groups in Trio and Wajana Indians from Surinam. Amer. J. Hum. Genet. 26:45-53.

Genz, T., Martin, J. P., and Cleve, H. (1977) Classification of α_1-antitrypsin (Pi) phenotypes by isoelectrofocusing. Hum. Genet. 38:325-332.

Germenis, A., Babionitakis, A., Kaloterakis, A., Filiotou, A., and Fertakis, A. (1983) Group-specific component and haptoglobin phenotypes in multiple myeloma. Hum. Hered. 33:188-191.

359

Germenis, A., Kolitsopoulos, A., Dimopoulos, A. M., and Fertakis, A. (1985) C3 polymorphism in Greece. Hum. Hered. 35:123-125.

Gershowitz, H., Layrisse, M., Layrisse, Z., Neel, J. V., Brewer, C., Chagnon, N., and Ayers, M. (1970) Gene frequencies and microdifferentiation among the Makiritare Indians. I. Eleven blood group systems and the ABH-Le secretor traits: A note on Rh gene frequency determinations. Amer. J. Hum. Genet. 22:515-525.

Gershowitz, H., Layrisse, M., Layrisse, Z., Neel, J. V., Chagnon, N., and Ayres, M. (1972) The genetic structure of a tribal population, the Yanomama Indians. II. Eleven blood-group systems and the ABH-Le secretor traits. Ann. Hum. Genet. 35:261-269.

Gershowitz, H. and Neel, J. V. (1978) The immunoglobulin allotypes (Gm and Km) of twelve Indian tribes of Central and South America. Amer. J. Phys. Anthrop. 49:289-302.

Geserick, G., Dufková, J., and Rose, M. (1969) Die Transferrintypen-Verteilung in einer Stichprobe der prager Bevölkerung. Acta Biol. Med. Germ. 22:637-642. Cited from Tills et al. (1983a).

Geserick, G., Marth, H., Schnitzler, St., Gogochia, S. D., Mirvis, A. B., and Annenkow, H. A. (1971) Pt-Typenverteilung bei einen Bevölkerungsisolat im abchasischen Swanetien (Grusinien). Acta Biol. Med. Ger. 26:411. Cited from Alper and Rosen (1976).

Geserick, G., Schnitzler, S., Marth, H., Gogochia, S. D., Annenkow, H. A., Kotrikadze, N. G., and Miruis, A. B. (1972) Verteilung der Serumgruppenmerkmale Hp, Gc, Tf, Lp, und Xh, in einen Bevölkerungsisolat im abchasischen Swanetien (Grusinien). Kriminal. Forens. Wiss. 8:105-111. Cited in Tills et al. (1983a).

Ghiselli, G., Gregg, R. E., Zech, L. A., Schaefer, E. J., and Brewer, H. B. Jr. (1982) Phenotype study of apolipoprotein E isoforms in hyperlipoproteinaemic patients. Lancet ii 405-407.

Ghosh, A. K. (1977a) Polymorphism of red cell glyoxalase 1 with special reference to South and Southeast Asia and Oceania. Hum. Genet. 39:91-95.

Ghosh, A. K. (1977b) The distribution of genetic variants of glyoxalase 1, esterase D, and carbonic anhydrase I and II in Indian populations. Indian J. Phys. Anthrop. & Hum. Genet. 3:73-83.

Ghosh, U., Banerjee, P. K., and Saha, N. (1980) Mitochondrial malic enzyme polymorphism in an Indian population. Hum. Hered. 30:159-160.

Ghosh, U., Banerjee, T., Banerjee, P. K., and Saha, N. (1981) Distribution of haemoglobin and glucose-6-phosphate dehydrogenase phenotypes among different caste groups of Bengal. Hum. Hered. 31:119-121.

Giblett, E. R. (1969) Genetic Markers in Human Blood, Blackwell, Oxford.

Giblett, E. R., Anderson, J. E., Chen, S. H., Teng, Y. S., and Cohen, F. (1974) Uridine monophosphate kinase: A new genetic polymorphism with possible clinical implications. Amer. J. Hum. Genet. 26:627-635.

Giblett, E. R., Anderson, J. E., Lewis, M., and Kaita, H. (1975) A new polymorphic enzyme, uridine monophosphate kinase: Gene frequencies and a linkage analysis. Cytogenet. & Cell Genet. 14:329-331.

Giblett, E. R. and Brooks, L. E. (1963) Haptoglobin sub-types in three racial groups. Nature 197:576-577.

Giblett, E. R., Chase. J., and Motulsky, A. G. (1957) Studies on anti-V, a recently discovered Rh antibody. J. Lab. Clin. Med. 49:433-439. Cited from Race and Sanger (1975).

Giblett, E. R., Motulsky, A. G., and Fraser, G. R. (1966) Population genetic studies in the Congo. IV. Haptoglobin and transferrin serum groups in the Congo and in other African populations. Amer. J. Hum. Genet. 18:553-558.

Gibney, S. F. A., Munroe, V., Nurse, G. T., and Schofield, E. C. (1981) Lactose absorption in a western Massim population. Ann. Hum. Biol. 8:477-480.

Gibson-Hill, M. H. H. (1953). Personal communication. Cited from Mourant et al. (1976).

Gilat, T., Kuhn, R., Gelman, E., and Mizrahy, O. (1970) Lactase deficiency in Jewish communities in Israel. Amer. J. Dig. Dis. 15:895-904.

Gilberg, A. and Persson, I. (1967) Serum protein types in polar Eskimos. Acta Genet. Stat. Med. 17:422-432.

Giles, C. M., Huth, M. C., Wilson, T. E., Lewis, H. B. M., and Grove, G. E. B. (1965a) Three examples of a new antibody, anti-Cs^a, which reacts with 98% of red cell samples. Vox Sang. 10:405-415.

Giles, C. M., Metaxas-Bühler, M., Romanski, Y., and Metaxas, M. N. (1967) Studies on the Yt blood group system. Vox Sang. 13:171-180.

Giles, E., Hansen, A. T., McCullough, J. M., Metzger, D. G., and Wolpoff, M. H. (1968) Hydrogen cyanide and phenylthiocarbamide sensitivity, mid-phalangeal hair and color blindness in Yucatán, Mexico. Amer. J. Phys. Anthrop. 28:203-212.

Giles, E., Ogan, E. and Steinberg, A. G. (1965b) Gamma-globulin factors (Gm and Inv) in New Guinea: Anthropological significance. Science 150:1158-1160.

Gill, P. and Sutton, J. G. (1984) α-L-Fucosidase polymorphism in human semen, blood, and vaginal fluid. Hum. Hered. 34:231-239.

Gitschier, J., Drayna, D., Tuddenham, E. G. D., White, R. L., and Lawn, R. M. (1985) Genetic mapping and diagnosis of haemophilia A achieved through a BclI polymorphism in the factor VIII gene. Nature. 314:738-740.

Glahs, G. (1974) Transferrinvarianten in der Wiener Bevölkerung. Med. Lab. 27:122-123. Cited from Tills et al. (1983).

Glasgow, B. G., Goodwin, M. J., Jackson, F., Kopec, A. C., Lehmann, H., Mourant, A. E., Tills, D., Turner, R. W. D., and Ward, M. P. (1968) The blood groups, serum groups and haemoglobins of the inhabitants of Lunana and Thimbu, Bhutan. Vox Sang. 14:31-42.

Godber, M., Kopec, A. C., Mourant, A. E., Teesdale, P., Tills, D., Weiner, J. S., El-Niel, H., Wood, C. H., and Barley, S. (1976) The blood groups, serum groups, red-cell isoenzymes and haemoglobins of the Sandawe and Nyaturu of Tanzania. Ann. Hum. Biol. 3:463-473.

Godber, M., Kopec, A. C., Mourant, A. E., Tills, D., and Lehmann, E. E. (1973) The hereditary blood factors of the Yemenite and Kurdish Jews. Phil. Trans. R. Soc. Lond. 266:169-184.

Goedde, H. W., Agarwall, D. P., Harada, S., Rothhammer, F., Whittaker, J. O., and Lisker, R. (1986a) Aldehyde dehydrogenase polymorphism in North American, South American, and Mexican Indian populations. Amer. J. Hum. Genet. 38:395-399.

Goedde, H. W. and Benkmann, H.-G. (1987) Personal communication.

Goedde, H. W., Benkmann, H.-G., Agarwal, D. P., Bienzle, U., Guggenmoos, R., Rosenkaimer, F., Hoppe, H. H., and Brinkmann, B. (1979a) Genetic studies in Cameroon: red cell enzyme and serum protein polymorphisms. Z. Morph. Anthrop. 70:33-40.

Goedde, H. W., Benkmann, H.-G., Agarwal, D. P., Hirth, L., Bienzle, U., Dietrich, M., Hoppe, H. H., Orlowski, J., Kohne, E., and Kleihauer, E. (1979b) Genetic studies in Saudi Arabia: Red cell enzyme, haemoglobin and serum protein polymorphisms. Amer. J. Phys. Anthrop. 50:271-278.

Goedde, H. W., Benkmann, H.-G., Agarwal, D. P., and Kroeger, A. (1977a) Genetic studies in Ecuador: Acetylator phenotypes, red cell enzyme and serum protein polymorphisms of Shuara Indians. Amer. J. Phys. Anthrop. 47:419-426.

Goedde, H. W., Benkmann, H.-G., Bienzle, U., and Bienzle, H. (1985a) Genetic markers in Liberia: Studies of Glo, AcP, EsD, 6-PGD, Ak, Sub PGM and α_1 at polymorphisms. Z. Morph. Anthrop. 75:349-354.

Goedde, H. W., Benkmann, H.-G., Flatz, G., Rahimi, A. G., Kaifie, S., and Delbrück, H. (1977b) Red cell enzyme polymorphisms in different populations of Afghanistan. Ann. Hum. Biol. 4:225-232.

Goedde, H. W., Benkmann, H.-G., Hirth, L., Rohde, R., Rougemont, A., and Delbrück, H. (1975) Phenotypes of Gc and Tf in leprosy patients of Mali and Ethiopia. Hum. Hered. 24:383-386.

Goedde, H. W., Benkmann, H.-G., Hussein, L., and el-Naggar, B. (1980) Genetic studies in Egypt: red cell enzyme and plasma protein polymorphisms. Z. Morph. Anthrop. 71:329-335.

Goedde, H. W., Benkmann, H.-G., Kriese, L., Bogdanski, P., Agarwal, D. P., Du, R., Chen, L., Cui, M., Yuan, Y., Xu, J., Li, S., and Wang, Y. (1984a) Aldehyde dehydrogenase isozyme deficiency and alcohol sensitivity in four different Chinese populations. Hum. Hered. 34:183-186.

Goedde, H. W., Benkmann, H.-G., Kriese, L., Bogdanski, P., Du, R., Chen, L., Cui, M., Yuan, Y., Xu, J., Li, S., and Wang, Y. (1984b) Population genetic studies in three Chinese minorities. Amer. J. Phys. Anthrop. 64:277-284.

Goedde, H. W., Benkmann, H.-G., Singh, S., Das, B. M., Chakravartti, M. R., Delbrück, H., and Flatz, G. (1972a) Genetic survey in the population of Assam. II. Serum protein and erythrocyte enzyme polymorphisms. Hum. Hered. 22:331-337.

Goedde, H. W., Czeizel, A., and Benkmann, H.-G. (1986b) Hungarian data. Personal communication.

Goedde, H. W., Flatz, G., Rahimi, A. G., Kaifie, S., Benkmann, H.-G., Kriese, G., and Delbrück, H. (1977c) The acetylator polymorphism in four populations of Afghanistan. Hum. Hered. 27:383-388.

Goedde, H. W., Hirth, L., Benkmann, H.-G., Pellicer, A., Pellicer, T., Stahn, M., and Singh, S. (1972b) Population genetic studies of red cell enzyme polymorphisms in four Spanish populations. Hum. Hered. 22:552-560.

Goedde, H. W., Hirth, L., Benkmann, H.-G., Pellicer, A., Pellicer, T., Stahn, M., and Singh, S. (1973) Population genetic studies of serum protein polymorphisms in four Spanish populations. Hum. Hered. 23:135-146.

Goedde, H. W. and Ohligmacher, H. (1965) Zur Problematik des Polymorphismus des Bitterschmeckens: Vergleichende Untersuchungen an Thioharnstoffderivaten und Anetholtrithion. Humangenetik 1:423-436.

Goedde, H. W., Paik, Y. K., Lee, C. C., and Benkmann, H.-G. (1986c) Korean data. Pers. comm.

Goedde, H. W., Rothhammer, F., Benkmann, H.-G., and Bogdanski, P. (1984c) Ecogenetic studies in Atacameño Indians. Hum. Genet. 67:343-346.

Goedde, H. W., Rothhammer, F., Benkmann, H.-G., and Bogdanski, P. (1985b) Genetic studies in Atacameño Indians: Serum protein and red-cell enzyme polymorphisms. Ann. Hum. Biol. 12:251-259.

Golan, R., Ben-Ezzer, J., and Szeinberg, A. (1977) Esterase D polymorphism in several population groups in Israel. Hum. Hered. 27:298-304.

Golan, R., Ben-Ezzer, J., and Szeinberg, A. (1979) Erythrocyte glyoxalase I polymorphism in several population groups in Israel. Hum. Hered. 29:57-60.

Golan, R., Ben-Ezzer, J., and Szeinberg, A. (1981) Phosphoglycolate phosphatase in several population groups in Israel. Hum. Hered. 31:89-92.

Goldschmidt, E. (1967) Summary and conclusions. 9th Int. Congr. Life Ass. Med., Tel-Aviv, pp. 200-206. Cited from Mourant et al. (1976).

Goldschmidt, E., Bayani-Sioson, P., Sutton, H. E., Fried, K., Sandor, A., and Block, N. (1962) Haptoglobin frequencies in Jewish communities. Ann. Hum. Genet. 26:39-45.

Gonzenbach, R., Hässig, A., and Rosin, S. (1955) Über post-transfusionelle Bildung von Anti-Lutheran-Antikörpern; Die Häufigkeit des Lutheran-Antigens Lu^a in der Bevölkerung Nord-, West- und Mitteleuropas. Blut. 1:272-274. Cited from Mourant et al. (1976).

Goodman, P. A., Yu, P-L., Azen, E. A., and Karn, R. C. (1985) The human salivary protein complex (SPC): A large block of related genes. Amer. J. Hum. Genet. 37:787-797.

Goti Iturriaga, J. L. (1966) Grupo ABO, factor Rh y sistema secretor-Lewis en vascos. Rev. Clin. Esp. 27: 30-40. Cited from Mourant et al. (1976).

Goti Iturriaga, J. L. and Velasco Alonso, R. (1965) Grupos sanguineos y úlcera péptica. Sustancias antigénicas ABH y Le^a en la úlcera péptica. Rev. Clin. Esp. 98:119-129. Cited from Mourant et al. (1976).

Govaerts, A., Massart, Th., Rivat, L., and Brocteur, J. (1972) Serological characteristics of a Bantu population. Histocompatibility Testing, Munksgaard, Copenhagen, pp. 409-413.

361

Greiner, J., Weber, F. J., Mauff, G., and Baur, M. (1980) Genetic polymorphisms of properdin factor B (Bf), the second component (C2), and the fourth component (C4) of complement in leprosy patients and healthy controls from Thailand. Immunobiol. 158:134-138.

Greuter, W., Hess, M., Renaud, N., Schmitter, M., and Bütler, R. (1963) Beitrag zur Genetik des Gm- und Gc-Serum-gruppensystems anhand von Untersuchungen an Schweizer Familien. Arch. Klaus-Stift. VererbForsch. 38:77-92. Cited from Mourant et al. (1976).

Griffiths, S. B. (1954) The distribution of the sickle-cell trait in Africa. S. Afr. J. Med. Sci. 19:56-57.

Grimmo, A. E. P. and Lee, S.-K. (1964a) A survey of blood groups in Hong Kong Chinese of Cantonese origin. Oceania 31:222-226. Cited from Mourant et al. (1976)

Grimmo, A. E. P. and Lee, S.-K. (1964b) Further blood groups of Hong Kong Chinese of Cantonese origin. Oceania 34:234-236. Cited from Mourant et al. (1976)

Grünwald, P. and Herman, C. (1963) Study of several gene frequencies in Yugoslav population. Nature 199:830-831.

Guasch, J. (1950) El factor Rh en España. Rev. Esp. Pediat. 6:387-390. Cited from Mourant et al. (1976).

Gudmand-Høyer, E., Dahlqvist, A., and Jarnum, S. (1969) Specific small-intestinal lactase deficiency in adults. Scand. J. Gastroent. 4:377-386.

Gudmand-Høyer, E. and Jarnum, S. (1969) Lactose malabsorption in Greenland Eskimos. Acta Med. Scand. 186:235-237.

Günther, A. (1982) Gene frequencies of human red cell phosphoglucomutase (PGM$_3$) in Western Germany (Düsseldorf region). Hum. Hered. 32:142-144.

Gürtler, H. (1970) Cited from Berg (1974).

Gürtler, H. (1971) Personal communication. Cited from Mourant et al. (1976).

Gutsche, B. B., Scott, E. M., and Wright, R. C. (1967) Hereditary deficiency of pseudocholinesterase in Eskimos. Nature 215:322-323.

Guttman, R., Guttman, L., and Rosenzweig, K. A. (1967) Cross-ethnic variation in dental, sensory and perceptual traits: A nonmetric multibivariate derivation of distances for ethnic groups and traits. Amer. J. Phys. Anthrop. 27:259-276.

Haas, E. J. C., Salzano, F. M., Araujo, H. A., Grossman, F., Barbetti, A., Weimer, T. A., Franco, M. H. L. P., Verruno, L., Nasif, O., Morales, V. H., and Arienti, R. (1985) HLA antigens and other genetic markers in the Mapuche Indians of Argentina. Hum. Hered. 35:306-313.

Habib, Z. (1983) Haptoglobin polymorphism in Egyptians. Ann. Hum. Biol. 10:385-388.

Habte, D., Sterky, G., and Hjalmarsson, B. (1973) Lactose malabsorption in Ethiopian children. Acta. Paediat. Scand. 62:649-654. Cited from Mourant et al. (1976).

Hackel, E., Hopkinson, D. A., and Harris, H. (1972) Population studies on mitochondrial glutamate oxaloacetate transaminase. Ann. Hum. Genet. 35:491-496.

Hainline, J., Clark, P., and Walsh, R. J. (1969) ABO, Rh and MNS blood typing results and other biochemical traits in the people of the Yap Islands. Archaeol. Phys. Anthrop. Oceania 4:64-71. Cited from Mourant et al. (1976).

Hamaguchi, H., Yamada, M., Noguchi, A., Fujii, K., Shibasaki, M., Mukai, R., Yabe, T., and Kondo, I. (1982a) Genetic analysis of human lymphocyte proteins by two-dimensional gel electrophoresis. 2. Genetic polymorphism of lymphocyte cytosol 64k polypeptide. Hum. Genet. 60:176-180.

Hamaguchi, H., Yamada, M., Shibasaki, M., and Kondo, I. (1982b) Genetic analysis of human lymphocyte proteins by two dimensional gel electrophoresis. 4. Genetic polymorphism of cytosol 100k polypeptide. Hum. Genet. 62:148-151.

Hamaguchi, H., Yamada, M., Shibasaki, M., Mukai, R., Yabe, T., and Kondo, I. (1982c) Genetic analysis of human lymphocyte proteins by two dimensional gel electrophoresis. 3. Frequent occurrence of genetic variants in some abundant polypeptides of PHA-stimulated peripheral blood lymphocytes. Hum. Genet. 62:142-147.

Hamamy, H. A. and Saeed, T. K. (1981) Glucose-6-phosphate dehydrogenase deficiency in Iraq. Hum. Genet. 58:434-435.

Hanngren, A., Borgå, O., and Sjöqvist, F. (1970) Inactivation of isoniazid (INH) in Swedish tuberculous patients before and during treatment with para-aminosalicylic acid (PAS). Scand. J. Resp. Dis. 51:61-69. Cited from Mourant et al. (1976).

Harada, S., Agarwal, D. P., and Goedde, H. W. (1978) Human liver alcohol dehydrogenase isoenzyme variations: Improved separation methods using prolonged high voltage starch-gel electrophoresis and isoelectric focusing. Hum. Genet. 40:215-220.

Harada, S., Agarwal, D. P., and Goedde, H. W. (1986) Genetic study of glutathione S-transferase in Japanese and its pharmacogenetic importance. 7th Int. Cong. Hum. Genet. Abstracts, Pt.II, p.431.

Harada, S., Itoh, M., and Misawa, S. (1975) Red cell uridine monophosphate kinase polymorphism in Japanese. Humangenetik 29:255-257.

Harada, S. and Misawa, S. (1976) Red cell glyoxalase I (EC 4.4.1.5) polymorphism in Japanese. Kyorin Med. J. 7:21-24.

Harada, S., Misawa, S., Agarwal, D. P., and Goedde, H. W. (1980) Liver alcohol dehydrogenase and aldehyde dehydrogenase in the Japanese: Isozyme variation and its possible role in alcohol intoxication. Amer. J. Hum. Genet. 32:8-15.

Harada, S. and Omoto, K. (1969) Electrophoretic variants of serum alpha-1-antitrypsin in Japan. Jpn. J. Hum. Genet. 14:248-249.

Harris, H. (1953-4) Family studies on the urinary excretion of β-aminoisobutyric acid. Ann. Eugen. 18:43-49. Cited from Mourant et al. (1976).

Harris, H. and Hopkinson, D. A. (1978) Handbook of Enzyme Electrophoresis in Human Genetics.

Supplement. North-Holland, Amsterdam.

Harris, H., Hopkinson, D. A. and Robson, E. R. (1974) The incidence of rare alleles determining electrophoretic variants: data on 43 enzyme loci in man. Ann. Hum. Genet. 37:237-253.

Harris, J. P., Tegoli, J., Swanson, J., Fisher, N., Gavin, J., and Noades, J. (1967) A nebulous antibody responsible for cross-matching difficulties (Chido). Vox Sang. 12:140-142.

Harrison, G. A., Küchemann, C. F., Moore, M. A. S., Boyce, A. J., Baju, T., Mourant, A. E., Godber, M. J., Glasgow, B. G., Kopec, A. C., Tills, D., and Clegg, E. J. (1969) The effects of altitudinal variation in Ethiopian populations. Phil. Trans. B.256:147-182. Cited from Mourant et al. (1976).

Hart, M. van der, Meike, M., Veer, M. V. D., and van Loghem, J. J. (1961) Ho and Lan - two new blood group antigens. Paper read at VIIIth Europ. Cong. Haemat. Cited from Mourant et al. (1976).

Hartmann, O., Heier, A. M., Kornstad, L., Weisert, O., and Örjasaeter, H. (1965) The frequency of the Lutheran blood group antigens, as defined by anti-Lu^a, in the Oslo population. Vox Sang. 10:234-238.

Harvey, R. G., and Giblett, E. R. (1968). Personal communication. Cited from Mourant et al. (1976).

Harvey, R. G., Godber, M. J., Kopec, A. C., Mourant, A. E., and Tills, D. (1969) Frequency of genetic traits in the Caribs of Dominica. Hum. Biol. 41:342-364.

Harvey, R. G., Tills, D., Mourant, A. E., Giblett, E. R., Cleve, H., Bearn, A. G., and McConnell, R. B. (1978) Blood groups, serum proteins and enzymes of the Ainu of Hokkaido. Hum. Biol. 50:425-450.

Hashem, N. and Khalifa, S. (1969) β-aminoisobutyric acid excretion patterns among Egyptians. Hum. Hered. 19:662-667.

Hashem, N., Khalifa, S., and, Nour, A. (1969) The frequency of isoniazid acetylase enzyme deficiency among Egyptians. Amer. J. Phys. Anthrop. 31:97-101.

Hässig, A. (1952) La répartition des génotypes rhésus en Suisse. Arch. Suisses Anthrop. Gén. 4:66-67. Cited from Mourant et al. (1976).

Hässig, A., Meyer, W., and Thommen, D. (1955) Zur klinischen Bedeutung des Lewis-Blutgruppensystems. Schweiz. med. Wschr. 85:786-787. Cited from Mourant et al. (1976).

Hauptmann, G., Wertheimer, E., Tongio, M. M., and Mayer, S. (1977) Bf polymorphism: Another variant (S0.8). Hum. Genet. 36:109-111.

Haverkorn, M. J. and Goslings, W. R. O. (1969) Streptococci, ABO blood groups and secretor status. Amer. J. Hum. Genet. 21:360-375.

Hawkins, B., Elliot, M., Kosasih, E. N., and Simons, M. J. (1973) Red cell genetic studies of the Toba Bataks of North Sumatra. Hum. Biol. Oceania 2:147-154.

Hawkins, B. R. and Simons, M. J. (1976) Blood group genetic studies in an urban Chinese population. Hum. Hered. 26:441-453.

Hayward, G. A. (1975) Human acetylation polymorphism in Polynesians. Proc. Univ. Otago Med. School 53:67-68. Cited from Evans (1986).

Heiken, A. (1962) Distribution of the Kell blood group factor K in the Swedish population. Acta Genet. Stat. Med. 12:352-358.

Heiken, A. (1965) A genetic study of the MNSs blood group system. Hereditas 53:187-211.

Heiken, A. (1967) Personal communication. Cited from Mourant et al. (1976).

Heiken, A., Balogun, R. A., Swan, T., and Rasmuson, M. (1974) Population genetic studies in Nigeria. Hereditas 76:117-135.

Heistö, H., van der Hart, M., Madsen, G., Moes, M., Noades, J., Pickles, M. M., Race, R. R., Sanger, R., and Swanson, J. (1967) Three examples of a new red cell antibody, anti-Co^a. Vox Sang. 12:18-24.

Helgeson, M., Swanson, J., and Polesky, H. F. (1970) Knops-Helgeson (Kn^a), a high-frequency erythrocyte antigen. Transfusion 10:137-138.

Henke, W. and Palsson, J. (1977) Homo 28:129. Cited from Walter (1981).

Henke, J., Schweitzer, H., Bär, W., Weidinger, S., Weissmann, J., and Baur, M. P. (1986) Extended polymorphism of the human esterase D isozyme system: description of a 'new' allele EsD*11. Hum. Genet. 73:89-90.

Hennig, W. and Hoppe, H. H. (1964) Häufigkeitsverteilung und Mutter/Kind-Kombinationen bei den Hp-, Gm- und Gc- Serumgruppen am Hamburger Material und ihre Brauchbarkeit im Blutgruppengutachten. Blut 10:361-367. Cited from Mourant et al. (1976).

Herz, B. and Bach, G. (1984) Arylsulfatase A in pseudodeficiency. Hum. Genet. 66:147-150.

Herzog, P., und Bohatová, U. (1969) Zur Populationsgenetik der sauren Phosphatase der erythrocyten (EC 3. 1. 3. 1): Phänotypen- und Allelhäufigkeiten in der CSSR. Humangenetik 7:183-184.

Herzog, P. and Bohatová, J. (1973) Phenotype and gene frequencies of adenosine deaminase in Prague. Humangenetik 17:173-174.

Herzog, P., Bohatová, J., and Drdová, A. (1970) Genetic polymorphisms in Kenya. Amer. J. Hum. Genet. 22:287-291.

Herzog, P., Bohatová, J., and Drdová, A. (1972) Contribution to the AK and 6-PGD polymorphism in Prague. Humangenetik 14:326.

Herzog, P. and Drdová, A. (1971) Contribution to the PGM$_3$ polymorphism phenotype frequencies in Prague. Humangenetik 13:64-65.

Herzog, P., Drdová, A., and Bohatová, J. (1976) Serum protein polymorphisms in North Vietnam. Hum. Hered. 26:203-206.

Herzog, P., Jezek, Z., Cerenshimid, O., Ochirvan, S., and Horejsi, J. (1978) Serum protein polymorphism in the Mongolian People's Republic. Acta Anthrop. 2:12-26.

Hevér, Ö. and Hajpál, A. (1978) The incidence of Hp^2 allele variants in Hungary. Hum. Hered. 28:100-103.

Hess, M. and Bütler, R. (1962) Untersuchungen über die Gc-Gruppen von Hirschfeld. Schweiz. Med. Wschr. 92:1351. Cited from Mourant et al. (1976).

Hewett-Emmett, D., Hanis, C. L., Bertin, T. K., Chakraborty, R., and Schull, W. J. (1986a) Mexican-

363

Americans in Starr county, Texas: Genetic markers, subtypes and disease. Amer. J. Hum. Genet.39:A237.

Hewett-Emmett, D., Hanis, C.L., Bertin, T. K., and Schull, W. J. (1986b) Personal communication.

Hewett-Emmett, D., Welty, R. J., and Tashian, R. E. (1983) A widespread silent polymorphism of human carbonic anhydrase III (31 Ile <-> Val): Implications for evolutionary genetics. Genetics 104:409-420.

Hiernaux, J. (1976) Blood polymorphism frequencies in the Sara Majingay of Chad. Ann. Hum. Biol. 2:127-140.

Higgs, D. R., Wainscoat, J. S., Flint, J., Hill, A. V. S., Thein, S. L., Nicholls, R. D., Teal, H., Ayyub, H., Peto, T. E. A., Falusi, A. G., Jarman, A. P., Clegg, J. B., and Weatherall, D. J. (1986) Analysis of the human α-globin gene cluster reveals a highly informative genetic locus. Proc. Natl. Acad. Sci. USA 83:5165-5169.

Hill, A. V. S., Bowden, D. K., Trent, R. J., Higgs, D. R., Oppenheimer, S. J., Thein, S. L., Mickleson, K. N. P., Weatherall, D. J., and Clegg, J. B. (1985) Melanesians and Polynesians share a unique α-thalassemia mutation. Amer. J. Hum. Genet. 37:571-580.

Hill, A. V. S., Bowden, D. K., and Weatherall, D. J., and Clegg, J. B. (1986) Chromosomes with one, two, three, and four fetal globin genes: Molecular and hematologic analysis. Blood. 67:1611-1618.

Hirschfeld, J. (1962) The Gc system: Immunoelectrophoretic studies of normal human sera with special reference to a new genetically determined serum system (Gc). Prog. Allergy 6:155-186.

Hirschfeld, J. (1968) Application of the Ag(x) antigen in medico-legal investigations. Vox Sang. 14:95-105.

Hirschfeld, J., Contu, L., Rittner, Ch., and Geserick, G. (1968) Inheritance of the Ag(x) and Ag(y) antigens. Vox Sang. 14:124-129.

Hirschfeld, J. and Okochi, K. (1967) Distribution of Ag(x) and Ag(y) antigens in some populations. Vox Sang. 13:1-3.

Hitzeroth, H. W. and Bender, K. (1980) Erythrocyte G-6-PD and 6-PGD genetic polymorphisms in South African Negroes with a note on G-6-PD and the malaria hypothesis. Hum. Genet. 54:233-242.

Hitzeroth, H. W., Bütler, R., and Bütler-Brunner, E. (1980) A population genetic study on the Ag polymorphism in South African Indians and Negroids. Hum. Hered. 30:94-103.

Hitzeroth, H. W., Skoda, U., Toit, E. du., and Mauff, G. (1986) The plasminogen polymorphism in South African populations: genetics and anthropogenetics. Hum. Genet. 74:341-345.

Ho, M. W., Povey, S., and Swallow, D. (1982) Lactase polymorphism in adult British natives: Estimating allele frequencies by enzyme assays in autopsy samples. Amer. J. Hum. Genet. 34:650-657.

Hobart, M. J. (1979) Genetic polymorphism of human plasminogen. Ann. Hum. Genet. 42:419-423.

Hobart, M. J., Vaz-Guedes, M. A., and Lachmann, P. J. (1981) Polymorphism of human C5. Ann. Hum. Genet. 45:1-4.

Hobart, M. J., Joysey, V., and Lachmann, P. J. (1978) Inherited structural variation and linkage relationships of C7. J. Immunogenet. 5:157-163.

Hobart, M. J., Lachmann, P. J., and Alper, C. A. (1974) Polymorphism of human C6. In: Protides of Biological Fluids, 22nd Colloquium (Peeters, H., ed.), Pergamon Press, New York, pp. 575-580.

Hodgkin, M. M., Eidus, L., and Bailey, W. C. (1979) Isoniazid phenotyping of black as well as white patients. Can. J. Physiol. Pharmacol. 57:760-763.

Holländer, L. (1951) Über die Blutgruppe Duffy und ihre Verteilung in Basel. Acta Haemat. 6:257-261. Cited from Mourant et al. 1976.

Hopkinson, D. A., Cook, P. J. L., and Harris, H. (1969) Further data on the adenosine deaminase (ADA) polymorphism and a report of a new phenotype. Ann. Hum. Genet. 32:361-367.

Hopkinson, D. A., Corney, G., Cook, P. J. L., Robson, E. B., and Harris, H. (1970) Genetically determined electrophoretic variants of human red cell NADH diaphorase. Ann. Hum. Genet. 34:1-10.

Hopkinson, D. A., Coppock, J. S., Mühlemann, M. F., and Edwards, Y. H. (1974a) The detection and differentiation of the products of the human carbonic anhydrase loci, CA_I and CA_{II}, using fluorogenic substrates. Ann. Hum. Genet. 38:155-162.

Hopkinson, D. A. and Harris, H. (1966) Rare phosphoglucomutase phenotypes. Ann. Hum. Genet. 30:167-181.

Hopkinson, D. A. and Harris, H. (1968) A third phosphoglucomutase in man. Ann. Hum. Genet. 31:359-367.

Hopkinson, D. A. and Harris, H. (1969) Red cell acid phosphatase, phosphoglucomutase, and adenylate kinase. In: Biochemical Methods in Red Cell Genetics (Yunis, J. J., ed.), Academic Press, New York, pp. 337-375.

Hopkinson, D. A., Peters, J., and Harris, H. (1974b) Rare electrophoretic variants of glycerol-3-phosphate dehydrogenase: evidence for two structural gene loci (GPD_1 and GPD_2). Ann. Hum. Genet. 37:477-484.

Hopkinson, D. A., Santisteban, I., Povey, S., and Smith, M. (1985) Biochemical genetic analysis of human and rodent aldehyde dehydrogenase (ALDH). Alcohol 2:73-78.

Hoppe, H. H. (1957) Die Häufigkeitsverteilung der Blutgruppensysteme ABO, MN, Rh, P und K in Hamburg. Blut 3:1-14. Cited from Mourant et al. (1976).

Horai, S. (1976) Genetic polymorphism of human serum factor B (Bf) in Japanese. Jpn. J. Hum. Genet. 21:177-186.

Horsfall, W. R., Lehmann, H., and Davis, D. (1963) Incidence of pseudocholinesterase variants in Australian aborigines. Nature 199:1115.

Hoste, B. (1979) Group-specific component (Gc) and transferrin (Tf) subtypes ascertained by isoelectric focusing. Hum. Genet. 50:75-79.

Hoste, B., Suys, J., and Mathy, M. R. (1984) Glyoxalase I polymorphism in Belgium. Hum. Hered. 34:192-193.

Huisman, T. H. J., Reese, A. L., Gardiner, M. B., Wilson, J. B., Lam, H., Reynolds, A., Nagle, S., Trowell, P., Yi-tao, Z.,Shu-zheng, H., Sukumaran, P. K., Miwa, S., Efremov, G. D., Petkov, G.,

364

Sciarratta, G. V. and Sansone, G. (1983) The occurrence of different levels of G_γ chain and of the $A_\gamma T$ variant of fetal hemoglobin in newborn babies from several countries. Amer. J. Hematol. 14:133-148.

Hummel, K., Pulverer, G., Schaal, K. P., and Weidtman, V. (1970) Häufigkeit der Sichttypen in den Erbsystemen Haptoglobin, Gc, saure Erythrocytenphosphatase, Phosphoglucomutase und Adenylatkinase sowie den Erbeigenschaften Gm(1), Gm(2), und Inv(1) bei Deutschen (aus dem Raum Freiburg i. Br. und Köln) und bei Türken. Humangenetik 8:330-333.

Hussein, L., Flatz, S. D., Kühnau, W., and Flatz, G. (1982) Distribution of human adult lactase phenotypes in Egypt. Hum. Hered. 32:94-99.

Hutz, M. H., Michelson, A. M., Antonarakis, S. E., Orkin, S. H., and Kazazian, H. H. Jr. (1984) Restriction site polymorphism in the phosphoglycerate kinase gene on the X chromosome. Hum. Genet. 66: 217-219.

Idle, J. R. and Smith, R. L. (1980) Inter-ethnic differences in the metabolic disposition of drugs and toxic substances. In: Toxicology in the Tropics (Smith, R. L. and Bababunmi, E. A., eds.), Taylor and Francis, London. Cited from Woolhouse (1986).

Ikemoto, S. (1983) New genetic markers,"Parotid types". Scientific American (Japanese version). December issue, pp.28-39.

Ikemoto, S., Minaguchi, K., Suzuki, K., and Tomita, K. (1977) New genetic marker in human parotid saliva (Pm). Science 197:378-379.

Ikin, E. W. (1963) The incidence of the blood antigens in different populations. Thesis-London. Cited from Mourant et al. (1976).

Ikin, E. W., Kopec, A. C., Mourant, A. E., Parkin, D. M., and Walby, J. A. E. (1952) Cited from Mourant et al. (1976).

Ikin, E. W., Lehmann, H., Mourant, A. E., and Thein, H. (1969) The blood groups and haemoglobins of the Burmese. Man, n.s. 4:118-122.

Ikin, E. W. and Mourant, A. E. (1952) The frequency of the Kidd blood-group antigen in Africans. Man 52:51. Cited from Mourant et al. (1976).

Ikin, E. W. and Mourant, A. E. (1962) A survey of some genetical characters in Ethiopian tribes. V. The blood groups of the Tigre, Billen, Amhara and other Ethiopian populations. Amer. J. Phys. Anthrop. 20:183-189.

Ikin, E. W., Prior, A. M., Race, R. R., and Taylor, G. L. (1939) The distribution of the A_1A_2BO blood groups in England. Ann. Eugen. 9:409-411.

Ikuramov, K. M., Dubrova, Y. E., Altukhov, Y. P., and Podogas, A. V. (1987) Population-genetic study of human differential fertility (examplified by habitual miscarriage). II. Distribution of genotypes and levels of heterozygosity according to a set of biochemical gene markers. Soviet Genetics 22:1073 -1078

Inaba, T., Jurima, M., Nakano, M., and Kalow, W. (1984) Mephenytoin and sparteine pharmacogenetics in Canadian Caucasians. Clin. Pharmacol. Ther. 36:670-676.

Iseki, S., Masaki, S., and Shibasaki, K. (1957) Studies on Lewis blood group system. ii. Distribution and heredity of Le^C blood group factor. Proc. Jpn. Acad. 33:686-691.

Ishii, C., Misawa, S., and Omoto, K. (1984) Polymorphism of γ-aminobutyric acid transaminase (GABAT) in Japanese. J. Anthrop. Soc. Nippon. 92:33-36.

Ishimoto, G. (1968) Urinary excretion of β-aminoisobutyric acid in Japanese-American hybrids. J. Anthrop. Soc. Nippon. 76:141-144. Cited from Mourant et al. (1976).

Ishimoto, G. (1969) Placental phosphoglucomutase in Japan. Jpn. J. Hum. Genet. 14:183-188.

Ishimoto, G. (1970) Further studies on the distribution of erythrocyte enzyme types in Japanese. Jpn. J. Hum. Genet. 15:26-34.

Ishimoto, G. (1975) Red cell enzymes. In: Anthropological and genetic studies on the Japanese (Watanabe, S., Kondo, S., and Matsunaga, E., eds.), Univ. Tokyo Press, Tokyo, pp.109-139.

Ishimoto, G. and Kuwata, M. (1974a) Electrophoretic variants of red cell phosphohexose isomerase in Japan. Jpn. J. Hum. Genet. 18:356-363.

Ishimoto, G. and Kuwata, M. (1974b) Studies on human glutamic-oxaloacetic transaminase variation, including enzyme patterns in some non-human primates. Jpn. J. Hum. Genet. 18:364-372.

Ishimoto, G. and Kuwata, M. (1974c) Red cell glutamic-pyruvic transaminase polymorphism in Japanese populations. Jpn. J. Hum. Genet. 18:373-377.

Ishimoto, G., Uemura, K., Toyomasu, T., and Watanabe, K. (1967) Population studies of serum protein variations in the Japanese. Cited from Omoto (1975).

Ishizaki, K., Noda, A., Ikenaga, M., Ida, K., Omoto, K., Nakamura, Y., and Matsubara, K. (1985) Restriction fragment length polymorphism detected by human salivary amylase cDNA. Hum. Genet. 71:261-262.

Isokoski, M., Sahi, T., Villako, K., and Tamm, A. (1981) Epidemiology and genetics of lactose malabsorption. Ann. Clin. Res. 13:164-168.

Javid, J. (1967) Haptoglobin 2-1 Bellevue, a haptoglobin beta-chain mutant. Proc. Natl. Acad. Sci. USA 57:920-924.

Jamil, T., Fisher, R. A., and Harris, H. (1975) Studies on the properties and tissue distribution of the isozymes of guanylate kinase in man. Hum. Hered. 25:402-413.

Jeannet, M., Schapira, M., Gervasoni, C., Metaxas-Bühler, M., Bütler, R., and van Loghem, E. (1972) Study of the HL-A system and other polymorphisms in the Tibetan population. In: Histocompatability Testing 1972, Munksgaard, Copenhagen, pp. 241-250.

Jenkins, T. (1965) Ability to taste phenylthiocarbamide among Kalahari Bushmen and Southern Bantu. Hum.

365

Biol. 37:371-374.

Jenkins, T. (1972) Genetic polymorphisms of man in southern Africa. M.D. Thesis, University of London. Cited from Nurse et al. (1985).

Jenkins, T., Beighton, P., and Steinberg, A. G. (1985) Serogenetic studies on the inhabitants of Tristan da Cunha. Ann. Hum. Biol. 12:363-371.

Jenkins, T., Dunn, D. S., Gibney, S. F. A., and Nurse, G. T. (1983) Serogenetic studies on the Daga of the interior of the mainland of Milne Bay Province, Papua New Guinea. Ann. Hum. Biol. 10:357-364.

Jenkins, T., Lane, A. B., Nurse, G. T., and Tanaka, J. (1975) Sero-genetic studies on the G/wi and G//ana San of Botswana. Hum. Hered. 25:318-328.

Jenkins, T., Lehmann, H., and Nurse, G. T. (1974) Public health and genetic constitution of the San ("Bushmen"): Carbohydrate metabolism and acetylator status of the !Kung of Tsumkwe in the north-western Kalahari. Br. Med. J. 1:23-26.

Jenkins, T., Gibney, S. F. A., Nurse, G. T., and Penketh, R. J. A. (1981) Persistent high intestinal lactase activity in Papua-New Guinea: Lactose absorption curves in two populations. Ann. Hum. Biol. 8:447-451.

Jenkins, T. and Nurse, G. T. (1976) Biomedical studies on the desert dwelling hunter-gatherers of Southern Africa. Prog. Med. Genet. n.s. 1:211-281.

Jenkins, T., Zoutendyk, A., and Steinberg, A. G. (1970) Gammaglobulin groups (Gm and Inv) of various Southern African populations. Amer. J. Phys. Anthrop. 32:197-218.

Jeremiah, S. and Povey, S. (1981) The biochemical genetics of human γ-aminobutyric acid transaminase. Ann. Hum. Genet. 45:231-236.

Johnson, A. M., Schmid, K., Alper, C. A., and Bissett, L. (1969) Inheritance of human α_1-acid glycoprotein (orosomucoid) variants. J. Clin. Invest. 48:2293-2299.

Johnston, F. E., Alarcon, O., Benedict, F., Dary, M., Galbraith, M., and Gindhart, P. S. (1973) Albumin Mexico (Al[Me]) in the Guatemalan Highlands. Amer. J. Phys. Anthrop. 38:27-29.

Johnston, F. E., Blumberg, B. S., Agarwal, S. S., Melartin, L., and Burch, T. A. (1969a) Alloalbuminemia in southwestern U.S. Indians: Polymorphism of albumin Naskapi and albumin Mexico. Hum. Biol. 41:263-270.

Johnston, F. E., Blumberg, B. S., Kensinger, K. M., Jantz, R. L., and Walker, G. F. (1969b) Serum protein polymorphisms among the Peruvian Cashinahua. Amer. J. Hum. Genet. 21:376-383.

Joó-Szabados, T. (1970) The frequency of Fy[a] and Fy[b] antigens in the population of Budapest. Hum. Hered. 20:436-437.

Jordal, K. (1958) The Lewis factors Le[b] and Le[x] and a family series tested by anti-Le[a], anti-Le[b] and anti-Le[x]. Acta Path. Microbiol. Scand. 42:269-284. Cited from Mourant et al. (1976).

Jørgensen, G., Dengler, H., und Hopper, U. (1965) Untersuchungen des β-Lipoproteinsystems nach Berg bei Gesunden, Kranken und Schwangern. Humangenetik 1:476-478.

Jørgensen, J., Drachmann, O., and Gavin, J. (1982) Duch, Dh[a]: A low frequency red cell antigen. Hum. Hered. 32:73-75.

Jurima, M., Inaba, T., and Kalow, W. (1985) Genetic polymorphism of mephenytoin p(4')-hydroxylation: difference between Orientals and Caucasians. Br. J. Clin. Pharmacol. 19:483-487.

Kaarsalo, E., Kortekangas, A. E., Tippett, P., and Hamper, J. (1962) A contribution to the blood group frequencies in Finns. Acta Path. Microbiol. Scand. 54:287-290. Cited from Mourant et al. (1976).

Kageoka, T., Satoh, C., Goriki, K., Fujita, M., Neriishi, S., Yamamura, K., Kaneko, J., and Masunari, N. (1985) Electrophoretic variants of blood proteins in Japanese. IV. Prevalence and enzymologic characteristics of glucose-6-phosphate dehydrogenase variants in Hiroshima and Nagasaki. Hum. Genet. 70:101-108.

Kaklamani, E. and Holborow, E. J. (1963) Secretor and Lewis gene frequencies in Blackfeet Indians. Vox Sang. 8:231-234.

Kalimanovska, V., Majkic-Singh, N., and Jelic-Ivanovic, Z. (1985) Polymorphism of red cell glyoxalase I in Serbia, Yugoslavia. Hum. Hered. 35:120-122.

Kalimanovska, V., Majkic-Singh, N., Stojanov, M., Grozdanic, V., Vucetic, G., Andelic, M., Gligorovic, V., and Tomasevic, R. (1983) Human red cell glutamic-pyruvic transaminase polymorphism in Serbia, Yugoslavia. Hum. Hered. 33:319-321.

Kalmus, H. (1957) Defective colour vision, P.T.C. tasting and drepanocytosis in samples from fifteen Brazilian populations. Ann. Hum. Genet. 21:313-317.

Kalmus, H., Amir, A., Levine, O., Barak, E., and Goldschmidt, E. (1961) The frequency of inherited defects of color vision in some Israeli populations. Ann. Hum. Genet. 25:51-55.

Kaloustian, V. M. D., Byrne, R., Young, W. J., and Childs, B. (1969) An electrophoretic method for detecting hypoxanthineguanine phosphoribosyl transferase variants. Biochem. Genet. 3:299-302.

Kamboh, M. I. and Ferrell, R. E. (1986a) Genetic studies of low abundance human plasma proteins. II. Population genetics of coagulation factor XIIIB. Amer J. Hum. Genet. 39:817-825.

Kamboh, M. I. and Ferrell, R. E. (1986b) Genetic studies of low abundance human plasma proteins. III. Polymorphism of the C1R subcomponent of the first complement component. Amer. J. Hum. Genet. 39:826-831.

Kamboh, M. I. and Ferrell, R. E. (1986c) Genetic studies of low abundance human plasma proteins. IV. Polymorphism of hemopexin. Personal communication.

Kamboh, M. I. and Ferrell, R. E. (1987a) Genetic studies of low abundance human plasma proteins. VIII. Inherited structural variation in antithrombin III. Ann. Hum. Genet. (in press).

Kamboh, M. I. and Ferrell, R. E. (1987b) Genetic studies of human apolipoproteins. I. Polymorphism of apolipoprotein A-IV. Amer. J. Hum. Genet. 41:119-127.

Kamboh, M. I. and Kirk, R. L. (1983) Distribution of transferrin (Tf) subtypes in Asian, Pacific and

Australian Aboriginal populations: Evidence for the existence of a new subtype Tf[C6]. Hum. Hered. 33:237-243.

Kamboh, M. I. and Kirk, R. L. (1984) Genetic studies of PGM1 subtypes: population data from the Asian-Pacific area. Ann. Hum. Biol. 11:211-219.

Kamboh, M. I. and Kirwood, C. (1984) Genetic polymorphism of thyroxin-binding globulin (TBG) in the Pacific area. Amer. J. Hum. Genet. 36:646-654.

Kamboh, M. I., Ramford, P. R., and Kirk, R. L. (1984) Population genetics of the vitamin D binding protein (GC) subtypes in the Asian-Pacific area: Description of new alleles at the Gc locus. Hum. Genet. 67:378-384.

Kamel, K., Chandy, R., Mousa, H., and Yunis, D. (1980) Blood groups and types, Hemoglobin variants, and G-6-PD deficiency among Abu Dhabians in the United Arab Emirates. Amer. J. Phys. Anthrop. 52:481-484.

Kamel, K., Umar, M., Ibrahim, W., Mansour, A., Gaballah, F., Selim, O., Azim, A., Hamza, S., Sabry, F., Moafy, N., El-Naggar, A., and Hoerman, K. (1975) Anthropological studies among Libyans: Erythrocyte genetic factors, serum haptoglobin phenotypes and anthropometry. Amer. J. Phys. Anthrop. 43:103-112.

Kang, Y. S., Cho, W. K., and Yurn, K. S. (1967a) Taste sensitivity to phenylthiocarbamide of Korean population. Eugen. Quart. 14:1-6.

Kang, Y. S. and Lee, C. C. (1973) The researches of the Korean population genetics. J. Nat. Acad. Sci. (Korea). 12:115-137. Cited from Tills et al. (1983a).

Kang, Y. S., Lee, S. W., Park, S., and Cho, W. K. (1967b) Colour blindness among Korean students. Eugen. Quart. 14:271-273.

Kaplanoglou, L. B. and Triantaphyllidis, C. D. (1982) Genetic polymorphisms in a North-Greek population. Hum. Hered. 32:124-129.

Karaphet, T. M., Sukernik, R. I., Osipova, L. P., and Simchenko, Y. B. (1981) Blood groups, serum proteins and red cell enzymes in the Nganasans (Tavghi) - Reindeer hunters from Taimir Peninsula. Amer. J. Phys. Anthrop. 56:139-145.

Karim, A. K. M. B., Elfellah, M. S., and Evans, D. A. P. (1981) Human acetylator polymorphism: estimate of allele frequency in Libya and details of global distribution. J. Med. Genet. 18:325-330.

Karlsson, S., Arnason, A., and Jensson, O. (1980a) GLO polymorphism in Iceland. Hum. Hered. 30:383-385.

Karlsson, S., Arnason, A., Thordarson, G., and Olaisen, B. (1980b) Frequency of Gc alleles and a variant Gc allele in Iceland. Hum. Hered. 30:119-121.

Karlsson, S., Skaftadóttir, I., Arnason, A., Thórdarson, G., and Jensson, O. (1983) Gc subtypes in Icelanders. Hum. Hered. 33:5-8.

Karn, R. C., Goodman, P. A., and Yu, P.-L. (1985) Description of a genetic polymorphism of a human proline-rich salivary protein, Pc, and its relationship to other proteins in the salivary protein complex (SPC). Biochem. Genet. 23:37-51.

Kataja, M. (1975) Simulation in paternity analysis. Ph.D. thesis. Insititute of Mathematics, Helesinki University of Technology.

Kattamis, C., Zannos-Mariolea, L., Franco, A. P., Liddell, J., Lehmann, H., and Davies, D. (1962) Frequency of atypical pseudo-cholinesterase in British and Mediterranean populations. Nature 196:599-600.

Kehr-Löke, E. R., Weyrauch, U., Ritter, H., and Goedde, H. W. (1966) Untersuchungen zur Populationsgenetik der Haptoglobin-Untergruppen. Blut 14:97-100. Cited from Mourant et al. (1976).

Kellermann, G. and Walter, H. (1970) Investigations on the population genetics of the α_1-antitrypsin polymorphism. Humangenetik 10:145-150.

Kellermann, G. and Walter, H. (1972) On the population genetics of the ceruloplasmin polymorphism. Humangenetik 15:84-86.

Kenrick, K. G. and Douglas, R. (1967) The distributions of Gc-types in selected populations from Southeast Asia, Polynesia and Australia. Acta Genet. Med. 17:518-523.

Kera, Y., Yamasawa, K., and Komura, S. (1983a) Genetic polymorphism of white blood cell glucose dehydrogenase in Japanese. Jpn. J. Hum. Genet. 28:29-34.

Kera, Y., Yamasawa, K., and Komura, S. (1983b) Phenotypic variation of human antithrombin III in normal plasma: Detection by isoelectric focusing. Jpn. J. Hum. Genet. 28:249-253.

Khérumian, R. and Moullec, J. (1956) Contribution à l'étude des groupes sanguins A_1A_2BO, Rh, MN et P dans l'Union française et les États associés. 2e Congr. Nat. Transfus. Sang., Bordeaux, 1956, pp. 501-504. Cited from Mourant et al. (1976).

Khérumian, R.., Moullec, J., and Desabie, J. (1958). Répartition des groupes sanguins ABO, A_1A_2, Rh et MN et de leurs constellations chez les étudiants de l'Université de Paris. Rev. Hémat. 13:231-238. Cited from Mourant et al. (1976).

Khérumian, R. and Pickford, R. W. (1959) Hérédité et fréquence des anomalies congénitales du sens cromatique. Paris: Vigot Frères. Cited from Post (1971).

Khoi, T. D., Glaise, D., LeTreut, A., Fauchet, R., Godin, Y., and LeGall, J. Y. (1979) Genetic polymorphism of α-L-fucosidase in Brittany (France). Hum. Genet. 51:293-296.

King, J. and Cook, P. J. L. (1981) Glucose dehydrogenase polymorphism in man. Ann. Hum. Genet. 45:129-134.

Kirjarinta, M., Fellman, J., Gustafsson, C., Keisala, E., and Eriksson, A. W. (1969) Two rare electrophoretic variants of erythrocyte enzymes in Finland. Scand. J. Clin. Lab. Invest., Suppl. 108, pp. 46.

Kirk, R. L. (1965) The distribution of genetic markers in Australian Aborigines. Occasional Papers in Aboriginal Studies (Australian Institute of Aboriginal Studies, Canberra) 4:1-67.

Kirk, R. L., Blake, N. M., Lai, L. Y. C., and Cooke, D. R. (1969) Population genetic studies in Australian aborigines of the Northern Territory: The distribution of some serum protein and enzyme groups among the Malag of Elcho Island. Archeol. Phys. Anthrop. Oceania 4:238-251.

Kirk, R. L., Blake, N. M., Moodie, P. M., and Tibbs, G. J. (1971a) Population genetic studies in Australian aborigines of the Northern Territory: The distribution of some serum protein and enzyme groups among the populations at various localities in the Northern Territory of Australia. Hum. Biol. Oceania 1:54-76.

Kirk, R. L., Blake, N. M., and Vos, G. H. (1971b) The distribution of enzyme group systems in a sample of South African Bantu. S. Afr. Med. J. 45:69-72.

Kirk, R. L., Cleve, H., and Bearn, A. G. (1963) The distribution of the group specific component (Gc) in selected populations in South and South East Asia and Oceania. Acta Genet. Stat. Med. 13:140-149.

Kirk, R. L., Keats, B., Blake, N. M., McDermid, E. M., Ala, F., Karini, M., Nickbin, B., Shabazi, H., and Kmet, J. (1977) Genes and people in the Caspian Littoral: A population genetic study in Northern Iran. Amer. J. Phys. Anthrop. 46:377-390.

Kirk, R. L. and Lai, L. Y. C. (1961) The distribution of haptoglobin and transferrin groups in South and South East Asia. Acta Genet. Stat. Med. 11:97-105.

Kirk, R. L., Lai, L. Y. C., Vos, G. H., Wickremasinghe, R. L., and Perera, D. J. B. (1962) The blood and serum groups of selected populations in South India and Ceylon. Amer. J. Phys. Anthrop. 20:485-497.

Kirk, R. L., McDermid, E. M., Blake, N. M., Gajdusek, D. C., Leyshon, W. C., and MacLennan, R. (1974) Blood group, serum protein and red cell enzyme groups of Amerindian populations in Colombia. Amer. J. Phys. Anthrop. 41:301-316.

Klasen, E. C., Bos, A., and Simmelink, H. D. (1982) PI (α_1-antitrypsin) subtypes: Frequency of PI*M4 in several populations. Hum. Genet. 62:139-141.

Klasen, E. C., Franken, C., Volkers, W. S., and Bernini, L. F. (1977) Population genetics of alpha$_1$-antitrypsin in the Netherlands. Hum. Genet. 37:303-313.

Klouda, P. T., Ollier, W. E. R., Al Hilali, A., and Bacchus, R. A. (1984) Properdin factor B (Bf) polymorphism in Saudi Arabs: High frequency of a "rare" allele Bf$^{S0.7}$. Hum. Hered. 34:269-272.

Knussmann, R. and Knussmann, R. (1976) Blood and serum protein groups of the Dama of south-west Africa. Hum. Hered. 26:34-42.

Kunstmann, G., Mauff, G., and Pulverer, G. (1980) C6 polymorphism and rare alleles in Western Germany. Immunobiol. 158:55-59.

Kobyliansky, E., Micle, S., Goldschmidt-Nathan, M., Arensburg, B., and Nathan, H. (1982) Jewish populations of the world: genetic likeness and differences. Ann. Hum. Biol. 9:1-34.

Kojima, T., Tanimoto, M., Kamiya, T., Obata, Y., Takahashi, T., Ohno, R., Kurachi, K., and Saito, H. (1987) Possible absence of common polymorphisms in coagulation factor IX gene in Japanese subjects. Blood 69:349-352.

Koppe, A. L., Walter, H., Chopra, V. P., and Bajatzadeh, M. (1970) Investigations on the genetics and population genetics of the β_2-glycoprotein I polymorphism. Humangenetik 9:164-171.

Kornstad, L. (1960) Absence of the Diego blood-group antigen in the Lapps. Nature 185:325.

Kornstad, L. (1961) Some observations on the Wright blood group system. Vox Sang. 6:129-135.

Kornstad, L. (1972) Distribution of the blood groups of the Norwegian Lapps. Amer. J. Phys. Anthrop. 36:257-266.

Kornstad, L. (1986) A rare blood group antigen, Ol^a (Oldeide), associated with weak Rh antigens. Vox Sang. 50:235-239.

Kornstad, L., Kout, M., Larsen, A. M. H., and Ørjasaeter, H. (1967) A rare blood group antigen, Jn^a. Vox Sang. 13:165-170.

Kornstad, L., Larsen, A. M. H., and Weisert, O. (1971) Further observations on the frequency of the Ny^a blood-group antigen and its genetics. Amer. J. Hum. Genet. 23:612-613.

Kornstad, L., Øyen, R., and Cleghorn, T. E. (1968) A new rare blood group antigen To^a (Torkildsen) and an unsolved factor Skjelbred. Vox Sang. 14:363-368.

Köster, B., Leupold, H., and Mauff, G. (1975) Esterase D polymorphism: High-voltage agarose gel electrophoresis and distribution of phenotypes in different European populations. Humangenetik 28:75-78.

Kout, M. (1962) The incidence of the C^w, M^g and Wr^a agglutinogens in the population of Prague. Vox Sang. 7:242-244.

Kouvatsi, A. and Triantaphyllidis, C. D. (1984) Enzyme polymorphism in placentae from northern Greece. Hum. Hered. 34:207-211.

Kouvatsi, A. and Triantaphyllidis, C. D. (1985) Placental alkaline phosphatase polymorphism in northern Greece. Hum. Hered. 35:259-262.

Kouvatsi, A. and Triantaphyllidis, C. D. (1987) GC and Tf subtypes in Greece. Hum. Hered. 37:62-64.

Kraus, A. P. and Neely, C. L., Jr. (1964) Human erythrocyte lactate dehydrogenase: Four genetically determined variants. Science 145:595-597.

Kreckel, P., Kühnl, P., and Scharrer, I. (1982a) Formal genetics and population data of the A and B subunits of the fibrin stabilizing factor (factor XIII)- evidence for a rare F XIII*QL variant and a new allele, F XIIIB*4. In: Factor XIII and Fibronectin (Egbring, R. and Klingemann, H. G., eds.), Medizinische Verlagsgesellschaft. pp. 81-89. Cited from Kamboh and Ferrell (1986a).

Kreckel, P., Kühnl, P., and Spielmann, W. (1982b) Human coagulation factor XIIIA (FXIIIA) phenotyping by immunofixation agarose gel electrophoresis (IAGE). Blut 44:309-314.

Kretchmer, N., Ransome-Kuti, O., Hurwitz, R., Dungy, C., and Alakija, W. (1971) Intestinal absorption of lactose in Nigerian ethnic groups. Lancet ii:392-395.

368

Kueppers, F. and Christopherson, M. J. (1978) Alpha$_1$-antitrypsin: Further genetic heterogeneity revealed by isoelectric focusing. Amer. J. Hum. Genet. 30:359-365.

Kueppers, F. and Harpel, B. M. (1979) Group-specific component (Gc) 'subtypes' of Gc1 by isoelectric focusing in US Blacks and Whites. Hum. Hered. 29:242-249.

Kueppers, F. and Harpel, B. M. (1980) Transferrin C subtypes in US blacks and whites. Hum. Hered. 30:376-382.

Kuhn, B., Bissbort, S., Kömpf, J., and Ritter, H. (1975) Red-cell uridine-5-monophosphate kinase (UMPK): Formal genetics, linkage analysis and population genetics from southwestern Germany. Humangenetik 28:255-258.

Kühnl, P., Langanke, U., Spielmann, W., and Neubauer, M. (1977a) Investigations of the polymorphism of sperm diaphorase in man. Hum. Genet. 40:79-86.

Kühnl, P., Schwabenland, R., and Spielmann, W. (1977b) Investigations on the polymorphism of glyoxalase I (EC 4.4.1.5) in the population of Hessen, Germany. Hum. Genet. 38:99-106.

Kühnl, P. and Spielmann, W. (1972) Untersuchungen zum C'3-polymorphismus (β_{1C}-globulin). Humangenetik 15:7-13.

Kühnl, P. and Spielmann, W. (1978) Transferrin: Evidence for two common subtypes of the Tf^C allele. Hum. Genet. 43:91-95.

Kühnl, P. and Spielmann, W. (1981) Recent developments in isoelectrofocusing techniques alpha-fucosidase (Fu), C6 and amylase 1 system. Arztl. Lab. 27:225-260.

Kühnl, P., Spielmann, W., and Loa, M. (1978) An improved method for the identification of Gc^1 subtypes (Group-Specific component) by isoelectric focusing. Vox Sang. 35:401-404.

Kühnl, P. and Tischberger, H. (1980) Amylase$_1$ polymorphism of human parotid saliva: Detection of a new allele, AMY$_1$^{5} by isoelectric focusing and AMY$_1$ population data from Germany. Electrophoresis 1:186-190.

Kukongviriyapan, V., Lulitanond, V., Areejitranusorn, C., Kongyingyose, B., and Laupattarakasem, P. (1984) N-acetyltransferase polymorphism in Thailand. Hum. Hered. 34:246-249.

Kunkel, H. G., Smith, W. K., Joslin, F. G., Natvig, J. B., and Litwin, S. D. (1969) Genetic marker of the γA2 subgroup of γA immunoglobulins. Nature 223:1247-1248.

Kunstmann, G., Mauff, G., and Pulverer, G. (1980) C6 polymorphism and rare alleles in western Germany. Immunobiology 158:55-59.

Küpfer, A. and Preisig, R. (1984) Pharmacogenetics of mephenytoin: a new drug hydroxylation polymorphism in man. Eur. J. Clin. Pharmacol. 26:753-759.

Kwok, K. Y. Y. and Lewis, W. H. P. (1981) Group-specific component (Gc) subtypes in the Chinese population of Hong Kong. Hum. Genet. 59:72-74.

La Du, B. N., Atkins, S., and Bayoumi, R. A-L. (1986) Analysis of the serum paraoxonase/arylesterase polymorphism in some Sudanese families. In: Ethnic differences in reactions to drugs and xenobiotics. (Kalow, W., Goedde, H. W., and Agarwal, D. P., eds.), Alan R. Liss, New York, pp 87-98.

Labie, D., Richin, C., Pagnier, J., Gentilini, M., and Nagel, R. L. (1984) Hemoglobins S and C in Upper Volta. Hum. Genet. 65:300-302.

Laet, M. de and van de Calseyde, P. (1935) Étude critique des méthodes de dépistage des daltoniens. Bull. Acad. Roy. Méd. Belg. 5:46. Cited from Post (1971).

Laha, P. K., Saha, N., and Bayoumi, R. A. (1979) Red cell glyoxalase I polymorphism among the selected tribes of the Sudan. Jpn. J. Hum. Genet. 24:259-264.

Lahav, M. and Szeinberg, A. (1972) Red-cell glutamic-pyruvic transaminase polymorphism in several population groups in Israel. Hum. Genet. 22:533-538.

Lai, L. Y. C. and Bloom, J. (1982) Genetic variation in Bougainville and Solomon Islands populations. Amer. J. Phys. Anthrop. 58:369-382.

Laird-Fryer, B., Dukes, C. V., Lawson, J., Moulds, J. J., Walker, E. M. Jr., and Glassman, A. B. (1983) Tc^a: A high-frequency blood group antigen. Transfusion 23:124-127.

Laisney, V., Van Cong, N., and Frezal, J. (1984) Human genes for glutathione S-transferases. Hum. Genet. 68:221-227.

Lamm, L. U. (1970) Family, population and mother-child studies of two phosphoglucomutase loci (PGM$_1$ and PGM$_3$). Hum. Hered. 20:292-304.

Lanchbury, J. S., Bernal, J. E., and Papiha, S. S. (1984) Genetic polymorphism of glutamate-pyruvate transaminase and glyoxalase I in Colombia. Hum. Hered. 34:222-225.

Larrick, J. W., Yost, J., Gourley, C., Buckley, C. E., Plato, C. C., Pandey, J. P., Burck, K. B., and Kaplan, J. (1985) Markers of genetic variation among the Waorani Indians of the Ecuadorian Amazon Headwaters. Amer. J. Phys. Anthrop. 66:445-453.

Larsen, B., Salimonu, L. S., Gow, C., and Marshall, W. H. (1981) Bf polymorphism: A very fast variant from Nigeria. Hum. Genet. 56:395-396.

Lasker, G. W., Mast, J., and Tashian, R. (1969) β-aminoisobutyric acid (BAIB) excretion in urine of residents of eight communities in the states of Michoacán and Oaxaca, Mexico. Amer. J. Phys. Anthrop. 30:133-136.

Laughlin, W. S. (1957) Cited from Mourant et al. (1976).

Laurell, C. B. and Niléhn, J. E. (1966) A new type of inherited serum albumin anomaly. J. Clin. Invest. 45:1935-1945.

Layrisse, M. (1958) Anthropological considerations of the Diego (Di^a) antigen. Amer. J. Hum. Genet. 16:173-186.

Leakey, T. E. B., Coward, A. R., Warlow, A., and Mourant, A. E. (1972) The distribution in human populations of electrophoretic variants of cytoplasmic malate dehydrogenase. Hum. Hered. 22:542-

551.

Lefèvre-Witier, P. and Vergnes, H. (1977) Enzyme polymorphisms of Ideles populations (Ahaggar, Algeria) and the Iwellemeden Kel Kummer Twaregs (Menaka, Mali). Hum. Hered. 27:454-469.

Lefranc, M. P., Chibani, J., Helal, A. N., Boukel, K., Segar, J., and Lefranc, G. (1981) Human transferrin (Tf) and group-specific component (Gc) subtypes in Tunisia. Hum. Genet. 59:60-63.

Lehmann, H., Ala, F., Hedeyat, S., Montazemi, K., Nejad, H. K., Lightman, S., Kopéc, A. C., Mourant, A. E., Teesdale, P., and Tills, D. (1973) XI. The hereditary blood factors of the Kurds of Iran. Phil. Trans. R. Soc. Lond. B. 266:195-205.

Lehmann, H. and Cutbush, M. (1952) Sub-division of some southern Indian communities according to the incidence of sickle-cell trait and blood groups. Trans. R. Soc. Trop. Med. Hyg. 46:380-383. Cited from Mourant et al. (1976).

Lehmann, H. and Kynoch, P. A. M. (1976) Human Haemoglobin Variants and Their Characteristics. North Holland, Amsterdam.

Lehmann, H., Sharih, A., Gilat, T., and Lenz, R. (1962) A survey of some genetical characters in Ethiopean tribes. II. Hemoglobin examinations. Amer. J. Phys. Anthrop. 20:174.

Leichter, J. and Lee, M. (1971) Lactose intolerance in Canadian west coast Indians. Amer. J. Dig. Dis. 16:809-813.

Levine, C., Cohen, T., Manny, N., Bar-Shany, S., and Moulds, J. J. (1985) Yt (Cartwright) blood groups among Israeli Jews. Transfusion 25:180.

Levine, M. H., von Hagen, V., Quilici, J. C., and Salmon, D. (1974) Anthropology of a Basque village: a new hemotypological study. Cah. Anthrop. Ecol. Hum. 2:159-171. Cited from Tills et al. (1983a).

Lewis, M. and Kaita, H. (1981) A 'new' low incidence "Hutterite" blood group antigen Waldner (Wd[a]). Amer. J. Hum. Genet. 33:418-420.

Lewis, M., Kaita, H., Allderdice, P. W., Bergren, M., and McAlpine, P. J. (1984) A 'new' low incidence red cell antigen, NFLD. Hum. Genet. 67:270-271.

Lewis, M., Kaita, H., McAlpine, P. J., Fletcher, J., and Moulds, J. J. (1978) A 'new' blood group antigen Fr[a]: Incidence, inheritance and genetic linkage analysis. Vox Sang. 35:251-254.

Lewis, W. H. P. (1973) Common polymorphism of peptidase A. Electrophoretic variants associated with quantitative variation of red cell levels. Ann. Hum. Genet. 36:267-271.

Lewis, W. H. P. and Harris, H. (1967) Human red cell peptidases. Nature 215:351-355.

Lewis, W. H. P. and Harris, H. (1969) Peptidase D (Prolidase) variants in man. Ann. Hum. Genet. 32:317-322.

Li, S., Wang, L., and Du, R. (1986) Polymorphism of human red cell glyoxalase I in six ethnic groups of China. Hum. Genet. 74:318-319.

Liddell, J., Brown, D., Beale, D., Lehmann, H., and Huntsman, R. G. (1964) A new haemoglobin J(alpha Oxford) found during a survey of an English population. Nature 204:269-270.

Lie, H. and Teisberg, P. (1973) Red cell acid phosphatase polymorphism in Norway. Hum. Hered. 23:257-262.

Lie-Injo, L. E. (1976) Genetic relationships of several aboriginal groups in Southeast Asia. In: The Origin of the Australians (Kirk, R. L. and Thorne, A. G., eds.), Australian Institute of Aboriginal Studies, Canberra, pp. 277-306.

Lie-Injo, L. E., Bolton, J. M., and Fudenberg, H. H. (1967) Haptoglobins, transferrins and serum gammaglobulin types in Malayan aborigines. Nature 215:777.

Lie-Injo, L. E., Ganesan, J., Herrera, A., and Lopez, C. G. (1978) α_1-antitrypsin variants in different racial groups in Malaysia. Hum. Hered. 28:37-40.

Lie-Injo, L. E., Solai, A., Herrera, A. R., Nicolaisen, L., Kan, Y. W., Wan, W. P., and Hasan, K. (1982) Hb Bart's level in cord blood and deletions of α-globin genes. Blood 59:370-376.

Lie-Injo, L. E. and Ti, T. S. (1964) Glucose-6-phosphate dehydrogenase deficiency in Malayans. Trans. Roy. Soc. Trop. Med. and Hyg. 58:500-502. Cited from Mourant et al. (1976).

Lie-Injo, L. E., Weitkamp, L. R., Kosasih, E. N., Bolton, J. M., and Moore, C. L. (1971) Unusual albumin variants in Indonesians and Malayan aborigines. Hum. Hered. 21:376-383.

Lightman, S. L., Carr-Locke, D. L., and Pickles, H. G. (1970) The frequency of PTC tasters and males defective in colour vision in a Kurdish population in Iran. Hum. Biol. 42:665-669.

Linnet-Jepsen, P., Galatius-Jensen, F., and Hauge, M. (1958) On the inheritance of the Gm serum group. Acta Genet. Stat. Med. 8:164-196.

Liotta, I., Purpura, M., Dawes, B. J., and Giles, C. M. (1970) Some data on the low frequency antigens Wr[a] and Bp[a]. Vox Sang. 19:540-543.

Lisker, R. and Giblett, E. R. (1967) Studies on several genetic hematological traits of Mexicans. XI. Red cell acid phosphatase and phosphoglucomutase in three Indian groups. Amer. J. Hum. Genet. 19:174-177.

Lisker, R., López-Habib, G., Daltabuit, M., Rostenberg, I., and Arroyo, P. (1974) Lactase deficiency in a rural area of Mexico. Amer. J. Clin. Nutr. 27:756-759.

Lisker, R., Loria, A., and Zárate, G. (1967) Studies on several genetic hematological traits of the Mexican population. XIII. Red cell and serum polymorphism in Spanish immigrants. Acta Genet. Stat. Med. 17:524-529.

Livingstone, F. B. (1985) Frequencies of Hemoglobin Variants. Oxford Univ. Press, New York.

Long, J. C., Naidu, J. M., Mohrenweiser, H. W., Gershowitz, H., Johnson, P. L., Wood, J. W., and Smouse, P. E. (1986) Genetic characterization of Gainj- and Kalam-speaking peoples of Papua New Guinea. Amer. J. Phys. Anthrop. 70:75-96.

Longster, G. and Giles, C. M. (1976) A new antibody specificity, anti-Rg[a], reacting with a red cell and serum antigen. Vox Sang. 30:175-180.

Lucarelli, P., Di Mino, M., Carapella, E., Agostino, R., and Palmarino, R. (1972) Red blood cell NADH diaphorase in Italian populations. Humangenetik 16:349-350.

Lucciola, L., Kaita, H., Anderson, J., and Emery, S. (1974) The blood groups and red cell enzymes of a sample of Cree Indians. Can. J. Genet. Cytol. 16:691-695.

Lugg, J. W. H. and Whyte, J. M. (1954) Taste thresholds for phenylthiocarbamide of some population groups. I. The thresholds of some civilized ethnic groups living in Malaya. Ann. Hum. Genet. 19:290-311.

Lukka, M., Ehnholm, C. and Kuusi, T. (1985) Phosphoglucomutase (PGM$_1$) subtypes in a Finnish population determined by isoelectric focusing in Agarose gel. Hum. Hered. 35:95-100.

Lukka, M., Turunen, P., Kataja, M., and Ehnholm, C. (1986) Group-specific component (Gc): Subtypes in the Finnish population. Hum. Hered. 36:299-303.

Lundsgaard, A. and Jensen, K. G. (1968) Two new examples of anti-Rd: A preliminary report on the frequency of the Rd (Radin) antigen in the Danish population. Vox Sang. 14:452-457.

MacDonald, J. L. (1975) C'3 polymorphism of human complement in North-East England. Hum. Hered. 25:393-397.

MacRoberts, M. H. (1964) Taste sensitivity to phenylthiocarbamide (P.T.C.) among the Papago Indians of Arizona. Hum. Biol. 36:28-31.

MacVie, S. I., Morton, J. A., and Pickles, M. M. (1967) The reactions and inheritance of a new blood group antigen, Sd^a. Vox Sang. 13:485-492.

Magnani, M., Cucchiarini, L., Stocchi, V., Stocchi, O., Carnevali, G., Dachà, M., Fornaini, G. (1982) Human erythrocyte galactokinase: A population survey. Hum. Hered. 32:274-279.

Majkic-Singh, N., Minic, M., Jelic, Z., Stojanov, M., Spasic, S., and Berkes, I. (1982) Human red cell adenylate kinase polymorphism in Serbija, Yugoslavia. Hum. Hered. 32:367-368.

Malaspina, P., Ciminelli, B. M., Pelosi, E., Santolamazza, P., Modiano, G., Santillo, C., Lofoco, G., Talone, C., Gatti, M., and Parisi, P. (1986) Colour blindness distribution in the male population of Rome. Hum. Hered. 36:263-265.

Malcolm, L. A., Woodfield, D. G., Blake, N. M., Kirk, R. L., and McDermid, E. M. (1972) The distribution of blood, serum protein and enzyme groups of Manus Island (Admiralty Islands, New Guinea). Hum. Hered. 22:305-322.

Mallinckrodt, M. G-von. (1978) Polymorphism of human serum paraoxonase. Hum. Genet. Suppl. 1:65-68.

Mann, I. and Turner, C. (1956) Color vision in native races in Australasia. Amer. J. Opthalmol. 41:797-800.

Manczak, M. (1984a) Polymorphism of C3 component of complement in the Polish population. I. Population and family studies. Arch. Immunol. Ther. Exp. 32:421-430.

Manczak, M. (1984b) BF group system in the Polish population. Arch. Immunol. Ther. Exp. 32:431-441.

Maranjian, G., Ikin, E. W., Mourant, A. E., and Lehmann, H. (1966) The blood groups and haemoglobins of the Saudi Arabians. Hum. Biol. 38:394-420.

Marengo-Rowe, A. J., Aviet, K., Godber, M. J., Kopec, A. C., Mourant, A. E., Tills, D., and Woodhead, B. J. (1974) The inherited blood factors of the inhabitants of southern Arabia. Ann. Hum. Biol. 1:311-326.

Margolis, E., Gurevitch, J., and Hermoni, D. (1960) Blood groups in Sephardic Jews. Amer. J. Phys. Anthrop. 18:197-199.

Marks, M. P., Jenkins, T., and Nurse, G. T. (1977) The red-cell glutamic-pyruvate transaminase, carbonic anhydrase I and II and esterase D polymorphisms in the Ambo populations of South West Africa, with evidence for the existence of an EsD^O allele. Hum. Genet. 37:49-54.

Marras, G., Zangani, P., and Stangoni, A. (1964) Il sistema P nei soggetti nati e residenti nella città e nella provincia di Sassari. Zacchia. 27:188-192. Cited from Mourant et al.1976(c)

Màrtensson, L. (1964) On the relationships between the γ-globulin genes of the Gm system. J. Exp. Med. 120:1169-1188.

Martin, J. P., Sesbove, R., Charlionet, R., Ropartz, C., and Pereira, M. T. (1976) Genetic variants of serum α_1-antitrypsin (Pi types) in Portuguese. Hum. Hered. 26:310-314.

Massi, G. and Vecchio, F. M. (1977) Alpha-1-antitrypsin phenotypes in a group of newborn infants in Somalia. Hum. Genet. 38:265-269.

Matousek, V. and Seemanová, E. (1973) Ucinky pokrevniho pribuzenstvi rodicu na potomstvo. Praha, Academia. Cited from Tills et al. (1983a).

Matson, G. A. (1938) Blood groups and ageusia in Indians of Montana and Alberta. Amer. J. Phys. Anthrop. 24:81-89.

Matson, G. A. (1940) Blood groups and ageusia in Indians of northern Alberta. Amer. J. Phys. Anthrop. 27:263-267.

Matson, G. A., Burch, T. A., Polesky, H. F., Swanson, J., Sutton, H. E., and Robinson, A. (1968) Distribution of hereditary factors in the blood of Indians of the Gila River, Arizona. Amer. J. Phys. Anthrop. 29:311-337.

Matson, G. A. and Swanson, J. (1959) Distribution of hereditary blood antigens among the Maya and non-Maya Indians in Mexico and Guatemala. Amer. J. Phys. Anthrop. 17:49-74.

Matson, G. A., Swanson, J., and Robinson, A. (1966) Distribution of hereditary blood groups among Indians in South America. III. In Bolivia. Amer. J. Phys. Anthrop. 25:13-33.

Matsumoto, H., Matsui, K., Ishida, N., Ohkura, K., and Teng, Y.-S. (1980) The distribution of Gc subtypes among the Mongoloid populations. Amer. J. Phys. Anthrop. 53:505-508.

Matsumoto, H., Miyazaki, T., and Fong, J. M. (1973) Further data on the Gm and Am allotypes of the Takasago in Taiwan. Japan J. Leg. Med. 27:273-277.

Matsunaga, E. (1962) The dimorphism in human normal cerumen. Ann. Hum. Genet. 25:273-286.

Matsunaga, E., Suzuki, T., Itoh, S., and Sugimoto, R. (1954) Individual difference of taste-ability for phenylthiocarbamide. Sapporo Med. J. 6:245-249.

Mattila, M. J. and Tiitinen, H. (1967) The rate of isoniazid inactivation in Finnish diabetic and non-diabetic patients. Ann. Med. Exp. Finn. 45:423-427.

Mauff, G., Gauchel, F. D., and Hitzeroth, H. W. (1976) Polymorphism of properdin factor B in South African Negroid, Indian and Coloured populations. Hum. Genet. 33:319-322.

Mauran-Sendrail, A., Bouloux, C., Gomila, J., and Langaney, A. (1975) Comparative study of haemoglobin types of two populations of eastern Senegal - Bedik and Niokholonko. Ann. Hum. Biol. 2:129-136.

Maxia, C., Cosseddu, G. G., Vona, G., and Floris, G. (1975) The sensitivity to PTC in 541 Sardinians. J. Hum. Evol. 4:281-286.

Mayes, J. S. and Guthrie, R. (1968) Detection of Heterozygotes for galactokinase deficiency in a human population. Biochem. Genet. 2:219-230.

Mayr, W. R., Mickerts, D., Pausch, V., and Pacher, M. (1970) Blutkörperchen- und Serummerkmale in Wien. Phänotypen- und Genfrequenzen. Mitt. Anthrop. Ges. Wien. 100:11-15. Cited from Tills et al. (1983a).

McAlpine, P. J., Chen, S.-H., Cox, D. W., Dossetor, J. B., Giblett, E., Steinberg, A. G., and Simpson, N. E. (1974) Genetic markers in blood in a Canadian Eskimo population with a comparison of allele frequencies in circumpolar populations. Hum. Hered. 24:114-142.

McConnell, R. B. (1969) Personal communication. Cited from Mourant et al. (1976).

McCurdy, P. R. and Mehmood, L. (1970) Red cell G6PD deficiency in Pakistan. J. Lab. Clin. Med. 76:943-948

McDermid, E. M. (1971a) Serum albumin variation in Indian populations. Vox Sang. 21:462-464.

McDermid, E. M. (1971b) Variants in human serum albumin and caeruloplasmin in populations from Australia, New Guinea, South Africa and India. Aus. J. Exp. Biol. Med. Sci. 49:309-312.

McDermid, E. M., Blake, N. M., Kirk, R. L., Kosasith, E. N., and Simons, M. J. (1973) The distribution of serum protein and enzyme groups among the Batak of Samosir Island (Sumatra, Indonesia). Humangenetik 17:351-356.

McDermid, E. M. and Vos, G. H. (1971a) Serum protein groups of South African Bantu. I. Albumin, caeruloplasmin, transferrin and haptoglobin. S. Afr. J. Med. Sci. 36:7-14.

McDermid, E. M. and Vos, G. H. (1971b) Serum protein groups of South African Bantu. II. α_1-antitrypsin, group specific component and further observations on haptoglobin and caeruloplasmin. S. Afr. J. Med. Sci. 36:63-68.

McLoughlin, K., Blake, N. M., and Hogan, P. F. (1982a) Blood group, red cell enzyme and serum protein types in the Buka Islanders, Papua New Guinea. Hum. Hered. 32:152-159.

McLoughlin, K., Blake, N. M., Korarome, J., and Alpers, W. (1982b) Blood group, red cell enzyme and serum protein types in an Asaro Village, Eastern Highlands, Papua New Guinea. Hum. Hered. 32:160-165.

McPherson, G. E., Wells, R. F., Hafleigh, E. B., and Grumet, F. C. (1979) A new low incidence red cell antigen, Pe. Vox Sang. 36:236-252.

Meera Khan, P. (1986) Red cell G6PD, GPX1, GLO1, LDH, PGP, and DIA2 polymorphisms. In: African Pygmies (Cavalli-Sforza, L. L., ed.), Academic Press, New York, pp. 273-306.

Meera Khan, P. and Doppert, B. A. (1976) Rapid detection of glyoxalase I (GLO) on cellulose acetate gel and the distribution of GLO variants in a Dutch population. Hum. Genet. 34:53-56.

Meera Khan, P., Verma, C., Wijnen, L. M. M., and Jairaj, S. (1984) Red cell glutathione peroxidase (GPX1) variation in Afro-Jamaican, Asiatic Indian, and Dutch populations. Is the GPX1*2 allele of "Thomas" variant an African marker? Hum. Genet. 66:352-355.

Meera Khan, P., Verma, C., Wijnen, L. M. M., Wijnen, J. Th., Prins, H. K., and Nijenhuis, L. E. (1986) Electrotypes and formal genetics of red cell glutathione peroxidase (GPX1) in the Djuka of Surinam. Amer. J. Hum. Genet. 38:712-723.

Melartin, L. (1967) Albumin polymorphism in man. Acta Path. Microb. Scand. Suppl. 191:1.

Menzel, H.-J., Kladetzky, R.-G., and Assmann, G. (1983) Apolipoprotein E polymorphism and coronary artery disease. Arteriosclerosis 3:310-315.

Menzel, H.-J., Kövary, P. M., and Assmann, G. (1982) Apolipoprotein A-IV polymorphism in man. 62:349-352.

Merritt, A. D., Rivas, M. L., Bixler, D., and Newell, R. (1973) Salivary and pancreatic amylase: Electrophoretic characterizations and genetic studies. Amer. J. Hum. Genet. 25:510-522.

Merton, B. B. (1958) Taste sensitivity to PTC in 60 Norwegian families with 176 children. Confirmation of the hypothesis of single gene inheritance. Acta Genet. Stat. Med. 8:114-128.

Mestriner, M. A., Simoes, A. L., and Salzano, F. M. (1980) New studies on the esterase D polymorphism in South American Indians. Amer. J. Phys. Anthrop. 52:95-101.

Metaxas, M. N. and Metaxas-Bühler, M. (1970) An agglutinating example of anti-Xg^a and Xg^a frequencies in 558 Swiss blood donors. Vox Sang. 19:527-529.

Metaxas, M. N. and Metaxas-Bühler, M. (1963) Studies on the Wright blood group system. Vox Sang. 8:707-716.

Metaxas, M. N., Metaxas-Bühler, M., and Romanski, J. (1966) Studies on the blood group antigen M^g. I. Frequency of M^g in Switzerland and family studies. Vox Sang. 11:157-169.

Migone, N., Feder, J., Cann, H., van West, B., Hwang, J., Takahashi, N., Honjo, T., Piazza, A., and Cavalli-Sforza, L. L. (1983) Multiple DNA fragment polymorphisms associated with immunoglobulin μ chain switch-like regions in man. Proc. Natl. Acad. Sci. USA 80:467-471.

Miller, S. A., Dykes, D. D., and Polesky, H. F. (1985) Gene frequency distribution of the B subunit of factor XIII (F XIIIB) in Minnesota Whites, Blacks and Amerindians. Electrophoresis 6:399-401.

Milner, A. E., Burnett, D., Rutter, J., and Bradwell, A. R. (1985) Detection of antithrombin III microheterogeneity. Thrombos. Res. 37:127-134.

Minaguchi, K., Ikemoto, S., Nakajima, I. and Susaki, K. (1976) Studies of genetic markers in human saliva. II. Frequencies of PR and Db systems from parotid saliva of Japanese in Tokyo. Bull. Tokyo Dent. Coll. 17:191-197. Cited from Pronk et al. (1984).

Misawa, S., Hayashida, Y., and Okochi, K. (1971) Distribution of Ag(x) and Ag(y) antigens in the Ainu. Jpn. J. Hum. Genet. 16:30-34.

Misawa, S. and Omoto, K. (1985) Blood groups of the Negrito in the Philippines. Personal communication.

Misawa, S., Omoto, K., Harada, S., Sumpaico, J. S., and Ogonuki, H. (1981) Population genetic studies of the Philippine Negritos. 9. Internationale Tagung der Gesellshaft für forensische Blutgruppenkunde.

Mitchell, R. J. (1976) ABH secretion in populations of the British Isles. Ann. Hum. Biol. 6:569-576.

Mitchell, R. J. (1977) Red-green colour blindness in the Isle of Man and Cumbria. Ann. Hum. Biol. 4:577-579.

Mitchell, R. J., Cook, R. M., and Sunderland, E. (1977) Phenylthiocarbamide (PTC) taste sensitivity in selected populations of the Isle of Man and Cumbria. Ann. Hum. Biol. 4:431-438.

Mitchell, R. J., Izatt, M. M., Sunderland, E., and Cartwright, R. A. (1976) Blood groups antigens, plasma protein and red cell isoenzyme polymorphisms in South-west Scotland. Ann. Hum. Biol. 3:157-171.

Mitchell, R. J., Tills, D., Warlow, A., Kopec, A. C., Sunderland, E., Mourant, A. E., and Marin, A. (1982) Genetic studies of the population of the Isle of Man. Ann. Hum. Biol. 9:57-68.

Miyano, M., Nanjo, K., Okai, K., Sowa, R., Nomura, Y., Kondo, M., Sanke, T., Kawa, A., Miyamura, K., Aiyathurai, E., Ferunando, R., and Vichayanrat, A. (1986) Properdin factor B frequencies in four Asian populations. Hum. Hered. 36:129-131.

Modiano, G., Bernini, L., Carter, N. D., Santachiara-Benerecetti, S. A., Detter, J. C., Baur, E. W., Paolucci, A. M., Gigliani, F., Morpurgo, G., Santolamazza, C., Scozzari, R., Terrenato, L., Meera Khan, P., Nijenhuis, L. E., and Kanashiro, V. K. (1972) A survey of several red cell and serum genetic markers in a Peruvian population. Amer. J. Hum. Genet. 24:111-123.

Modiano, G., Terrenato, L., Novelletto, A., Shrestha, S., Parajuli, M., Luzzatto, L., Morpurgo, G., Purpura, M., Colombo, B., Santachiara, S. A., and Brega, A. (1986) Decreased malaria morbidity and relevant polymorphisms in the Tharu people of Nepal. 7th Int. Cong. Hum. Genet. Abstracts. Pt.II. p.433.

Modrzewska, K. (1958) Taste sensitivity to phenylthiocarbamide in the Polish population. (In Polish) Przegl. Antrop. 24:540-564. Cited from Mourant et al. (1976).

Moharram, I. (1942) The blood group factor P in Egypt. Lab. Med. Progr. 3:1-8.Cited from Mourant et al. (1976).

Moharram, I. (1943) The group properties in the saliva of Egyptian population. Lab. Med. Progr. 4:1-13.

Mohr, J. (1951-2) Taste sensitivity to phenylthiourea in Denmark. Ann. Eugen. 16:282-286.

Molthan, L. and Moulds, J. (1978) A new antigen, McC^a(McCoy), and its relationship to Kn^a(Knops). Transfusion 18:566-568.

Monn, E. (1969a) Red cell phosphoglucomutase (PGM) types of Norwegian Lapps. Characteristic gene frequencies and variant types. Hum. Hered. 19:264-273.

Monn, E. (1969b) Human red cell phosphoglucomutase (PGM) types in Norway. Hum. Hered. 19:274-282.

Monn, E., Berg, K., Reinskou, T., and Teisberg, P. (1971) Serum protein polymorphisms among Norwegian Lapps. Hum. Hered. 21:134-139.

Monn, E. and Christiansen, R. O. (1972) Guanylate kinase in man - multiple molecular forms. Hum. Hered. 22:18-27.

Monn, E. and Gjønnaess, H. (1971) Placenta phosphoglucomutase types in Norway. Hum. Hered. 21:254-262.

Moore, J. M., Deutsch, H. F., and Ellis, F. R. (1973) Human carbonic anhydrases. IX. Inheritance of variant erythrocyte forms. Amer. J. Hum. Genet. 25:29-35.

Moore, J. M., Funakoshi, S., and Deutsch, H. F. (1971) Human carbonic anhydrases. VII. A new C type isozyme in erythrocytes of American Negroes. Biochem. Genet. 5:497-504.

Moores, P. and Brain, P. (1968) Lewis groups and secretor status in Natal Bantu. Transfusion 8:283-288.

Moral, P. and Panadero, A. M. (1983) Haptoglobin subtypes in Barcelona (Spain). Hum. Hered. 33:192-194.

Morel, P. A. and Hamilton, H. B. (1979) Ok^a: An erythrocytic antigen of high frequency. Vox Sang. 36:182-185.

Morganti, G., Beolchini, P. E., Bütler, R., Brunner, E., and Vierucci, A. (1970) Contribution to the genetics of serum β-lipoproteins in man. IV. Evidence for the existence of the Ag$^{al/d}$ and Ag$^{c/g}$ loci, closely linked to the Ag$^{x/y}$ locus. Humangenetik 10:244-253.

Moro-Furlani, A. M., Turner, V. S., and Hopkinson, D. A. (1980) Genetical and biochemical studies on human phosphoserine phosphatase. Ann. Hum. Genet. 43:323-333.

Mortensen, J. P. and Lamm, L. U. (1981) Quantitative differences between complement factor B phenotypes. Immunology 42:505-511.

Morton, N. E. and Yamamoto, M. (1973) Blood groups and haptoglobins in the Eastern Carolines. Amer. J. Phys. Anthrop. 38:695-698.

Motulsky, A. G. and Morrow, A. (1968) Atypical cholinesterase gene E$_1$^{a}: Rarity in Negroes and most Orientals. Science. 159:202-203.

Motulsky, A. G., Stransky, E., and Fraser, G. R. (1964) Glucose-6-phosphate dehydrogenase (G6PD) deficiency, thalassemia, and abnormal haemoglobins in the Philippines. J. Med. Genet. 1:102-106.

Moulinier, J. P. (1958) Iso-immunisation maternelle antiplaquettaire et purpura néo-natal. Le système de groupe plaquettaire <<duzo>>. Proc. 6th Congr. Europ. Soc. Haematology, Karger, Basel, pp.817-820.

Moullec, J. (1963) Les groupes Gc. Étude de 221 donneurs de sang parisiens. Rev. Franç. Étud. Clin. Biol. 8:910-912. Cited from Mourant et al. (1976).

Moullec, J. (1967) Cited from Mourant et al. (1976).

Moullec, J., Ruffié, J., Matte, C., Audran, R., and Noël, M. (1961) Observations sur la rériques (haptoglobines, transferrines, groupes Gm) dans quelques populations. Second Int. Congr. Hum. Genet., Rome 2:762-765.

Mourant, A. E., Godber, M. J., Kopec, A. C., Lehmann, H., Steele, P. R., and Tills, D. (1969) The hereditary blood factors of some populations in Bhutan. Anthropologist (Sp. Vol.):29-43.

Mourant, A. E., Kopec, A. C., and Domaniewska-Sobczak, K. (1976) The Distribution of the Human Blood Groups and Other Polymorphisms, 2nd ed., Oxford University Press, Oxford.

Mourant, A. E., Tills, D., Kopec, A. C., Warlow, A., Teesdale, P., Booth, P.B., and Hornabrook, R. W. (1981) Red cell antigen, serum protein, and red cell enzyme polymorphisms in inhabitants of the Jimi valley, Western Highlands, New Guinea. Hum. Genet. 59:77-80.

Mourant, A. E., Tills, D., Kopec, A. C., Warlow, A., Teesdale, P., Booth, P. B., and Hornabrook, R. W. (1982) Red cell antigen, serum protein and red cell enzyme polymorphisms in Eastern Highlanders of New Guinea. Hum. Hered. 32:374-384.

Mowbray, S., Watson, B., and Harris, H. (1972) A search for electrophoretic variants of human adenine phosphoribosyl transferase. Ann. Hum. Genet. 36:153-162.

Moya, J. (1971) Los grupos sanguineos de los sistemas Kell y Duffy en los vascos. 1e Sem. Antrop. Vasca. Bilbao. pp. 553-561. Cited from Mourant et al. (1976).

Mueller, W. H. and Weiss, K. M. (1979) Colour-blindness in Colombia. Ann. Hum. Biol. 6:137-145.

Mukherjee, B. N. and Das, S. K. (1970) The haptoglobin and transferrin types in West Bengal and a case of haptoglobin "Johnson". Hum. Hered. 20:209-214.

Mukherjee, B. N., Das, S. K., Basak, S. N., and Kellermann, G. (1974) The distribution of some erythrocyte and serum enzyme group systems in the Mahishyas and the Muslims of 24-Parganas, West Bengal. Hum. Biol. 46:425-433.

Mukherjee, B. N., Das, S. K., Malhotra, K. C., Kate, S. L., Mutalik, G. S., Sainani, G. S., and Bhidya, S. (1978) Placental enzyme polymorphism among Maharashtrians: Alkaline phosphatase and lactate dehydrogenase. Ann. Hum. Biol. 5:435-440.

Mueller-Eckhardt, C., Mueller-Eckhardt, G., Willen-Ohff, H., Horz, A., Küenzlen, E., and Schendel, D. J. (1985) Immunogenicity of and immune response to the human platelet antigen Zwa is strongly associated with HLA-B8 and DR3. Tissue Antigens 26:71-76.

Müller, U., Ananthakrishnan, R., Walter, H., and Berg, K. (1974) Studies on the geographic distribution of the human serum β-lipoprotein antigen Ag(x). Hum. Hered. 24:458-462.

Murray, J. (1946) The incidence of Rh types in the British population. Brit. J. Exp. Path. 27:102-110.

Murray, J., Donovan, M., Sadler, E., Hornung, S., Anderson, J., Buetow, K., Giblett, E.,and Motulsky, A. (1984a) Association of DNA and protein polymorphisms at the human plasminogen locus. Amer. J. Hum. Genet. 36(4):176S.

Murray, J., Mills, K. A., Demopulos, C. M., Hornung, S., and Motulsky, A. G. (1984b) Linkage disequilibrium and evolutionary relationships of DNA variants (restriction enzyme fragment length polymorphisms) at the serum albumin locus. Proc. Natl. Acad. Sci. USA 81:3486-3490.

Murray, J. C., Watanabe, K., Tamaoki, T., Hornung, S., and Motulsky, A. (1985) RFLPs for the human alphafetoprotein (AFP), at 4q11-4q13. Nucleic Acids Res. 13:6794.

Murty, J. S. and Vijayalaxmi, C. (1974) Frequency of colour-blindness in Hyderabad school children. Ann. Hum. Biol. 1:225-228.

Mya-Tu, M. and Ma Than Saw. (1970) A note on the ABH secretion among the Burmese. Union of Burma J. Life Sci. 3:101. Cited from Mourant et al. (1976).

Mya-Tu, M., May-May-Yi, and Thin-Thin-Hlaing (1971) Blood groups of the Burmese population. Hum. Hered. 21:420-430. Cited from Mourant et al. (1976).

Naidu, J. M., Babu, V. R., and Veerraju, P. (1978) The incidence of colour-blindness among the tribal populations of Andhra Pradesh. Ann. Hum. Biol. 5:159-163.

Naidu, J. M., Mohrenweiser, H. W., and Neel, J. V. (1985) A sero-biochemical genetic study of Jalari and Brahmin caste populations of Andhra Pradesh, India. Hum. Hered. 35:148-156.

Nakajima, H. (1958) Studies on the S-T blood-type. IV. Classification of the S-T blood type. Acta Crim. Japon.24(4) Suppl. pp. 16-43 (in Japanese). Cited from Akaishi and Kudo (1975).

Nakajima, H., Ikemoto, S., Tokunaga, E., and Furuhata, T. (1965) Further investigation of the Webb (Wb) blood antigen among the Japanese. Proc. Japan Acad. 41:86-87. Cited from Watanabe et al. (1975).

Nakajima, H. (1973a) Blood and serum groups of Koreans in Seoul. Cited from Tills et al. (1983a).

Nakajima, H. (1973b) Cited from Akaishi, S. and Kudo, T. (1975)

Nakajima, H. and Ito, K. (1978) An example of anti-Jra causing hemolytic disease of the newborn and frequency of Jra antigen in the Japanese population. Vox Sang. 35:265-267.

Nakajima, H., Kuniyuki, M., and Takahara, N. (1967a) An infrequent blood antigen 'Tsunoi (Ts)' defined by the serum from a patient with acquired haemolytic anemia. Jpn. J. Hum. Genet. 12:187-189.

Nakajima, H. and Moulds, J. J. (1980) Doa (Dombrock) blood group antigen in the Japanese. Tests on further population and family samples. Vox Sang. 38:294-296.

Nakajima, H. and Ohkura, K. (1971) The distribution of several serological and biochemical traits in East Asia. III. The distribution of gamma-globulin [Gm(1), Gm(2), Gm(5), and Inv(1)] and Gc groups in Taiwan and Ryukyu. Hum. Hered. 21:362-370.

Nakajima, H., Ohkura, K., Ørjasaeter, H., and Kornstad, L. (1967b) The Nya blood group antigen among Japanese, Ryukyuan and Chinese and the mountainous aborigines in Taiwan. Jpn. J. Hum. Genet. 11:263-265.

Nakamura, K., Goto, F., Ray, W. A., McAllister, C. B., Jacqz, E., Wilkinson, G. R., and Branch, R. A. (1985) Interethnic differences in genetic polymorphism of debrisoquine and mephenytoin hydroxylation between Japanese and Caucasian populations. Clin. Pharmacol. Ther. 38:402-408.

Nakamura, S. and Abe, K. (1982a) Genetic polymorphism of coagulation factor XIIIB subunit in Japanese. Ann. Hum. Genet. 46:203-207.

Nakamura, S. and Abe, K. (1982b) Genetic polymorphism of human plasminogen in the Japanese population: New plasminogen variants and relationship between plasminogen phenotypes and their biological activities. Hum. Genet. 60:57-59.

Nakamura, S. and Abe, K. (1985) Genetic polymorphism of human factor I (C3b inactivator). Hum. Genet. 71:45-48.

Nakamura, S., Ooue, O., and Abe, K. (1984a) Genetic polymorphism of the seventh component of complement in a Japanese population. Hum. Genet. 66:279-281.

Nakamura, S., Ooue, O.,and Abe, K. (1986a) Genetic polymorphism of human complement component C81 in the Japanese population. Hum. Genet. 72:344-347.

Nakamura, S., Ohue, O., and Abe, K. (1986b) Genetic polymorphism of coagulation factor XIIIB subunit in the Japanese population: description of three new rare alleles. Hum. Genet. 73:183-185.

Nakamura, S., Ooue, O., Akiyama, K., and Abe, K. (1984b) Genetic polymorphism of complement C6 and haplotype analysis between C6 and C7 in a Japanese population. Hum. Genet. 68:138-141.

Nakashima, K. (1974) Further evidence of molecular alteration and aberration of erythrocyte pyruvate kinase. Clinica Chimica Acta 55:245-254.

Neel, J. V. (1978) Rare variants, private polymorphisms, and locus heterozygosity in Amerindian populations. Amer. J. Hum. Genet. 30:465-490.

Neel, J. V., Ferrell, R. E., and Conard, R. A. (1976) The frequency of 'rare' protein variants in Marshall Islanders and other Micronesians. Amer. J. Hum. Genet. 28:262-269.

Neel, J. V., Gershowitz, H., Mohrenweiser, H. W., Amos, B., Kostyu, D. D., Salzano, F. M., Mestriner, M. A., Lawrence, D., Simoes, A. L., Smouse, P. E., Oliver, W. J., Spielman, R. S., and Neel, J.V.,Jr. (1980) Genetic studies on the Ticuna, an enigmatic tribe of Central Amazonas. Ann. Hum. Genet. 44:37-54.

Neel, J. V., Gershowitz, H., Spielman, R. S., Migliazza, E. C., Salzano, F. M., and Oliver, W. J. (1977a) Genetic studies of the Macushi and Wapishana Indians. II. Data on 12 genetic polymorphisms of the red cell and serum proteins: Gene flow between the tribes. Hum. Genet. 37:207-219.

Neel, J. V., Tanis, R. J., Migliazza, E. C., Spielman, R. S., Salzano, F., Oliver, W. J., Morrow, M., and Bachofer, S. (1977b) Genetic studies of the Macushi and Wapishana Indians. I. Rare genetic variants and a "Private polymorphism" of esterase A. Hum. Genet. 36:81-107.

Nei, M. and Imaizumi, Y. (1966) Genetic structure of human populations. I. Local differentiation of blood group gene frequencies in Japan. Heredity 21:9-35.

Nelson, R. L., Povey, M. S., Hopkinson, D. A., and Harris, H. (1977a) Electrophoresis of Human L-glutamate dehydrogenase: Tissue distribution and preliminary population survey. Biochem. Genet. 15:87-91.

Nelson, R. L., Povey, S., Hopkinson, D. A., and Harris, H. (1977b) Detection after electrophoresis of enzymes involved in ammonia metabolism using L-glutamate dehydrogenase as a linking enzyme. Biochem. Genet. 15:1023-1035.

Nemoto, H. and Murao, M. (1961) A genetic study of colour blindness. Jpn. J. Hum. Genet. 6:165-173.

Neppert, J. (1980) Blood group antibody anti-M^g and isoagglutinin frequency in the Republic of Liberia, West Africa. Transfusion 20:448-449.

Nerell, G. (1963) Secretors of ABH antigen in a central Swedish population. Ann. Hum. Genet. 27:119-123.

Nerstrøm, B. (1965) Gc-serumtypesystemet og dets anvendelse i retsmedicin. Thesis - Copenhagen University. Cited from Berg (1974).

Neumann, S. and Walter, H. (1968) Frequencies of pseudocholinesterase variants in Icelanders, Greeks and Pakistanis. Nature 219:950.

Nevanlinna, H. R. (1966). Cited from Race and Sanger (1975).

Nevanlinna, H. R. (1972) The Finnish population structure: A genetic and genealogical study. Hereditas 71:195-236.

Nevanlinna, H. R. (1980) Genetic markers in Finland. Haematologia 13:65-74.

Nevo, S. and Cleve, H. (1983) Gc subtypes in the Middle East: Report on an Arab Moslem population from Israel. Amer. J. Phys. Anthrop. 60:49-52.

Nicholls, E. M., Lewis, H. B. M., Cooper, D. W., and Bennett, J. H. (1965) Blood group and serum protein differences in some Central Australian aborigines. Amer. J. Hum. Genet. 17:293-307.

Nickel, B. E. and McAlpine, P. J. (1982) Extension of human acid α-glucosidase polymorphism by isoelectric focusing in polyacrylamide gel. Ann. Hum. Genet. 46:97-103.

Nielsen, J. C. (1961) Studies on the inheritance of the Gm groups. Proc. 2nd Intl. Congr. Hum. Genet. II:766-770.

Niessner, H. and Beutler, E. (1974) Starch gel electrophoresis of phosphofructokinase in red cells. Biochem. Med. 9:73-76.

Nijenhuis, L. E. (1956) Blood group frequencies in French Basques. First Int. Congr. Hum. Genet., Kopenhagen, iii: 375-379. Cited from Mourant et al. (1976).

Nijenhuis, L. E. (1961) Blood group frequencies in the Netherlands, Curaçao, Surinam and New Guinea. Thesis-Amsterdam. Cited from Mourant et al. (1976).

Nilsson, L.-O. and Eriksson, A. W. (1972) Screening for haemoglobin and lactate dehydrogenase variants in the Icelandic, Swedish, Finnish, Lappish, Mari and Greenland Eskimo populations. Hum. Hered. 22:372-379.

Nishigaki, I., Itoh, T., Suzuki, H., and Fujiki, N. (1980) Genetic studies of red cell glutamic-pyruvic transaminase in some Japanese populations. Hum. Hered. 30:33-38.

Nishigaki, I., Omoto, K. and Juji, T. (1981) Genetic polymorphism of the A subunit of human coagulation factor XIII in Japanese. Jpn. J. Hum. Genet. 26:237-241.

Nishimukai, H., Kitamura, H., Sano, Y., and Tamaki, Y. (1985) C3 variants in Japanese. Hum. Hered. 35:69-72.

Nishimukai, H. and Tamaki, Y. (1986) Genetic polymorphism of the seventh component of complement: A new variant. Vox Sang. 51:60-61.

Niswander, J. D., Brown, K. S., Iba, B. Y., Leyshon, W. C., and Workman, P. L. (1970) Population studies on Southwestern Indian tribes. I. History, culture, and genetics of the Papago. Amer. J. Hum. Genet. 22:7-23.

Niu, K., Chen, L., and Du, R. (1984) The genetic polymorphisms of α_1-antitrypsin (α_1-AT) in Dong, Hui, Bai and Tujia populations. Ann. Report Inst. Genet. Academ. Sinica, p. 65.

Niu, K. and Du, R. (1984). The phenotype distributions of haptoglobin (HP) in Dong, Hui, Bai and Tujia ethnic groups. Ann. Report Inst. Genet. Academ. Sinica, p. 65.

Noades. J., Gavin, J., Tippett, P., Sanger, R., and Race, R. R. (1966) The X-linked blood group system Xg. Tests on British, Northern American, and Northern European unrelated people and families. J. Med. Genet. 3:162-168.

Noel, E. P., Sampson, L., Pepper, B. M., and Farid, N. R. (1980) Polymorphism of the second component of complement (C2) in Graves' disease. Hum. Hered. 30:245-247.

Novelo, G., Salazar, M. M., and Kornhauser, T. (1964) Caracteristicas de los aglutinogenos de grupos sanguineos de los indigenas mexicanos. 9th. Congr. Int. Soc. Blood Transfus., Mexico, 1962. Bibl. Haemat. 19:256-262.

Nurse, G. T. and Jenkins, T. (1975) The Griqua of Campbell, Cape Province, South Africa. Amer. J. Phys. Anthrop. 43:71-78.

Nurse, G. T. and Jenkins, T. (1977) Health and the Hunter-Gatherer. Monographs in Human Genetics. Vol. 8, Karger, Basel.

Nurse, G. T., Jenkins, T., Africa, B. J., and Stellmacher, F. F. (1982) Sero-genetic studies on the Basters of Rehoboth, South West Africa/Namibia. Ann. Hum. Biol. 2:157-166.

Nurse, G. T., Jenkins, T., Santos David, J. H., and Steinberg A. G. (1979) The Njinga of Angola: a serogenetic study. Ann. Hum. Biol. 4:337-348.

Nurse, G. T., Lane, A. B., and Jenkins, T. (1976) Sero-genetic studies on the Dama of south west Africa. Ann. Hum. Biol. 3:33-50.

Nurse, G. T., Weiner, J. S., and Jenkins, T. (1985) The Peoples of Southern Africa and Their Affinities. Clarendon Press, Oxford.

Nussbaum, R. L., Crowder, W., E., Nyhan, W. L., and Caskey, C. T. (1983) A three-allele restriction-fragment-length polymorphism at the hypoxanthine phosphoribosyltransferase locus in man. Proc. Natl. Acad. Sci. USA 80:4035-4039.

Oettgen, P., Antonarakis, S. E., and Karathanasis, S. K. (1986a) PvuII polymorphic site upstream to the human ApoCIII gene. Nucleic Acids Res. 14:5571.

Oettgen, P., Antonarakis, S. E., and Karathanasis, S. K. (1986b) BglII polymorphic site downstream to the human apolipoprotein AIV (apoAIV) gene. Nucl. Acids Res. 14:7138.

Ohayon, E., De Mouzon, A., Hauptmann, G., Klein, J., Abbal, M., Constans, J., Mayer, S., and Ducos, J. (1980) High frequency of the properdin factor Bf F1 and its linkage to HLA in French Basques. J. Immunogenet. 7:441-445.

Ohkura, K., Miyashita, T., Nakajima, H., Matsumoto, H., Matsumoto, K., Rahabar, S., and Hedayat, S. (1984) Distribution of polymorphic traits in Mazandaranian and Guilanian in Iran. Hum. Hered. 34:27-39.

Ohta, Y. (1963) An investigation of abnormal hemoglobins in southern Japan. I. A case of thalassemia minor discovered on Amami Islands. Jpn. J. Hum. Genet. 8:227-238.

Ojikutu, R. O., Nurse, G. T., and Jenkins, T. (1977) Red cell enzyme polymorphisms in the Yoruba. Hum. Hered. 27:444-453.

Okubo, Y., Nagao, N., Tomita, T., Yamaguchi, H., and Moulds, J. J. (1986a) The first examples of the Gy(a-), Hy- phenotype and anti-Gy^a found in Japan. Transfusion 26:214-215.

Okubo, Y., Yamaguchi, H., Seno, T., Araki, Y., Noguchi, M., Shioda, K., Takai, M., and Daniels, G. L. (1984a) The rare red cell phenotype Lan negative in Japanese. Transfusion 24:534-535.

Okubo, Y., Yamaguchi, H., Seno, T., Kikuchi, M., Abe, S., Ishuijima, A., and Daniels, G. L. (1984b) The rare red cell phenotype Gerbich negative in Japanese. Transfusion 24:274-275.

Okubo, Y., Yamaguchi, H., Seno, T., Miyata, Y., Moulds, M. K., and Moulds, J. J. (1986b) The first example of anti-Ul^a and and Ul (a+) red cells found in Japan. Transfusion 26:215.

Ökte, M. (1960) Colorblindness in Turkey (in Turkish). Türk. Oftal. Kong. Bül. 344-347. Cited from Post (1971).

Olaisen, B. (1975) Distribution of GPT types in Norway. Hum. Hered. 25:20-29.

Olaisen, B., Siverts, S., Toias, G. D., Janassen, R., and Teisberg, P. (1983) 10th Int. Congr. Soc. Forens. Haemogenet., Munich, Abstracts vol., pp.459-469. Cited from Miller et al. (1985).

Olaisen, B. and Teisberg, P. (1972) Erythrocyte alanine aminotransferase polymorphism in Norwegian Lapps. Hum. Hered. 22:380-386.

Olaisen, B., Teisberg, P., and Jonassen, R. (1976a) ESD polymorphism in Norway. Hum. Genet. 34:63-64.

Olaisen, B., Teisberg, P., and Jonassen, R. (1976b) GLO polymorphism in Norway. Hum. Hered. 26:454-457.

Olaisen, B., Teisberg, P., Gedde-Dahl, T., Jr., and Thorsby, E. (1978) Genetic polymorphism of the second component of human complement (C2). Hum. Genet. 42:301-305.

Olving, J. H., Teisberg, P., and Olaisen, B. (1980) Polymorphism of the sixth component of complement (C6) in Norwegian Lapps. Hum. Hered. 30:211-214.

Omoto, K. (1970) The distribution of polymorphic traits in the Hidaka Ainu. I. Defective colour vision,

PTC taste sensitivity and cerumen dimorphism. J. Fac. Sci. Univ. Tokyo, Sec.V, 3:337-355.

Omoto, K. (1972) The distribution of red cell adenosine deaminase phenotypes in Oceania. Jpn. J. Hum. Genet. 16:166-169.

Omoto, K. (1973) Observation of some polymorphic traits among the Australian aborigines. J. Anthrop. Soc. Nippon 81: 61-67.

Omoto, K. (1975) Serum protein groups. In: Anthropological and genetic studies on the Japanese (Watanabe, S., Kondo, S., and Matsunaga, E., eds.), Univ. Tokyo Press, Tokyo, pp.141-162.

Omoto, K., Aokoi, K., and Harada, S. (1975) Polymorphism of esterase D in some population groups in Japan. Hum. Hered. 25:378-381.

Omoto, K. and Blake, N. M. (1972) Distribution of genetic variants of erythrocyte phosphoglycerate kinase (PGK) and phosphohexose isomerase (PHI) among some population groups in South-east Asia and Oceania. Ann. Hum. Genet. 36:61-67.

Omoto, K. and Harada, S. (1968) Red cell and serum protein types in the Ainu population of Shizunai, Hokkaido. Proc. 8th Intl. Congr. Anthrop. Ethnol. Sci. 1:206-209.

Omoto, K. and Harada, S. (1972) The distribution of polymorphic traits in the Hidaka Ainu. II. Red cell enzyme and serum protein groups. J. Fac. Sci., Univ. Tokyo, Sec. V, 4:171-211. Cited from Ishimoto (1975).

Omoto, K., Misawa, S., Harada, S., Sumpaico, J. S., Medado, P. M., and Ogonuki, H. (1978) Population genetic studies of the Philippine Negritos. I. A pilot survey of red cell enzyme and serum protein groups. Amer. J. Hum. Genet. 30:190-201.

Omoto, K., Ueda, S., Goriki, K., Takahashi, N., Misawa, S., and Pagaran, I. G. (1981) Population genetic studies of the Philippine Negritos. III. Identification of the carbonic anhydrase-1 variant with CA_1 Guam. Amer. J. Hum. Genet. 33:105-111.

Ørjasaeter, H., Kornstad, L., and Heier Larsen, A. M. (1966) A contribution to the blood group frequencies in Tibetans. Vox Sang. 11:726-729.

Oya, M., Shibata, R., Kido, A., and Komatsu, N. (1985) Placental-soluble aconitase polymorphism in Japanese. Hum. Hered. 35:346-348.

Pagnier, J., Dunda-Belkhodja, O., Zohoun, I., Teyssier, J., Baya, H., Jaeger, G., Nagel, R. L., and Labie, D. (1984) α-Thalassemia among sickle cell anemia patients in various African populations. Hum. Genet. 68:318-319.

Palmhert-Keller, R., Nurse, G. T., and Jenkins, T. (1983) Sero-genetic studies on the Caucasoids of South West Africa/Namibia. Hum. Hered. 33:79-87.

Pals, G. and Pronk, J. C. (1979) Genetic variation in parotid basic proteins (Pb) in the Bozo (Mali, West Africa). Hum. Genet. 49:355-359.

Pálsson, J. und Walter, H. (1967) Untersuchungen zur Populationsgenetik von Island, insbesondere der Region Dalasysla. Humangenetik 4:352-361.

Pálsson, J. O. P., Walter, H., and Bajatzadeh, M. (1970) Serogenetical studies in Ireland. Hum. Hered. 20:231-239.

Papiha, S. S. (1973) Haptoglobin and abnormal haemoglobin types in Sinhalese and Punjabis. Hum. Hered. 23:147-153.

Papiha, S. S. (1981) Genetic diversity of complement system: A study of C_3 and Bf variations in human populations. Bionature 1:29-34.

Papiha, S. S. and Al-Agidi, S. K. (1976) Esterase D and superoxide dismutase polymorphisms in Iraq. Hum. Hered. 26:394-400.

Papiha, S. S., Bernal, J. E., Roberts, D. F., Habeebullah, C. M., and Mishra, S. C. (1979) C'3 polymorphism in some Indian populations. Hum. Hered. 29:193-196.

Papiha, S. S. and Chhaparwal, B. C. (1973) Serum protein and red cell enzyme polymorphism in two religious groups of Madhya Pradesh, India. IXth Int. Cong. Anthrop. Ethno. Sc., Chicago, USA.

Papiha, S. S., Constans, J., White, I., and McGregor, I. A. (1985) Group-specific component (Gc) subtypes in Gambian and Transkeian populations: a description of a new variant. Ann. Hum. Biol. 12:17-26.

Papiha, S. S. and Nahar, A. (1977) The world distribution of the electrophoretic variants of the red cell enzyme esterase D. Hum. Hered. 27:424-432.

Papiha, S. S., Roberts, D. F., Ali, S. G. M., and Islam, M. M. (1975) Some hereditary blood factors of the Bengali Muslim of Bangladesh. (Red cell enzymes, haemoglobins, and serum proteins.) Humangenetik 28:285-293.

Papiha, S. S., Roberts, D. F., Mukherjee, D. P., Singh, S. D., and Malhotra, M. (1978) A genetic survey in the Bhil tribe of Madhya Pradesh, Central India. Amer. J. Phys. Anthrop. 49:179-186.

Papiha, S. S., Roberts, D. F., and Rahimi, A. G. (1977a) Genetic polymorphisms in Afghanistan. Ann. Hum. Biol. 4:233-241.

Papiha, S. S., Roberts, D. F., and Shah, K. C. (1977b) Genetic variants of cytoplasmic malate dehydrogenase (MDH:EC:1.1.1.37) in populations in England and the Indian subcontinent. A new S-MDH variant. Hum. Genet. 36:73-79.

Papiha, S. S., Roberts, D. F., Shah, K. C., and Shah, A. C. (1981) A genetic study of some Gujarat populations. Acta Anthropogenet. 5:23-40.

Papiha, S. S., Roberts, D. F., White, I., Chahal, S. M. S., and Asefi, J. A. (1982a) Population genetics of the group specific component (Gc) and phosphoglucomutase (PGM_1) studied by isoelectric focusing. Amer. J. Phys. Anthrop. 59:1-7.

Papiha, S. S., Roberts, D. F., Wig, N. N., and Singh, S. (1972) Red cell enzyme polymorphisms in Punjabis in north India. Amer. J. Phys. Anthrop. 37:293-299.

Papiha, S. S. and Rodger, R. S. C. (1986) C3 and Bf complement types in chronic renal failure. Hum.

Genet. 72:260-261.

Papiha, S. S., Seyedna, Y., and Sunderland, E. (1982b) Phosphoglucomutase (PGM) and group-specific component (Gc), isoelectric focusing sub-types among Zoroastrians of Iran. Ann. Hum. Biol. 9:571-574.

Papiha, S. S. and Wastell, H. J. (1974) Transferrin variants in the Indian subcontinent. Humangenetik 21:69-73.

Parikh, N. P., Baxi, A. J., and Jhala, H. I. (1969) Blood groups, abnormal haemoglobins and other genetical characters in three Gujarati-speaking groups. Hum. Hered. 19:486-498.

Parisi, H. A., Lanara, E. C., and Triantaphyllidis, C. D. (1980) Protein and enzyme polymorphisms in affective disorders in northern Greece. Hum. Hered. 30:181-184.

Park, K. S., Tokunaga, K., and Omoto, K. (1985) Genetic polymorphisms of human complement components BF and C2 in Korean: Population and association studies. Jpn. J. Hum. Genet. 30:9-14.

Parr, C. W. (1966) Erythrocyte phosphogluconate dehydrogenase polymorphism. Nature 210:487-489.

Parr, C. W., Bagster, I. A., and Welch, S. G. (1977) Human red cell glyoxalase I polymorphism. Biochem. Genet. 15:109-133.

Partanen, J., and Koskimies, S. (1986) Human MHC class III genes, Bf, and C4: Polymorphism, complotypes and association with MHC class I genes in the Finnish population. Hum. Hered. 36: 269-275.

Pascali, V. L. and Auconi, P. (1983) Transferrin: Common and rare variants in Italy. Evidence for the existence of the rare Tf^{C}6 among Caucasians. Hum. Genet. 64:232-234.

Pascali, V. L. and de Mercurio, D. (1981) Determination of alpha-1-antitrypsin subtypes in the population of Rome: A study in ultrathin-layer isoelectric focusing. Hum. Hered. 31:296-298.

Pascasio, F. M., Bias, W. B., Manipol, V., and Campos, P. C. (1974) Genetic marker systems in Philippine Negritos. Birth Defects, Orig. Artic. Ser. 10:220-225.

Paslier, D. L., Rochu, D, and Lucotte, G. (1985) Pst I polymorphism of the antithrombin III gene in a French population. Vox Sang. 49:168-170.

Pausch, V., Weirather, M., Dub, E., Göbel, J., and Mayr, W. R. (1985) No association between GLO I and Hp in the Austrian population. Hum. Hered. 35:111-112.

Pena-Yanez, A., Pena Angulo, J. F., and Juarez Fernandez, C. (1971) Malabsorcion de la lactosa en estudiantes espanoles. I. Tolerancia intestinal al la sobrecarga oral de lactosa. Rev. Esp. Enferm. Apar. Dig. 35:925-938. Cited from Flatz (1986).

Penketh, R. J. A., Gibney, S. F. A., Nurse, G. T., and Hopkinson, D. A. (1983) Acetylator phenotypes in Papua New Guinea. J. Med. Genet. 20:37-40.

Pereira, T. M. and Manso, C. (1975) Immunoglobulin allotypes in Portugal. Humangenetik. 27:137-140.

Persson, I. (1968) The distribution of serum types in West Greenland Eskimos. Acta Genet. Stat. Med. 18:261-270.

Persson, I., Melartin, L., and Gilberg, A. (1971) Alloalbuminemia. A search for albumin variants in Greenland Eskimos. Hum. Hered. 21:57-59.

Peters, J., Hopkinson, D. A., and Harris, H. (1973) Genetic and non-genetic variation of triose phosphate isomerase isozymes in human tissues. Ann. Hum. Genet. 36:297-312.

Petersen, G. M., Rotter, J. I., Cantor, R. M., Field, L. L., Greenwald, S., Lim, J. C. T., Roy, C., Schoenfeld, V., Lowden, A., and Kaback, M. M. (1983) The Tay-Sachs disease gene in North American Jewish populations: Geographic variations and origin. Amer. J. Hum. Genet. 35:1258-1269.

Petrakis, N. L. (1969) Dry cerumen - a prevalent genetic trait among American Indians. Nature 222:1080-1081.

Petrakis, N. L. (1971) Cerumen genetics and human breast cancer. Science 173:347-349.

Petrakis, N. L., Pingle, U., Petrakis, S. J., and Petrakis, S. L. (1971) Evidence for a genetic cline in earwax types in the Middle East and Southeast Asia. Amer. J. Phys. Anthrop. 35:141-144.

Petrucci, R. and Congedo, P. (1983) Genetic studies of Gc (vitamin D binding globulin) polymorphism in the population of Latium (Italy). J. Hum. Evol. 12:439-441.

Petrucci, R., Leonardi, A., and Battistuzzi, G. (1982) The genetic polymorphism of Δ -aminolevulinate dehydrase in Italy. Hum. Genet. 60:289-290.

Pflugshaupt, R., Scherz, R., and Bütler, R. (1978) Polymorphism of several red cell enzymes in the Swiss population. Acta Anthropogenet. 2:1-11.

Pflugshaupt, R., Scherz, R., Steinegger, L., and Bütler, R. (1975) The frequency of the polymorphisms of C3 in the Swiss population and some remarks on the identification of rare phenotypes. In: Protides of the Biological Fluids: 22nd Colloquim, (Peeters, H., ed.), Pergamon Press, New York, pp. 559-562.

Piazza, A., van Loghem, E., de Lange, G., Curtoni, E. S., Ulizzi, L., and Terrenato, L. (1976) Immunoglobulin allotypes in Sardinia. Amer. J. Hum. Genet. 28:77-86.

Pierce, J. A., Eradio, B., and Dew, T. A. (1975) Antitrypsin phenotypes in St. Louis. JAMA 231:609-612.

Pieters, J. J. L. and van Rens, R. (1973) Lactose malabsorption and milk tolerance in Kenyan school age children. Trop. Geogr. Med. 25:365-371.

Pinder, L. B., Farr, D. E., and Woodfield, D. G. (1984) Milne, a new low-frequency antigen. Vox Sang. 47:290-292.

Pinder, L. B., Staveley, J. M., Douglas, R., and Kornstad, L. (1969) Pt^a- a new private antigen. Vox Sang. 17:303-305.

Pirastu, M., Lee, K. Y., Dozy, A. M., Kan, Y, W., Stamatoyannopoulos, G., Hadjiminas, M. G., Zachariades, Z., Angius, A., Furbetta, M., Rosatelli, C., and Cao, A. (1982) Alpha-thalassemia in two Mediterranean populations. Blood 60:509-512.

Plato, C. C. and Cruz, M. (1966) Blood group and haptoglobin frequencies of the Trukese of Micronesia. Acta Genet. Stat. Med. 16:74-83.

Plato, C. C., Cruz, M. T., and Kurland, L. T. (1964) Frequency of glucose-6-phosphate dehydrogenase deficiency, red-green colour blindness and Xg[a] blood-group among Chamorros. Nature 202:728.

Playfer, J. R., Eze, L. C., Bullen, M. F., and Evans, D. A. P. (1976) Genetic polymorphism and interethnic variability of plasma paroxonase activity. J. Med. Genet. 13:337-342.

Polesky, H. F. and Swanson, J. L. (1966) Studies on the distribution of the blood group antigen Do[a] (Dombrock) and the characteristics of anti-Do[a]. Transfusion 6:268-270.

Polunin, I. and Sneath, P. H. A. (1953) Studies of blood groups in Southeast Asia. J. R. Anthrop. Inst. 83:215-251.

Pongpaew, P. and Schelp, F. P. (1980) Alpha-1-protease inhibitor phenotypes and serum concentrations in Thailand. Hum. Genet. 54:119-124.

Pons, J. (1955) Taste sensitivity to phenylthiourea in Spaniards. Hum. Biol. 27:153-160.

Popivanov, I., Boev, P., and Karamihova-Tsacheva, L. 1965. The distribution of secreting types among the Bulgarians. C. R. Acad. Bulg. Sci. 18:955-958. Cited from Mourant et al. 1976.

Porck, H. J., Fleming, A. F., and Frants, R. R. (1984) Distribution of genetic variants of transcobalamin II in Nigerian Black populations. Amer. J. Hum. Genet. 36:710-717.

Post, R. H. (1962) Population differences in red and green color vision deficiency: A review, and a query of selection relaxation. Eugen. Quart. 9:131-146.

Post, R. H. (1971) Possible cases of relaxed selection in civilized populations. Humangenetik 13:253-284.

Potrafki, B. G., Hellenbroich, H., und Pulverer, G. (1972) Polymorphismus der erythrocytären NADH-Diaphorase in der westdeutschen Bevölkerung. Humangenetik 15:182-185.

Povey, S., Corney, G., and Harris, H. (1975) Genetically determined polymorphism of a form of hexokinase, HKIII, found in human leucocytes. Ann. Hum. Genet. 38:407-415.

Povey, S., Corney, G., Lewis, W. H. P., Robson, E. B., Parrington, J. M., and Harris, H. (1972) The genetics of peptidase C in man. Ann. Hum. Genet. 35:455-465.

Próchnicka, B. (1968) The transferrin group system in the Polish population. Acta Med. Polon. 9: 263-279.

Prochownik, E. V., Antonarakis, S., Bauer, K. A., Rosenberg, R. D., Fearon, E. R., and Orkin, M. D. (1983) Molecular heterogeneity of inherited antithrombin III deficiency. New Eng. J. Med. 308:1549-1552.

Pronk, J. C., Frants, R. R., Jansen, W., Eriksson, A. W., and Tonino, G. J. M. (1982) Evidence for duplication of the human salivary amylase gene. Hum. Genet. 60:32-35.

Pronk, J. C., Jansen, W. M., Pronk, A., Pol, C. F. A. M. v.d., Frants, R. R., and Eriksson, A. W. (1984) Salivary protein polymorphism in Kenya: Evidence for a new AMY1 allele. Hum. Hered. 34:212-216.

Przybylski, Z., Dobosz, T., and Stawarz, M. (1982) Genetic polymorphism of FUC in Polish population. (in Polish) Z. Rechtsmed. 89:21-24. Cited from Gill and Sutton (1984).

Püschel, K., Schmidt, M., Söder, R., and Brinkmann, B. (1979) Population genetic data on properdin factor B (Bf) in northern Germany. Amer. J. Phys. Anthrop. 50:247-250.

Quilici, J. C. (1968) Les altiplanides du corridor interandin. Étude hémotypologique. Toulouse, Centre Régional de Transfusion sanguine et d'Hématologie, pp. 103. Cited from Mourant et al. (1976).

Race, R. R. and Sanger, R. (1968) Blood Groups in Man, 5th ed. Oxford, Blackwell Scientific Publications.

Race, R. R. and Sanger, R. (1975) Blood Groups in Man, 6th ed. Oxford, Blackwell Scientific Publications.

Rahimi, A. G., Delbrück, H., Haeckel, R., Goedde, H. W., and Flatz, G. (1976) Persistence of high intestinal lactase activity (lactose tolerance) in Afghanistan. Hum. Genet. 34:57-62.

Rahimi, A. G., Goedde, H. W., Flatz, G., Kaifie, S., Benkmann, H.-G., and Delbrück, H. (1977) Serum protein polymorphisms in four populations of Afghanistan. Amer. J. Hum. Genet. 29:356-360.

Ram Kumar, N. S. and Rao, P. R. (1982) Some polymorphic enzymes from Andhra Pradesh. Hum. Hered. 32:121-123.

Ramsay, G. and Salamon, D. J. (1986) Frequency of PL[A1] in blacks. Transfusion 26:531-532.

Ranford, P. R., Kirk, R. L., and Zimmet, P. (1982) Distribution of complement factors Bf, C2, and C6 in Western Pacific. Acta Anthropogenet. 6:23-32.

Rantala, H., Finni, K., and Similä, S. (1982) Alpha-1-antitrypsin: The PiM subtypes and serum concentrations in Finnish newborns. Hum. Hered. 32:228-232.

Ranzani, G., Antonini, G., and Santachiara-Benerecetti, A. S. (1979) Red cell glyoxalase I polymorphism in Italians. Hum. Hered. 29:261-264.

Ranzani, G., Bertolotti, E., and Santachiara-Benerecetti, A. S. (1977) The polymorphism of red cell uridine monophosphate kinase in two samples of the Italian population. Hum. Hered. 27:332-335.

Ranzani, G. N., Brdicka, R., Antonini, G., Pardini, R., and Santachiara-Benerecetti, A. S. (1985) Electrophoretic subtyping of phosphoglucomutase locus 1 (PGM_1) polymorphism in the Italian and Czechoslovakian populations. Hum. Hered. 35:273-278.

Rao, P. R., Sampath, K., Char, N., Theophilus, J., Parasa, L., and Hussain, S. (1985) Incidence of C_5 isozyme of serum cholinesterase (E2 locus) in populations of Andhra Pradesh, South India. Hum. Hered. 35:126-128.

Rapley, S., Robson, E. B., Harris, H., and Maynard Smith, S. (1967) Data on the incidence, segregation and linkage relations of the adenylate kinase (AK) polymorphism. Ann. Hum. Genet. 31:237-242.

Rasch, L. H. (1960) Die Verteilung der Blutgruppen und Untergruppen in Oberbayern unter Einschluß von München. Blut 6:257-260.

Ratanaubol, K. and Ratanasirivanich, P. (1971) Xg blood groups of Thais. Nature 229:430

Raum, D., Marcus, D., and Alper, C. A. (1980) Genetic polymorphism of human plasminogen. Amer. J. Hum. Genet. 32:681-689.

379

Raum, D., Spence, M. A., Balavitch, D., Tideman, S., Merritt, A. D., Taggart, R. T., Petersen, B. H., Day, N. K., and Alper, C. A. (1979) Genetic control of the eighth component of complement. J. Clin. Invest. 64:858-865.

Rausen, A. R., Rosenfield, R. E., Alter, A. A., Hakim, S., Graven, S. N., Apollon, C. J., Dallman, P. R., Dalziel, J. C., Konugres, A. A., Francis, B., Gavin, J., and Cleghorn, T. E. (1967) A 'new' infrequent red cell antigen, Rd (Radin). Transfusion 7:336-342.

Ray, A. K. (1969) Color blindness, culture, and selection. Soc. Biol. 16:203-208.

Reddy, A. P., Mukherjee, B. N., Malhotra, K. C., Walter, H., and Sauber, P. (1984) Transferrin subtyping by isoelectric focusing in three West Bengal populations, India. Z. Morph. Anthrop. 74:345-349.

Reddy, V. and Pershad, J. (1972) Lactase deficiency in Indians. Amer. J. Clin. Nutr. 25:114-119.

Reinskou (1967) Unpublished data. Cited from Tills et al. (1983a).

Reinskou (1974) Unpublished data. Cited from Berg (1974).

Rex, D. K., Bosron, W. F., Smialek, J. E., and Li, T.-K. (1985) Alcohol and aldehyde dehydrogenase isoenzymes in North American Indians. Alcoholism 9:147-152.

Rex-Kiss, B. (1967) Das MN-Ss Blutgruppensystem und ihre Brauchbarkeit in der medizinischen Vaterschaftsbegutachtung. (In Hungarian). Morph. és. Ig. Orv. Szemle. 7:214-220. Cited from Tills et al. (1983a).

Rex-Kiss, B. (1970) Rh-Bestimmungen und ihre Anwendung zur Klärung strittiger Vaterschaft in Ungarn. Z. Rechtsmed. 67:319-323. Cited from Tills et al. (1983a).

Rex-Kiss, B. Szabó, R., and Hartmann, E. (1973) Ergebnisse der im Bezirk von Ráckeve (Komitat Pest, Ungarn) vorgenommenen Blutgruppenuntersuchungen. Ann. Immun. Hung. 17:169-181. Cited from Tills et al. (1983a).

Reys, L., Manso, C., and Stamatoyannopoulos, G. (1970) Genetic studies on southeastern Bantu of Mozambique. i. Variants of glucose-6-phosphate dehydrogenase. Amer. J. Hum. Genet. 22:203-215.

Rife, D. C. (1953) An investigation of genetic variability among Sudanese. Amer. J. Phys. Anthrop. 11:189-202.

Ritter, H. (1969). Personal communication. Cited from Mourant et al. (1976).

Ritter, H., Friedrichson, U., and Schmitt, J. (1974) Genetic variation of mannose phosphate isomerase in man. Humangenetik 22:261-262.

Ritter, H. and Wendt, G. G. (1971a) Population genetics of phosphoglucose isomerase (EC:5.3.1.9). Humangenetik 13:356-357.

Ritter, H. and Wendt, G. G. (1971b) Indophenol oxidase variability. Humangenetik 14:72.

Rittner, C., Hargesheimer, W. and Mollenhauer, E. (1984) Population and formal genetics of the human C81 (α-γ) polymorphism. Hum. Genet. 67:166-169.

Roberts, D. F. (1967) Red/green color blindness in the Niger Delta. Eugen. Quart. 14:7-13.

Roberts, D. F., and Al-Agidi, S. K. (1979) The C3 component of complement: Gene frequency variation in Iraq. Amer. J. Phys. Anthrop. 50:511-514.

Roberts, D. F. and Boyo, A. E. (1962) Abnormal haemoglobins in childhood among the Yoruba. Hum. Biol. 34:20-37.

Roberts, D. F., Creen, C. K., and Abeyaratne, K. P. (1972a) Blood groups of the Sinhalese. Man 7:122-127. Cited from Tills et al. (1983a).

Roberts, D. F., Papiha, S. S., and Abeyaratne, K. P. (1972b) Red cell enzyme polymorphisms in Ceylon Sinhalese. Amer. J. Hum. Genet. 24:181-188.

Roberts, D. F., Papiha, S. S., Creen, C. K., Chhaparwal, B. C., and Mehta, S. (1974) Red cell enzyme and other polymorphic systems in Madhya Pradesh, Central India. Ann. Hum. Biol. 1:159-174.

Roberts, D. F., Papiha, S. S., Rao, G. N., Habeebullah, C. M., Kumar, N., and Murty, K. J. R. (1980) A genetic study of some Andhra Pradesh populations. Ann. Hum. Biol. 7:199-212.

Roberts, D. F., Papiha, S. S., and Ssebabi, E. C. T. (1977) Red cell isoenzymes in east Africa. Hum. Biol. 49:301-308.

Robson, E. B., Glen-Bott, A. M., Cleghorn, T. E., and Harris, H. (1964) Some rare haptoglobin types. Ann. Hum. Genet. 28:77-86.

Robson, E. B. and Harris, H. (1966) Further data on the incidence and genetics of the serum cholinesterase phenotype C_5+. Ann. Hum. Genet. 29:403-408.

Robson, E. B. and Harris, H. (1967) Further studies on the genetics of placental alkaline phosphatase. Ann. Hum. Genet. 30:219-232.

Rodriguez-Córdoba, S., Bootello, A., and Arnaiz-Villena, A. (1981) Bf polymorphism and its relationship with HLA antigens in a sample of the Spanish population: High Bf F1 frequencies. Tissue Antigens 17:231-237.

Rogde, S., Mevåg, B., Teisberg, P., Gedde-Dahl Jr., T., Tedesco, F., and Olaisen, B. (1985) Genetic polymorphism of complement component C8. Hum. Genet. 70:211-216.

Rokala, D. A., Polesky, H. F., and Matson, G. A. (1977) The genetic composition of reservation populations: The Blackfeet reservation, Montana, U.S.A. Hum. Biol. 49:19-29.

Ropartz, C., Rivat, L., Rousseau, P. Y., and Fine, J. M. (1965) Myélomes, maladies de Waldenström et groupes de gamma-globulines Gm et Inv. Rev. Fr. Étud. Clin. Biol. 10:507-513. Cited from Steinberg and Cook (1981).

Ropartz, C., Rivat, L., Rousseau, P. Y., Lauridsen, B., and Persson, I. (1970) A survey of 9 Gm-factors, the Inv and the Isf systems in Danes. Hum. Hered. 20:456-461.

Ropartz, C., Rousseau, P. Y., Rivat, L., Baitsch, H., Ritter, H., Pinkerton, F. J., et Mermod, L. E. (1964) Les groupes de gamma-globulines Gm et Inv parmi la population d'Honolulu (Hawaii). Acta Genet. Stat. Med. 14:25-35.

Rosenfield, R. E., Haber, G. V., Kissmeyer-Nieelsen, F., Jack, J. A., Sanger, R., and Race, R. R. (1960) Ge, a very common red-cell antigen. Brit. J. Haemat. 6:344-349.

380

Rosenkranz, W., Hadorn, B., Müller, W., Heinz-Erian, P., Hensen, Ch., and Flatz, G. (1982) Distribution of human adult lactase phenotypes in the population of Austria. Hum. Genet. 62:158-161.

Rothhammer, F., Goedde, H. W., Llop, E., Acuña, M., and Carvajal, P. (1984) Erythrocyte and HLA antigens of Atacameño Indians. Amer. J. Phys. Anthrop. 65:243-247.

Rowe, G. P. and Hammond, W. (1983) A new low-frequency antigen, Hg[a] (Hughes). Vox Sang. 45:316-319.

Roy, M. N. and Catterjea, J. B. (1965) Some observations on the secretion of blood group substances in the body fluids of man. J. Indian Med. Assn. 45:413-417.

Rudduck, C., Beckman, L., Franzén, G., and Lindström, L. (1985) C3 and C6 complement types in schizophrenia. Hum. Hered. 35:255-258.

Rudduck, C., Franzén, G., Hansson, A., and Rorsman, B. (1984) Properdin factor B (Bf) types in schizophrenia. Hum. Hered. 34:331-333.

Russell, S. and Russell, D. W. (1971) Isoniazid acetylator phenotyping of Amharas in Ethiopia. Bull. World. Hlth. Org. Notes. 11pp. stencilled. Cited from Mourant et al. (1976).

Rychkov, Y. G. and Sheremeteva, V. A. (1972) Population genetics of peoples in the north of Pacific Ocean basin: The problems of their history and adaptation. III. Populations of Asian Eskimos and Chukchi of Bering Sea coast. Vop. Anthrop. 42:3-30.

Sadre, M. and Karbasi, K. (1979) Lactose intolerance in Iran. Amer. J. Clin. Nutr. 32:1948-1954.

Saha, N. (1981) Erythrocyte glutathione reductase polymorphism in a Sudanese population. Hum. Hered. 31:32-34.

Saha, N. and Banerjee, B. (1973) Xg[a] blood group in Chinese, Malays and Indians in Singapore. Vox Sang. 24:542-544.

Saha, N. and Banerjee, B. (1986) A study of some blood genetic characteristics of Bedouin and non-Bedouin Arabs of Jordan. Hum. Hered. 36:276-280.

Saha, N., Bayoumi, R. A., El Sheikh, F. S., Samuel, A. P. W., El Fadil, I., El Houri, I. S., Sebai, Z. A., and Sabaa, H. M. A. (1980) Some blood genetic markers of selected tribes in western Saudi Arabia. Amer. J. Phys. Anthrop. 52:595-600.

Saha, N., Bhattacharyya, S. P., Yeoh, S. C., Chua, S. P. K., and Ratnam, S. S. (1987) Glucose dehydrogenase polymorphism among ethnic groups of Singapore - with report of two additional alleles (GDH[4] and GDH[5]). Amer. J. Hum. Genet. 40:126-130.

Saha, N., Jeremiah, S. J., and Povey, S. (1978a) Further data on mitochondrial malic enzyme in man. Hum. Hered. 28:421-425.

Saha, N., Kirk, R. L., Shanbhag, S., Joshi, S. R., and Bhatia, H. M. (1976) Population genetic studies in Kerala and the Nilgiris (South West India). Hum. Hered. 26:175-197.

Saha, N., Samuel, A. P. W., Omer, A., Ahmed, M. A., Hussein, A. A., and Gaddoura, E. N. (1978b) A study of some genetic characteristics of the population of the Sudan. Ann. Hum. Biol. 5:569-575.

Sahi, T. (1974) Lactose malabsorption in Finnish-speaking and Swedish-speaking populations in Finland. Scand. J. Gastroenterol. 9:303-308.

Salák, J. and Palousová, Z. (1971) On the phenotype distribution of red cell acid phosphatase in Czechoslovakia: the district of Ceské Budejovice. Humangenetik 13:247-249.

Saleh, H., Davrinche, C., Charlionet, R., and Rivat, C. (1986). Genetic variants of factor B in a population of Jordan. Hum. Hered. 36:405-407.

Salmon, C., Salmon, D., Liberge, G., André, R., Tippett, P., and Sanger, R. (1961) Un nouvel antigène de groupe sanguin érythrocytaire présent chez 80% des sujets de race blanche. Nouv. Rev. Franc. Hémat. 1:649-661. Cited from Mourant et al. (1976).

Salzano, F. M. (1972) Visual acuity and color blindness among Brazilian Cayapo Indians. Hum. Hered. 22:72-79.

Salzano, F. M., Gershowitz, H., Junqueira, P. C., Woodall, J. P., Black, F. L., and Hierholzer, W. (1972a) Blood groups and H-Le[a] salivary secretion of Brazilian Cayapo Indians. Amer. J. Phys. Anthrop. 36:417-426.

Salzano, F. M., Gershowitz, H., Mohrenweiser, H., Neel, J. V., Smouse, P. E., Mestriner, M. A., Weimer, T. A., Franco, M. H. L. P., Simoes, A. L., Constans, J., Oliveira, A. E., and de Melo E Freitas, M. J. (1986) Gene flow across tribal barriers and its effect among the Amazonian Içana river Indians. Amer. J. Phys. Anthrop. 69:3-14.

Salzano, F. M., Jacques, S. M. C., Franco, M. H. L. P., Hutz, M. H., Weimer, R. S. S., and da Rocha, F. J. (1980) The Caingang revisted: Blood genetics and anthropometry. Amer. J. Phys. Anthrop. 53:513-524.

Salzano, F. M., Mohrenweiser, H., Gershowitz, H., Neel, J. V., Mestriner, M. A., Simoes, A. L., Constans, J., and de Melo E Freitas, M. J. (1984) New studies on the Macushi Indians of northern Brazil. Ann. Hum. Biol. 11:337-350.

Salzano, F. M., Neel, J. V., Weitkamp, L. R., and Woodall, J. P. (1972b) Serum proteins, hemoglobins and erythrocyte enzymes of Brazilian Cayapo Indians. Hum. Biol. 44:443-458.

Salzano, F. M. and Shreffler, D. C. (1966) The Gc polymorphism in the Caingang Indians of Brazil. Acta Genet. Stat. Med. 16:242-247.

Salzano, F. M., Steinberg, A. G., and Tepfenhart, M. A. (1973) Gm and Inv allotypes of Brazilian Cayapo Indians. Amer. J. Hum. Genet. 25:167-177.

Salzano, F. M., Weimer, T. A., Franco, M. H. L. P., Mestriner, M. A., Simoes, A. L., Constans, J., and de Melo E Freitas, M. J. (1985) Population structure and blood genetics of the Pacaás Novos Indians of Brazil. 12:241-249.

Salzano, F. M., Woodall, J. P., Black, F. L., Weitkamp, L. R., and Franco, M. H. L. P. (1974) Blood groups, serum proteins and hemoglobins of Brazilian Tiriyo Indians. Hum. Biol. 46:81-87.

Samloff, M., Liebman, W. M., Glober, G. A., Moore, J. O., and Indra, D. (1973) Population studies of pepsinogen polymorphism. Amer. J. Hum. Genet. 25:178-180.

Samuel, A. P. W., Saha, N., Acquaye, J. K., Omer, A., Ganeshaguru, K., and Hassounh, E. (1986) Association of red cell glucose-6-phosphate dehydrogenase with haemoglobinopathies. Hum. Hered. 36:107-112.

Sandler, S. G., Kravitz, C., Sharon, R., Hermoni, D., Ezekiel, E., and Cohen, T. (1979) The Duffy blood group system in Israeli Jews and Arabs. Vox Sang. 37:41-46.

Sanger, R., Walsh, R. J., and Kay, M. P. (1951) Blood types of natives of Australia and New Guinea. Amer. J. Phys. Anthrop. 9:71-78.

Sangiorgi, S., Mochi, M., Beretta, M., Prosperi, L., Costantino, G., and Romeo, G. (1982) Genetic and demographic characterization of a population with high incidence of fucosidosis. Hum. Hered. 32:100-105.

Sanpitak, N., Delbrück, H., Muangintra, J., Winyar, B., and Flatz, G. (1972) Polymorphism of erythrocyte phosphoglucomutase, adenylate kinase and adenosine deaminase in northern Thailand. Humangenetik 14:330-332.

Santachiara-Benerecetti, A. S. (1986) A study of 17 enzyme markers in African Pygmies. In: African Pygmies (Cavalli-Sforza, L. L., ed.), Academic Press, New York, pp. 247-272.

Santachiara-Benerecetti, A. S., Bauer, E. W., Beretta, M., Ranzani, G., Morpurgo, G., Carter, N. D., D'Udine, B., Ranjit, S. K., and Modiano, G. (1976) A study of several genetic biochemical markers in Sherpas with description of some variant phenotypes. Hum. Hered. 26:351-359.

Santachiara-Benerecetti, A. S., Beretta, M., Negri, M., Ranzani, G., Antonini, G., Barberio, C., Modiano, G., and Cavalli-Sforza, L. L. (1980) Population genetics of red cell enzymes in Pygmies: A conclusive account. Amer. J. Hum. Genet. 32:934-954.

Santachiara-Benerecetti, A. S. and Modiano, G. (1964) The frequencies of haptoglobin and transferrin types in some villages of the Milan province. Acta Genet. Stat. Med. 14:36-40.

Santachiara-Benerecetti, A. S. and Modiano, G. (1969) Studies on African Pygmies. II. Red cell phosphoglucomutase studies in Babinga Pygmies: A common PGM_2 variant allele. Amer. J. Hum. Genet. 21:315-321.

Santachiara-Benerecetti, A. S., Ranzani, G. N., and Antonini, G. (1977) Studies on African Pygmies: V. Red cell acid phosphatase polymorphism in Babinga Pygmies. High frequency of ACP^R allele. Amer. J. Hum. Genet. 29:635-638.

Santolamazza, C., Benincasa, A., and Scozzari, R. (1986) Phosphoglycolate phosphatase polymorphism: Gene frequencies in three Italian samples. Hum. Hered. 36:281-285.

Santoro, C., Olivetti, E., and Carbonara, A. O. (1983) Distribution of haptoglobin subtypes in continental Italy and Sardinia. Hum. Hered. 33:195-198.

Satoh, C., Takahashi, N., Asakawa, J., Masunari, N., Fujita, M., Goriki, K., Hazama, R., and Iwamoto, K. (1984a) Electrophoretic variants of blood proteins in Japanese. I. Phosphoglucomutase-2 (PGM2). Jpn. J. Hum. Genet. 29:89-104.

Satoh, C., Takahashi, N., Kaneko, J., Kimura, Y., Fujita, M., Asakawa, J., Kageoka, T., Goriki, K., and Hazama, R. (1984b) Electrophoretic variants of blood proteins in Japanese. II. Phosphoglucomutase-1 (PGM1). Jpn. J. Hum. Genet. 29:287-310.

Satoh, C., Takahashi, N., Kimura, Y., Miura, A., Kaneko, J., Fujita, M., and Toyama, M. (1986) Electrophoretic variants of blood proteins in Japanese. VII. Cytoplasmic glutamate-oxaloacetate transaminase (GOT1). Jpn. J. Hum. Genet. 31:1-14.

Sawhney, K. S., Sunderland, E., and Woolley, V. (1984) Genetic polymorphisms in the Kuwaiti Arabs. Hum. Hered. 34:303-307.

Say, B., Kiran, O., Altay, C., and Berkel, I. (1966) The incidence of PTC taste sensitivity in the population of Ankara. Turk. J. Pediat. 8:171-175. Cited from Mourant et al. (1976).

Say, B., Ozand, P., Berkel, I., and Cevik, N. (1965) Erythrocyte glucose-6-phosphate dehydrogenase deficiency in Turkey. Acta Pediat. Scand. 54:319-324. Cited from Mourant et al. (1976).

Sayek, I., Karahasanoglu, A. M., and Özand, P. (1967) Pseudocholinesterases. iii. The presence of pseudo-cholinesterase variants in a Turkish population. Turk. J. Pediat. 9:8-12.

Scacchi, R., Corbo, R. M., Calzolari, E., Laconi, G., Palmarino, R., and Lucarelli, P. (1985) Human placental glucose dehydrogenase: IEF polymorphism in two Italian populations and enzyme activity in the six common phenotypes. Hum. Hered. 35:349-352.

Schamaun, O., Olaisen, B., Teisberg, P., Gedde-Dahl, Jr., T., and Ehnholm, C. (1985) Genetic studies of apolipoprotein A-IV by two-dimensional electrophoresis. In: Protides of the Biological Fluids. vol. 33. (Peeters, H., ed.) Pergamon, New York, pp.471-474.

Schanfield, M. S., Gergely, J., and Fudenberg, H. H. (1975a) Immunoglobulin allotypes in European populations. I. Gm and Km (Inv) allotypic markers in Hungarians. Hum. Hered. 25:370-377.

Schanfield, M. S., Gershowitz, H., Hong, K.-J., and Shim, B.-S. (1972) Studies on the immunoglobulin allotypes of Asiatic populations. III. Gm and Inv allotypes among random Koreans. Hum. Hered. 22:144-148.

Schanfield, M. S., Herzog, P., and Fudenberg, H. H. (1975b) Immunoglobulin allotypes in European populations. II. Gm, Am, and Km (Inv) allotypic markers in Czechoslovakia. Hum. Hered. 25:382-389.

Scherz, R., Fràter-Schröder, M., Steinegger, L., Pflugshaupt, R., and Bütler, R. (1982a) Transcobalamin II polymorphism in the Swiss population - in application to paternity testing. Hum. Hered. 32:289-292.

Scherz, R., Pflugshaupt, R., and Bütler, R. (1977) Genetic polymorphism of glycine-rich β-glycoprotein in the Swiss and Italian populations. Hum. Hered. 27:143-146.

Scherz, R., Pflugshaupt, R., and Bütler, R. (1981a) Isoelectric focusing of human red cell phosphoglucomutase (PGM_1): Phenotype distribution in the Swiss population - Rare phenotypes. Hum. Hered. 31:187-190.

Scherz, R., Pflugshaupt, R., and Bütler, R. (1981b) Phosphoglycolate phosphatase (PGP, EC 3.1.3.18) polymorphism in the Swiss population. Its application in paternity testing. Forensic Sci. Int. 18:267. Cited from Caeiro and Rey (1985).

Scherz, R., Pflugshaupt, R., Bütler, R., and Peyretti, F. (1982b) Genetic polymorphism of glycine-rich beta-glycoprotein in the Italian population. Hum. Hered. 32:11-14.

Scherz, R., Reber, B., Pflugshaupt, R., and Bütler, R. (1985) Genetic polymorphism of human serum transferrin in the Swiss population: Evidence for three "new" transferrin-variants. Electrophoresis 6:569-571.

Schiff, F. (1940) Racial differences in frequency of the 'secreting factor'. Amer. J. Phys. Anthrop. 27:255-262.

Schlesinger, D. (1968) The Inv(1) factor in the Polish population. Arch. Immunol. Ther. Exp. 16:742-746. Cited from Steinberg and Cook (1981).

Schlesinger, D. (1971) Hp subgroups in the Polish population. Arch. Immun. Ther. Exp. 23:339-344. Cited from Tills et al. (1983a).

Schmechta, H. and Geserick, G. (1972) Populationsgenetische Untersuchungzum $Alpha_1$-Antitrypsin-Polymorphismus (Pi-System) in der Berliner Beölkerung. Dtsch. Gesundheitswes. 27:1089-1091. Cited from Tills et al. (1983a).

Schmidt, I. (1936) Ergebnis einer Massenuntersuchung des Farbensinns mit dem Anomaloscop. Z. Bahnärzte 31:44. Cited from Post (1971).

Schmidtke, J., Pape, B., Krengel, U., Langenbeck, U., Cooper, D. N., Breyel, E., and Mayer, H. (1984) Restriction fragment length polymorphisms at the human parathyroid hormone gene locus. Hum. Genet. 67:428-431.

Schmitt, J. and Ritter, H. (1974) Genetic variation of aconitate hydratase in man. Humangenetik 22:263-264.

Schneider, P., Ananthakrishnan, R., Walter, H., Xirotiris, N., and Abele, R. (1975) Enzyme polymorphisms and haemoglobin variants in Greeks. Humangenetik 27:217-222.

Scott, E. M. (1973) Genetic disorders in isolated populations. Arch. Environ. Health 26:32-35.

Scott, E. M., Duncan, I. W., Ekstrand, V., and Wright, R. C. (1966) Frequency of polymorphic types of red cell enzymes and serum factors in Alaskan Eskimos and Indians. Amer. J. Hum. Genet. 18:408-411.

Scott, E. M. and Wright, R. C. (1978) Polymorphism of red cell enzymes in Alaskan ethnic groups. Ann. Hum. Genet. 41:341-346.

Scott, E. M. and Wright, R. C. (1980) Genetic polymorphism of rhodanese from human erythrocytes. Amer. J. Hum. Genet. 32:112-114.

Scott, J., Knott, T. J., Priestley, L. M., Robertson, M. E., Mann, D. V., Kostner, G., Miller, G. J., and Miller, N. E. (1985) High-density lipoprotein composition is altered by a common DNA polymorphism adjacent to apoprotein AII gene in man. Lancet. i :771-773.

Scott-Emuakpor, A. B., Uviovo, J. E., and Warren, S. T. (1975) Genetic variation in Nigeria. I. The genetics of phenylthiourea tasting ability. Hum. Hered. 25:360-369.

Scozzari, R., Trippa, G., Barberio, C., and Menini, P. (1975) Red cell glutamic-pyruvic transaminase gene frequencies in the region of the Po Delta (Ferrara, Northern Italy). Humangenetik 26:147-150.

Sebastio, G., Riccio, A., Verde, P., Scarpato, N., and Blasi, F. (1985) BamHI RFLP linked to the human urokinase gene. Nucleic Acids Res. 13:5404.

Sebetan, I. M., Akaishi, S., Matsumoto, H., and Toyomasu, T. (1982) Genetic variants of the human diaphorase DIA_3 in Japanese: Report of a new rare allele, $DIA_3{}^4$. Jpn. J. Hum. Genet. 27:313-318.

Sebetan, I. M., Hiraiwa, K., and Akaishi, S. (1985) Genetic polymorphism of transferrin in Egyptians: Analysis by two electrofocusing methods with description of unusual B variant. Jpn. J. Hum. Genet. 30:15-20.

Segal, I., Gagjee, P. P., Essop, A. R., and Noormohamed, A. M. (1983) Lactase deficiency in the South African black population. Amer. J. Clin. Nutr. 38:901-905.

Séger, J. (1971) Fréquences des types de phosphatase acide, phosphoglucomutase et adénylate kinase dans la population parisienne. Rev. Franç. Transfus. 14:393-399. Cited from Tills et al. (1983a).

Séger, J. (1977) Fréquences dans la population parisienne des gènes correspondant au polymorphisme de l'adénosine déaminase, de la 6 phosphogluconate deshydrogenase et de l'esterase D. Rev. fr. Transf. Immuno-Hématol. 4:575-583. Cited from Vergnes et al. (1980b).

Sen, D. K. (1960) Blood groups and haemoglobin variants in some upper castes of Bengal. J. R. Anthrop. Inst. 90:161-172.

Seniwiratne, B., Thambipillai, S., and Perera, H. (1977) Intestinal lactase deficiency in Ceylon (Sri Lanka). Gastroenterology 72:1257-1259.

Seno, T., Yamaguchi, H., Okubo, Y., Sunni, R., and Green, C. A. (1983) Os^a, a 'new' low-frequency red cell antigen. Vox Sang. 45:60-61.

Serafini, N. A., Serra, A., Fagiolo, E., and Schinco, G. (1968) Haptoglobin phenotype and gene frequencies in the population of Rome. Acta Genet. Stat. Med. 18:458-467.

Seth, S. (1968) The Lewis blood groups and their correlation with ABO(H) group specific substances. Z. Morph. Anthrop. 59:232-237.

Seyfried, H., Frankowska, K., and Giles, C. M. (1966) Further examples of anti-Bu^a found in immunized donors. Vox Sang. 11: 512-516.

Shaker, Y., Onsi, A., and Aziz, R. (1966) The frequency of glucose-6-phosphate dehydrogenase deficiency in the newborns and adults in Kuwait. Amer. J. Hum. Genet. 18:609-613.

Shapiro, M. (1953) Blood groups and skin colour. J. Forens. Med. 1:2-10. Cited from Mourant et al. (1976).

Shapiro, M. (1964) Serology and genetics of a 'new' blood factor: hr^H. J. Forens. Med. 11:52-66. Cited from Race and Sanger (1975).

Sharma, J. C. (1959) Blood and PTC taste studies in Punjabis and the effect of age and certain eating habits on taste threshold. Anthropologist 6(1&2):40-46.

Sheehy, T. W. and Anderson, P. R. (1965) Disaccharidase activity in normal and diseased small bowel. Lancet ii:1-4.

Shibata, K. (1983) Haptoglobin, group-specific component, transferrin and α_1-antitrypsin subtypes and new variants in Japanese. Jpn. J. Hum. Genet. 28:17-27.

Shibata, Y., Matsuda, I., Miyaji, T., and Ichikawa, Y. (1986) Yuk[a], a new platelet antigen involved in two cases of neonatal alloimmune thrombocytopenia. Vox Sang. 50: 177-180.

Shih, L. Y. and Hsia, D. Y. Y. (1969) The distribution of genetic polymorphisms among Chinese in Taiwan. Hum. Hered. 19:227-233.

Shim, B. and Bearn, A. G. (1964) The distribution of haptoglobin subtypes in various populations, including subtype patterns in some nonhuman primates. Amer. J. Hum. Genet. 16:477-483.

Shimizu, K., Keino, H., Mizutani, A., Itoh, T., and Nishigaki, I. (1985) Red cell NADH diaphorase variants in Japanese. Hum. Hered. 35:212-217.

Shinoda, T. (1970) Polymorphism of red cell adenosine deaminase in the Japanese population. Jpn. J. Genet. 45:147-152.

Shinoda, T. and Matsunaga, E. (1970) Studies on polymorphic types of several red cell enzymes in a Japanese population. Jpn. J. Hum. Genet. 15:133-143.

Shokeir, M. H. K. and Shreffler, D. C. (1970) Two new ceruloplasmin variants in Negroes - Data on three populations. Biochem. Genet. 4:517-528.

Shrivastava, P. K. (1969) Molecular heterogeneity of human serum albumin. Arch. Immunol. Ther. Exp. (Warsz) 17:747.

Shulman, N. R., Moor-Jankowski, J., and Hiller, M. C. (1965) Platelet and leukocyte isoantigens common to man and other animals. Histocompatibility Testing 1965, Munksgaard, Copenhagen, pp.113-123.

Silvestroni, E., Bianco, I., Graziani, B., and Carboni, C (1978) First premarital screening of thalassaemia carriers in intermediate schools in Latium. J. Med. Genet. 15:202-207.

Simmons, R. T. (1966) The blood group genetics of Easter Islanders (Pascuense), and other Polynesians. In: The Norwegian archaeological expedition to Easter Island and the East Pacific (Heyerdahl, T. and Ferdon, E. N., eds.), Stockholm, ii, Report 15:333-343. Cited from Mourant et al. (1976).

Simmons, R. T. (1970) The apparent absence of the Diego (Di[a]) and the Wright (Wr[a]) blood group antigens in Australian aborigines and in New Guineans. Vox Sang. 19:533-536.

Simmons, R. T. and Albrey, J. A. (1963) A 'new' blood group antigen Webb (Wb) of low frequency found in two Australian families. Med. J. Aust. i:8-10. Cited from Race and Sanger (1975).

Simmons, R. T. and Cooke, D. R. (1969) Population genetic studies in Australian aborigines of the Northern Territory. Blood group genetic studies in the Malag of Elcho Island. Archaeol. Phys. Anthrop. Oceania 4:252-259.

Simmons, R. T., Graydon, J. J., and Semple, N. M. (1953) A further blood genetical survey in Micronesia: Palauans, Trukese and Kapingas. Med J. Aust. ii:589-596. Cited from Mourant et al. (1976).

Simmons, R. T., Graydon, J. J., and Semple, N. M. (1954) A blood group genetical survey in Australian aborigines. Amer. J. Phys. Anthrop. 12:599-606.

Simmons, R. T., Graydon, J. J., Semple, N. M., Birdsell, J. B., Milbourne, J. D., and Lee, J. R. (1952) A collaborative genetical survey in Marshall Islanders. Amer. J. Phys. Anthrop. 10:31-54.

Simmons, R. T., Graydon, J. J., Semple, N. M., and Fry, E. I. (1955) A blood group genetical survey in Cook Islanders, Polynesia, and comparisons with American Indians. Amer. J. Phys. Anthrop. 13:667-690.

Simmons, R. T., Tindale, N. B., and Birdsell, J. B. (1962) A blood group genetical survey in Australian aborigines of Bentinck, Mornington and Forsyth Islands, Gulf of Carpentaria. Amer. J. Phys. Anthrop. 20:303-320.

Simmons, R. T. and Were, S. O. M. (1955) A "new" family blood group antigen and antibody (By) of rare occurrence. Med. J. Aust. ii:55-58. Cited from Race, R. R. and Sanger, R. (1975).

Simpson, N. E. (1968) Genetics of esterase in man. Ann. N. Y. Acad. Sci. 151:699-709.

Singh, S., Jensen, M., Goedde, H. W., Lehmann, W., Pyröälä, K., and Eriksson, A. W. (1971) Pseudocholinesterase polymorphism among Lapp populations in Finland. Humangenetik 12:131-135.

Singh, S., Sareen, K. N., and Goedde, H. W. (1974a) Investigation of some biochemical genetic markers in four endogamous groups from Punjab (N. W. India). I. Protein and enzyme polymorphisms in serum. Humangenetik 21:341-346.

Singh, S., Sareen, K. N., and Goedde, H. W. (1974b) Investigation of some biochemical genetic markers in four endogamous groups in Punjab (N. W. India). II. Red cell enzyme polymorphisms. Humangenetik 22:133-138.

Singh, S., Saternus, K., Münsch, H., Altland, K., Goedde, H. W., and Eriksson, A. W. (1974c) Pseudocholinesterase polymorphism among Ålanders (Finno-Swedes), Maris (Cheremisses, USSR), and Greenland Eskimos, and the segregation of some E_1 and E_2 locus types in Finnish Lapp families. Hum. Hered. 24:352-362.

Siniscalco, M. (1967). Cited from Tills et al. (1983a).

Siniscalco, M., Filippi, G., Latte, B., Piomelli, S., Rattazzi, M., Gavin, J., Sanger, R., and Race, R. R. (1966) Failure to detect linkage between Xg and other X-borne loci in Sardinians. Ann. Hum. Genet. 29:231-252.

Sistonen, P., Nevanlinna, H. R., Virtaranta-Knowles, K., Pirkola, A., Leikola, J., Kekomäki, R., Gavin, J. and Tippet, P. (1981) Ne[a], a new blood group antigen in Finland. Vox Sang. 40:352-357.

Skeller, E. (1954) Anthropological and ophthalmological studies on the Angmagssalik Eskimos. Meddeleser om Grønlande. 107:1-231. Cited from Post (1962).

Skov, F. (1972) A new rare blood group antigen, Je[a]. Vox Sang. 23:461-463.

Slaughter, C. A., Hopkinson, D. A., and Harris, H. (1975) Aconitase polymorphism in man. Ann. Hum. Genet. 39:193-202.

Smith, D. S., Stratton, F., Johnson, T., Brown, R., Howell, P., and Riches, R. (1969) Haemolytic disease of the newborn caused by anti-Lan antibody. Brit. Med. J. 3:90-92.

Smith, M., Hopkinson, D. A., and Harris, H. (1971) Developmental changes and polymorphism in human alcohol dehydrogenase. Ann. Hum. Genet. 34:251-271.

Smith, M., Hopkinson, D. A., and Harris, H. (1972) Alcohol dehydrogenase isozymes in adult human stomach and liver: evidence for activity of the ADH3 locus. Ann. Hum. Genet. 35:243-253.

Smith, S. E. and Tun, K. (1968) Inactivation of isoniazid in Burmese subjects. Nature 217:1273.

Socha, J. and Kaczera, Z. (1968) Studies on the gamma-globulin group systems in the Polish population. Folia Biol. Kraków 16:145-165.

Socha, J., Ksiazyk, J., Flatz, G., and Flatz, S. D. (1984) Prevalence of primary adult lactose malabsorption in Poland. Ann. Hum. Biol. 11:311-316.

Solaas, M. H. (1970) Frequency of the Ag(x) antigen in a Norwegian population sample. Hum. Hered. 20:290-291.

Sørensen, S. A. (1972) Adenylate kinase, adenosine deaminase and phosphoglucomutase phenotypes in a Danish population. Hum. Hered. 22:362-371.

Sørensen, S. A. (1973) Human red cell acid phosphatase polymorphism. Hum. Hered. 23:470-481.

Soulier, J. P., and Patereau, C. (1976) Groupage et détection d'anticorps dans le système PLA (pla-quettaire. Rev. Fr. Transfus. Immunohématol. 19: 431-448. Cited from Ramsey and Salamon (1986).

Sowers, M. F. and Winterfeldt, E. (1975) Lactose intolerance among Mexican Americans. Amer. J. Clin. Nutr. 28:704-705.

Spanidou, E. P. and Petrakis, N. L. (1972) Lactose intolerance in Greeks. Lancet ii:872-873.

Spedini, G. (1960) Ricerche sull'antigene Diego. Riv. Antrop. 47:249-255. Cited from Mourant et al. (1976).

Spedini, G., Capucci, E., Crosti, N., Danubio, M. E., and Romagnoli, S. (1982) Erythrocyte glyoxalase I and superoxide dismutase polymorphisms in the Mbugu and some other populations of the Central African Republic. Hum. Hered. 32:253-258.

Speiser, P. (1958) Krankheiten und Blutgruppen; Ueber Beziehung zwischen den Genitalkarzinomen bzw. Mammakarzinomen bei Frauen, Blutgruppen und Rh_0-(D)-Faktor. Wien, klin. Wschr. 70:315-316. Cited from Mourant et al. (1976).

Spielmann, W. (1966) Die Verteilung der Kidd-Gruppen in der hessischen Bewvölkerung. Folia Haemat. Lpz. 85:292-295. Cited from Mourant et al. (1976).

Spitsyn, V. A., Irisova, O. V., Annenkov, G. A., and Filippov, I. K. (1978) Some genetic characters of blood serum of the Karamojo (North-east Uganda). Hum. Biol. 50:229-234.

Ssebabi, E. C. T. and Roberts, D. F. (1975) East African frequencies in some lesser known blood group systems. Man 10:524-529. Cited from Tills et al. (1983a).

Stamatoyannopoulos, G., Thomakos, A., and Giblett, E. R. (1975) Red cell enzyme polymorphisms in the Greek populations. Humangenetik 27:23-30.

Steegmüller, H. (1975) On the geographical distribution of pseudocholinesterase variants. Humangenetik 26:167-185.

Steinberg, A. G. (1966) Letters to the Editor: Correction of previously published Gm(c) phenotypes of Africans and Micronesians. Amer. J. Hum. Genet. 18:109.

Steinberg, A. G. (1973) The Gm and Inv allotypes of some Ashkenazic Jews living in Northern U.S.A. Amer. J. Phys. Anthrop. 39:409-412.

Steinberg, A. G. and Cook, C. E. (1981) The Distribution of the Human Immunoglobulin Allotypes. Oxford Univ. Press, Oxford.

Steinberg, A. G., Damon, A., and Bloom, J. (1972a) Gammaglobulin allotypes of Melanesians from Malaita and Bougainville, Solomon Islands. Amer. J. Phys. Anthrop. 36:77-84.

Steinberg, A. G., Gajdusek, D. C., and Alpers, M. (1972b) Genetic studies in relation to Kuru. V. Distribution of human gamma globulin allotypes in New Guinea populations. Amer. J. Hum. Genet. 24:s95-s110.

Steinberg, A. G., Jenkins, T., Nurse, G. T., and Harpending, H. C. (1975) Gammaglobulin groups of the Khoisan peoples of southern Africa: evidence for polymorphism for a $Gm^{1,5,13,14,21}$ haplotype among the San. Amer. J. Hum. Genet. 27:528-542.

Steinberg, A. G. and Kageyama, S. (1970) Further data on the Gm and Inv allotypes of the Ainu; confirmation of the presence of a $Gm^{2,17,21}$ phenogroup. Amer. J. Hum. Genet. 22:319-325.

Steinberg, A. G. and Kirk, R. L. (1970) Gm and Inv types of Aborigines in the Northern Territory of Australia. Archaeol. Phys. Anthrop. Oceania 5:163-172.

Steinberg, A. G. and Morton, N. E. (1973) Immunoglobulins in the eastern Carolines. Amer. J. Phys. Anthrop. 38:699-702.

Steinberg, A. G., Stauffer, R., Blumberg, B. S., and Fudenberg, H. (1961) Gm phenotypes and genotypes in U.S. whites and Negroes; in American Indians and Eskimos; in Africans; and in Micronesians. Amer. J. Hum. Genet. 13:205-213.

Steinberg, A. G., Tiilikainen, A., Eskola, M. R., and Eriksson, A. W. (1974) Gammaglobulin allotypes in Finnish Lapps, Finns, Åland Islanders, Maris (Cheremis), and Greenland Eskimos. Amer. J. Hum. Genet. 26:223-243.

Steiner, E., Iselius, L., Alván, G., Lindsten, J., and Sjöqvist, F. (1985) A family study of genetic and environmental factors determining polymorphic hydroxylation of debrisoquine. Clin. Pharmacol. Ther. 38:394-401.

Stewart, R. E. and Lovrien, E. W. (1971) Haemopexin in human serum: A search for genetic polymorphism. Ann. Hum. Genet. 35:19-24.

Stoffersen, E. and Jørgensen, K. A. (1980) C3 polymorphism in patients with chronic uremia. Hum. Hered. 30:46-49.

Stoffersen, E., Jørgensen, K. A., Nymand, G., and Dyerberg, J. (1982) C3 polymorphism in Greenland Eskimos. Hum. Hered. 32:49-51.

Strange, R. C., Faulder, C. G., Davis, B. A., Hume, R., Brown, J. A. H., Cotton, W., and Hopkinson, D. A. (1984) The human glutathione S-transferases: studies on the tissue distribution and genetic variation of the GST1, GST2 and GST3 isozymes. Ann. Hum. Genet. 48:11-20.

Sugita, H. and Takahama, K. (1983) Red cell glyoxalase II type in a Japanese population. Jpn. J. Hum. Genet. 28:201-203.

Sukernik, R. I., Karaphet, T. M., and Osipova, L. P. (1978) Distribution of blood groups, serum markers and red cell enzymes in two human populations from northern Siberia. Hum. Hered. 28:321-327.

Sukernik, R. I., Lemza, S. V., Karaphet, T. M., and Osipova, L. P. (1981) Reindeer Chukchi and Siberian Eskimos: Studies on blood groups, serum proteins and red cell enzymes with regard to genetic heterogeneity. Amer. J. Phys. Anthrop. 55:121-128.

Sukernik, R. I., Osipova, L. P., Karaphet, T. M., and Abanina, T. A. (1980) Studies on blood groups and other genetic markers in Forest Nentzi: Variation among the subpopulations. Hum. Genet. 55:397-404.

Sunahara, S., Urano, M. and Ogawa, M. (1961) Genetical and geographic studies on isoniazid inactivation. Science 134:1530-1531.

Sunderland, E. (1971). Personal communication. Cited from Mourant et al. (1976).

Sunderland, E., Sawhney, K. S., and Bethell, I. L. (1979) Plasma protein and red cell enzyme groups among the Nepalese. Hum. Hered. 29:14-26.

Sunderland, E., Tills, D., Bouloux, C., and Doyl, J. (1971). Cited from Mourant et al. (1976).

Sung, J. L. and Shih, P. L. (1972) The jejunal disaccharidase activity and lactose intolerance in Chinese adults. Asian J. Med. 8:149-151. Cited from Flatz (1986).

Sussman, L. N., Meyer, L. H., and Conard, R. A. (1959) Blood groupings in Marshallese. Science 129:644-645.

Sussman, L. N., Meyer, L. H., Conard, R. A., and Smith, H. (1960) Blood grouping in Marshallese. 8th Intl. Congr. Hemat., Tokyo, pp. 1456-1458. Cited from Mourant et al. (1976).

Suzuki, K., Matsui, K., and Matsumoto, H. (1986) FXIIIA polymorphism in a Japanese population: Occurrence of FXIIIA*4 allele. Electrophoresis 7:289-290.

Svanda, M., Prochazka, R., Kout, M., and Giles, C. M. (1970) A case of haemolytic disease of the newborn due to a new red cell antigen, Zd. Vox Sang. 18:366-369.

Svensson, M. and Wetterling, G. (1982) Identification of PGM$_1$ (phosphoglucomutase EC 2.7.5.1) by isoelectric focusing in a Swedish population. Hum. Hered. 32:357-361.

Swallow, D. M., Corney, G., and Harris, H. (1975) Acid α-glucosidase: A new polymorphism in man demonstrable by 'affinity' electrophoresis. Ann. Hum. Genet. 38:391-406.

Swanson, J., Mary, Z., and Polesky, H. F. (1967) A new public antigenic determinant Gy^a (Gregory). Transfusion 7:304-306.

Szaloky, A., Sijpesteijn, N. K., and van der Hart, M. (1973) A new blood group antigen, Hov. Vox Sang. 24:535-541.

Szathmary, E. J. E. (1983) Dogrib Indians of the Northwest Territories, Canada: genetic diversity and genetic relationship among subarctic Indians. Ann. Hum. Biol. 10:147-162.

Szathmary, E. J. E., Cox, D. W., Gershowitz, H., Rucknagel, D. L., and Schanfield, M. S. (1974) The northern and southeastern Ojibwa: Serum proteins and red cell enzyme systems. Amer. J. Phys. Anthrop. 40:49-65.

Szathmary, E. J. E., Ferrell, R. E., and Gershowitz, H. (1983) Genetic differentiation in Dogrib Indians: serum protein and erythrocyte enzyme variation. Amer. J. Phys. Anthrop. 62:249-254.

Szathmary, E. J. E., Mohn, J. F., Gershowitz, H., Lambert, R. M., and Reed, T. E. (1975) The northern and southeastern Ojibwa: Blood group systems and the causes of genetic divergence. Hum. Biol. 47:351-368.

Szeinberg, A. (1974) Investigation of genetic polymorphic traits in Jews. In: Genetic Polymorphisms and Diseases in Man (Ramot, B., ed.), Academic Press, New York, pp. 45-54.

Szeinberg, A., Bar-Or, R., and Sheba, C. (1961) Distribution of isoniazid inactivator types in various Jewish groups in Israel. 2nd Int. Cong. Hum. Genet., Rome, pp.110-112.

Szeinberg, A., Pipano, S., Assa, M., Medalie, J., and Neufeld, H. N. (1972) High frequency of atypical pseudocholinesterase gene among Iraqi and Iranian Jews. Clin. Genet. 3:123-127.

Szeinberg, A., Pipano, S., Rozansky, Z., and Ravia, N. (1971) Frequency of red cell adenosine deaminase phenotypes in several population groups in Israel. Hum. Hered. 21:357-361.

Szeinberg, A. and Tomashevsky-Tamir, S. (1971) Red cell adenylate kinase and phosphoglucomutase polymorphisms in several population groups in Israel. Hum. Hered. 21:289-296.

Taggart, R. T., Karn, R. C., Merritt, A. D., Yu, P. L., Conneally, P. M. (1979) Urinary pepsinogen isozymes: A highly polymorphic locus in man. Hum. Genet. 52:227-238.

Takahara, S., Ogura, Y., Mitani, Y., Kasai, H., Kawasaki, K., Kuroda, Y., and Ohkura, K. (1973) A report of field survey on acatalasaemia and hypocatalasaemia in 1970 and 1971. Okyama I. Z. 85:607-613 (in Japanese). Cited from Ishimoto (1975).

Tan, S. G. (1976) Human saliva esterases: Genetic studies. Hum. Hered. 26:207-216.

Tan, S. G. and Ashton, G. C. (1976) An autosomal glucose-6-phosphate dehydrogenase (hexose-6-phosphate dehydrogenase) polymorphism in human saliva. Hum. Hered. 26:113-123.

Tan, S. G., Gan, Y. Y., Asuan, K., and Abdullah, F. (1981) Gc subtyping in Malaysians and in Indonesians from North Sumatra. Hum. Genet. 59:75-76.

Tan, S. G., Gan, Y. Y., and Asuan, K. (1982) Transferrin C subtyping in Malaysians and in Indonesians from North Sumatra. Hum. Genet. 60:369-370.

Tan, S. G. and Teng, Y. S. (1978) Saliva acid phosphatases and amylase in Senoi and aboriginal Malays and superoxide dismutase in various racial groups of peninsular Malaysia. Jpn. J. Hum. Genet. 23:133-138.

Tan, S. G. and Teng, Y. S. (1979) Saliva acid phosphatases in Malaysians: Report of a new variant. Hum. Hered. 29:61-63.

Tandon, R. K., Goel, U., Mukherjee, S. N., Pandey, S. C., and Lal, K. (1977) Lactose intolerance during pregnancy in different Indian communities. Indian. J. Med. Res. 66:33-38.

Tandon, R. K., Joshi, Y. K., Singh, D. S. Narendranathan, M., Balakrishnan, V., and Lal, K. (1981) Lactose intolerance in north and south Indians. Amer. J. Clin. Nutr. 34:943-946.

Tanis, R. J., Neel, J. V., Dovey, H., and Morrow, M. (1973) The genetic structure of a tribal population, the Yanomama Indians. IX. Gene frequencies for 18 serum protein and erythrocyte enzyme systems in the Yanomama and five neighboring tribes: nine new variants. Amer. J. Hum. Genet. 25:655-676.

Tanis, R. J., Ueda, N., Satoh, C., Ferrell, R. E., Kishimoto, S., Neel, J. V., Hamilton, H. B., and Ohno, N. (1978) The frequency in Japanese of genetic variants of 22 proteins. IV. Acid phosphatase, NADP-isocitrate dehydrogenase, peptidase A, peptidase B, and phosphohexose isomerase. Ann. Hum. Genet. 41:419-428.

Tashian, R. E. (1969) The estrases and carbonic anhydrases of human erythrocytes. In: Biochemical Methods in Red Cell Genetics (Yunis, J. J., ed.) Academic Press, New York, pp. 307-336. Cited from Harris and Hopkinson (1978).

Tashian, R. E., Brewar, G. J., Lehmann, H., Davies, D. A., and Rucknagel, D. L. (1967) Further studies on the Xavante Indians. V. Genetic variability in some serum and erythrocyte enzymes, hemoglobin and urinary excretion of β-aminoisobutyric acid. Amer. J. Hum. Genet. 19:524-531.

Tashian, R. E. and Carter, N. D. (1976) Biochemical genetics of carbonic anhydrase. In: Advances In Human Genetics, Vol. 7 (Harris, H. and Hirschhorn, K., eds.), Plenum Press, New York, pp. 1-56.

Tashian, R. E., Kendall, A. G., and Carter, N. D. (1980) Inherited variants of red cell carbonic anhydrases. Hemoglobin 4:635-651.

Tchen, P., Bois, E., Lanset, S., and Feingold, N. (1981) Blood group antigens in the Emerillon, Wayampi, and Wayana Amerindians of French Guiana. Hum. Hered. 31:47-53.

Tchen, P., Bois, E., Séger, J., Grenand, P., Feingold, N., and Feingold, J. (1978) A genetic study of two French Guiana Amerindian populations. I. Serum proteins and red cell enzymes. Hum. Genet. 45:305-315.

Tedesco, T. A., Miller, K. L., Rawnsley, B. E., Mennuti, M. T., Spielman, R. S., and Mellman, W. J. (1975) Human erythrocyte galactokinase and galactose-1-phosphate uridyltransferase: A population survey. Amer. J. Hum. Genet. 27:737-747.

Teisberg, P. (1971a) The distribution of C'3 types in Norway. Hum. Hered. 21:154-161.

Teisberg, P. (1971b) C'3 types of Norwegian Lapps. Hum. Hered. 21:162-167.

Teisberg, P. and Olaisen, B. (1977) Properdin factor B (Bf) polymorphism in Norway. Vox Sang. 32:52-55.

Teng, Y.-S. (1981) Stomach aldehyde dehydrogenase: Report of a new locus. Hum. Hered. 31:74-77.

Teng, Y.-S., Anderson, J. E. and Giblett, E. R. (1975) Cytidine deaminase: A new genetic polymorphism demonstrated in human granulocytes. Amer. J. Hum. Genet. 27:492-497.

Teng, Y.-S., Jehan, S., and Lie-Injo, L. E. (1979) Human alcohol dehydrogenase ADH_2 and ADH_3 polymorphisms in ethnic Chinese and Indians of West Malaysia. Hum. Genet. 53:87-90.

Teng, Y.-S. and Lie-Injo, L. E. (1977) Erythrocyte superoxide dismutase in different racial groups in Malaysia. A variant in a Filipino. Hum. Genet. 36:231-234.

Teng, Y.-S. and Tan, S. G. (1979) Acid α-glucosidase in Malaysians. Hum. Hered. 29:2-4.

Teng, Y.-S. and Tan, S. G. (1982) Subtyping of properdin factor B (Bf) by isoelectrofocusing. Hum. Hered. 32:362-366.

Teng, Y.-S., Tan, S. G., Lopez, C. G., Ng, T., and Lie-Injo, L. E. (1978a) Genetic markers in Malaysians: Variants of soluble and mitochondrial glutamic oxaloacetic transaminase and salivary and pancreatic amylase, phosphoglucomutase III and saliva esterase polymorphisms. Hum. Genet. 41:347-354.

Teng, Y.-S., Tan, S. G., Ng, T., and Lopez, C. G. (1978b) Red cell glyoxalase I and placental soluble aconitase polymorphisms in the three major ethnic groups of Malaysia. Jpn. J. Hum. Genet. 23:211-215.

Terry, M. C. (1950) Taste-blindness and diabetes in the colored population of Jamaica. J. Hered. 41:306-307. Cited from Mourant et al. (1976).

Than-Than-Sint and Mya-Tu, M. (1973) Haptoglobin and transferrin distribution in the Burmese. Hum. Hered. 23:267-269.

Thieme, F. P. (1952) The geographic and racial distribution of ABO and Rh blood types and tasters of PTC in Puerto Rico. Amer. J. Hum. Genet. 4:94-112.

Thorsby, E., Colombani, J., Dausset, J., Figueroa, J., and Thorsby, A. (1972) HL-A, blood group and serum type polymorphism of natives of Easter Island. Histocompatibility Testing, Münksgaard, Copenhagen, pp. 287-297.

Thymann, M. (1978) Identification of a new serum protein polymorphism as transferrin. Hum. Genet. 43:225-229.

Thymann, M. (1981) Gc subtypes determined by agarose isoelectrofocusing: Distribution in Denmark and application to paternity cases. Hum. Hered. 31:214-221.

Thymann, M. (1986) Distribution of alpha-1-antitrypsin (Pi) phenotypes in Denmark determined by separator isoelectric focusing in agarose gel. Hum. Hered. 36:19-23.

387

Tiitinen, H., Mattila, M. J., and Eriksson, A. W. (1967) Isoniazid inactivation in Finns and Lapps. Bull. Europ. Soc. Hum. Genet. 1:77-78.

Tiilikainen, A., Rouslahti, E., Seppälä, A., Mårtensson, L., and van Loghem, E. (1969) Low frequency of genetic factors Gm(s) and Di(a) in Finland. Hum. Hered. 19:180-184.

Tills, D. (1977) Red cell and serum proteins and enzymes of the Irish. Ann. Hum. Biol. 4:35-42.

Tills, D., van den Branden, J. L., Clements, V. R., and Mourant, A. E. (1971) The distribution in man of genetic variants of 6-phosphogluconate dehydrogenase. Hum. Hered. 21:305-308.

Tills, D., Harvey, R. G., Warlow, A., Kopec, A. C., Sutter, D., Hauge, M., Simonsen, H. J. and Marin, A (1985) Blood groups, serum proteins and enzymes of the Faroe Islanders. J. Hum. Evol. 14:725-738.

Tills, D., Kopec, A. C., Fox, R. F., and Mourant, A. E. (1979) The inherited blood factors of some Northern Nigerians. Hum. Hered. 29:172-176.

Tills, D., Kopec, A. C., and Tills, R. E. (1983a) The Distribution of the Human Blood Groups and Other Polymorphisms, Supplement 1, Oxford University Press, Oxford.

Tills, D., Kopec, A. C., Warlow, A., Barnicot, N. A., Mourant, A. E., Marin, A., Bennett, F. J., and Woodburn, J. C. (1982a) Blood group, protein, and red cell enzyme polymorphisms of the Hadza of Tanzania. Hum. Genet. 61:52-59.

Tills, D., Teesdale, P., and Mourant, A. E. (1977a) Blood groups of the Irish. Ann. Hum. Biol. 4:23-34.

Tills, D., Warlow, A., Kopéc, A. C., Fridriksson, S., and Mourant, A. E. (1982b) The blood groups and other hereditary blood factors of the Icelanders. Ann. Hum. Biol. 9:507-520.

Tills, D., Warlow, A., Lord, J. M., Suter, D., Kopec, A. C., Blumberg, B. S., Hesser, J. E., and Economidou, I. (1983b) Genetic factors in the population of Plati, Greece. Amer. J. Phys. Anthrop. 61:145-156.

Tills, D., Warlow, A., Mourant, A. E., Kopéc, A. C., Edholm, O. G., and Garrard, G. (1977b) The blood groups and other hereditary blood factors of Yemenite and Kurdish Jews. Ann. Hum. Biol. 4:259-274.

Tipler, T. D., Dunn, D. S., and Jenkins, T. (1982) Phosphoglucomutase first locus polymorphism as revealed by isoelectric focusing in southern Africa. Hum. Hered. 32:80-93.

Tippett, P. (1967) Genetics of the Dombrock blood group system. J. Med. Genet. 4:7-11.

Tiwari, S. C. (1961) The frequencies of serum haptoglobin and transferrin types in a Punjabi population. Anthropologist 8:23-28.

Tiwari, S. C. (1966) The blood groups of the Tibetans. In: Human adaptability to environments and physical fitness (Malhotra, M. S., ed.), Madras, India, pp. 281-289. Cited from Mourant et al. (1976).

Tiwari, S. C. (1969) The incidence of colour-blindness among the Tibetans. J. Génét. Hum. 17:95-98.

Tiwari, S. C. and Bhasin, M. K. (1967) Taste sensitivity to phenylthiourea among the Newar population groups in Nepal. Acta Genet. Med. Gemell. 16(suppl.):57-67.

Tokunaga, K., Araki, C., Juji, T., and Omoto, K. (1981) Genetic polymorphism of the complement C2 in Japanese. Hum. Genet. 58:213-216.

Tokunaga, K., Dewald, G., Omoto, K., and Juji, T. (1986) Family study on the polymorphisms of the sixth and seventh components (C6 and C7) of human complement: Linkage and haplotype analyses. Amer. J. Hum. Genet. 39:414-419.

Tokunaga, K., Omoto, K., Akaza, T., Akiyama, N., Amemiya, H., Naito, S., Sasazuki, T., Satoh, H, and Juji, T. (1985) Haplotype study on C4 polymorphism in Japanese. Associations with MHC alleles, comlotypes, and HLA-complement haplotypes. Immunogenet. 22:359-365.

Toncheva, D. (1986) Variants of glucose-6-phosphate dehydrogenase in a Vietnamese population. Hum. Hered. 36:348-351.

Toncheva, D., and Tzoneva, M. (1984) Genetic polymorphism of G6PD in a Bulgarian population. Hum. Genet. 67:340-342.

Tongmao, Z. (1983) Genetic polymorphisms of C3 and Bf in the Chinese population. Hum. Hered. 33:36-38.

Torrinha, J. A. F. (1967) Haptoglobin frequencies in the North of Portugal. Acta Genet. Stat. Med. 17:74-76.

Torrinha, J. A. F. (1969) Frequencia dos genotipos do sistema Gc nos portugueses. Medico, Porto 51:719-720. Cited from Mourant et al. (1976).

Toyomasu, T., Sakakibara, S., Kagamiyama, H., and Matsumoto, H. (1984) Genetic polymorphism of mitochondrial glutamate-oxaloacetate transaminase in Japanese. Hum. Genet. 66:90-91.

Trent, R. J., Mickleson, K. N. P., Wilkinson, T., Yakas, J., Bluck, R., Dixon, M., Liley, A. W., and Kornenberg, H. (1985) α globin gene rearrangements in Polynesians are not associated with malaria. Amer. J. Hematology. 18:431-433.

Trincao, C. and Cordeiro, F. N. (1962) Thalassaemia in Portugal. Proc. 8th Congr. Eur. Soc. Haematol. S. Karger, New York. Pt. 2. No. 307a. Cited from Livingstone (1985).

Tsipouras, P., Børresen, A-L, Dickson, L. A., Berg, K., Prockop, D, J., and Ramirez, F. (1984) Molecular heterogeneity in the mild autosomal dominant forms of osteogenesis imperfecta. Amer. J. Hum. Genet. 36:1172-1179.

Tsuchiya, S., Yamanouchi, Y., Onuki, M., Yamakawa, K., Miyazaki, R., Taya, T., Kondo, I., Ohnuki, M., and Hamaguchi, H. (1985) Frequencies of apolipoproteins E5 and E7 in apparently healthy Japanese. Jpn. J. Hum. Genet. 30:271-278.

Tuchinda, S., Rucknagel, D. L., Na-Nakorn, S., Wasi, P. (1968) The Thai variant and the distribution of alleles of 6-phosphogluconate dehydrogenase and the distribution of glucose 6-phosphate dehydrogenase deficiency in Thailand. Biochem. Genet. 2:253-264.

Turner, B. M., Turner, V. S., Beratis, N. G., and Hirschhorn, K. (1975) Polymorphism of human α fucosidase. Amer. J. Hum. Genet. 27:651-661.

Turowska, B., Marek, Z., and Jaegermann, K. (1977) Blood group and serum protein investigations in a south Polish population. Acta Anthropogenet. 1:9-17.

Ueda, N., Satoh, C., Tanis, R. J., Ferrell, R. E., Kishimoto, S., Neel, J. V., Hamilton, H. B., and Baba, K. (1977) The frequency in Japanese of genetic variants of 22 proteins. II. Carbonic anhydrase I and II, lactate dehydrogenase, malate dehydrogenase, nucleoside phosphorylase, triose phosphate isomerase, haemoglobin A and haemoglobin A_2. Ann. Hum. Genet. 41:43-52.

Umetsu, K., Ikeda, N., Kashimura, S., and Suzuki, T. (1985) Orosomucoid (ORM) typing by print lectinofixation: a new technique for isoelectric focusing. Two common alleles in Japan. Hum. Genet. 71:223-224.

Umetsu, K., Kashimura, S., Ikeda, N., and Suzuki, T. (1984) A new α_2HS-glycoprotein allele (AHS*5*) in two Japanese families. Hum. Genet. 68:264-265.

Umnova, M. A. (1959) Factor P of the human blood and its distribution among the Moscow population. In Russian. Soviet Anthrop. 3:99-105. Cited from Mourant et al. (1976).

Umnova, M. A., Prokop, O., Piskunov, T. M., Samusova, G. C., Ichalovskaya, T. A., and Prozorovskaya, G. P. (1964) Blood group distribution among the Moscow population. 7th Int. Congr. Anthrop. Ethnol. Sci., Moscow, 1:496-501. Cited from Mourant et al. (1976).

Undevia, J. V., Kirk, R. L., and McDermid, E. M. (1973) Serum protein systems among Parsis and Iranis in Bombay. Hum. Hered. 23:492-498.

Uotila, L. (1984) Polymorphism of red cell S-formylglutathione hydrolase in a Finnish population. Hum. Hered. 34:273-277.

Vaccaro, A. M., Mandara, I., Muscillo, M., Ciaffoni, F., de Pellegrin, S., Benincasa, A., Novelleto, A., and Terrenato, L. (1984) Polymorphism of erythrocyte galactose-1-phosphate uridyl-1-transferase in Italy: Segregation analysis in 693 families. Hum. Hered. 34:197-206.

Valls, A. (1974) Primeros datos del sistema Kidd de antigenos ertrocittarios en la poblacion española. Miscelánea Alcobé. January, 67-72. Cited from Tills et al. (1983a).

Valls, A. (1975) Seroantropologiá de la población española. Rev. Univ. Complutense, Madrid. 24:111-139. Cited from Tills et al. (1983a).

Valls Medna, A. (1958) Estudio antropogenético de la capacidad gustativa para la feniltiocarbamida. Madrid, Inst. Sahagún, Thesis-Madrid. Cited from Mourant et al. (1976).

van den Branden, J. L., Clements, V. R., Mourant, A. E., and Tills, D. (1971) The distribution in human populations of genetic variants of adenosine deaminase. Hum. Hered. 21:60-62.

van der Does, J. A., D'Amaro, J., Meerakhan, P., Bernini, L. F., van Loghem, E., Nijenhuis, L., van der Steen, G., van Rood, J. J., and Rubinstein, P. (1972) HL-A typing in Chilean Aymara Indians. Histocompatibility Testing 1972, Munksgaard, Copenhagen, pp.391-395.

van der Weerdt, C. M. (1965) The platelet agglutination test in platelet grouping. Histocompatibility Testing 1965, Munksgaard, Copenhagen, pp. 161-166.

van Loghem, E., Chandanayingyong, D., and Douglas, R. (1975) Immunoglobulin genetic markers in the Thai population. J. Immunogenet. 2:141-145.

van Loghem, E. Jr., Dorfmeijer, H., and van der Hart, M. (1959) Serogical and genetical studies on a platelet antigen (Zw). Vox Sang. 4:161-169.

van Loghem, E., Wang, A. C., and Shuster, J. (1973) A new genetic marker of human immunoglobulins determined by an allele at the α2 locus. Vox Sang. 24:481-488.

von dem Borne, A. E. G. K., van Riesz, E., Verheugt, F. W., A., ten Cate, J. W., Koppe, I. G., Engelfriet, C. P., and Nijenhuis, L. E. (1980) Bak[a], a new platelet-specific antigen involved in neonatal allo-immune thrombocytopenia. Vox Sang. 39:113-120.

Vandeville, D., Martin, J. P., Lebreton, J. P., and Ropartz, C. (1972) Le système Pi dans les populations normandes et amérindiennes. Rev. Franc. Transfus. 15:213-218.

Vaz-Guedes, M. A., Hobart, M. J., and Lachmann, P. J. (1978) Absence of variation in human C5. J. Immunogenet. 5:279-282.

Vega, C. E. (1975) Aportación al conocimento del sistema de grupos sanguineos Xg(a) en dos poblaciones españolas. IV Congr. Nac. Genet. Hum. Zaragoza. Cited from Tills et al. (1983a).

Vergnes, H. and Cabannes, R. (1976) Polymorphism of erythrocyte and serum enzyme systems in the Gagu of the Ivory coast. Ann. Hum. Biol. 3:423-429.

Vergnes, H., Constans, J., Quilici, J. C., Lefèvre-Witier, P., Sevin, J., and Stevens, M. (1980a) Study of red blood cell and serum enzymes in five Pyrenean communities and in a Basque population sample. Hum. Hered. 30:171-180.

Vergnes, H., Meyer, S., Weil, D., Goudemand, J., Brevière, D., Sevin, J., and Constans, J. (1980b) Erythrocyte glyoxalase I and esterase D polymorphisms in four French populations. Hum. Hered. 30:232-236.

Vergnes, H., Quilici, J. C., and Constans, J. (1976a) Serum and red cell enzyme polymorphisms in six Amerindian tribes. Ann. Hum. Biol. 3:577-585.

Vergnes, H., Quilici, J. C., Gherardi, M., and Bejarano, G. (1976b) Serum and red cell enzyme variants in an Amerindian tribe. Hum. Hered. 26:252-262.

Vergnes, H. and Sevin, J. (1981) Isoelectric focusing of human phosphoglucomutase: Data on the distribution of variant phenotypes in three population samples from southwest France. Hum. Hered. 31:156-160.

Vernon, P. E. and Straker, A. (1943) Distribution of colour-blind men in Great Britain. Nature 152:690.

Vierucci, A. (1965) Gm groups and anti-Gm antibodies in children with Cooley's anaemia. Vox Sang. 10:82-93.

Vincent-Viry, M., La Du, B. N., LePage, L., and Mikstacki, T. (1986) Distribution des differents phenotypes de la paraoxonase dans une population Francaise. Ann. Biologie Clinique. (in press).

Virtaranta-Knowles, K. and Nevanlinna, H. R. (1982) Red cell glyoxalase I polymorphism in the Finnish population. Hum. Hered. 32:285-288.

389

population. Hum. Hered. 32:285-288.

Vogel, F., Krüger, J., Chakravartti, M. R., Flatz, G., and Ritter, H. (1971) Inv phenotypes and quantitative gamma globulin determinations in leprosy patients and control populations from India and Thailand. Humangenetik. 12:35-41.

Vos, G. H. and Comley, P. (1967) Red cell and saliva studies for the evaluation of ABH and Lewis factors among the Caucasian and aboriginal populations of western Australia. Acta Genet. Stat. Med. 17:495-510.

Vos, G. H. and Kirk, R. L. (1961) Di^a, Js^a and V blood groups in South and Southeast Asia. Nature 189:321-322.

Vos, G. H., Kirk, R. L., and Steinberg, A. G. (1963) The distribution of the gamma globulin types Gm(a), Gm(b), Gm(x) and Gm-like in South and Southeast Asia and Australia. Amer. J. Hum. Genet. 15:44-52.

Vyas, G. N., Bhatia, H. M., Sukumaran, P. K., Balakrishnan, V., and Sanghvi, L. D. (1962) Study of blood groups, abnormal hemoglobins and other genetical characters in some tribes of Gujarat. Amer. J. Phys. Anthrop. 20:255-265.

Waaler, G. H. M. (1927) Über die Erblichkeitsverhältnisse der verschiedenen Arten von angeborener Rotgrünblindheit. Acta Ophthal. (Kbh) 5:309-345. Cited from Post (1971).

Wainscoat, J. S., Hill, A. V. S., Boyce, A. L., Flint, J., Hernandez, M., Thein, S. L., Old, J. M., Lynch, J. R., Falusi, A. G., Weatherall, D. J., and Clegg, J. B. (1986) Evolutionary relationships of human populations from an analysis of nuclear DNA polymorphisms. Nature. 319:491-493.

Wallis, S. C., Donald, J. A., Forrest, L. A., Williamson, R., and Humphries, S. E. (1984) The isolation of a genomic clone containing the apolipoprotein CII gene and the detection of linkage disequilibrium between two common DNA polymorphisms around the gene. Hum. Genet. 68:286-289.

Walter, H. (1981) On the population genetics of Iceland - A Review. Coll. Anthropol. 5:155-177.

Walter, H., Ananthakrishnan, R., and Tsacheva, L. (1972a) Serum protein groups in Bulgaria. Hum. Hered. 22:529-532.

Walter, H., Arndt-Hanser, A., Raffa, M. A., and Gumbel, B. (1975) On the distribution of some genetic markers in Libya. Humangenetik 27:129-136.

Walter, H, Dannewitz, A., Veerraju, P., and Goud, J. D. (1984) Gc subtyping in south Indian tribal and caste populations. Hum. Hered. 34:250-254.

Walter, H., Hilling, M., Brachtel, R., and Hitzeroth, H. W. (1979) On the population genetics of β_2-glycoprotein I. Hum. Hered. 29:236-241.

Walter, H., Kannapinn, G., Dannewitz, A., Rickards, O., de Stefano, G. F. (1986) On the variability of Gc subtypes in Italy. Hum. Hered. 36:50-53.

Walter, H., Kellermann, G., Bajatzadeh, M., Krüger, J., and Chakravartti, M. R. (1972b) Hp, Gc, Cp, Tf, Bg and Pi phenotypes in leprosy patients and healthy controls from West Bengal (India). Humangenetik 14:314-325.

Walter, H., Mukherjee, B. N., Gilbert, K., Lindenberg, P., Dannewitz, A., Malhotra, K. C., Das, B. M., and Deka, R. (1986) Investigations on the variability of haptoglobin, transferrin and Gc polymorphisms in Assam, India. Hum. Hered. 36:388-396.

Walter, H., Neumann, S., Backhausz, R., and Nemeskéri, J. (1965) Populationsgenetische Untersuchungen über die Pseudocholinesterase-varianten bei Ungarn und Deutschen. Humangenetik 1:551-556.

Walter, H. and Palsson, J. (1973) The incidence of some genetic markers in Ireland. In: Genetic Variation in Britain (Roberts, D. F. and Sunderland, E., eds.), Taylor and Francis, London, pp. 207-220.

Wang, L. and Cavalli-Sforza, L. L. (1986) Transport proteins in Pygmies. In: African Pygmies (Cavalli-Sforza, L. L., ed.), Academic Press, New York, pp. 291-303.

Wang, Y., Mao, Z., Chen, L., Cui, M., Xu, J., and Du, R. (1983) Colorblindness in twelve minor ethnic groups of China. Chinese J. Med. (in Chinese) 63:339-341.

Wang, Y., Yan, Y., Xu, J., Du, R., Flatz, S. D., Kühnau, W., and Flatz, G. (1984) Prevalence of primary adult lactose malabsorption in three populations of northern China. Hum. Genet. 67:103-106.

Watanabe, S., Kondo, S., and Matsunaga, E. (1975) Anthropological and Genetic Studies on the Japanese. Univ. Tokyo Press, Tokyo.

Wardell, M. R., Suckling, P. A., and Janus, E. D. (1982) Genetic variation in human apolipoprotein E. J. Lip. Res. 23:1174-1182.

Weidinger, S., Cleve, H., Schwarzfischer, F., Postel, W., Weser, J., and Görg, A. (1984a) Transferrin subtypes and variants in Germany: Further evidence for a Tf null allele. Hum. Genet. 66:356-360.

Weidinger, S., Schwarzfischer, F., Burgemeister, R., and Cleve, H. (1984b) Two new Bf S subtypes revealed by isoelectric focusing and immunofixation. Hum. Genet. 68:90-92.

Weidinger, S., Schwarzfischer, F., Burgemeister, R., and Cleve, H. (1984c) Genetic alpha-2-HS glycoprotein phenotypes demonstrated by isoelectric focusing and immunofixation. In: Electrophoresis '84 (Neuhoff, V., ed.), Verlag Chemie, Weinheim. pp. 487-490.

Weidinger, S., Scwarzfischer, F., and Cleve, H. (1983) Antithrombin III: A new polymorphism revealed by isoelectrofocusing and immunofixation. In: Electrophoresis' 82 (Stathakos, D., ed.), Walter de Gruyer & Co., Berlin. pp.761-765.

Weiner, J. S. and Zoutendyk, A. (1959) Blood-group investigation on Central Kalahari Bushmen. Nature 183:843-844.

Weissmann, J. and Reuter, W. (1981) Properdin factor B polymorphism in Portugal. Hum. Hered. 31:370-372.

Weissmann, J., Vollmer, M., and Pribilla, O. (1982) Survey of the distribution of adenosine deaminase and superoxide dismutase markers in different populations. Hum. Hered. 32:344-356.

Weitkamp, L. R. (1974) The contribution of variations in serum albumin to the characterization of human populations. In: Genetic Polymorphisms and Diseases in Man (Ramot, B., ed.), Academic Press, New York, pp. 112-122.

Weitkamp, L. R. (1976) Linkage of GLO with HLA and Bf. Effect of population and sex on recombination frequency. Tissue Antigens 7:273-279.

Weitkamp, L. R., Arends, T., Gallango, M. L., Neel, J. V., Schultz, J., and Shreffler, D. C. (1972) The genetic structure of a tribal population, the Yanomama Indians. III. Seven serum protein systems. Ann. Hum. Genet. 35:271-279.

Weitkamp, L. R. and Buck, A. A. (1972) Phenotype frequencies for four serum proteins in Afghanistan: Two "new" albumin variants. Humangenetik 15:335-340.

Weitkamp, L. R. and Neel, J. V. (1970) Gene frequencies and microdifferentiation among the Makiritare Indians. III. Nine erythrocyte enzyme systems. Amer. J. Hum. Genet. 22:533-537.

Weitkamp, L. R. and Neel, J. V. (1972) The genetic structure of a tribal population, the Yanomama Indians. IV. Eleven erythrocyte enzymes and summary of protein variants. Ann. Hum. Genet. 35:433-444.

Welch, S. (1974) Red cell esterase D polymorphism in Gambia. Humangenetik 21: 365-367.

Welch, S. G., Barry, J. V., Dodd, B. E., Griffiths, P. D., Huntsman, R. G., Jenkins, G. C., Lincoln, P. J., McCathie, M., Mears, G. W., and Parr, C. W. (1973) A survey of blood group, serum protein and red cell enzyme polymorphisms in the Orkney Islands. Hum. Hered. 23:230-240.

Welch, S. G., and Lee, J. (1974) The population distribution of genetic variants of human esterase D. Humangenetik 24:329-331.

Welch, Q. B. and Lie-Injo, L. E. (1972) Serum albumin variants in three Malaysian racial groups. Hum. Hered. 22:503-507.

Welch, Q. B., Lie-Injo, L. E., and Bolton, J. M. (1971) Adenylate kinase and malate dehydrogenase in four Malaysian racial groups. Humangenetik 14:61-63.

Welch, Q. B., Lie-Injo, L. E., and Ganesan, J. (1975a) Erythrocyte adenosine deaminase in Malaysians. Hum. Hered. 25:69-72.

Welch, S. G. and Mears, G. W. (1972) Genetic variants of human indophenol oxidase in the Westray Island of the Orkneys. Hum. Hered. 22:38-41.

Welch, S. G., Mills, P. R., and Gaensslen, R. E. (1975b) Phenotypic distributions of red cell glutamate-pyruvate transaminase (E.C. 2.6.1.2) isoenzymes in British and New York populations. Humangenetik 27:59-62.

Welty, R. G. and Hewett-Emmett, D. (1986) Personal communication.

Whitehouse, D. B., Hopkinson, D. A., Hill, A. V. S., and Bowden, D. K. (1985) Analysis of genetic variation in two human thyroxine-binding plasma proteins by immunodetection after isoelectric focusing. Ann. Hum. Genet. 49:259-265.

Whitehouse, D. B. and Putt, W. (1983) Immunological detection of the sixth complement component (C6) following flat bed polyacrylamide gel isoelectric focusing and electrophoretic transfer to nitrocellulose filters. Ann. Hum. Genet. 47:1-8.

Whittaker, M. (1968) Frequency of atypical pseudocholinesterase in groups of individuals of different ethnographical origin. Acta Genet. Stat. Med. 18:567-572.

Whittaker, M., and Britten, J. J. (1985) Plasma cholinesterase variants. Hum. Hered. 35:364-368.

Whittaker, M. and Lowe, R. F. (1976) The cholinesterase variants found in some African tribes living in Rhodesia. Hum. Hered. 26:380-393.

Whittaker, M. and Reys, L. (1975) Plasma cholinesterase studies on south-eastern Bantu of Mozambique. Hum. Hered. 25:296-301.

Wickremasinghe, R. L., Ikin, E. W., Mourant, A. E., and Lehmann, H. (1963) The blood groups and haemoglobins of the Veddahs of Ceylon. J. R. Anthrop. Inst. 93:117-125.

Wiebecke, D., Spielmann, W., and Seidl, S. (1967) Ein Beitrag zur Populationsgenetik des Gm- and Inv-Systems. Klin. Wschr. 45:736-737. Cited from Steinberg and Cook (1981).

Wiener, A. S. (1969) Problems and pitfalls in blood grouping tests for non-parentage. i Distribution of the blood groups. Amer. J. Clin. Path. 51:9-14. Cited from Mourant et al. (1976).

Wierst, B. V., Blake, N. M., Kirk, R. L., Jacobs, D. S., and Johnson, D. G. (1973) Genetic variation at the third locus of phosphoglucomutase in placentas from Australia and Papua New Guinea. Aust. J. Expt. Biol. & Med. Sci. 51:857-860.

Willcox, M., Beckman, G., and Beckman, L. (1986) Serum protein polymorphisms in a Liberian population. Hum. Hered. 36:54-57.

Willcox, M. C., and Beckman, L. (1981) Haemoglobin variants, β-thalassaemia and G-6-PD types in Liberia. Hum. Hered. 31:339-347.

Willis, M. F. and Booth, P. B. (1968) Takuu and Nukumanu atolls, Bougainville District, Territory of Papua and New Guinea. Blood groups and other genetic data. Archaeol. Phys. Anthrop. Oceania 3:55-63.

Wilson, D. E., Povey, S. and Harris, H. (1976) Adenylate kinases in man: evidence for a third locus. Ann. Hum. Genet. 39:305-313.

Windhof, O. and Walter, H. (1983) Blood group, serum protein and red cell enzyme polymorphisms in Filipinos. Hum. Hered. 33:357-364.

Wing, J. P. (1974) Blood protein polymorphisms in Jewish populations. Hum. Hered. 24:323-344.

Winichagoon, P., Higgs, D. R., Goodbourn, S. E. Y., Clegg, J. B., Weatherall, D. J., and Wasi, P. (1984) The molecular basis of α-thalassaemia in Thailand. EMBO J. 3:1813-1818.

Winship, P. R., Anson, D. S., Rizza, C. R., and Brownlee, G. G. (1984) Carrier detection in haemophilia B using two further intragenic restriction fragment length polymorphisms. Nucl. Acids. Res. 12:8861-8872.

391

Wolanski, N., Nahar, R. A., and Roberts, D. F. (1983) Genetic studies in Poland. Hum. Hered. 33:270-276.

Won, C. D., Shin, H. S., Kim, S. W., Swanson, J., and Matson, A. (1960) Distribution of hereditary blood factors among Koreans residing in Seoul, Korea. Amer. J. Phys. Anthrop. 18:115-124.

Woodfield, D. G., Scragg, R. F. R., Blake, N. M., Kirk, R. L., and McDermid, E. M. (1974) Distribution of blood, serum protein and enzyme groups among the Fuyuge speakers of the Goilala Sub-District. Hum. Hered. 24:507-519.

Woolhouse, N. M. (1986) The debrisoquine/sparteine oxidation polymorphism: evidence of genetic heterogeneity among Ghanaians. In: Ethnic Differences in Reactions to Drugs and Xenobiotics (Kalow, W., Goedde, H. W., and Agarwal, D. P., eds.), Alan R. Liss, New york, pp.189-206.

Workman, P. L., Niswander, J. D., Brown, K. S., and Leyshon, W. C. (1974) Population studies on southwestern Indian tribes. IV. The Zuni. Amer. J. Phys. Anthrop. 41:119-132.

Wurzel, H. A. and Haesler, W. E. (1968) The Yt blood groups of American Negroes. Vox Sang. 15:304-305.

Wüst, H. (1971) Further studies on the adenosine deaminase (ADA) polymorphism in Austria. Vox Sang. 21:443-446.

Xu, J., Cui, M., Li, S., Chen, L., Du, R., Goedde, H. W., Benkmann, H.-G., Kriese, L., and Bogdanski, P. (1986) Polymorphisms of Pi, Hp, ADA and AK in Mongolian, Korean and Zhuang populations of China. Ann. Hum. Biol. 13:245-251.

Xu, J. and Du, R. (1984a) The genetic investigation of red cell acid phosphatase in Dong, Hui, and Bui ethnic groups. Ann. Report Inst. Genet. Academ. Sinica, p. 57.

Xu, J. and Du, R. (1984b) A genetic study on the gene frequency distribution of red cell glutamic-pyruvic transaminase in Hui, Dong, and Bai ethnic groups. Ann. Report Inst. Genet. Academ. Sinica, p. 57

Xu, J. and Du, R. (1984c) The genetic polymorphism of red cell esterase D in the ethnic groups of Bai, Dong and Hui. Ann. Report Inst. Genet. Academ. Sinica, p. 58.

Xu, J. and Du, R. (1984d) The phenotype distribution and gene frequency of 6-phosphogluconate dehydrogenase (6-PGD) in Hui, Dong and Bai ethnic groups. Ann. Report Inst. Genet. Academ. Sinica, p. 59.

Xu, J., Mao, Z., Li, S., Cui, M., Wang, Y., Chen, L., and Du, R. (1982) Distribution of taste sensitivity of phenylthiocarbamide (PTC) of various nationalities in China. Acta Genet. Sinica 9:308-314.

Xu, Y. -K. and Ng, W. G. (1983) Polymorphism of erythrocyte galactose-1-phosphate uridyltransferase among Chinese. Hum. Genet. 63:280-282.

Yamamoto, M. and Fu, L. (1973) Red cell isozymes in the eastern Carolines. Amer. J. Phys. Anthrop. 38:703-708.

Yamamoto, M., Wada, T., Watanabe, T., Kanazawa, H., Saito, R., Kondo, M., Hosokawa, K., Masuda, M., Nakai, T., and Fujiki, N. (1972) Genetic polymorphisms in four isolated communities in Kinki district. Jpn. J. Hum. Genet. 17:273-285.

Yang, S. Y., Coleman, P., Ochs, H. D., and Dupont, B. (1981) Inheritance and genetic linkage of transcobalamin II. Hum. Genet. 57:307-311.

Yenchitsomanus, P-T, Summers, K. M., Bhatia, K. K., Cattani, J., and Board, P. G. (1985) Extremely high frequencies of α-globin gene deletion in Madang and on Kar Kar Island, Papua New Guinea. Amer. J. Hum. Genet. 37:778-784.

Yenchitsomanus, P., Summers, K. M., Board, P. G., Bhatia, K. K., Jones, G. L., Johnston, K., and Nurse, G. T. (1986) Alpha-thalassemia in Papua New Guinea. Hum. Genet. 74:432-437.

Yin, S.-J., Bosron, W. F., Li, T.-K., Ohnishi, K., Okuda, K., Ishii, H., and Tsuchiya, M. (1984) Polymorphism of human liver alcohol dehydrogenase: Identification of ADH$_2$ 2-1 and ADH$_2$ 2-2 phenotypes in the Japanese by isoelectric focusing. Biochem. Genet. 22:169-180.

Ying, Q., Zhang, M., Liang, C., Chen, L., Chen, L., Huang, Y., Wang, R., Zhang, N., Li, H., Liu, S., and Gao, E. (1985a) Alpha-1-antitrypsin types in five Chinese national minorities. Hum. Genet. 71:225-226.

Ying, Q., Zhang, M., Liang, C., Liu, X., Huang, Y., Wang, R., Zhang, N., Chen, L., Chen, L., Yu, N., and Muo, X. (1985b) Geographical variability of alpha-1-antitrypsin alleles in China: A study on six Chinese populations. Hum. Genet. 69:184-187.

York, L. J., Marshall, W. H., and Huang, S. N. (1986) Polymorphism of the seventh complement component, C7, in Chinese. Hum. Hered. 36:261-262.

Yuan, Y., Du, R., Chen, L., Xu, J., Cui, M., Wang, Y., Li, S., Goedde, H. W., Benkmann, H.-G., Kriese, L., and Bogdanski, P. (1984a) Distribution of eight blood- group systems and ABH secretion in Mongolian, Korean, and Zhuang nationalities in China. Ann. Hum. Biol. 11:377-388.

Yuan, Y., Du, R., and Hao, L. (1984b) Distribution of Lewis, ABO, MN, Rh, P blood group systems and ABH secretion in Han nationality of north China (in Chinese). Acta Anthrop. Sinica 3:158-164.

Yuan, Y., Du, R., and Li, C. (1985) A survey on red cell blood groups of Hui ethnic group in Ningxia (in Chinese). Acta Anthrop. Sinica. 4:385-393.

Yuan, Y., Jin, F., Du, R., Long, Y., and Cai, R. (1984c) Distribution of nine blood group systems and ABH secretion in the Dong nationality (in Chinese). Acta Anthrop. Sinica 3:277-284.

Yuan, Y., Xu, J., Ai, S., Jin, F., and Du, R. (1984d) A study of red cell blood groups in Uygur of Xinjiang (in Chinese). Chinese J. Hematology. 5:305-310.

Yuan, Y., Xu, J., Zhang, Z, and Du, R. (1982) Distribution of Kell, Kidd, Diego, Duffy, Lutheran, and Xg blood group systems in Han nationality of north China (in Chinese). Acta Genet. Sinica 9:395-401.

Yuasa, I., Ikebuchi, J., Suenaga, K., and Ito, K. (1986a) Phosphoglucomutase-1 subtypes: Polymorphic occurrence of PGM*7+ and geographical variation in Japan. Hum. Hered. 36:233-237.

Yuasa, I., Saneshige, Y., and Okada. K. (1983a) Geographical cline of allele frequency of group-specific component (Gc) in the Japanese populations: An analysis of data obtained by immunoelectrophoresis. Jpn. J. Hum. Genet. 28:255-261.

Yuasa, I., Saneshige, Y., Okamoto, N., Ikawa, S., Hikita, T., Ikebuchi, J., Inoue, T., and Okada, K. (1983b) Distribution of Hp, Tf, Gc, and Pi polymorphisms in a Nepalese population. Hum. Hered. 33:302-306.

Yuasa, I., Suenaga, K., Gotoh, Y., Ito, K., and Yokoyama, N. (1984a) Gc types in western Japan: Report of a new variant Gc 1C35. Hum. Hered. 34:174-177.

Yuasa, I., Suenaga, K., Gotoh, Y., Ito, K., Yokoyama, N., and Okada, K. (1984b) PI (α_1-antitrypsin) polymorphism in the Japanese: Confirmation of PI*M4 and description of new PI variants. Hum. Genet. 67:209-212.

Yuasa, I., Taira, T., Suenaga, K., Ito, K., and Okada, K. (1985a) Determination of α_2HS-glycoprotein phenotypes by isoelectric focusing and immunoblotting: Polymorphic occurrence of HSGA*5 in Okinawa. Hum. Genet. 70:32-34.

Yuasa, I., Tamaki, N., Suenaga, K., Ito, K., Inoue, T., and Okada, K. (1985b) Reliable phenotyping of esterase D by low voltage isoelectric focusing: Evidence for the new variant ESD Yamaguchi. Electrophoresis 6:588-592.

Yuasa, I., Umetsu, K., Suenaga, K., and Robinet-Levy, M. (1986b) Orosomucoid (ORM) typing by isoelectric focusing: evidence for structural loci ORM1 and ORM2. Hum. Genet. 74:160-161.

Yvart, J., Gerbal, A. and Salmon, C. (1974) A new "private" antigen: Hey. Vox Sang. 26:41-44.

Zarinah, K. H., Abdullah, F., and Tan, S. G. (1984) Genetic markers in a Malaysian population: variants of uridine monophosphate kinase (UMPK), phosphoglycolate phosphatase (PGP) and pancreatic amylase (AMY2). Ann. Hum. Biol. 11:533-536.

Zeng, Z., Tokunaga, K., Omoto, K., and Du, C. (1986) Genetic polymorphisms of complement C6 and C7 in two Chinese populations. Jpn. J. Hum. Genet. 31:263-271.

Zhang, M., Ying, Q., Wang, L., and Liang, C. (1985) Frequencies of ceruloplasmin alleles in a chinese population. Hum. Hered. 35:117-119.

Zhao, T. (1983) Genetic polymorphisms of C3 and Bf in the Chinese population. Hum. Hered. 33:36-38.

Zhao, H., Chen, L., and Du, R. (1984) The distribution of phosphoglucomutase$_1$ (PGM$_1$) polymorphism in Dong, Hui, and Bai ethnic groups. Ann. Report Inst. of Genetics. Academ. Sinica. p. 66.

Zhao, H. and Du, R. (1984a) The distribution of adenylate kinase (AK) phenotype in the ethnic groups of Dong, Hui, and Bai ethnic groups. Ann. Report Inst. Genet. Academ. Sinica, p.67.

Zhao, H. and Du, R. (1984b) The distribution of adenosine deaminase (ADA) phenotypes in Dong, Hui, and Tujia ethnic groups. Ann. Report Inst. Genet. Academ. Sinica, p.68.

Zlotogora, J., Bach, G., Barak, Y., and Elian, E. (1980) Metachromatic leukodystrophy in the Habbanite Jews: High frequency in a genetic isolate and screening for heterozygotes. Amer. J. Hum. Genet. 32:663-669.